Springer Series in Synergetics

Editor: Hermann Haken

Synergetics, an interdisciplinary field of research, is concerned with the cooperation of individual parts of a system that produces macroscopic spatial, temporal or functional structures. It deals with deterministic as well as stochastic processes.

W. Tschacher G. Schiepek E. J. Brunner (Eds.)

Self-Organization and Clinical Psychology

Empirical Approaches
to Synergetics in Psychology

With 156 Figures

Springer-Verlag

Berlin Heidelberg New York
London Paris Tokyo
Hong Kong Barcelona
Budapest

Dr. Wolfgang Tschacher
Sozialpsychiatrische Universitätsklinik, Murtenstraße 21
CH-3010 Bern, Switzerland

Priv.-Doz. Dr. Günter Schiepek
Lehrstuhl für Klinische Psychologie
Universität Bamberg, Markusplatz 3
W-8600 Bamberg, Fed. Rep. of Germany

Professor Dr. Ewald Johannes Brunner
Universität Tübingen, Institut für Erziehungswissenschaft I
Münzgasse 22–30, W-7400 Tübingen 1, Fed. Rep. of Germany

Series Editor:
Professor Dr. Dr. h. c. Hermann Haken
Institut für Theoretische Physik und Synergetik der Universität Stuttgart
Pfaffenwaldring 57/IV, W-7000 Stuttgart 80, Fed. Rep. of Germany and
Center for Complex Systems, Florida Atlantic University
Boca Raton, FL 33431, USA

ISBN-13: 978-3-642-77536-9   e-ISBN-13: 978-3-642-77534-5
DOI: 10.1007/978-3-642-77534-5

Library of Congress Cataloging-in-Publication Data. Self-organization and clinical psychology: empirical approaches to synergetics in psychology / [edited by] W. Tschacher, G. Schiepek, E.J. Brunner. p. cm. – (Springer series in   synergetics; v. 58)   Includes bibliographical   references   and index.   1. Clinical psychology–Philosophy–Congresses. 2. System theory–Congresses. 3. Psychotheraphy–Philosophy–Congresses. 4. Self-organizing systems–Congresses. I. Tschacher, Wolfgang. II. Schiepek, Günter. III. Brunner, Ewald Johannes. IV. Series. RC467.E46 1992 616.89'001–dc20 92–19036

Typesetting: Camera ready by authors / editors

54/3140-5 4 3 2 1 0 – Printed on acid-free paper

Foreword

The present book edited by Tschacher, Schiepek and Brunner signals a change of paradigms in psychology. Psychology and its related disciplines deal with probably the most complex system of our world, namely the human being. In order to cope with its tasks, psychology has developed a number of basic ideas. During this development some of its ideas have been based on general concepts originally developed in other disciplines. Among these concepts are those of cybernetics and of autopoiesis. Cybernetics originated from engineering and deals with the problem of control. An often quoted example is that of the control of the temperature by an automatic device. The other field that has found considerable interest in psychology is that of autopoiesis, which is concerned with the maintenance of biological structures. Examples are provided by the problem of how a cell maintains itself by adequate control mechanisms and metabolism or how a whole species maintains itself.

Over the last twenty years or so the new discipline of synergetics has evolved. This interdisciplinary field of research is concerned with systems composed of many parts that may produce spatial, temporal or functional structures. In contrast to cybernetics and autopoiesis, which, in a way, deal with the problem of *maintaining* the state of a system, synergetics deals with the emergence of new kinds of states or structures, i.e. synergetics is concerned with *changes*. It has found both a rigorous mathematical foundation as well as many applications which range from physics and chemistry to biology and economics. For instance, in biology synergetics has been usefully applied to evolution, morphogenesis, movement coordination, perception, and other problems.

The word synergetics is taken from Greek and means cooperation. At a superficial level this may mean cooperation between different disciplines or it may mean that specific phenomena are produced by cooperation of parts of a system. In the present context, which is the one used throughout the Springer Series in Synergetics, synergetics is meant in a more precise sense. It means the application of a specific methodology. It has been shown by synergetics that when systems, or even a whole discipline, change their states, so-called critical fluctuations occur caused by the destabilization of the former state of the system. It is the view of the authors of this volume, and also my personal opinion, that psychology is presently undergoing such a change of paradigms and I believe this change is well-mirrored by the articles of this volume. But quite evidently, we are only at the beginning of a fascinating new development which will have its impact on neighboring disciplines of psychology also. One can hardly deny that the paradigm shift alluded to above will have its impact on the theoretical foundation of therapy. When a patient comes to a therapist, he or she is certainly not interested in the *maintenance* of his or her state, but he or she is highly interested in the help of the therapist in *changing* his or her state.

It is here in particular that the results of synergetics come in.

Stuttgart 1991   *Hermann Haken*

Preface

Clinical psychology deals with systems that change and evolve in time, i.e. with dynamical systems. There are several ways of modelling the mental and social systems that form and influence a psychotherapeutical context – in each case we have to consider phenomena of stability and sensitivity with respect to environmental influences, of the spontaneous formation of patterns and of rapid switching between such patterns. Clinical practice often has to cope with resistance to interventions, and with relapses to problem behavior after seemingly successful therapy or training; the behavior patterns that emerged from the system result from the interaction of many components. In most cases, however, patterns cannot be predicted from what we know of the single components (the persons, the $S - R$ links, the schemata etc. depending on the conceptualization). Attempts to reduce this problem to ever smaller problems, which may then be solved in some kind of basic research effort, remain futile.

In this situation there is growing promise that the difficulties can be surmounted by an interdisciplinary effort, namely that of synergetics and dynamical systems theory. There are basically two reasons for our expectation that clinical psychology and related disciplines may be enriched in this way. The first reason is given above; we doubt that reductionism will be of much further use in explaining complex behavior; microscopic theories will not add up to elucidate what happens in the multi-faceted natural systems of our field. The second reason is that we do not want to throw out the baby with the bathwater – systemicity does not mean that systems cannot be studied or modelled with the help of precise methods; we therefore aim to foster empirical research and mathematical models.

This volume gives a comprehensive overview of recent developments in the field of self-organization theory in clinical psychology; adjacent fields like psychiatry and medicine, cognitive psychology, social psychology and social science are also addressed. Accordingly, the contributors to this book come from various disciplines – from psychology, psychiatry, physics, mathematics, medicine, and sociology.

Many of the contributions are based on talks given at two symposia ("Herbst-akademie" – autumn academy) in October 1990 and October 1991. Both autumn academies were organized in Bamberg, Bavaria, by Günter Schiepek with Wolfgang Tschacher and Ewald Johannes Brunner. In this series of symposia we cooperate with Hermann Haken, founder of synergetics, in order to advance the evolving field of synergetics in psychology and other sciences that study behavior and mind. Additionally, an informal "society for synergetics in human sciences" was founded during the most recent symposium. We hope that these activities will turn out to provide incentives and crystallizing kernels for the further interdisciplinary integration of empirical and theoretical research in our field. This is the intention of this volume as well.

Let us briefly summarize the *contents* of the book: Schiepek and Tschacher's article introduces synergetic modelling to clinical psychology. A general overview of the application of synergetics in psychology is given in Haken's contribution.

The issue of deterministic chaos is seen from different angles by several authors: Kratky gives an introduction to chaos in which self-organization is stressed rather than disorder; an der Heiden reviews empirical research on chaotic systems found in connection with dynamical diseases; Ambühl et al. report dimension analyses of the time course of schizophrenia; Steitz et al. discuss options and limitations of chaos methodology when applied to time series in psychology.

Self-organization in cognitive order formation and its importance to psychotherapy is treated by Kruse et al. Schaub and Schiepek present a computer simulation of depression with propositions extracted from a case study. Kriz proposes to simulate family process with the help of equations taken from population dynamics. Schiepek and Schoppek also model time courses of schizophrenia by difference equations. The question of whether psychiatric disorders are dynamical diseases is considered in Emrich and Hohenschutz.

The article of Ackermann et al. uses multivariate time series models in analyzing data from psychotherapeutic processes. Schiepek et al. give a general account of the application of synergetics in psychotherapy. Schneider reflects passages of therapy session transcripts with a model of change in relationship schemata and Znoj proposes a connectionist model of such schemata. Nissen discusses a short sequence from a group therapy using the concept of interpenetration of social and mental systems. Reiter develops the concept of clinical constellation as appropriate especially for clinical practice.

A more philosophical understanding is found in Rössler's formal approach to interactional bifurcations. Schöppe and Brunner derive prerequisites for paradoxical interventions from the theory of operationally closed systems.

Tschacher et al. put forward theoretical considerations, a simulation model and a pilot study on self-organization in small groups. Schiepek and Reicherts report on results gained from the application of the "systems game" methodology. Bergold describes the system of psycho-social and psychiatric health care facilities of an urban area. Krohn et al. view research teams as central agents of the recursive process of scientific knowledge production.

Our thanks go to several persons and institutions. First of all we are indebted to Prof. Hermann Haken for the invaluable support and inspiration during recent years. We especially thank Dr. Angela Lahee from Springer-Verlag Heidelberg for knowledgeable suggestions and corrections to all the texts in this book. The sponsors of the autumn academies also contributed significantly – the Stiftung Volkswagenwerk provided a grant for the symposium of October 1990; the University of Bamberg and the City of Bamberg through its mayor supported and encouraged both symposia by the provision of financial means and amenities. We are also grateful to Martina Weisensel, Veronika Seifert, and Heike Mayer who assisted in organizing the autumn academies.

Tübingen and Bamberg,
Spring 1992

Wolfgang Tschacher
Günter Schiepek
Ewald Johannes Brunner

Contents

Part III   **Self-Organizing Processes in Psychotherapy**

Part IV   **Studies of Social and Mental Self-Organization**

List of Contributors

Klaus Ackermann
  Universität Tübingen, Psychologisches Institut,
  Gartenstr. 29, W-7400 Tübingen 1, Fed. Rep. of Germany

Brigitte Ambühl
  Sozialpsychiatrische Universitätsklinik, Murtenstr. 21,
  CH-3010 Bern, Switzerland

Jarg B. Bergold
  Freie Universität Berlin, Psychologisches Institut,
  Habelschwerdter Allee 45, 1000 Berlin 33, Fed. Rep. of Germany

Ewald Johannes Brunner
  Universität Tübingen, Institut für Erziehungswissenschaft I,
  Münzgasse 22–30, W-7400 Tübingen 1, Fed. Rep. of Germany

Luc Ciompi
  Sozialpsychiatrische Universitätsklinik, Murtenstr. 21,
  CH-3010 Bern, Switzerland

Rudolf Dünki
  Psychiatrische Universitätsklinik, Lenggstr. 31,
  CH-8029 Zürich 8, Switzerland

Hansjörg Ebell
  Universität München, Klinikum Großhadern,
  Marchioninistr. 15, W-8000 München 70, Fed. Rep. of Germany

Hinderk M. Emrich
  Wissenschaftskolleg zu Berlin,
  Wallotstr. 19, 1000 Berlin 33, Fed. Rep. of Germany

Bernd Fricke
  Universität Bamberg, Psychologische Forschungs- und Beratungsstelle,
  Markusplatz 3, W-8600 Bamberg, Fed. Rep. of Germany

Vladimir Gheorghiu
  Universität Bremen, Institute for Cognitive Psychology,
  W-2800 Bremen 33, Fed. Rep. of Germany

Hermann Haken
   Institut für Theoretische Physik und Synergetik, Universität Stuttgart,
   Pfaffenwaldring 57/4, W-7000 Stuttgart 80, Fed. Rep. of Germany

Uwe an der Heiden
   Institut für Mathematik, Zentrum für Nichtlineare Dynamik,
   Universität Witten/Herdecke, Postfach 6260
   W-5810 Witten-Annen, Fed. Rep. of Germany

Charlotte Hohenschutz
   Max Planck Institut für Psychiatrie, Kraepelinstr. 10,
   W-8000 München 40, Fed. Rep. of Germany

Peter Kaimer
   Universität Bamberg, Psychologische Forschungs- und Beratungsstelle,
   Markusplatz 3, W-8600 Bamberg, Fed. Rep. of Germany

Karl W. Kratky
   Institut für Experimentalphysik der Universität Wien,
   Boltzmanngasse 5, A-1090 Wien, Austria

Jürgen Kriz
   Fachbereich Psychologie, Fach: Klinische Psychologie,
   Universität Osnabrück, Knollstr. 15,
   W-4500 Osnabrück, Fed. Rep. of Germany

Wolfgang Krohn
   Universität Bielefeld, Center for Science Studies,
   Postfach 8640, W-4800 Bielefeld, Fed. Rep. of Germany

Peter Kruse
   Universität Bremen, Institute for Cognitive Psychology,
   W-2800 Bremen 33, Fed. Rep. of Germany

Günter Küppers
   Universität Bielefeld, Center for Science Studies,
   Postfach 8640, W-4800 Bielefeld, Fed. Rep. of Germany

Bernd Nissen
   Freie Universität Berlin, Psychologisches Institut,
   Habelschwerdter Allee 45, 1000 Berlin 33, Fed. Rep. of Germany

Wolf Nowack
   Universität Bielefeld, Abteilung für Psychologie,
   Postfach 8640, W-4800 Bielefeld, Fed. Rep. of Germany

Boris Pavleković
   Universität Bremen, Institute for Cognitive Psychology,
   W-2800 Bremen 33, Fed. Rep. of Germany

Michael Reicherts
   University of Fribourg, Psychologisches Institut,
   Route des Fougères, CH-1700 Fribourg, Switzerland

Ludwig Reiter
   Universitätsklinik für Tiefenpsychologie und Psychotherapie,
   Lazarettgasse 14, A-1090 Wien, Austria

Dirk Revenstorf
   Universität Tübingen, Psychologisches Institut,
   Gartenstr. 29, W-7400 Tübingen 1, Fed. Rep. of Germany

Otto E. Rössler
   Universität Tübingen, Institut für Physikalische und Theoretische Chemie,
   Auf der Morgenstelle, W-7400 Tübingen, Fed. Rep. of Germany

Harald Schaub
   Universität Bamberg, Lehrstuhl Psychologie II,
   Markusplatz 3, W-8600 Bamberg, Fed. Rep. of Germany

Günter Schiepek
   Universität Bamberg, Lehrstuhl Psychologie III,
   Markusplatz 3, W-8600 Bamberg, Fed. Rep. of Germany

Henri Schneider
   Stapferstr. 8, CH-8006 Zürich, Switzerland

Arno Schöppe
   Universität der Bundeswehr Hamburg, Fachbereich Pädagogik,
   Holstenhofweg 85, W-2000 Hamburg 70, Fed. Rep. of Germany

Wolfgang Schoppek
   Universität Bayreuth, Lehrstuhl für Psychologie,
   Geschwister-Scholl-Platz 3, W-8580 Bayreuth, Fed. Rep. of Germany

Michael Stadler
   Universität Bremen, Institute for Cognitive Psychology,
   W-2800 Bremen 33, Fed. Rep. of Germany

Arno Steitz
   Universität Tübingen, Institut für Theoretische Astrophysik,
   Auf der Morgenstelle, W-7400 Tübingen 1, Fed. Rep. of Germany

Uta Streit
Universität Tübingen, Psychologisches Institut,
Gartenstr. 29, W-7400 Tübingen 1, Fed. Rep. of Germany

Felix Tretter
Bezirkskrankenhaus Haar, W-8013 Haar, Fed. Rep. of Germany

Wolfgang Tschacher
Sozialpsychiatrische Universitätsklinik,
Murtenstr. 21, CH-3010 Bern, Switzerland

Ilse M. Zalaman
Universität Tübingen, Psychologisches Institut,
Gartenstr. 29, W-7400 Tübingen 1, Fed. Rep. of Germany

Hansjörg Znoj
Universität Bern, Department of Clinical Psychology,
Gesellschaftsstr. 49, CH-3012 Bern, Switzerland

Part I

Synergetics in Psychology:
Basic Issues

Application of Synergetics to Clinical Psychology

Günter Schiepek and Wolfgang Tschacher

With 8 Figures

The present contribution shows that clinical psychology as a theoretical and applied science is essentially concerned with phenomena of autonomous order formation and transformation within complex, dynamic systems. Evidence for this is provided by examples derived from clinical assessment, epidemiology, etiology research, psychotherapy and the dynamics of multipersonal systems. It is described how the theoretical and methodological instruments provided by synergetics may be employed in analyzing, explaining and perhaps even modifying such processes of self-organization. The concepts utilized in synergetics are explained and the synergetic systems concept is distinguished from various other conceptualizations (such as interactional constellations of several individuals or the systems concept of autopoiesis).

1. Evolution of Order of Psychosocial Phenomena

1.1 Why Synergetics in Clinical Psychology ?

There is no doubt that synergetics has become one of the most encompassing theoretical conceptualizations in modern science since it was introduced more than two decades ago (e.g. Haken & Graham 1971). It is a scientific concept which keeps its promise of offering an integrating and general approach. The precise mathematical formulation of its theoretical assumptions (Haken 1983a, b; 1990a; see Haken 1988 for macroscopic synergetics) forms the core of this approach. A large number of areas of application (domains), beyond the original experimental paradigm, laser physics, were identified after these central formulations had been accordingly extended and specified, thus fulfilling the requirements of a structuralistic view of theorizing (e.g. Stegmüller 1973). Synergetics has been applied to scientific fields such as thermodynamics, fluid dynamics, phenomena of collective and dynamic order formation in chemistry, biochemistry, meteorology, biology (e.g. morphogenesis and population dynamics), neurobiology, cognitive science, sociology, and economics. Even the basic disciplines of psychology have already begun to model perception, cognition and psychological development, all of which exhibit characteristics of complexity and discontinuous dynamics (phase transitions), using synergetic conceptualizations (see Haken, this volume; contributions to Haken & Stadler 1990, e.g. Stadler & Kruse 1990; Kruse & Stadler 1990; Bischof 1990). As has been pointed out (Stadler &

3

Springer Series in Synergetics, Vol. 58   Self-Organization and Clinical Psychology
Editors: W. Tschacher, G. Schiepek, and E.J. Brunner   © Springer-Verlag Berlin Heidelberg 1992

Kruse 1990), there are obvious parallels to German Gestalt psychology (see Köhler 1920; 1940; Wertheimer 1912; 1923; and others).

However, apart from certain exceptions (Kriz 1990), synergetic concepts have not been introduced into clinical psychology. In general, this discpline finds it hard to integrate systems theory and ideas of self-organization, as can be observed from the latest textbooks (Reinecker 1990a; Baumann & Perrez 1990). Nevertheless the fact that there is no lack of differing concepts, this must not necessarily be a disadvantage. On the contrary: First of all, the utilization of theories from other disciplines for heuristic purposes does not guarantee better quality, as one might have to "bend and distort" one's own field of interest to make it fit the theory in question. Secondly, synergetic terminology which uses terms such as "control parameters" or "slaving principle" poses the danger of inviting technical thinking to come in through the systemic back door (through which it had not been expected) and enter the psychosocial practice once again. This would be even more regrettable, as academic psychotherapy research has only just begun to abandon with great effort technologically oriented concepts of intervention with the help of the approaches of community psychology and heuristic concepts of therapy (Caspar & Grawe 1989). Thirdly, it is evident that it is not possible to apply a certain theory to every field of interest. There certainly are areas of research in clinical psychology which, at first glance, deal neither with phenomena of dynamics nor with order formation. These include classification, epidemiology, outcome evaluation and cost-effectiveness evaluations.

However, if one takes a closer look at clinical psychology, it soon becomes evident that much of the work in clinical research and practice is aimed at the development and change, as well as the identification and recognition of structures of order within complex, dynamic systems.

1.2 Classification and Clinical Assessment

The very first step, psychodiagnostic classification, the controversial assignment of human suffering and interpersonal problems to nosological entities, itself constitutes a process of identifying complex patterns, or, in other words, of pattern recognition. This is substantiated by findings indicating how quickly this process of diagnostic pattern-identification takes place (see Blaser 1977). Obviously only a few pieces of information may already suffice as the basis for a symmetry breaking in favor of a certain order parameter (in this case a certain discrete diagnostic category), and further information is utilized merely to confirm or differentiate the decision made. Principles governing Gestalt recognition and associative memory processes contribute to rendering fragmentary patterns complete and to producing the most concise pattern possible (potential valley). Haken's postulate that "the order parameters have features required by gestalt theory" (1990b, p. 11) might also apply to processes of clinical assessment, and thus present an alternative to regarding clinical assessment as guided by criteria of inclusion and exclusion as well as by decisions following the guidelines of flow charts (compare DSM III and

4

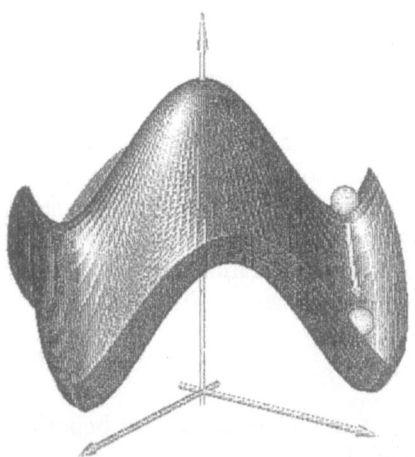

Fig. 1: Visualization of the potential landscape for two prototype patterns and two features (from Haken 1990b, p. 13).

DSM III-R, APA 1987): The process of clinical assessment being governed by principles inherent to the synergetic computer. In this view, diagnosticians' decisions are assumed to be based on standard examples or ideal types stored as prototype pattern $\underline{v}_u$, rather than on lists of criteria. A test pattern $\underline{q}(0)$ containing (possibly incomplete) initial information is transformed during the process of recognizing complex patterns, until it coincides with the most similar of the stored prototype patterns $\underline{v}_u$

$$\underline{q}(0) \longrightarrow \underline{q}(t) \longrightarrow \underline{v}_u. \tag{1}$$

When potential landscapes are utilized as a means of representation, the process described above is illustrated by that of a ball rolling into a potential valley (Fig. 1).

The basic equation representing this process of pattern recognition is given as

$$\dot{\underline{q}} = \sum_u \lambda_u \, \bar{A} \, \underline{v}_u \, (\underline{v}_u^+ A \, \underline{q}) + N \, (\underline{v}_u^+ A \, \underline{q}) + \underline{F}(t), \tag{2}$$

where

$\underline{q}$ is the state vector which encodes a specific feature

λ_u represents u different attention parameters designating attention biases which may be caused by a specific sensitivity to certain phenomena, by emotions (such as avoidance or fear of proximity) or by certain diagnostic interests (e.g. due to a   specific focus of research)

A   is an adaptation operator which influences $\underline{q}$ and leads to its matching with $\underline{v}_u^+$

$\underline{v}_u$ represents u different prototype patterns with the external product of vectors $\underline{v}_u \cdot \underline{v}_u^+$ serving as a learning matrix ($\underline{v}_u^+$ is the vector adjoint to $\underline{v}_u$)

5

N is a nonlinear function N $(\underline{v}_u^+ A\underline{q})$ which serves to differentiate between different patterns and their degree of saturation and

$\underline{F}(t)$ represents internal and external fluctuations of the system.

Thus, this equation covers basic psychological aspects (such as the individual's learning history and hypothesis formation, his ability to differentiate, as well as motivational biases of perception), as also discussed in the empirical literature on cognitive science, perception psychology and clinical assessment.

The order parameters ξ_u of pattern recognition processes thus described can be defined by means of the scalar product

$$\xi_u(t) = (\underline{v}_u^+ A\underline{q}) \tag{3}$$

with the order parameters representing specific patterns of behavior, perception and cognition on a macroscopic level.

The analogy between pattern formation and pattern recognition is the basis for the development of the synergetic pattern recognition computer: "...in pattern formation when a part of a system is in an ordered state, it can generate its corresponding order parameter which in turn forces the rest of the system into the totally ordered state. Similarly, in pattern recognition a set of known features can generate their corresponding order parameter which in turn forces the rest of the system into the total state that represents the total pattern in the sense of associative memory" (Haken 1990b, p. 11).

This approach may also be employed to explain clinical assessment processes taking place within a very short time on the basis of incomplete information (compare above; Blaser 1977). Also, it would be possible to generate further hypotheses, for example concerning phenomena such as multistability within certain stages of diagnosticians' clinical assessment.

A very well-known empirical finding concerns the α-error of the re-diagnosis of patients who have already been diagnosed previously. As shown by Rosenhan (1973), diagnostic labels are hardly ever abandoned, even though the required criteria are not or no longer observable. Thus, patients who had been diagnosed as "schizophrenic" were labeled "remitted schizophrenic" on being discharged from the clinic, although their diagnoses had not been justified at any time. This phenomenon may be understood in analogy to that of hysteresis as a desynchronisation of the phase transitions on the way from percept a to percept b as opposed to that from percept b to percept a (compare Schwegler 1978; Haken 1990b). This effect is illustrated in Fig. 2.

Looking at the row of pictures from left to right, one first sees a face, which is then deformed in an indefinite manner until one suddenly perceives the figure of a girl. However, if one begins to look at the pictures starting from the right-hand side, the percept of the figure of the girl persists far beyond that point where it had become a girl when looking at the pictures in the opposite direction. The field of synergetics enables a mathematical formulation of this delayed switching from order parameter ξ_1 to ξ_2 as opposed to that from ξ_2

to ξ_1 (see Haken 1990b, p. 22ff.), an effect which is in accordance with assumptions of the 'hypothesis theory' of perception (e.g. Bruner & Postman 1949; Lilli 1978).

1.3 Analytic Epidemiology

Another field of clinical psychology where questions are usually not approached from the perspective of self-organization is that concerned with epidemiology. According to the traditional approach, the prerequisites of descriptive epidemiological research are the ability to clearly define and classify the disorder in question, coupled with minute and painstaking field investigations, especially with regard to primary surveys. However, questions of pattern formation within complex systems comprising large numbers of individuals and possible sources of influence come into play as soon as the next step towards analytical and interpretative epidemiology is to be taken. At any rate, it is the task of analytic epidemiology to identify and explain the conditions for the occurrence of temporal and spatial clusters of certain mental disorders or their irregular distribution within a multidimensional space of possible causal factors. Apart from the covariation of the distribution of mental disorders with factors such as social stratum, family environment, residential area, biopsychological stress factors and age, there are also temporal changes in the frequency of occurrence, which often seem to develop in a discontinuous manner. Thus, Reinecker (1990b, p. 33) mentions drastic changes in the frequency of occurrence of several different disorders within only a few years or decades. Examples of this are the change in the rate of alcoholism after the Second World War, changes regarding the diagnosis of "hysteria" since the beginning of the century, the introduction of the diagnoses of "panic attacks" or "bulimia" into the DSM III (APA 1987), etc.

All in all, the occurrence of such patterns points to the conclusion that the factors involved are correlated rather than independent of each other. They exhibit cooperative behavior — thus violating the central limit value theorem of probability theory (see Haken 1983a, chap. 2) — suggesting synergetic explanations for these phenomena of pattern formation. It is feasible that synergetic modeling could give a far better account of the complexity of these phenomena than is possible by simply referring to general presumptions such as "social causation" or "social drift" (see Dohrenwend & Dohrenwend 1981).

1.4 Etiological Research

Another field within clinical psychology where researchers have tried to manage without self-organizational models is that concerned with etiology. Rather, it has been the aim to identify individual causal factors which lead to mental disorder becoming overt when affecting a vulnerable person. This kind of approach was then progressively called "multifactorial", "interactive" or "transmissive", while still implicitly starting out from the principle of strong causality, e.g. similar causal factors leading to similar effects. It is only on the basis of this principle that etiological research concerned with attributing a certain configuration of psychosocial problems to a certain configuration of causal factors makes any sense. However, questions concerning the problem of causality usually remain unsolved. As has already been demonstrated by the critics of classical methodology, neither the usual correlational study nor the cross-sectional study at two or more points in time provides an adequate answer to questions of causality. But even when the highly praised prospective longitudinal design is employed, it is not at all possible to include every relevant factor, which means that an arbitrary number of rival explanations will remain untested depending on which theoretical approach has been chosen. It should be the main focus of research, that is to say of theoretical conceptualization and empirical survey, to examine how the relevant causal factors interact in the course of time and lead to specific configurations of psychosocial problems or disorders under specific, varying initial conditions. However, when the complex dynamics are taken into account, this again modifies the specific meaning of the individual factors. Moreover, even those etiological models promising the consideration of processes in an integrating or interactive way do not go beyond simply enumerating subprocesses, postulating inadequately specified interactive processes or relating subprocesses in such a way (e.g. assuming positive feedback loops only) that it seems implausible from a systems-theoretical point of view. This argument however also applies to those recent models which are adequately elaborated regarding the individual factors (e.g. Lewinsohn et al. 1985, with regard to the etiology of depression; see Hautzinger & de Jong-Meyer 1990 for a survey).

Mental suffering or "disorder" has been conceived of in terms of highly structured coherent states which impair and determine an individual's mental and social being since as far back as the classical phenomenological approach to psychopathology; put differently, these states are seen as considerably reducing the individual's "degrees of freedom". Very often we can observe discontinuous transitions between different pathological states or between states of health and illness which resemble those nonlinear phase transitions observable within highly complex physical systems (compare Haken 1989). This has been demonstrated, for example, with regard to uni- or bipolar cyclic depression (e.g. Wehr & Goodwin 1979) and schizophrenia. Strauss (1989) emphasizes the discontinuity in the evolution of patient functioning over time and describes several temporal patterns found in schizophrenia (e.g. "woodshedding", "the low turning point", "oscillating levels of function"; see also Schiepek, Schoppek & Tretter, this volume). It therefore seems obvious to attempt the utilization of synergetic concepts with respect to etiological research and dynamic psychopathology (Schmid 1991). After all, synergetics presents an elaborate theory of disorder-to-order as well as order-to-order transitions within highly complex nonlinear systems. (With regard to clinical psychology, these are biopsychosocial systems.) Moreover, this view is supported by the fact that, as mentioned above, the principle of strong

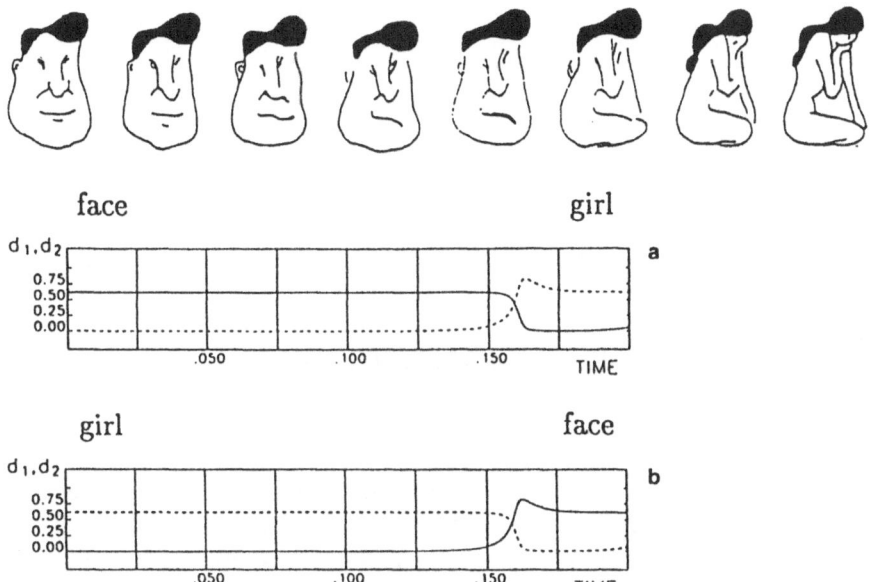

face   girl

girl   face

Fig. 2: Visualization of the hysteresis present in the perception of a man's face or a girl.

(a) Transition from the perception of a man's face to that of a girl.

(b) Transition from the perception of a girl to that of a man's face. Note the difference in the switching between figures (a) and (b). (from Haken 1990b, p. 23/24).

causality obviously does not apply with respect to etiological processes, something which is indicated, e.g. by the unspecific effects of factors such as traumatic childhood experiences or life events.

Different initial conditions and different degrees of stress may result in similar pathological states (see for example Akiskal & McKinney 1975, regarding the concept of the "final common pathway" of the evolution of depression), while only minimum fluctuations within an individual's intrapsychic or environmental conditions may also lead to quite different effects (see the concept of "critical instability", Schiepek & Schaub 1991; Schaub & Schiepek, this volume).

Thus, two characteristic features of strange attractors (see Tschacher 1990) are combined with regard to etiological processes: the postulate of convergence (attraction) and the postulate of divergence (expansion), resulting in the crucial dependence of the evolution of such processes upon their initial conditions (Fig. 3).

Considering these premises, one may argue that the unspecific effects of life events (see various contributions to Katschnig 1980, especially Brown, Harris & Peto 1980) are no artefact of inadequate research methods but are rather a substantial feature of the nonlinear dynamics of etiological processes.

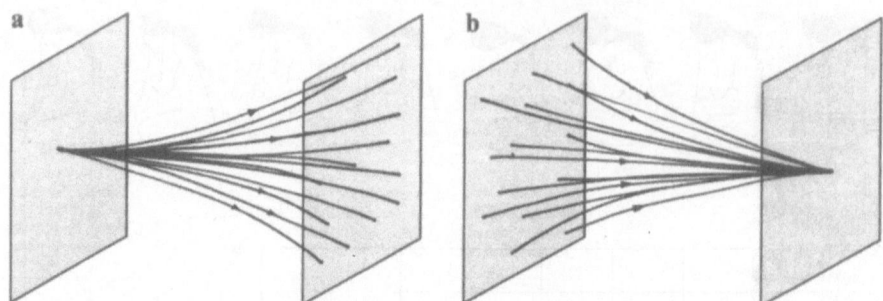

Fig. 3: Beyond the principle of strong causality: Convergence (a) and divergence (b) in the realm of "chaotic" causality.

1.5 Psychotherapy

The application of the concept of nonlinear phase transitions seems especially appropriate with regard to processes taking place within psychotherapy. They may be considered to represent nonlinear phase transitions between different biopsychosocial states of order. This view is supported by both practical experience and detailed analysis, indicating that therapeutic change obviously does not take place in the form of a continuous, incremental process, as may be suggested for example by the model of Skinner's shaping. On the contrary, we can observe the occurrence of sudden inspiration, "aha-experience" or the restructuring of problems — phenomena similar to the oscillating perception of ambiguous patterns — both during problem-solving and within psychotherapeutic processes (Stadler & Kruse 1990; Kruse & Stadler 1990; Mahoney 1980). Therapists are acquainted with the phenomenon of casual remarks or ideas being picked up by clients only after a substantial delay, and it often seem to be these fluctuations which then trigger the jump into a modified pattern of experience and behavior. This may be considered an effect similar to that described by Zeigarnik, who postulates a period of latency or distraction to be necessary after several trials of learning in order that the individual be enabled to spontaneously achieve a higher level of functioning than before with regard to the pattern of behavior to be acquired. It is the occurrence of phenomena such as those described above which supports the postulate of nonlinear processes being effective within psychotherapy. The transfer of methods applied in chaos research to the analysis of psychotherapeutic processes is therefore quite obvious (Fuhriman & Burlingame 1991; Burlingame et al. 1991). However, this requires a sufficient amount and range of quantitative, time-serial data, collected by employing an instrument which will provide measurement of adequately high resolution. Our research group at the University of Bamberg is currently preparing such time series by subdividing the therapy transcripts and video recordings of two completed therapies (twelve sessions each) into ten-second units and coding the interactional plans (see Caspar 1989) of both client and therapist. The time-series analysis of these data will allow us to ascertain their correlational dimension, and thus obtain empirical evidence for the chaotic dynamics effective within therapeutic processes.

The process of acquiring complex patterns of behavior was conceptualized as a process of "pattern formation" or, put differently, as a phase transition between different patterns of behavior by Kelso (1990), thereby presenting a synergetic learning theory. Apart from taking into account the intrinsic dynamics of learning, he especially succeeds in allowing for the consideration of intentional behavior, which, after all, represents an important aspect of human learning. Learning processes are considered as intentional, if information about the behavioral pattern to be learned is able to modify the intrinsic dynamics of the previous pattern, thus enabling a phase transition to take place (compare the purpose of goal identification and anticipation within psychotherapy). This implies that intentions act in the same space of collective variables as that in which the intrisic patterns are measured. "Intentional information defines an attractor in that space and is meaningful to the extent that it attracts the system toward an intended behavioral pattern. At the same time, intentions are restricted by the intrinsic dynamics, in that the ability to perform a particular pattern is influenced by the relative stability of the available patterns. In short, intentions parameterize the dynamics but are in turn constrained by the dynamics" (Kelso 1990, p. 261).

Kelso (1990) and co-workers have carried out numerous studies aimed at investigating the phase transitions between different rhythmic movements (e.g. parallel as opposed to antiphase finger movement). He concludes that in the case of intended behavior not corresponding to the intrinsic dynamics of the system — which is the usual case as far as learning processes are concerned — the question of which attractor will determine overt behavior depends on the intensity of the behavioral information given. "1. If behavioral information is weak or absent, the system's behavior is determined primarily by the intrinsic dynamics; 2. If behavioral information specifying a required behavioral pattern is strong, the system will possess a stable state at that required pattern" (Kelso 1990, p. 262).

Yet it is questionable with regard to therapeutic processes whether goal-oriented behavioral information alone is sufficient to effect a phase transition of the behavior of a biopsychosocial system into a desired state of attraction. Rather, one has to expect bistable or even multistable system behavior once the system begins to develop one or more attractors (or potential valleys) in addition to the attractor determining behavior so far (cf. Schiepek, Fricke & Kaimer, this volume, Fig. 1). We may also recognize such multistabilities during practical therapeutic work. Clients sometimes seem to process their experiences within quite different mental and emotional states, or "states of mind" (see Horowitz 1987). They may also repress or dissociate insights they have already gained, or reverse behavioral changes and pick them up again at a later date to work on them at an even more profound level. Planned, goal-oriented development is therefore rather unlikely during periods of psychosocial instability, because new attractors which determine behavior have not yet evolved; also, behavioral information determining emotional and social processes is still being worked out during a process of co-evolution on the basis of the intrinsic emotional dynamics of the client. It may be assumed that these processes are very sensitive to external fluctuations and take place at an unconscious rather than a conscious level (see Kruse & Stadler 1990; Revenstorf 1985, 1990).

At any rate, the synergetic concept of phase transitions between biopsychosocial states of order encourages researchers to take a closer look at characteristic features and processes of stability and instability within psychotherapy. The identification and empirical description of patterns or states of order within biopsychosocial systems is a fundamental prerequisite of this kind of research. We are currently undertaking efforts to this end by employing the concept of "lifestyle scenario" and measures proposed for its analysis (see Schiepek, Fricke & Kaimer, this volume).

Another systematic approach to pattern recognition in this difficult area of observation, termed "configurational analysis", has been developed by Horowitz (1987). "The approach is to segmentalize behavior and experience into states that recur; each "state of mind" is then explained in terms of the schemata that organize it, and the transition of states is described in terms of how representations and expressions of meaning are regulated. The stability of a given state will be explained essentially by a person's self-concept and inner model of relationship to others. The transitions from one state to another will be examined primarily in terms of events, including internal emotions and motives and the processing or avoidance of information related to these events, emotions and motives" (Horowitz 1987, p. 1). It can be expected that the concept of "states of mind", together with the corresponding configurational analysis and the synergetic concept of order-to-order transitions, will have a stimulating impact on therapy research.

1.6 Multipersonal Systems

In the field concerned with the dynamics of multipersonal systems (e.g. couples, families, groups, institutions) — in contrast to those areas of research in clinical psychology mentioned so far — phenomena of spontaneous order formation have always been regarded as presenting plausible explanations. It is a common experience that a group to which people are arbitrarily assigned develops a characteristic life of its own after only a short period of time. On Saturdays one may observe phenomena of coherence, together with a distinct reduction of degrees of freedom, among the spectators of soccer games. This example also shows that spontaneous, collective order formation (self-organization) usually prevails against (attempts at) organization by others (e.g. security guards) — something which becomes only too evident when one takes into consideration the catastrophic events in soccer stadiums in the past few years.

If we are concerned with a group consisting of a moderate number of people who did not know each other before, we can observe spontaneous processes of structure formation and identity development. The group will evolve to offer specific role positions and relations as well as its own we-feeling. We all know from experience that amazing things happen within such "systems": Dynamic processes and overall qualities may evolve which cannot simply be explained by regarding them as the sum of individual behavior or individual affective-cognitive schemata (Tschacher, Brunner & Schiepek, this volume). The term designating this process of *bifurcating* into the regime of newly developed attractors is that of "emergence" or, put differently, *emergent order formation*.

However, processes of emergent order formation also occur when people are not associated as a group, but rather widely dispersed (locally distributed), as within a society. An example of this is the evolution of public opinion (see Weidlich & Haag 1983). The changing of collective "modes" of behavior and attitudes regarding health care is of special importance for health psychology as well as clinical psychology, as these disciplines are concerned with information about, and the prevention of, mental and somatic disorder.

Processes occurring at the level of social systems are not necessarily more complicated than those occurring at the level of individual mental systems — it is merely a question of different levels of resolution. It is a well-known phenomenon in synergetics that the self-organizing, spontaneous activities of a system (pattern formation) occur along with a substantial reduction of degrees of freedom regarding the behavior of its elements, or, as Willke (1983) puts it: "The whole is *less* than the sum of its parts". A vivid example of this may be seen in the restrictions imposed on the possible behavior of family members and their relations towards each other by rigid family structures (compare Minuchin 1974). Apart from stable patterns, however, we may also expect chaotic dynamic processes to occur within social systems. As shown by Kratky (this volume), chaotic dynamics may exhibit rather different numbers of degrees of freedom, and it is possible to distinguish between low-dimensional and high-dimensional chaos. At any rate, chaotic dynamics represent a specific kind of order, one which can be distinguished from disorder or white noise by means of the identified dimensionality. The properties of information reduction (convergence, attraction) and information production (divergence) characteristic of strange attractors was already mentioned in Sect. 1.4. At the present time, however, empirical evidence is too sparse to answer the question as to whether — apart from physical, chemical, biological and cognitive systems (regarding the latter, see Nicolis 1986; Freeman 1990) — chaos also plays a part in the evolution of psychosocial systems, not only in a metaphorical but also in a mathematical sense (for an analysis of the correlational dimension of a group-therapy process, see Burlingame et al. 1991).

However, phenomenological evidence is available: The development of interpersonal relationships is often dependent upon *crucial initial conditions*, e.g. subtle nuances conveyed within the first impression, tiny gestures, minute bits of preliminary information. This *butterfly effect of social life* may be especially noticeable at the beginning of a love affair. Every smile and eye contact, every sign and its possible meaning may have farreaching consequences, may initiate a development, may invite or reject — a field of great insecurity (compare the figure of "Que faire" in Roland Barthes' "Fragment d'un discours amoureux" 1977). Over and above the special case of a beginning romance, every interpersonal encounter is characterized by double contingency (Luhmann 1984). This is a highly sensitive and instable state of symmetry, in which even minute fluctuations may suffice to break the symmetry of communication in favor of specific paths of communication, in which the partner may join again (see Markowitz 1991 with regard to the problem of double contingency).

Another indication of the chaotic features of social life can be concluded from the phenomenon of the *self-similarity* of social patterns. We often recognize similar patterns of interaction at different levels of communication (communication, meta-communication, meta-meta-communication etc.). An example: During a group-therapy session, the participants work on a problematic situation described by one of the members. The situation is illustrated by means of a group sculpture. The client explains how as a boy he was always driven to the wall for tiny reasons, how he had to appease his dominant accusers as a guilty child. While taking part in the sculpture he again makes excuses addressed towards his parents and significant others. When the sculpturing is finished, the actors are asked to report their observations, thoughts and feelings in order to be able to leave their roles. While doing so, a pattern of interaction develops which is similar to that represented in the sculpture. The main actor makes excuses — this or that wasn't that way, he only wanted to ..., he was sorry that, but ..., etc. It seems as if the accounts of the other members of the group have an invasive effect upon the client. The therapist then tells the group about his observations concerning the group processes here-and-now. But even after his intervention the process is repeated in an analogous way. Implicitly, the client makes excuses for constantly making excuses, and the rest of the group tries to explain his behavior in a reproachful way. Further attempts at clarification fail in the same way and the group dynamics become more and more entangled until finally all the participants feel guilty, angry and exhausted.

This is to show that meta-communication is not to be considered a universal remedy or cure-all; often it leads to nothing but the exact pattern from which it originally started.

We may recognize phenomena of scale-invariant self-similarity not only when analyzing the recursive loops of communication systems, but also when taking a closer look at the lifestyle of an individual across different time-scales. The specific patterns of an individual's lifestyle can be found on a large as well as a small scale. On a large scale, this may be represented as the way in which one makes difficult decisions or the avoidance of decision-making, the way in which a person arranges his or her relationships, or the way in which one meets others in passing or in which a person uses his or her time (Ciompi, personal communication; compare the concept of the lifestyle scenario, Schiepek, Fricke & Kaimer, this volume).

1.7 Conclusions

Each of the previous examples shows that the nonlinear processes and phenomena of self-organization occur everywhere within the traditional areas of the research and practice of clinical psychology. Clinical psychology can therefore be regarded as a theoretical and applied science, the genuine task of which is to analyze, and act within, complex dynamic systems, be they of biological, psychological or social nature. Normally, biological, social and psychological processes will be intricately interwoven, so that clinical psychology deals with biopsychosocial systems.

In order to gain an understanding of the dynamics of evolution of such systems, theories of nonlinear systems and especially synergetic conceptualizations will be necessary in the future. It is these theories which provide universally applicable scenarios for the explanation of the evolution of systems from disorder to self-organized order and from self-organized equilibrium conditions, via periods of instability, to new and more complex states of order. Generally speaking, this happens when certain parameters are changed which regulate the distance of the system from the point of thermodynamic equilibrium, and thus the extent of nonlinearity.

These transitions are characterized by bifurcations. Once the system has approached this point, minute fluctuations may "throw the switch", leading to one specific state of equilibrium or another. At points of bifurcation the attractors governing the system behavior are "flat" (compare the effects of critical slowing down), so that it is random fluctuations which effect the decisive symmetry breaking. Thus, the history of a nonlinear system is governed by an interplay of chance and necessity or, as proposed by evolution theory, variation and selection.

It should be clear by now that the synergetic approach to phenomena treated by clinical psychology neither leads to physicalist reductionism nor means mere metaphorical thinking. On the contrary, it presents a consistently *empirical* approach which sets constructive scientific endeavor against the renunciation of empirical analysis to which systemic approaches were dedicated because of their misinterpretation of constructivism (compare various contributions to this volume). Psychological synergetics can be regarded as an approach which allows the construction of adequate, sufficiently complex and scientifically fruitful models for the phenomena with which clinical psychology is concerned, models which, moreover, can be integrated into a coherent, interdisciplinary theoretical framework.

2. Synergetic Modeling of Complexity and Dynamic Processes Within Clinical Psychology

As described earlier, it is a central concern of clinical psychology to explain processes of order formation and transformation within complex dynamic systems, and also to promote such processes in specific ways, as in the case of psychotherapy.

We are thus faced with the question as to what can be determined to be a "system" or, in other words, what is an adequate, empirical conceptualization of a system, enabling us to apply this construct in an adequate way to empirical analysis. The following first describes different conceptualizations of this construct, in order to allow a distinction to be made from the conceptualization proposed here.

2.1 What is Regarded as a "System"?

There are various different normative definitions in clinical psychology as to what is regarded as a system. How this concept is defined is determined by specific conditions with respect to the person using this concept, such as his/her theoretical background, which discipline he/she is committed to, or the intended area of application. In principle, the conceptualization of the target of research as a system, together with the decision as to where the boundaries of the system are considered to be, depends upon the researcher's own disposition, although this freedom may be restricted by common institutional, methodological or theoretical agreements. It is the usual practice to declare as environment that part of the target of research considered to be outside the delimitations of the system, and to relate variations within the environment to system behavior. Incidentally, this corresponds to traditional experimental procedure.

The different varieties of the concept of "system" each supply a formal framework which must then be filled with actual content. The concept alone does not explain anything. It seems necessary here to point out that a semantic connotation has been attached to the term "system" (in the literature on family therapy or systemic therapy, for example) suggesting that the mere utilization of this term to designate families or other multipersonal constellations would present an explanation.

2.1.1 Systems as Interactional Constellations of Several Individuals

It has become especially popular within the family therapy approach to define a family or a group of persons as a system. The individuals (e.g. family members) are considered the elements, and the relationships or communicative interactions between individuals or subgroups (e.g. parent subsystem, sibling subsystem) the relations of the system. The system dynamics are considered the result of interpersonal communication and are interpreted in terms of dynamic relationship processes.

This conceptualization is supported by its face validity and simple descriptiveness, which are both given for assessable face-to-face interaction within families or small groups only. Modeling approaches such as structural diagrams (Minuchin) or family genograms or sculptures (Satir) are adequate for systems of this size. As soon as one is concerned with organizations, however, one is forced for reasons of practicability to abstract from individuals — there are too many. (Organigrams show functional units instead of individuals.) Yet it is an advantage of this individual-centered conceptualization that phenomena of communicative emergence are considered without losing sight of the individual person/personality and his/her psychological characteristics. This is of special importance for clinical and especially therapeutic application. If phenomena of transpersonal or interpersonal emergence were acknowledged, this sociopsychological systems concept could be linked with psychological data or theories. Thus, an unstable state of floating between two different risks would result: Either the main responsibility for complexity would be attributed to the elements of the system, that is to say to the individual person and his/her intrapsychic characteristics, thereby making the systems concept seem artificial, or, on the other hand, individual characteristics would be treated only as an epiphenomenon of characteristics of communication.

As could be expected, this instability exists only in theory. In reality, however, the symmetry is broken in favor of one of the two possibilities. Within family or systemic therapy, intrapsychic and biological processes were neglected while interpersonal processes were emphasized (see Minuchin or early publications by Selvini-Palazzoli and her team, for example), at least for as long as family therapy approaches were still developing their identity, thus rendering systems theory a rather naive communications theory. However, the examples mentioned in Sects. 1.2-6 should make it clear that a reduction of systems theory to an interpersonal communications theory cannot be the aim of clinical psychology. It is an absolute necessity that intrapsychic and biological processes be taken into consideration when one is concerned with biopsychosocial systems.

Another problem inherent to the systems concept described above has been pointed out by Kriz (1985). The relations of a system thus defined, i.e. communication, not only relate individuals, but are also concerned with how these individuals themselves relate to each other and, in turn, how individuals relate to this relationship, etc. It is easy to see that the problem of iterated recursive relations can be solved much better by applying Luhmann's (1984) systems-theoretical approach, which refrains from the naive assumption that communication refers to individual persons.

2.1.2 The Systems Concept of Autopoiesis

A recent and highly elaborated systems concept has been developed on the basis of the concept of autopoiesis and introduced to biological (an der Heiden, Roth & Schwegler 1985), psychological (Schiepek 1990; 1991, chap.

IV) and social (Luhmann 1984; Teubner 1990; Willke 1989) theories of self-organization. According to this conceptualization, a system is conceived of as an entity which produces and reproduces its components, its structures and its autonomous boundary by means of the self-referential operation of its components (detailed descriptions of this approach are given in the literature quoted). Within the framework of a general theory of self-referential systems (TSS), this systems concept allows an explanation of psychological and social processes (of structure formation and transformation, for example), and therefore its application within clinical psychology should be emphasized. In addition to the synergetic approach, the theory of self-referential systems offers the second main systems-theoretical approach to clinical psychology and is aimed at a consistent and coherent, but primarily qualitative, theorizing. However, this means that, in contrast to synergetics, which originates from formal mathematical modeling, operationalization and empirical analysis are much more difficult. By defining the basic components of a system (e.g. biomolecules, affective-cognitive units, communication) a distinction is made between qualitatively different (biological, psychological, social) types of systems. Following this distinction, the interaction of the components is analyzed. Synergetics, on the other hand, proceeds from the assumption of interactions taking place at a high degree of resolution between biological, psychological and social processes on a microscopic level to create a coherent pattern on a macroscopic level characterized by biological, psychological and social phenomena. Moreover, the synergetic approach primarily treats processes of order transformation (i.e. disorder-to-order and order-to-order transitions), while the theory of self-referential systems is primarily concerned with questions of how the organization and boundaries of a system are maintained (see Maturana 1982). This can be understood in the light of the fact that its origin in biology resulted in the investigation of internal processes maintaining the existence of autopoietic systems becoming the main focus of attention. It has been only in the second place and, more importantly, outside the domain of biology that questions of self-organization (disorder-to-order transitions) have been raised (see Teubner 1990). However, it has been possible to further elaborate the theory of self-referential systems with regard to psychological and social systems and make it a fruitful theory of self-organization. As far as biological systems are concerned, it is the hypercycle theory proposed by Manfred Eigen which presents a genuine theory of self-organization or, put differently, a theory of the evolution of life.

Despite their similarities, we plead for a strict distinction between the theory of self-referential systems, on the one hand, and synergetics, on the other, in order to avoid confusion with respect to terminology and content. Keeping these theories apart also means that they can enter into fruitful scientific competition with respect to rival explanations of the phenomena of self-organization.

2.1.3 The Systems Concept of Synergetics

The systems concept of synergetics cannot be described by means of a few single key words, as it comprises a whole model of the self-organization of complex systems (see below, Fig. 4). In the first place, synergetics assumes that a self-organizing system is constituted on a microscopic level by a large number of components, the behavior of which is characterized by a large number of degrees of freedom. To give an example, the molecules of a gas can be regarded as solid bodies moving about in an ideal Newtonian world and colliding every now and then in an elastic way. It is possible to draw up equations with respect to physical, chemical and biological systems describing such microscopic dynamics on the basis of substantiated assumptions about the laws governing the behavior and interactions of the microscopic components of the system in question. This allows a modeling of the microscopic dynamics, i.e. the behavior and interaction of a large number of components. Knowledge about the order parameters active on the macroscopic level thus results from knowledge about the laws effective on the microscopic level (Haken 1983a).

For various reasons, such a microscopic modeling has not been possible up to now, and might even be quite impossible in principle, in those biopsychosocial systems in which we are interested within the scope of psychological and psychiatric research. Neither field nor laboratory research enables an understanding that is in any way complete of the high degree of complexity of the components and processes which interact in multifarious ways, are linked up or act in parallel on the biopsychosocial microscopic level. (Examples of laboratory research concerned with processes on the microscopic level are the evoked-potentials research within neurophysiology (e.g. Birbaumer & Schmid 1990) and the approach of Ekman & Friesen (1978) aimed at analyzing microscopic movements of emotion expression.)

Here it is necessary to insert a word of caution about the risk of being misunderstood: We do not intend to support a simple, reductionist view by reducing psychological phenomena to the microscopic material components of a physical or chemical nature, such as atoms or molecules. Rather, *the microscopic level of psychological synergetics is conceived of as the virtual horizon of the maximum degree of resolution attainable in the course of the analysis of interacting biological, psychological and social processes*. This also means that psychological synergetics is an interdisciplinary approach, even at the microscopic level of analysis. This microscopic level is regarded as the hypothetical limit of the maximum degree of resolution. Thus the postulate of this biopsychosocial microscopic level can best be characterized as a basic axiom (see Tschacher 1990).

In psychology, therefore, we shall prefer to rely on a different approach of modeling designated by Haken (1988) as the "second foundation of synergetics". This approach is based upon the phenomenological identification of patterns generated by a system through self-organization. This pattern, representing the dynamics of the macroscopic variables of the system in question, presents the empirically accessible grounds for analysis. These empirically obtained data are to be explained, i.e. they form the explanandum. We have only conceptual assumptions at our disposal about the explanans, which must be reconstructed by means of a macroscopic model comprising the identification of the relevant *order parameters* (or macroscopic variables) of the pattern, their interactions and autorecursive relations, as well as the influence of specific control parameters.

Thus we postulate the existence of a macroscopic level defined by the order parameters or macroscopic variables of the system dynamics. We are concerned with a phenomenological level of observation and a description of variables commonly analyzed in psychological, psychophysiological and psychiatric research (including the corresponding constructs). Which variables are focused upon depends on the specific system in question. In synergetic terms, the order parameters on the macroscopic level represent the collective modes of behavior of a large number of individual processes and components. Just like the microscopic processes, the order parameters can be of biological, psychological or social nature, thus confirming practical experience regarding psychological and psychiatric phenomena which are often located at a right angle with respect to the boundaries of established disciplines. A detailed example illustrating these ideas is given by Schiepek, Schoppek and Tretter (this volume). In the analysis of schizophrenic processes presented there, control parameters (such as the metabolism of neurotransmitters) and order parameters (such as stress or expressed emotions) stand for biological, cognitive, affective, and interpersonal processes.

The existence of a specific relation between the macroscopic and the microscopic level renders the former the preferred level for methodological and theoretical analysis. Under certain conditions determined by the control parameters of the system the behavior of the microscopic components becomes coherent and organized. If such coherent behavior occurs without being explicitly imposed by the system's environment, this is termed *self-organization*. The process of self-organization distinctly reduces the degrees of freedom of the behavior of the microscopic components. This phenomenon enables us to describe the system's behavior completely by identifying the collective variables responsible for the coherent system dynamics, instead of having to draw up a list of the behavior of the system's components, which would in fact be infinite. The resulting emergent states of order (or, put differently, the collective modes of the hypercomplex behavior of the microscopic level) in turn "enslave" by these very modes or order parameters the behavior of the microscopic components. This observable coherent system behavior, resulting from processes of self-organization, will be termed "*pattern*" here. Let us add that one basic precondition for self-organization to occur is the openness of a system to an exchange of energy (and, in some cases, matter) with its environment. In the case of psychological self-organization this also includes sensory and social stimulation by the environment. If coherence and order reach a very high degree at the observable macroscopic level, then the phenomenon of self-organization possesses a high degree of face validity. It was the occurrence of simple equilibrium conditions within multicomponent systems which stimulated research into self-organization which all began with the research paradigms of the laser, the Bénard instability and the Belousov-Zhabotinsky reaction. Standing behavior patterns are observable once a certain mode of behavior prevails, thus drastically reducing the complexity of the system behavior. Dynamic behavior at the macroscopic level, on the other hand, may seem unpredictable or in disorder, i.e. chaotic and turbulent: If the system is far from thermodynamic equilibrium and there is a high degree of matter or energy throughput while certain control parameters exhibit high values, then the emergent order parameters (variables) alternate in chaotic ways. Thus chaos, especially low-dimensional chaos, may indicate the existence of a self-organizing system. Just as is possible with regard to the dynamics of a stable periodic or complex periodic system, the macroscopic dynamics of a system showing this type of chaos can be modeled by means of a nonlinear system of variables.

To summarize, it can be stated that the order parameters evolving out of a complex microscopic level through self-organization and their interaction can be considered as a system of variables and modeled accordingly. Thus, systems are defined as systems of variables by the macroscopic approach of psychological synergetics, precisely as suggested by Schiepek (1986; 1991) with respect to the method of idiographic systems modeling. Neither method takes as a starting point any ontological determination of biopsychosocial systems — neither for the microscopic nor for the macroscopic level. Rather, it is a vital characteristic of (psychological) synergetics that it starts out from an inter-disciplinary and "structuralist" point of view, postulating that similar structures and dynamics will evolve out of very different substrata under certain given conditions. Fig. 4 shows a graphic summary of the ideas presented in this section, Fig. 5 gives an example (comp. also Fig. 8.).

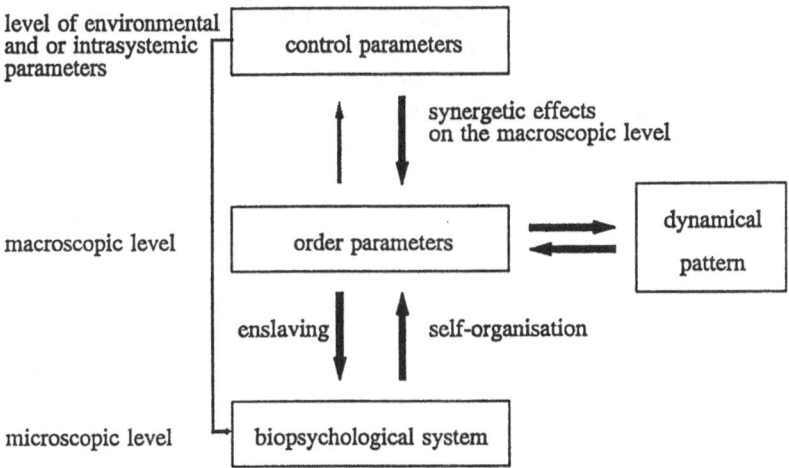

Fig. 4: The synergetic model of self-organization.

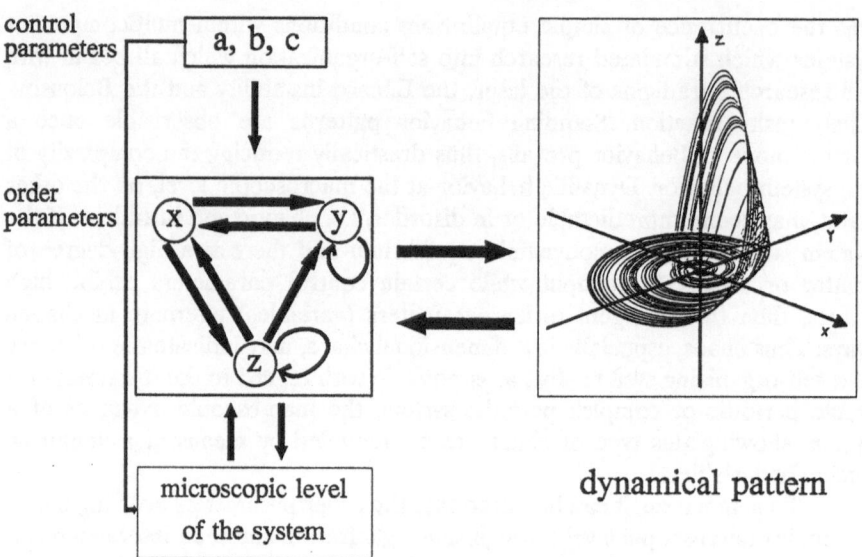

control parameters

a, b, c

order parameters

x → y
y → (loop)
z → (loop)

microscopic level of the system

dynamical pattern

Fig. 5: An example of the interactions between macroscopic variables and the pattern of the pertaining system dynamics (here: Rössler attractor, phase space diagram).

2.2 The Problem of Determining Patterns, Variables, and the Macroscopic Model

It is usually assumed that the components constituting a system, the laws governing the components' behavior, and particularly the boundaries of the system are known when physical systems are analyzed. When the Bénard instability is investigated, for example, a fluid is poured into a flat tray of a round or rectangular shape which is then heated from below. Thus, both system and relevant environment are well-defined. Even the process of pattern recognition by the synergetic computer starts out from a clearly defined, digitalized reference pattern with given boundaries. However, the processes of pattern identification and explanation with which clinical psychology and psychiatry are confronted represent a very different situation.These fields of application offer neither fixed anchor points nor well-defined limits as to the target of investigation. The phenomena we are dealing with here can be considered "fuzzy sets". Regarding schizophrenia research, for example, what is to be determined as the relevant "pattern"? Is it the pathological features of a patient described in a qualitative fashion? The temporal evolution of the disorder? The peculiarities of interpersonal communication within the patient's family? The dynamics of the interpersonal relations within and between the psychosocial institutions taking care of the patient? The pattern of the patient's EEG? The dynamic patterns of specific biochemical values? Or something else? Which of the patterns is chosen surely depends on the researcher's interests and theoretical background, as well as the extent to which his/her method of approach allows, or inhibits, the identification of a certain pattern, and also on the conciseness and range of dynamic or structural patterns. Taking these preconditions into account, it is plain to see that it is hardly possible to determine a pattern eo ipso, that is, independent of those variables or order parameters applied in its description and explication — which are, of course, in turn dependent on the theory preferred. It is in fact a process of coevolution: The explanans, the pattern in question, and the explanandum, the model of the nonlinear interrelations between the relevant macroscopic variables or order parameters and the control parameters, determine each other. This process of coevolution is based on the reciprocal specification of (a) the patterns by means of adequate methods, especially the collecting of time-serial data, and (b) the model applied. As already proposed by Schiepek (1986, pp. 118 ff.) with respect to the method of idiographic systems modeling, this can be viewed as a process of "dialectic problem solving" (Dörner 1976). If one takes into consideration the fact that the clinical approach to determining "patterns" — whether the intention be of a scientific, practical or therapeutic nature — always takes place in the context of concrete action and communication, of data collection, and of purposeful interaction, it becomes clear that this dialectic process of problem-solving takes on the form of a perception-action cycle (Turvey, Carello & Nam-Gyoon Kim 1990). This perception-action cycle illustrates how perception and performatory action combine in a synergistic way (see Fig. 6).

In the terminology of yet another approach, the process of the reciprocal "determination" of pattern and model is considered as the

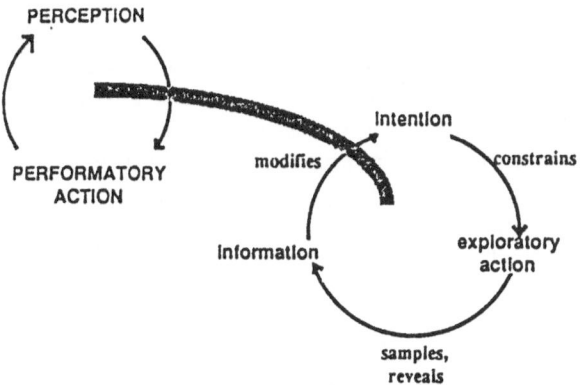

Fig. 6: The perception-action cycle. The main cycle revolves around performatory activity. Nested within the main cycle is a cycle that revolves around exploratory activity, such as adjusting and moving organs of sensitivity. In the course of several perception-performatory activity cycles one can imagine many more perception-exploratory activity cycles. Perception-exploratory activity cycles comprise intentional states guiding exploratory states, sampling informational states, and modifing intentional states. (from Turvey, Carello & Nam-Gyoon 1990, p. 279)

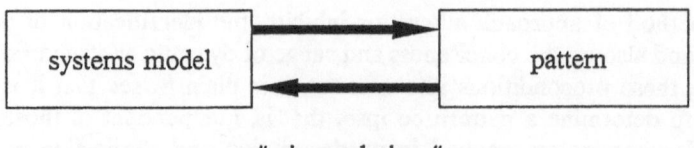

" eigensolution "
Fig. 7: Pattern identification (generation of data, pattern recognition) and
model building (hypothesizing, explanation) as a recursively interrelated
process, at best converging towards a stable attractor ("eigensolution").

generation of an "eigensolution" (von Foerster 1985). In Krohn and Küppers'
(1989) view, the production of scientific evidence is the result of an
"eigensolution" which is generated during the research process via the
recursive interrelations existing between our collecting data, hypothesizing or
theorizing (interpreting data), and developing appropriate methods. This
approach leads one to see the generating of an "eigensolution" of pattern and
model determination as the result of a process of cognitive and social self-
organization (see Fig. 7).

Arriving at an "eigensolution" is facilitated by three conditions. First,
biopsychosocial systems themselves reduce complexity via self-organization by
creating order parameters which render structural and dynamic patterns more
concise. Second, psychologists and psychiatrists, whose profession it is not only
to recognize patterns but also to supplement missing links in order to get a
grasp of the overall patterns, usually have an eye for relevant patterns, which is
trained through daily practical work. Third, there are different methods
available for developing models of complex dynamic processes. These are (a)
the collection of multiple time-serial data by means of adequate measures; (b)
the analysis of these data by means of adequate mathematical methods such as
auto- and cross-correlational analysis, ARIMA models, Fourier analysis and,
especially, the analysis of dimensionality (e.g. Grassberger & Procaccia 1983;
Steitz et al., this volume) which provides information about the presence of
chaotic processes and their fractal dimensionality, which in turn indicates how
many variables (dimensions) are required in order to generate a specific
(chaotic) time series; and (c) the building of idiographic or theoretical systems
models (Schiepek 1986) which provide the necessary basis for generating
mathematical models.

As has been shown, it is quite possible to develop explanatory models
for those types of patterns with which we are concerned in psychological
synergetics, namely dynamic patterns in the form of time-serial or process
data. Factors which contribute to a reduction in the usefulness of the data
collected are noise and error of measurement, an inadequate length of of time-
series as well as an inadequate level of resolution of the measuring
instruments. Given the declared empirical approach of psychological
synergetics, the amount of work to be done on methodology in order to
achieve its aim of overcoming these restrictions should not be underestimated.

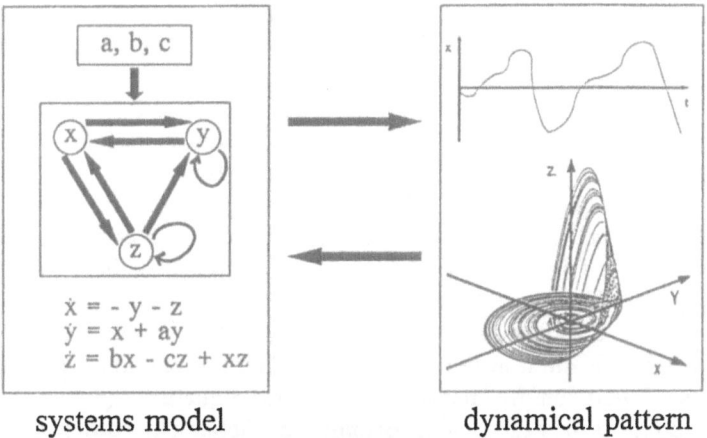

$$\dot{x} = - y - z$$
$$\dot{y} = x + ay$$
$$\dot{z} = bx - cz + xz$$

systems model   dynamical pattern

Fig. 8: The relation of systems model and pattern at the macroscopic level. The systems model is illustrated as an interrelated structure of the macroscopic variables and as a system of equations; the dynamic pattern is given as (x;t) diagram and as phase space illustration. The illustration shows the strange Rössler attractor with its pertinent system of equations. It contains 3 variables and one quadratic nonlinearity (see Rössler 1976; Nicolis & Prigogine 1987).

Fig. 8 illustrates the relationship between the dynamic pattern and its model, comprising the system's macroscopic variables (order parameters) and their interactions and autorecursive interrelations. These are given in the form of a graphic structural model, on the one hand, and as a system of difference equations, on the other. The example given here shows the system of variables pertinent to the Rössler attractor (Rössler 1976). It consists of three variables (x, y, z) and three control parameters a, b, and c. This system can be fully described by means of the three interconnected equations given. The solution resulting from these equations when specific initial conditions and parameter values are inserted shows the system dynamics (as to the theory of nonlinear dynamic systems see Bergé, Pomeau & Vidal 1984; Thompson & Stewart 1986; Schuster 1988). Any modification of the values of the initial conditions or the parameters may decisively determine the course of the resulting dynamics. There is a great chance of creating a system showing chaotic dynamics for larger systems of variables: Only three variables together with one nonlinearity (Poincaré-Bendixon theorem) will suffice in the case of continuous maps, as is shown for example by the chaotic Rössler system. Systems of one or two variables are sufficient to create chaotic dynamics in the case of discrete maps (as an example for a one-dimensional system see the logistic map; for two-dimensional systems see the Hénon map).

The systems concept of interrelated variables offers a decisive advantage in opening the way towards dynamic considerations since the dimension of time is introduced into the analysis. This presents a sharp contrast to static, structural systems concepts. Furthermore, it is convenient to conceptualize systems via interrelated variables, as mathematical procedures are available for the description, analysis, simulation and classification of the behavior of these systems. A translation of terms used in cybernetics is also possible. Input-output relations may be converted into functions, and concepts such as feedback control system, equilibrium conditions, stability and instability can also be adapted to the theory of dynamic systems.

In addition, the number of variables, their initial values, and the type of interactions between them can be determined specifically in each case so as to take into account theories and empirical evidence for the scientific field to the greatest possible extent (compare the method of idiographic systems modeling, Schiepek 1986). It is as a result of these possibilities that the conceptualization of systems via variables is often used in simulations. The present volume gives several examples: Simulation by means of difference equations (Schiepek, Schoppek & Tretter; Kriz; Tschacher, Brunner & Schiepek), differential equations (an der Heiden), and production systems (Schaub & Schiepek). Another possible way of simulating the dynamics of complex systems lies in applying the approach of connectionism, which is a result of the research on artificial intelligence (Haken, this volume; Znoj, this volume; Schaub & Schiepek, this volume). The connectionist approach goes beyond that of viewing systems as constituted by macroscopic variables, in as far as the former explains properties of pattern formation (self-organization) out of a complex microscopic level.

To summarize, the following steps are to be taken in modeling within the framework of psychological synergetics:

1) Choosing the desired area of investigation (e.g. intrafamiliar communication, development of intragroup structures, dynamics of etiological processes, therapy process research, epidemiology, perception, coordination of perception and action, etc.), along with the qualitative identification of patterns in this field. The laws of gestalt perception naturally play an important part here (see Stadler & Kruse 1986).

2) Method-guided description and, if possible, a collection of quantitative data about the dynamic pattern (via time series or coding schemes, for example).

3) Systems modeling. This implies (a) selecting adequate macroscopic variables or order parameters and (b) stating hypotheses concerning the interactions and autorecursive interrelations of the variables. The hypotheses must, of course, be based on a profound knowledge of empirical evidence and theories concerning the area of investigation. A detailed description of the method of systems modeling is given by Schiepek (1986).

4) Simulation by means of systems of nonlinear equations.

5) Testing of the model and comparison with empirical data. Possible ways of testing the model include examinations of its internal logical consistency and specific modifications of the simulated system

corresponding to the hypotheses   applied (for example by modifying certain parameter values in accordance with   prognoses stemming from certain hypotheses). This enables the system's   behavior, as it results from the simulation, to be assessed on the basis of   theoretical assumptions and with reference to the behavior of the actual (empirical) system under comparable conditions.

Simulations have to prove their worth, especially when compared to empirical systems. Further mathematical analyses, such as auto- or cross-correlograms, Fourier analysis (power spectra), and the determination of indicators of the chaotic dimensionality within the pattern in question (correlational dimension, Lyapunov exponent, Kolmogorov entropy), also serve to provide a comparison of the time-serial data obtained via simulation and through empirical data collection.

It is now the task of clinical psychology and psychiatry, utilizing the methods devised by psychological synergetics, to obtain a more profound understanding of self-organization and nonlinear dynamic processes effective in different systems.

Acknowledgement. We would like to thank Dipl.-Psych. Martina Weisensel (Staatl. gepr. Übersetzerin) for the translation of the present contribution.

References

Akiskal HS, McKinney WT (1975) Overview of Recent Research in Depression, Integration of ten Conceptual Models into a Comprehensive Clinical Frame. Arch Gen Psychiatry 32: 285-295

American Psychiatric Association (1987) Diagnostic and Statistical Manual of Mental Disorders. Third Edition. APA, Washington DC

Barthes R (1977) Fragments d'un discours amoureux. Éditions du Seuil, Paris

Baumann U, Perrez M (Hrsg) (1990) Klinische Psychologie. 2 Bde. Huber, Bern

Bergé P, Pomeau Y, Vidal C (1984) L'Ordre dans le Chaos. Hermann, Paris

Birbaumer N, Schmid RF (1990) Biologische Psychologie. Springer, Berlin

Bischof N (1990) Phase Transitions in Psychoemotional Developement. In: Haken H, Stadler M (Eds) Synergetics of Cognition. Springer Series in Synergetics, Vol. 45. Springer, Berlin, pp 361-378

Blaser A (1977) Klinische Urteilsbildung. Huber, Bern

Brown GW, Harris TO, Peto J (1980). Die Kausalbeziehung zwischen lebensverändernden Ereignissen und psychischen Störungen. In: Katschnig H (Hrsg) Sozialer Streß und psychische Erkrankung. Urban & Schwarzenberg, München, S 214-237

Bruner JS, Postman L (1949) Perception, Cognition, and Behavior. Journal of Personality 18: 14-31

Burlingame GM, Fuhriman A, Barnum K, Lyman R (1991) Deterministic Chaos in Group Psychotherapy Interaction: An Illustration. Paper presented at the Society for Psychotherapy Research 22nd Annual Meeting, July, 1991, Lyon, France

Caspar F (1989) Beziehungen und Probleme verstehen. Eine Einführung in die psychotherapeutische Plananalyse. Huber, Bern

Caspar F, Grawe K (1989) Weg vom Methoden-Monismus in der Psychotherapie. Bulletin der Schweizer Psychologen 1989, Nr. 3: 6-19

Dohrenwend BS, Dohrenwend BP (1981) Socioenvironmental Factors, Stress and Psychopathology. American Journal of Community Psychiatry 9): 128-159

Dörner D (1976) Problemlösen als Informationsverarbeitung. Kohlhammer, Stuttgart

Ekman P, Friesen WV (1978) Manual for the Facial Action Coding System. Consulting Psychologists Press, Palo Alto

Foerster H von (1985) Sicht und Einsicht, Vieweg, Braunschweig

Freeman WJ (1990) On the Problem of Anomalous Dispersion in Chaoto-Chaotic Phase Transitions in Neural Masses, and Its Significance for the Management of Perceptual Information in Brains. In: Haken H, Stadler M (Eds) Synergetics of Cognition. Springer Series in Synergetics, Vol. 45. Springer, Berlin, pp 126-143

Fuhriman A, Burlingame GM (1991) Chaos: Emerging Developmental Patterns in Process Research. Paper presented at the Society for Psychotherapy Research 22nd Annual Meeting, July, 1991, Lyon, France

Grassberger P, Procaccia I (1983) Measuring the Strangeness of Strange Attractors. Physica 9 D: 189-208

Haken H (1983a) Synergetics. An Introduction. Springer Series in Synergetics, Vol.1. Springer, Berlin (3.Aufl.)

Haken H (1983b) Advanced Synergetics (Instability Hierarchies of Self-Organizing Systems and Devices). Springer Series in Synergetics, Vol. 45. Springer, Berlin (3. Aufl)

Haken H (1988) Information and Self-Organization: A Macroscopic Approach to Complex Systems. Springer Series in Synergetics, Vol. 40. Springer, Berlin

Haken H (1989) Synergetik: Vom Chaos zur Ordnung und weiter ins Chaos. In: Gerok W (Hrsg) Ordnung und Chaos in der unbelebten und belebten Natur. S Hirzel, Wissenschaftliche Verlagsgesellschaft, Stuttgart, S 65-75

Haken H (1990a) Synergetik. Eine Einführung. Springer, Berlin

Haken H (1990b) Synergetics as a Tool for the Conceptualization and Mathematization of Cognition and Behavior - How Far Can We Go? In: Haken H, Stadler M (Eds) Synergetics in Cognition. Springer Series in Synergetics, Vol. 45. Springer, Berlin, S 2-31

Haken H, Graham R (1971) Synergetik - Die Lehre vom Zusammenwirken. Umschau 6: 191

Haken H, Stadler M (Eds) (1990) Synergetics of Cognition. Springer Series in Synergetics, Vol. 45. Springer, Berlin

Hautzinger M, de Jong-Meyer R (1990) Depressionen. In: Reinecker H (Hrsg) Lehrbuch der Klinischen Psychologie. Hogrefe, Göttingen, S 126-165

an der Heiden U, Roth G, Schwegler H (1985) Die Organisation der Organismen: Selbstherstellung und Selbsterhaltung. Funkt Biol Med 5: 330-346

Horowitz MJ (1987) States of Mind. Plenum Press, New York

Katschnig H (Hrsg) (1980) Sozialer Streß und psychische Erkrankung. Urban & Schwarzenberg, München

Kelso J (1990) Phase Transitions: Foundation of Behavior. In: Haken H, Stadler M (Eds) Synergetic of Cognition. Springer Series in Synergetics, Vol. 45. Springer, Berlin, pp. 249-268

Köhler W (1920) Die physischen Gestalten in Ruhe und im stationären Zustand. Vieweg, Braunschweig

Köhler W (1940) Dynamics in Psychology. Liveright, New York

Kriz J (1985) Grundlegende Aspekte systemischer Therapien. In: Kriz J Grundkonzepte der Psychotherapie. Urban & Schwarzenberg, München, S 227-241

Kriz J (1990) Synergetics in Clinical Psychology. In: Haken H, Stadler M (Eds) Synergetics of Cognition. Springer Series in Synergetics, Vol. 45. Springer, Berlin, pp 393-404

Krohn W, Küppers G (1989) Die Selbstorganisation der Wissenschaft. Suhrkamp, Frankfurt

Kruse P, Stadler M (1990) Stability and Instability in Cognitive Systems: Multistability, Suggestion, and Psychosomatic Interaction. In: Haken H, Stadler M (Eds) Synergetics of Cognition. Springer Series in Synergetics, Vol. 45. Springer, Berlin, pp. 201-215

Lewinsohn PM, Hoberman H, Teri L, Hautzinger M (1985) An Integrative Theory of Depression. In: Reiss S, Bootzin RR (Eds) Theoretical Issues in Behavior Therapy. Academic Press, Orlando, pp 331-359

Lilli W (1978) Die Hypothesentheorie der sozialen Wahrnehmung. In: Frey D (Hrsg) Kognitive Theorien der Sozialpsychologie. Huber, Bern, S 17-46

Luhmann N (1984) Soziale Systeme. Grundriß einer allgemeinen Theorie. Suhrkamp, Frankfurt am Main

Mahoney MJ (Eds) (1980) Psychotherapy Process. Current Issues and Future Directions. Plenum Press, New York

Markowitz J (1991) Referenz und Emergenz: Zum Verhältnis von psychischen und sozialen Systemen. Systeme (Interdisziplinäre Zeitschrift für systemisch orientierte Forschung und Praxis in den Humanwissenschaften) 5: 22-46

Maturana HR (1982) Erkennen: Die Organisation und Verkörperung von Wirklichkeit. Vieweg, Braunschweig

Minuchin S (1974) Families and Family Therapy. Harvard University Press, Cambridge

Nicolis G, Prigogine I (1987) Die Erforschung des Komplexen. Piper, München

Nicolis JS (1986) Dynamics of Hierarchical Systems. An Evolutionary Approach. Springer Series in Synergetics, Vol. 25. Springer, Berlin

Reinecker H (1990a) Lehrbuch der Klinischen Psychologie. Hogrefe, Göttingen

Reinecker H (1990b) Forschungsprobleme in der Klinischen Psychologie. In: Reinecker H (Hrsg) Lehrbuch der Klinischen Psychologie. Hogrefe, Göttingen, S 25-44

Revenstorf D (1985) Nonverbale und verbale Informationsverarbeitung als Grundlage psychotherapeutischer Intervention. Hypnose und Kognition 2(2): 13-35

Revenstorf D (1990) Zur Theorie der Hypnose. In: Revenstorf D (Hrsg) Klinische Hypnose. Springer, Berlin, S 79-99

Rössler OE (1976) An Equation for Continous Chaos. Phys Lett 57A: 397-398

Rosenhan DL (1973) On Being Sane in Insane Places. Science 179: 250-258

Schiepek G (1986) Systemische Diagnostik in der Klinischen Psychologie. Psychologie Verlags Union, München Weinheim

Schiepek G (1990) Selbstreferenz in psychischen und sozialen Systemen. In: Kratky K, Wallner F (1990) Grundprinzipien der Selbstorganisation. Wissenschaftliche Buchgesellschaft, Darmstadt, S 182-200

Schiepek G (1991) Systemtheorie der Klinischen Psychologie. Vieweg, Braunschweig

Schiepek G, Schaub H (1991) Als die Theorien laufen lernten... Computersimulation einer Depressionsentwicklung. In: Schiepek G: Systemtheorie der Klinischen Psychologie. Vieweg, Braunschweig, S 221-305

Schmid GB (1991) Chaos Theory and Schizophrenia: Elementary Aspects. Psychopathology 24: 185-198

Schuster HG (1988) Deterministic Chaos. An Introduction. VCH Verlagsgesellschaft, Weinheim

Schwegler H (1978) Sind Katastrophen vorhersehbar? - Geometrische Modelle für sprunghafte Veränderungen in Natur und Gesellschaft. Jahrbuch der Wittheit zu Bremen XXII: 181-198

Stadler M, Kruse P (1986) Gestalttheorie und die Theorie der Selbstorganisation. Gestalttheory 8: 75-98

Stadler M, Kruse P (1990) The Self-Organization Perspective in Cognition Research: Historical Remarks and New Experimental Approaches. In: Haken H, Stadler M (Eds) Synergetics of Cognition. Springer Series in Synergetics, Vol. 45. Springer, Berlin, pp 32-52

Stegmüller W (1973) Theorie und Erfahrung, Zweiter Halbband: Theorienstrukturen und Theoriendynamik. Springer, Berlin

Strauss JS (1989) Intermediäre Prozesse in der Schizophrenie: zu einer neuen dynamisch orientierten Psychiatrie. In: Böker W, Brenner HD (Hrsg) Schizophrenie als systemische Störung. Huber, Bern, S 35-50

Teubner G (1990) Hyperzyklus in Recht und Organisation. Zum Verhältnis von Selbstbeobachtung, Selbstkonstitution und Autopoiese. In: Krohn W, Küppers G (Hrsg) Selbstorganisation: Aspekte einer wissenschaftlichen Revolution. Vieweg, Braunschweig, S 231-264

Thompson J MT, Stewart HB (1986) Nonlinear Dynamics and Chaos. Geometrical Methods for Engineers and Scientists. Wiley, Chichester

Tschacher W (1990) Interaktion in selbstorganisierenden Systemen. Asanger, Heidelberg

Turvey MT, Carello C, Nam-Gyoon K (1990) Links Between Active Perception and the Control of Action. In: Haken H, Stadler M (Eds) Synergetics of Cognition. Springer Series in Synergetics, Vol. 45. Springer, Berlin, pp 269-295

Wehr TA, Goodwin FK (1979) Rapid Cycling in Manic-Depressives Induced by Tricyclic Antidepressants. Arch Gen Psychiatry 36: 555-559

Weidlich W Haag G (Eds) (1983) Concepts and Models of a Quantitative Sociology. Springer Series in Synergetics, Vol. 14. Springer, Berlin

Wertheimer M (1912) Experimentelle Studien über das Sehen von Bewegung. Z Psychol 61: 165-292

Wertheimer M (1923) Untersuchungen zur Lehre von der Gestalt II. Psychol Forsch 4: 301-350

Willke H (1983) Methodologische Leitfragen systemtheoretischen Denkens. Annäherung an das Verhältnis von Intervention und System. Z system Ther 1(2): 23-37

Willke H (1989) Systemtheorie entwickelter Gesellschaften. Juventa, München

Synergetics in Psychology

Hermann Haken

With 25 Figures

Abstract. The interdisciplinary field of synergetics may be considered as a strategy for dealing with complex systems. It focusses its attention on those situations where systems change their behavior qualitatively. The basic concepts of synergetics, such as stability and instability, order parameters, control parameters, the slaving principle, hysteresis, critical fluctuations, etc. are illustrated by means of examples taken from physics (lasers and fluid dynamics) and physiology (finger movements). It is shown how the concepts of synergetics may be used as metaphors at the conceptual level, but that they allow us also to devise computer models for phenomena treated by psychology, e.g. visual perception, including those of ambiguous pictures, where contact with Gestalt theory can also be made. As is shown in particular, the change of rather unspecific control parameters can induce a qualitative change in the behavior of a system.

1. Relating Synergetics to Psychology – What Kind of Endeavor?

Psychology deals with probably the most complex system in the world, namely the human being. In order to cope with its problems, psychology has developed its own repertoire of methods, concepts, theories, and so on. (For references consult the articles in this volume.) But the whole problem is so difficult that it represents an enormous challenge over and over again. Thus we are, perhaps, led to the question of whether complex systems exist also in other fields of science, and how one deals with complex systems there. In the next step of such an endeavor we have to carefully check the extent to which the results of other disciplines can be transferred to and used by psychology. Over the past twenty years a new field of science has emerged that deals with complex systems from a unifying point of view. This field, called synergetics, considers systems composed of many parts that may interact with each other (Haken, 1970; 1983; 1987; 1991a). Synergetics focusses its attention on those situations where the macroscopic behavior of such systems change qualitatively, or, in other words, where new macroscopic qualities of a system emerge. Or in still other words, synergetics studies transitions between behavioral patterns of a complex system. Though synergetics originated from physics, it is by no means a physical theory that tries to reduce complex systems or phenomena in the animate or inanimate world to the laws of physics. Rather the physical systems, which were studied first, allowed us to unearth a number of general principles that are common to a great variety of complex systems. In this

Springer Series in Synergetics, Vol. 58   Self-Organization and Clinical Psychology
Editors: W. Tschacher, G. Schiepek, and E.J. Brunner   © Springer-Verlag Berlin Heidelberg 1992

way it became possible to establish far reaching analogies between the behavior of quite different systems that may belong to a variety of disciplines, such as physics, chemistry, electrical engineering, computer science, biology and economics, to mention some examples. Our general principles are based on mathematical relationships or mathematical structures. In the context of this article, however, we shall not dwell on the mathematical aspects but rather try to elucidate the basic ideas by means of explicit examples, some of which are taken from physics. In psychology, qualitative changes of behavior are a well-known experience from every day life. Such changes (or transitions) may occur in facial expressions, gestures, actions, moods, perceptions, psychotic states, behavior of groups, families, etc. Furthermore all these phenomena are linked to a material substrate – the brain – which is a huge network of neurones, i.e. a complex system by itself. I hope these remarks will help to convince the reader that synergetics may be a subject worthy of study by psychologists.

2. Synergetics – Its Models, Paradigms, Concepts, and Computer Simulations and Realizations

In the following we wish to show that synergetics can be viewed in a variety of ways each of which may provide a link to psychology. First of all, there are a number of rather simple systems that have acquired the character of paradigms for qualitative changes of behavior in complex systems. Among these systems are the light source "laser", spontaneous formation of movement patterns in fluid dynamics and spontaneous transitions in human finger movement. These examples allow us to establish and explain basic concepts of synergetics, such as stability and instability, order parameters, the slaving principle, non-equilibrium phase transitions, hysteresis, etc. Finally we shall see how these concepts can be used to simulate human perception by computers or to model the neural network of a brain. Thus synergetics may serve as a tool for psychology at several levels of approach: at the conceptual level, at the level of models, and at the level of concrete computer simulations of behavior.

We trust that the reader will not be too bored if we begin with some examples from physics. We present these examples because the conclusions that can be drawn are cogent and show that a number of typical processes observed in psychology can be found at a rather elementary level. The most outstanding example in this respect is the laser light source (cf. Haken, 1970).

2.1 The Laser Paradigm

This example shows how the cooperation of rather simple elements can produce quite different patterns of macroscopic behavior, where sharp transitions between one kind of behavior and another occur. Let us consider the gas laser, where in a glass tube a gas consisting of atoms (or molecules) is enclosed. When an electric current is sent through the gas, its individual atoms can be energetically excited.

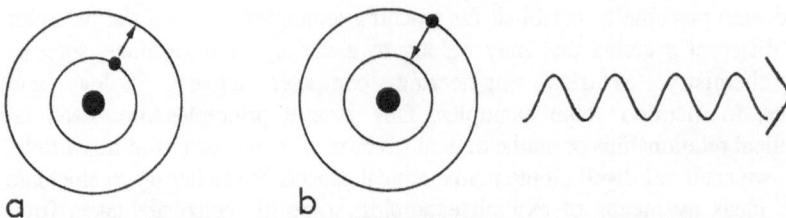

Fig. 1. (a) When an electron orbiting around a nucleus in an atom is excited, it jumps from its inner orbit to an outer orbit which is energetically higher; (b) The electron may spontaneously go from the outer orbit to the inner orbit whereby it gives away its energy to the light field, i.e. the electron emits a light wave

More precisely speaking, an electron is brought from its lower orbit in an atom to an energetically higher orbit (Fig. 1). As is shown in quantum mechanics, the electron may then spontaneously return to its lower orbit giving away its energy to the light field, i.e. a light wave is emitted. It is as if we are throwing a pebble into water where a water wave is generated. In a normal lamp such emission processes take place again and again. These processes are entirely uncorrelated; it is as if we are throwing a handful of pebbles into water such that a rather chaotic water surface emerges (Fig. 2a). We speak of microscopically chaotic light. A laser differs from a normal lamp in two essential ways:

a) At the end faces of the glass tube mirrors are mounted. By means of these the light emitted from the atoms is reflected rather often and can now act on the atoms in a way we shall discuss immediately.

b) The electric current is considerably increased.

When the atoms are excited more rapidly by increasing the electric current, suddenly an entirely new phenomenon occurs, namely the emitted light wave becomes entirely ordered (Fig. 2b). This phenomenon can be understood only if we assume that the electrons of the individual atoms, which emit the light waves, become highly correlated in their motion. How is this achieved? The underlying mechanism is this: When a light wave hits an energetically excited electron, it causes this electron to transfer its energy to that light wave so that the wave is amplified. When more and more excited electrons are hit in this way, this specific light wave is amplified more and more: a light avalanche evolves. In general, different excited electrons may emit originally different kinds of waves (they differ by their wavelengths). Each of these waves competes with all the other waves for the energy stored in the excited electron. Eventually one specific wave wins this competition: This is the ordered (or "coherent") laser light wave. Because this ordered light wave describes the order of the system, it is called "the order parameter". When atoms are excited by the current, or when new atoms are brought into the gas tube, these new atoms are subjected to the ordering action of the light wave and are, in this way, enslaved by it. After being enslaved, they support the laser wave. When the laser wave is perturbed for a moment, for instance by an additional light field added to it from the outside, it adjusts much

34

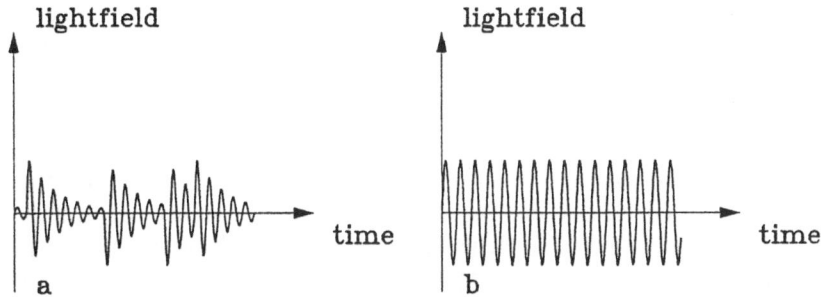

Fig. 2. (a) In the case of an ordinary lamp, the light field, which is plotted here against time, consists of individual wave tracks; (b) In the case of the laser the light field consists of an infinitely long well-ordered coherent wave

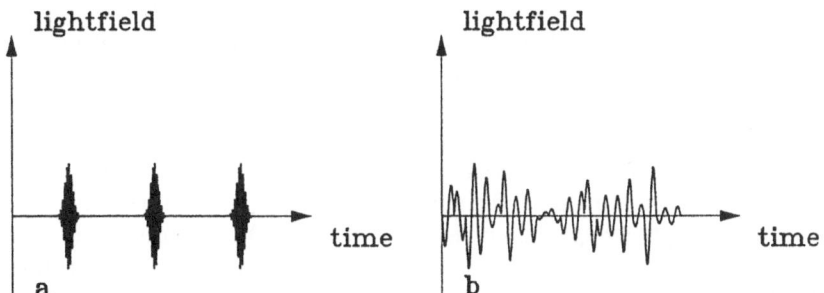

Fig. 3. (a) At a higher pump power the laser light may consist of ultra-short regular light flashes; (b) Under different conditions (bad mirrors) and high pumping the light field may show deterministic chaos

more slowly than the individual electrons would do after a perturbation of their movements. This is a general feature of order parameters: they relax (adjust) much more slowly than the enslaved subsystems. In the laser, the principle of circular causality holds: On the one hand, the order parameter (laser light wave) enslaves the subsystems (electrons), but on the other hand the action of the latter enables the existence of the order parameter. An anthropomorphic example may help to illustrate the situation. In a nation, the national language plays the role of an order parameter that lives much longer than the individuals. When a baby is born it is subjected to the language. He or she learns it and, eventually, carries the language further. Quite clearly, an individual may be subjected to several order parameters, such as family, school, language, culture, religion, political party, and so on. In the case of the laser it is remarkable that when we increase the current further and further, suddenly the coherent laser wave is replaced by a new type of light, namely regular very short light flashes (Fig. 3a). Under still other conditions the laser may show a macroscopic irregular behavior which is nowadays described as deterministic chaos (Fig. 3b). In such a case few order parameters, say three, cooperate, but by their interaction an entirely irregular emission is achieved. A number of concepts result from the laser example:

1. A system of individual active elements can become entirely ordered in its action by means of self-organization, i.e. the order is not imposed on the system from the outside. Instead, the change of a rather unspecific control parameter (in the laser case it is the size of the electric current), induces a self-ordering of the system (in the present case the motion of the electrons).

2. The huge number of individual variables or degrees of freedom of the individual electrons in the case of a lamp, where all the electrons act independently of each other, is replaced by a single variable, namely the order parameter, i.e. the laser light wave. We thus see that the behavior of a complex system may be governed by only few variables, namely the order parameters.

3. The order parameters become apparent when a system changes its macroscopic behavior qualitatively, i.e. for instance at the transition from the microscopically chaotic emission of a lamp (Fig. 2a) to the highly ordered emission of a laser (Fig. 2b).

As we have seen above, typical phenomena occur when the control parameter is changed so that the transition from the lamp to the laser light occurs. They can be best visualized by comparing the behavior of the order parameter with a particle moving in a "potential landscape". In the case of the lamp, this landscape is shown in Fig. 4. It must be noted that the individual acts of excitation of the atoms by the electric current occur at random. This means that the light field describing the order parameter is receiving random pushes all the time. In our visualization, the ball moving in the valley of Fig. 4 is pushed all the time to the left or to the right in a random sequence. When the energy input by the current is increased, the landscape is deformed to a very flat potential as shown in Fig. 5. Here the so-called *instability* sets in. If we still stick to the picture of a ball hit by random pushes, we immediately see that the pushes have now a bigger effect because the valley is very flat and the restoring force is quite small. As a consequence, the position of the ball will fluctuate quite strongly, or, in other words, the order parameter shows *critical fluctuations*. When the ball is brought away from its

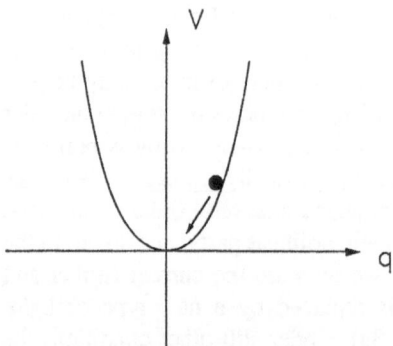

Fig. 4. Visualization of the behavior of the order parameter. The size of the order parameter is interpreted as the position q of a ball which may slide down the slope of a hill in a potential landscape. Quite evidently, the ball will always return to its equilibrium position at q = 0, i.e. the order parameter vanishes

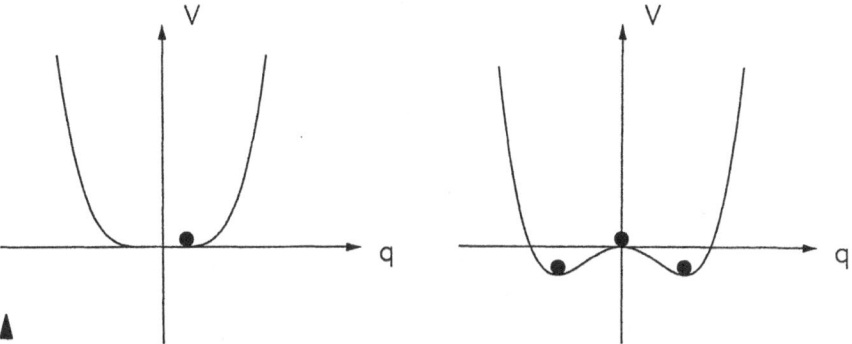

Fig. 5. When the control parameter, in the present case the electric current, is enhanced, the potential landscape of Fig. 4 is deformed and the phenomena of critical fluctuations and critical slowing down, as described in the text, occur

Fig. 6. When the control parameter exceeds a critical value, the potential landscape acquires two minima as shown

equilibrium position, it will relax to it only very slowly because the bottom of the valley is very flat. We speak in this case of *critical slowing down*. Eventually, when the control parameter, namely the power input into the system, is increased still more, new minima occur (Fig. 6). The former minimum becomes unstable and a slight push on the ball will cause it to move to the left or to the right valley. One speaks of a *symmetry breaking instability* because the symmetry of the potential landscape must be violated by the system itself when trying to realize the new state which is visualized here as the position at the bottom of one of the valleys. As a consequence we may state that close to instability points, when an equilibrium state becomes unstable, in general one or even more new minima occur. A random push, i.e. a microscopic event may drive the system into a new state that is macroscopic. The phenomena shown by the order parameter as described by Figs. 4-6 are well known from physical systems in thermal equilibrium, such as ferromagnets where the magnetization spontaneously emerges when the temperature (as control parameter) is lowered. These phenomena are typical for a phase transition. Because a laser is a system away from equilibrium, we call these transitions nonequilibrium phase transitions.

2.2 The Fluid Dynamics Paradigm

Our second example is taken from fluid dynamics. Again, a famous example is provided by the so-called Bénard instability (Bénard, 1900). A liquid, e.g. silicon oil, in a vessel is heated from below. When the temperature difference between the lower and upper surface exceeds a critical value, suddenly the formerly homogeneous fluid becomes ordered, e.g. in the form of hexagons or rolls. The hexagons or rolls are large compared to the dimensions of the individual atoms or molecules of the liquid. The mechanism can be visualized as follows: When the fluid is heated from below, its volume elements expand and become specifically

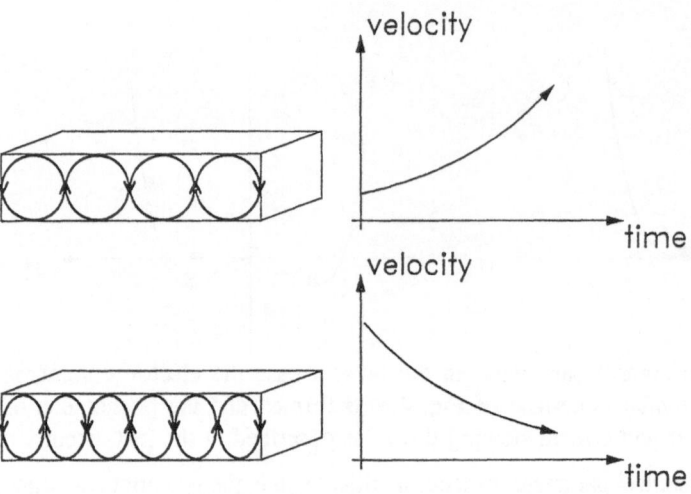

Fig. 7. Left: Two different configurations of movement of the fluid in the form of rolls; Right: The increase or decrease of velocity of the rolls in the course of time depending on the shape of the rolls

lighter. Thus they want to rise. On the other hand, the still cold volume elements at the top of the liquid tend to move downwards. It is as if there are two crowds of people at a broad staircase, one crowd at the bottom trying to move upwards, the other one at the top trying to move downwards. Quite often in life a rather chaotic movement starts. In nature the liquid tests, so to speak, various configurations which can also be shown mathematically in detail. When it tests the configurations of Fig. 7, it turns out that one configuration can exploit the offered energy in a better fashion so that the velocity of the rotation of the rolls increases in the course of time. In the other configuration, the usage of the offered energy is much worse, and because of internal friction the velocity eventually dies out. There may be cases in which different configurations grow but at a different speed. In such a case some kind of Darwinian selection occurs. The configuration with the biggest growth rate wins and then forces all the parts of the fluid into its specific motion. The winning configuration is again described by its order parameter which enslaves the individual parts of the system. Competition need not always happen between order parameters: cooperation can occur also. For instance, hexagons in the liquid (Fig. 8) are generated by the cooperation of three order parameters, each governing a specific system of roll motions.

In a number of cases, because of symmetry, order parameters may have the same growth rate so that they are not distinguishable by their energy consumption. Such a situation arises when the fluid is enclosed in a circular vessel. Then all directions of roll systems are possible each one being governed by a specific order parameter. Which order parameter and its corresponding roll system is realized depends on the initial conditions imposed on the system. This is clearly seen in Fig. 9, where in the left column first a specific upwelling roll is prescribed and then by self-organization the liquid completes this pattern to the full roll pattern

Fig. 8. The Bénard experiment. This figure shows a liquid in a circular vessel which is heated from below. In the center of each honeycomb cell the fluid rises, cools down, and sinks down at the borders of each cell

governed by a specific order parameter (Bestehorn and Haken, 1989). In the middle column, the direction of the roll initially given is another one and now a new order parameter is called upon. Finally, in the third column the fluid is exposed to a conflict situation because here two rolls are initially given at different directions. Eventually the roll pattern that was initially somewhat stronger than the other one wins the competition. As we shall see below, this behavior can be interpreted as the action of some kind of associative memory and the mechanism described with respect to the fluid can be exploited in a realization of pattern recognition by the synergetic computer (see below).

2.3 The Finger Movement Paradigm

The concepts of order parameters etc. have been applied to modeling a number of physiological phenomena. A by now well-known example is Kelso's experiment (Kelso, 1984). He asked test persons to move their fingers slowly in parallel which the persons could do quite easily. When the test persons were asked to move their fingers more rapidly, they could first follow but then beyond a critical speed the finger movement changed quite involuntarily to a new mode of behavior, namely to a symmetric one (Fig. 10). This transition was conceived as a phase transition described above (Haken et al., 1985; Schöner et al., 1986). The relative phase between the fingers, i.e. the difference between the angles of the two fingers, can easily be identified as an order parameter, whereas, quite evidently, the speed of movement plays the role of the control parameter. Based on rather simple arguments taken from synergetics and from symmetry considerations one can construct a sequence of potential curves which are analogous to those of Figs. 4-6 and are adapted to the present situation (Fig. 11). Quite clearly, a number of

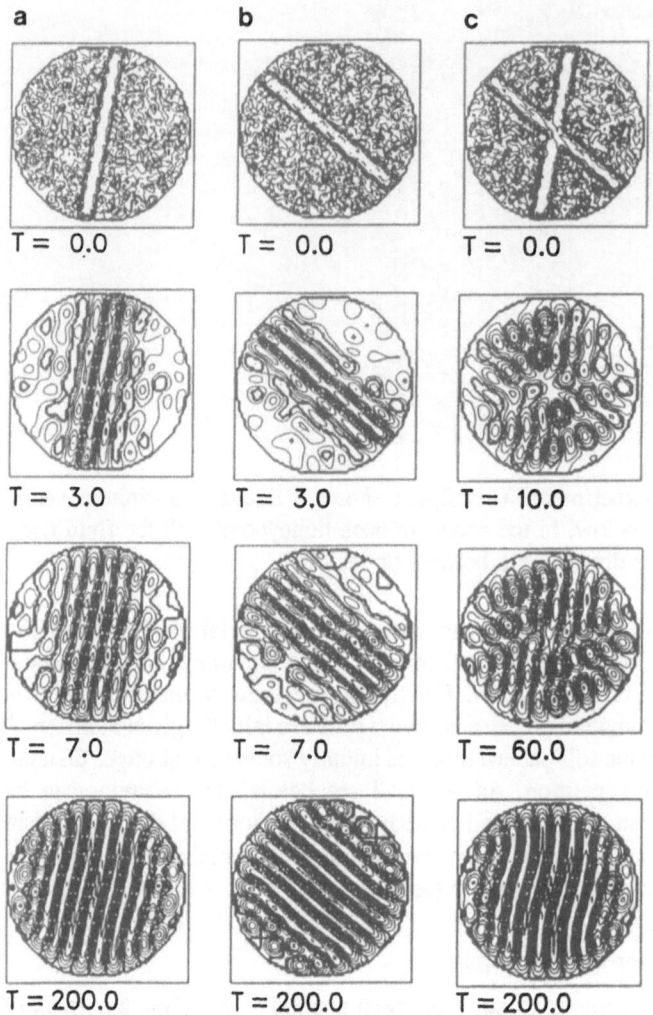

a	b	c
T = 0.0	T = 0.0	T = 0.0
T = 3.0	T = 3.0	T = 10.0
T = 7.0	T = 7.0	T = 60.0
T = 200.0	T = 200.0	T = 200.0

Fig. 9. A computer simulation of a liquid in a circular vessel heated from below when different initial states are prescribed; Left column: A single roll is initially given and develops into a full roll pattern; Middle column: A roll in a different direction is prescribed and develops into a full roll pattern in its original orientation; Right column: A conflict situation for the fluid, where two rolls are prescribed, the one being by about 10 percent bigger than the other. The fluid eventually develops a roll pattern in the direction of the initially stronger roll

predictions can be made based on this simple representation. When we start at the upper left corner of Fig. 11, the parallel position of the fingers corresponds to the position of the ball in the upper minimum. When the frequency is increased, this minimum becomes flatter and flatter so that critical fluctuations can be expected in analogy to the considerations of Fig. 5. Finally, the ball falls down to the middle

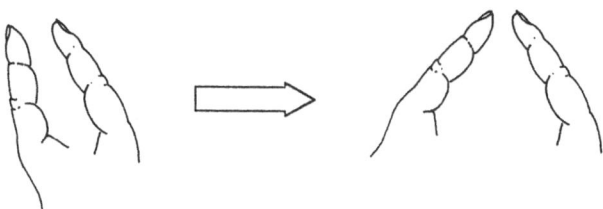

Fig. 10. Transition of a parallel movement of fingers into a symmetric movement

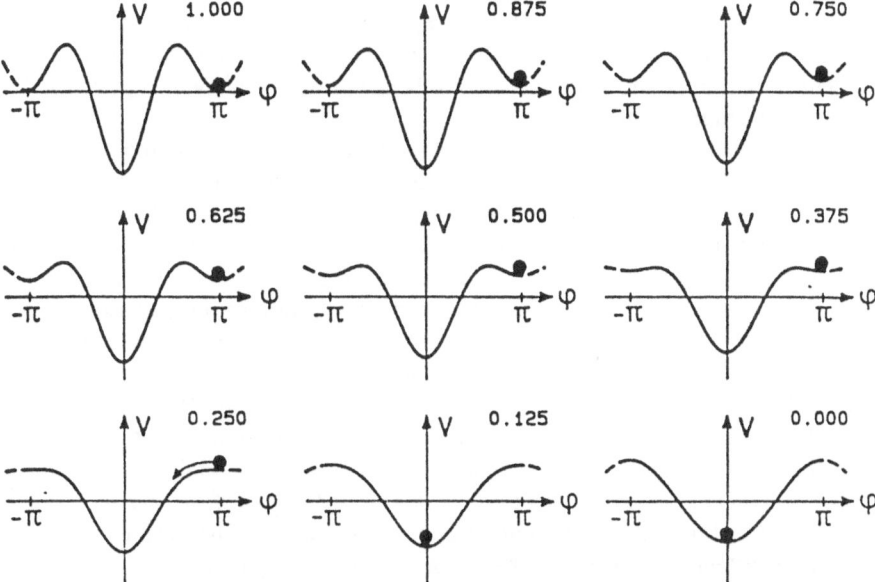

Fig. 11. Visualization of the behavior of the relative phase between the fingers by means of a potential landscape. The potential is plotted versus the relative angle. From upper left to lower right corner the potential is deformed as a consequence of the increasing frequency of the finger movement as a control parameter

valley where it adopts a stable position. We may also do an experiment in the reverse order: We let the test persons start from a symmetric finger movement at a high frequency and then ask them to lower the frequency. When we adopt the model of Fig. 11, the symmetric movement corresponds to the ball lying at the bottom of the potential. When the potential is deformed by starting from the right lower corner in Fig. 11 to the left upper corner, the ball will certainly stay in this position. This means that we expect test persons to move their fingers in the symmetric fashion even if the frequency of finger movement is lowered. This is precisely found in Kelso's experiments as well as the critical fluctuations mentioned above. In the remainder of this article we shall be concerned with some applications of synergetics to psychology.

3. Synergetics as Metaphor, Tool and Model for Processes Dealt with by Psychology

Synergetics can be applied to psychology at a variety of levels, and we want to discuss a few typical examples. In traditional linear science it has not been understandable that one and the same system (in our case the brain) can produce qualitatively quite different kinds of behavior. Let us take the somewhat extreme (because rare) example of multiple personalities where one and the same person exhibits different personalities. In traditional linear science a possible explanation could have only been that different regions of the brain become different personalities. The nonlinear system of the laser has shown that one and the same system can exhibit quite different kinds of behavior. Each behavior is governed by one or few order parameters.

How can switches between different kinds of behavior be caused? Does this require a complicated mechanism? The laser paradigm tells us that the change of even a single, rather unspecific control parameter may cause a qualitative change of laser light. In psychiatry it is known that a single drug may cause a change of behavior and even the microscopic events are known. For example it is well known that Haloperidol blocks Dopamin 2 receptors which, in turn, may cause changes in behavior, such as a considerable reduction of some psychotic states and even their replacement by a quiet state. Because the drug acts in its first place unspecifically, the occurrence of side-effects also becomes conceivable. The same remarks hold for caffeine which blocks serotonin receptors. It is more than speculation to think that the action of drugs taken by addicts and that of endorphines is based on the same principles. A good deal of the mystery of the action of drugs (i.e. matter) on our psyche can be removed, because in nonlinear complex systems changes of one or a few control parameters may induce qualitative changes of macroscopic behavior just as it was the case in the laser. The general paradigm evolving from our results is that of "indirect influence", namely one does not act on behavior directly (for instance by telling the person how to behave) but by changing (control) parameters. Not only the concept of control parameters (for instance concentration of drugs) may prove useful in psychiatry but also that of order parameters. As it appears a number of mental diseases are accompanied by a decrease of a repertoire of behavioral patterns, or, in other words, by a decrease of the number of available order parameters. A further exploitation of the laser paradigm might help to unearth reasons for the decrease of available order parameters (which can supplement, but not replace research on the neuronal level).

Let us discuss some further applications of synergetics to psychology and psychotherapy in one of their proper domains, namely to help a patient (or client) to change his or her attitudes, perception of the environment, kinds of behavior, etc. A consideration of the laser paradigm (Figs. 4-6) may be helpful. When a transition of a system from one state of behavior to another is desirable, the old state must first be destabilized. This is a well-known method in psychotherapy, and applies to both individual and group or family therapy. But in the destabilized state, "critical fluctuations" and "critical slowing down" must be expected and be taken into account. This means for instance that a new stable state is not immediately reached but its achievement may require a considerable amount of time. Furthermore the destabilization of the old state is not enough, but some help must be provided to determine into which new direction the system has to go. This is clearly established by the example of the two minima of Fig. 6. Quite evidently, here a random push must be avoided because one minimum may represent an unwanted state. Rather some well-defined initial state or push must be given to the system which then can achieve a new state. When several members of a group are involved, in general, it will not be sufficient to provide only one member of the group with this new impetus but all the members must be brought, so to speak, into the new coherent state. If the minima are still flat, i.e. the system is not well stabilized, random pushes may easily throw the system out into an unwanted state (including the former destabilized state). In the case of the laser, the depth of the minima depends on the pump power, or, more generally speaking, it depends on the effect of the surroundings on the order parameter. It may be expected that the concepts of synergetics and their models, e.g. of finger movements, etc., may be useful in order to model transitions between mental states or states of behavior. A warning should be added on the range of applicability of synergetics. While synergetics has revealed quite far-reaching analogies between different systems, it is not a unique way to derive from macroscopic behavior the detailed processes at the microscopic level. In general, different microscopic models may produce the same macroscopic behavior or transitions between the macroscopic states. This means for example that the knowledge of the behavior of a person alone will not allow us to construct a unique microscopic model of his or her brain. It is definitely here where the cooperation of all the disciplines involved in a specific problem must cooperate. On the other hand, as we shall demonstrate below, synergetics will allow us to construct brain models that capture some of the behavior shown by humans.

4. Gestalt Psychology

In order to illustrate in how far the methods and concepts of synergetics can be applied to psychological phenomena, we quote the example of Gestalt psychology founded by Köhler (1920, 1940), Wertheimer (1912, 1923) and others. One of the prominent features of synergetics is its consideration of the emergence of new qualities, or, in other words, the validity of the slogan "the whole is more than the sum of its parts". This was clearly put forward by the Gestalt psychologists, who, at their time, had to fight structuralism, in which the attempt was made to explain the properties of the whole out of the properties of the individual parts. A beautiful example of the synergetic view is provided by a painting by Arcimboldo in Fig. 12. Here at first sight we all recognize a face, but when we look more closely, we realize that the face is composed of fruits and vegetables. Other concepts found in Gestalt psychology are those of bistability of perception (Fig. 13), (Stadler and

Fig. 12. A painting by Arcimboldo

44

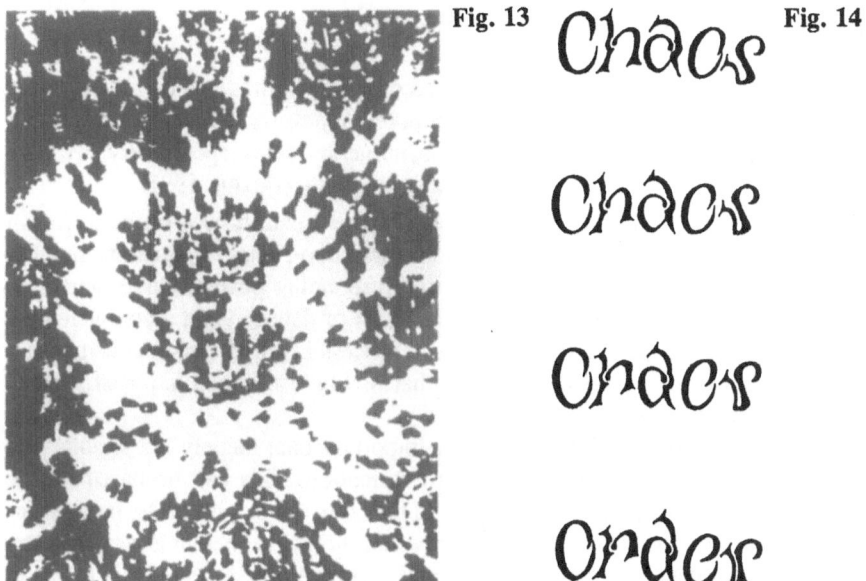

Fig. 13   Fig. 14

Fig. 13. A physicist will interpret this as a microscopic array of elementary magnets in a ferromagnet, other people will interpret it as groups at a stock market

Fig. 14. When we read these words in the top-down direction, the switch from CHAOS to ORDER will occur only after the third word. When we read the same words in the bottom-up direction, the switch occurs from ORDER to CHAOS after the third word from below, i.e. the switching occurs at different positions depending on the history; this effect is called hysteresis

Kruse, 1990) and of hysteresis (Fig. 14). Both phenomena are well known for synergetic systems where we just quote the bistable potential of Fig. 6. The concept of *Prägnanz* of Gestalt theory requires a further study by synergetics. We further quote a number of features of the concept of Gestalt. Gestalt is independent of its position and orientation in space and of its size and it remains intact in spite of deformations, noise, or, say, frequency filtering. These invariance properties are maintained, as we shall see, by order parameters. The question, of course, arises of whether the analogies, such as those of bistability, are at the conceptual level or can be carried still further. In fact, what we have in mind is the realization of pattern recognition tasks by computers. We shall be concerned with them in the next section.

5. Synergetic Computer for Pattern Recognition

To get at least a feeling of how pattern recognition by humans or animals may work, one may try to develop adequate computer models (or "brain models") which in our case are based on the concepts of synergetics. More specifically, we base pattern recognition on the following three ingredients (Haken, 1991b):

1. Pattern recognition is based on associative memory. An example for an associate memory is provided by a telephone dictionary, where a telephone number is provided if we know the name of the corresponding person. Quite generally, associative memory completes an incomplete set of data to yield a complete one.

2. We conceive pattern recognition as a process in which one or several order parameters move in a potential, such as that of Fig. 15, which is constructed in analogy to Fig. 6.

3. The third assumption is the most important one, namely we assume that pattern recognition is nothing but pattern formation. This can be based on the following argument where we refer again to the laser. When part of the laser atoms are in an ordered state so that they produce a well-defined coherent laser wave, this laser wave acting as an order parameter may enslave the rest of the laser atoms to form a total state in a well-ordered fashion (compare Fig. 16). Similarly when we look at a figure, we perceive a number of its features. These features may then generate their order parameter which, in turn, forces all the lacking features into the whole picture.

These concepts may be cast into a rigorous mathematical form which can be implemented on a computer. The total state of the system describing the grey values of the pixels of a picture (Fig. 17) undergoes a dynamic process which is determined by the following parts: There are a number of prototype patterns stored in the computer, which may be for instance faces together with letters encoding the corresponding family names (Fig. 18). These are the quantities to be learnt by the computer. They form the so-called learning matrix. Each learning matrix is multiplied by an attention parameter. Finally, there are terms facilitating the strict discrimination between different patterns and terms yielding a saturation so that the

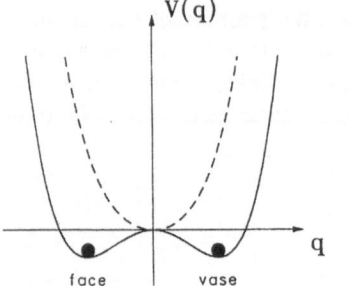 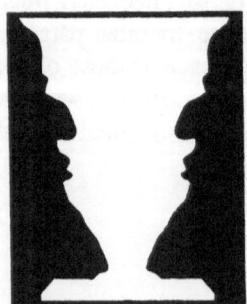

Fig. 15. Right: The well-known ambiguous Fig. vase/faces; Left: Visualization of the potential landscape with two minima corresponding to the interpretation vase or face

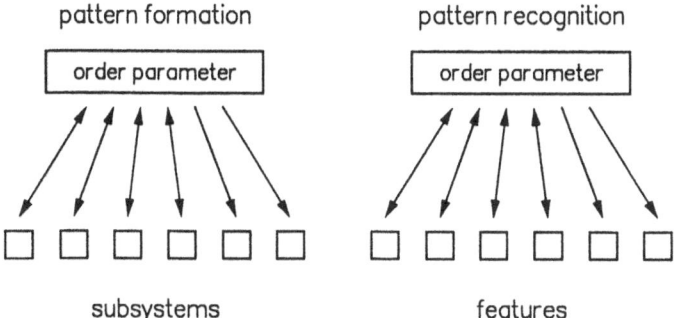

pattern formation   pattern recognition

order parameter   order parameter

subsystems   features

Fig. 16. The analogy between pattern formation and pattern recognition; Left: When part of the subsystems are in an ordered state, they generate their order parameter which then enslaves the rest of the system and produces a totally ordered pattern; Right: When some features of a pattern are given, they generate their corresponding order parameter which then restores the whole recognition pattern

Fig. 17. In this digitized pattern a grey value is attributed to each pixel

Fig. 18. Examples of prototype patterns encoded in the computer

excitation or the height of the grey values cannot exceed a critical value. The dynamics of the grey value distribution is now constructed in such a way that once a set of grey values is given as initial set, the grey values undergo a dynamics so that, eventually, one of the original prototype pattern is completed. This occurs in

Fig. 19. The recognition of a face and its name out of an initially given set of features

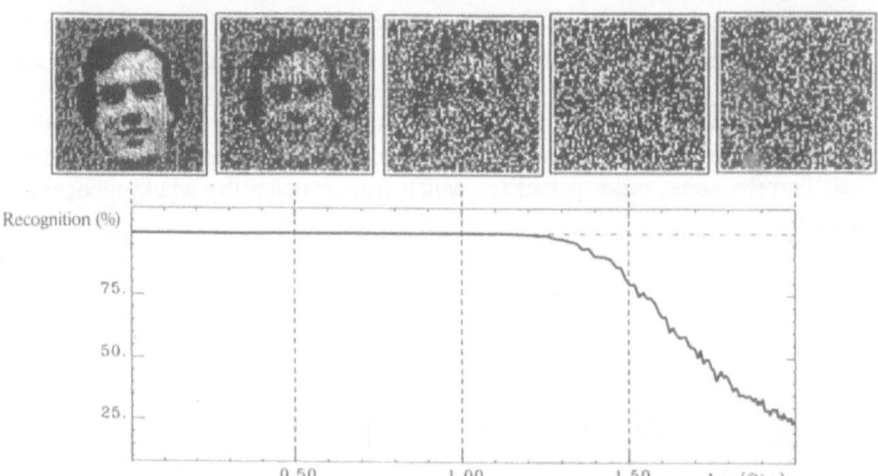

Fig. 20. The synergetic computer recognizes a face even if there is a strong superposition of noise on it. The noise level increases from the left to the right. In the lower part of the figure the percentage of recognized figures is plotted versus the noise level

Fig. 21. Examples of deformed faces that were recognized by the synergetic computer

complete analogy to Fig. 9. Fig. 18 showed a set of typical prototype patterns stored in the computer. Fig. 19 shows the recall of a complete face and letter out of part of a face. Fig. 20 shows that the procedure is stable with respect to noise, Fig. 21 shows that the procedure allows for the recognition of deformed faces. Finally, Fig. 22 and Fig. 23 show that the computer can recognize high and low frequency band-filtered faces. Here frequency refers to the spatial frequency. It is well known that painters occasionally also present objects in one of the two

Fig. 22. The synergetic computer recognizes high frequency filtered faces. The figure shows examples of the degree of filtering

Fig. 23. Same as Fig. 22 but with respect to low frequency filtered faces

fashions. Our computer experiments are in good agreement with psychophysical experiments done by O'Toole et al. (1988). Our computer program allows for practical applications, e.g. it enables the computer to distinguish between facial expression. Figs. 24 and 25 show that the computer can distinguish in a very fine way between facial expressions. Actually, the computer can also categorize but we shall not dwell on this problem here. As may be shown in detail, the prototype patterns correspond to the order parameters, while the attention parameters correspond to the control parameters.

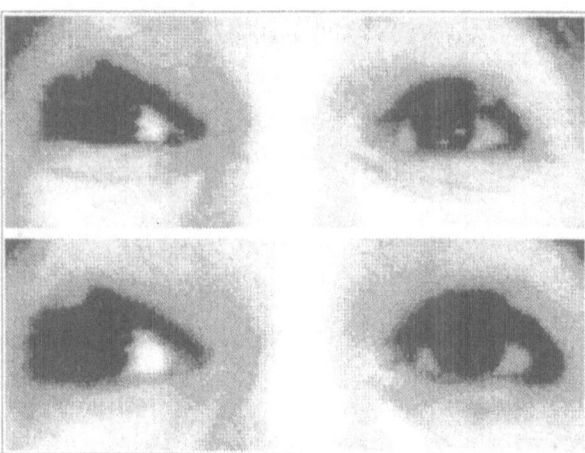

Fig. 24. Example of facial expressions that were recognized by the synergetic computer. This work is based on an unpublished paper by Haken, Hönlinger, and Vanger

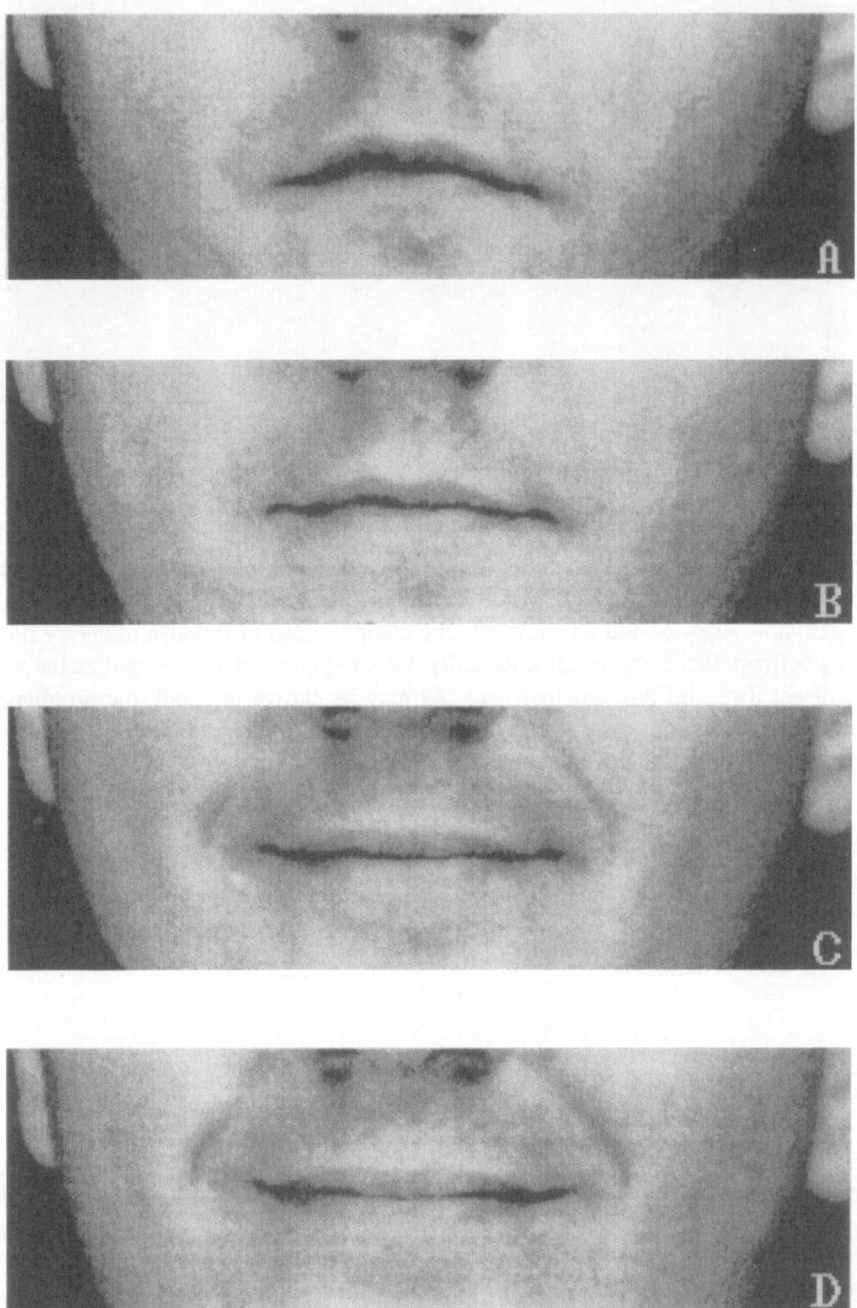

Fig. 25. Same as Fig. 24 but with different parts of a face

6. Recognition of Ambiguous Patterns

The recognition of ambiguous patterns, such as that of Fig. 15, has intrigued psychologists since a long time (Fisher, 1967; Gombrich, 1973; Gräser, 1977; Pöppel, 1982; Kawamoto and Anderson, 1985; Rubin, 1921; Jastrow, 1900; Wallach and Austin, 1954; Botwinick, 1961; Bugelsky and Alampay, 1961; Marbe, 1893; Eichler, 1930; Köhler, 1940, 1920; Orbach et al., 1963; Stadler and Erke, 1968; Kruse, 1988; Attneave, 1971). As is well known, a number of phenomena occur in the recognition of ambiguous patterns, the most prominent being the phenomenon of oscillation. One percept cannot be held indefinitely, it is rather replaced by the other one. Then the latter fades away and is replaced by the former, etc. Köhler (1940) had suggested that such an effect can be explained by a saturation of attention where he made a rather specific physical model, which is probably obsolete nowadays but is still valid as far as its basic concept is concerned. We developed a detailed mathematical theory of the saturation of attention parameters and also of bias effects (Ditzinger and Haken, 1989, 1990, see also Haken, 1991b). In this way we can model a number of experimentally observed phenomena, such as the increase of duration of one percept when the bias of the first recognition of the correponding percept is increased. The theory also covers a number of effects, such as fluctuations of the duration of the reversion times.

These computer results are certainly not based on models of individual neurones and their properties, rather they may be considered as a phenomenological theory of the behavior of large ensembles of neurones. But our model allows us to establish relationships between the experimentally found parameters and those of the model, and to establish relationships between such parameters, e.g. in the case of bias and duration of recognized pattern. (For details cf. Ditzinger and Haken, 1989, see also Haken, 1991b).

7. Conclusion and Outlook

We have tried to give an outline of basic paradigms of synergetics, its concepts and models. In Sect. 6 we formulated a model that can be checked directly experimentally where even the parameters are subject to experimental verification. Quite clearly, the thus elaborated model may serve as a paradigm for mental functions at a higher cognitive or behavioral level. In this way we hope that a way may be paved to the conceptualization and modeling of complex human behavior which may be further traced back to mental states or states of the brain. In our interpretation, behavioral patterns, mental states, and brain states are governed by order parameters or hierarchies thereof and it will be a prominent task of future research to identify the relevant order parameters, their transitions, and their dynamics.

Depending on the number of order parameters, a variety of phenomena may occur. If there is only one order parameter, the discussion connected with Figs. 4-6 holds. If there are two order parameters, they may relax towards a so-called fixed point, i.e. a stationary state of the system, or the system may undergo oscillations, or, in technical terms, it may have limit cycles. When three order parameters are present, besides fixed points, limit cycles and generalized limit cycles in the form of so-called tori representing quasi-periodic motion, deterministic chaos may occur. The most characteristic feature of deterministic chaos is the following: Even if the initial conditions of the same system differ only slightly, in the course of time the so-called trajectories of the system may diverge strongly so that a prediction of the behavior of the system over a longer period of time becomes impossible. A now famous example is the impossibility of giving a reliable weather forecast for more than a few days. A quantitative measure for the speed of divergence of the trajectories is provided by the so-called Lyapunov exponents. It is, of course, tempting to apply the concepts of chaos theory to phenomena in psychology. An important question is whether it is possible to determine the dynamics of the order parameters by means of measured time series. For instance, in EEGs one may derive the electric potential as a function of time. A basic difficulty arises here because the number of measured quantities, e.g. the electric potential, is, in general, smaller than the number of order parameters. In this case a number of methods have been proposed and have been applied to extract nevertheless the form of the so-called chaotic attractor.

We shall not dwell here further on this fascinating but difficult problem and rather refer the reader to a recent article of ours where, in the case of the EEG of an epileptic seizure, it was possible not only to reconstruct the chaotic attractor but also to write down the detailed order parameter equations (Friedrich et al., 1991).

References

ATTNEAVE, F. 1971. Multistability in Perception. Sci. Am. **225**, 62-71.

BÉNARD, H. 1900. Les Tourbillons Cellulaires dans une Nappe Liquide. Rev. Gen. Sci. Pures Appl. **11**, 1261, 1309.

BESTEHORN, M. and HAKEN, H. 1989. Associative Memory of a Fluid, unpublished.

BOTWINICK, J. 1961. Husband and Father-in-law: A Reversible Figure. Am. J. Psychol. **74**, 312-313.

BUGELSKI, B.R. and ALAMPAY, D.A. 1961. The Role of the Frequency in Developing Perceptual Sets. Can. J. Psychol. **15**, 205-211.

DITZINGER, T. and HAKEN, H. 1989. Oscillations in the Perception of Ambiguous Patterns. Biol. Cybern. **61**, 279.

DITZINGER, T. and HAKEN, H. 1990. The Impact of Fluctuations on the Recognition of Ambiguous Patterns. Biol. Cybern. **63**, 453-456.

EICHLER, W. 1930. Der Rhythmische Wechsel in der Auffassung räumlich-zweideutiger geometrischer Figuren. Z. Sinnesphysiol. **61**, 154-193.

FISHER, GH. 1967. Measuring Ambiguity. Am. J. Psychol. **80**, 541-547.

FRIEDRICH, R., FUCHS, A., HAKEN, H. 1991. Spatio-Temporal EEG Pattern. In: H. Haken and H.P. Koepchen (Eds.): Synergetics of Physiological Rhythms. Berlin: Springer.

GOMBRICH, EH. 1973. Illusion in Nature and Art. In: R.L. Gregory and E.H. Gombrich (Eds.): Illusion in Nature and Art. London: Duckworth.

GRÄSER, H. 1977. Spontane Reversionsprozesse in der Figuralwahrnehmung. Trier: Dissertation.

HAKEN, H. (Ed.) 1970. Laser Theory. Berlin: Springer.

HAKEN, H. 1983. Advanced Synergetics. Instability Hierarchies of Self-Organizing Systems and Devices. Berlin: Springer.

HAKEN, H. 1987. Synergetic Computers for Pattern Recognition and Associative Memory. In: H. Haken (Ed.): Computational Systems, Natural and Artificial. Berlin: Springer.

HAKEN, H. 1991a. Erfolgsgeheimnisse der Natur. Synergetik: Die Lehre vom Zusammenwirken. Frankfurt: Ullstein. Engl. transl.: The Science of Structure: Synergetics.

HAKEN, H. 1991b. Synergetic Computers and Cognition. Berlin: Springer.

HAKEN, H., KELSO, J., BUNZ, H. 1985. A Theoretical Model of Phase Transitions in Human Hand Movements. Biol. Cybern. 51, 347.

JASTROW, J. 1900. Fact and Fable in Psychology. New York: Houghton Mifflin.

KAWAMOTO, AH. and ANDERSON, JA. 1985. A Neural Network Model of Multistable Perception. Acta Psychol. 59, 35-65.

KELSO, J. 1984. Phase transitions and Critical Behavior in Human Bimanual Coordination. Am. J. Physiol.: Reg. Integ. Comp. 15, R1000-R1004.

KÖHLER, W. 1920. Die physischen Gestalten in Ruhe und im stationären Zustand. Braunschweig: Vieweg.

KÖHLER, W. 1940. Dynamics in Psychology. New York: Liveright.

MARBE, K. 1893. Die Schwankungen der Gesichtsempfindungen. Phil. Stud. 8, 615-637.

ORBACH, J., EHRLICH, D., HEATH, H.A. 1963. Reversibility of the Necker Cube. I: An Examination of the Concept of "Satiation of Orientation". Percept Mot. Skills 17, 439-458.

O'TOOLE, AJ., MILLWARD, R.B., ANDERSON, J.A. 1988. A Physical System Approach to Recognition Memory for Spatially Transformed Faces. Neural Networks, Vol. I, No. 3.

PÖPPEL, E. 1982. Lust und Schmerz. Neuronale Grundlagen menschlichen Verhaltens. Berlin: Severin u. Siedler.

RUBIN, E. 1921. Visuell wahrgenommene Figuren. Copenhagen: Gyldendalske.

SCHÖNER, G., HAKEN, H., KELSO, J. 1986. A Stochastic Theory of Phase Transitions in Human Hand Movements. Biol. Cybern. 53, 247.

STADLER, M. and ERKE, H. 1968. Über einige periodische Vorgänge in der Figuralwahrnehmung. Vision Res. 8, 1081-1092.

STADLER, M. and KRUSE, P. 1990. The Self-Organization Perspective in Cognition Research: Historical Remarks and New Experimental Approaches.

In: H. Haken and M. Stadler (Eds.): Synergetics of Cognition. Berlin: Springer.

WALLACH, H. and AUSTIN, P. 1954. Recognition and Localization of Visual Traces. Am. J. Psychol. 67, 338-340.

WERTHEIMER, M. 1912. Experimentelle Studien über das Sehen von Bewegung. Z. Psychol. 61, 165-292.

WERTHEIMER, M. 1923. Untersuchungen zur Lehre von der Gestalt II. Psychol. Forsch. 4, 301-350.

Chaos in Health and Disease
– Phenomenology and Theory

Uwe an der Heiden

With 17 Figures

Abstract. After presenting empirical data concerning apparent chaos in health and disease the modern concept of "deterministic chaos" is discussed and developed in some detail. This includes the description of a system in state space, the notions of chaotic or strange attractors, and of fractal dimensions. Subsequently, two main sources of chaos in biological systems are discussed. One of them, called "circular organization", involves nonlinear interactions, feedback, and time delays. The other consists in the coupling of several oscillators. These basic mechanisms are elaborated in the form of mathematical models for epileptic seizures and for the endocrine system. The last section is devoted to the detection and quantification of chaos in empirical data with special application to the heart beat.

1. Introduction. Homeostasis Versus Dynamics

In particular since the work of Bertalanffy it has been emphasized that a basic principle of biological organization is *homeostasis*. This means that the organism as a whole as well as its parts, e.g. the organs and other functional units, at least under healthy conditions operates in such a way as to permanently reestablish a state of equilibrium. It was realized that these equilibria of the organism and its parts are dynamic equilibria (Bertalanffy's "Fließgleichgewicht") in the sense that, as equilibria far from thermodynamic equilibrium, they are maintained under permanent exchange of matter and energy with the environment.

Without denying that the concept of homeostasis has certainly a deep rooted basis in biological organization, a too narrow fixation of this concept can and has masked the fact that under closer inspection it turns out that the organism and many of its functional units are never at an equilibrium in a strict sense. In contrast, on a variety of different time scales they reveal a nonequilibrium dynamics, which not only becomes apparent in development but also in the adult life, and is marked by a large manifold of different time patterns.

One such typical time pattern consists in rhythmic activity. And indeed, in recent decades substantial research, mainly under the banner of chronobiology, has been directed towards the investigation of biological oscillations, e.g. circadian rhythms, physiological rhythms such as the heart beat, breathing, menstrual cycles, and even rhythms in population dynamics.

However, most if not all of these rhythms have been considered as strictly periodic, i.e. after a definite fixed time interval, called the period, the time pattern

55

Springer Series in Synergetics, Vol. 58   Self-Organization and Clinical Psychology
Editors: W. Tschacher, G. Schiepek, and E.J. Brunner   © Springer-Verlag Berlin Heidelberg 1992

is exactly or nearly exactly repeated again and again. Limit cycles became an object of central interest.

Only in recent years, parallel to corresponding developments in mathematics and physics, has it been discovered that most of these rhythms are not strictly periodic, and that moreover, there are a great many rhythms which show nearly no repetition and have a rather irregular appearance. Whereas previously these irregularities would have been attributed to some noise or random purturbation of an otherwise simple signal, the *modern theory of deterministic chaos* has opened a new perspective and a much deeper insight into the origin of many of these seemingly "lawless" temporal fluctuations. Similar insight is to be expected with regard to spatial or even spatio-temporal patterns, however here the theory is less advanced at present.

In the following section we give some examples of apparently chaotic time patterns. In subsequent sections we shall develop and apply some of the theoretical approaches and models.

One of the conclusions will be that biological organization is not bound to rigid, always identical structures, and that chaotic dynamics might enhance the flexibility of organs and other functional units under varying environmental conditions. In fact, there are many different kinds of chaos, and this article will argue against the simple associations of healthy and regular on the one side and ill and chaotic on the other.

2. Some Examples of Irregular Time Patterns Under Healthy and Pathological Conditions

It has long been assumed that a very steady and regular heart beat is a sign of healthiness. However, closer inspection shows that variability of the interbeat time interval (R-R interval), even under constant environmental conditions, is not only characteristic of healthy and not too old persons but that a lack of variability can be an indication of some malfunction. This has been shown by the work of A. Goldberger (1985, 1987), G. Mayer-Kress (1986, 1988), A. Babloyantz and A. Destexhe (1987), and others. Fig. 1.1a shows a continuous record across one day of the R-R intervals of a healthy person. More precisely vertically plotted values are the average of the interbeat intervals over ten minutes. These ten minute averages have been determined every ten minutes, such that Fig. 1a altogether shows 144 values. Despite of the smoothing effect of averaging, a substantial variability and irregularity is evident. Fig. 1.1b shows the corresponding plot of a patient during stationary treatment after a heart attack. The pattern is again apparently irregular; however, the irregularity seems to be of another type, and the amplitude of the variation is reduced. We will return to the discussion of heart rhythms in a later section after the concept of the dimension of a chaotic attractor has been introduced.

Fig. 1.2a shows a record over 84 days of the concentration of blood cells, again from a healthy person. In contrast, Fig. 1.2b contains the corresponding plot

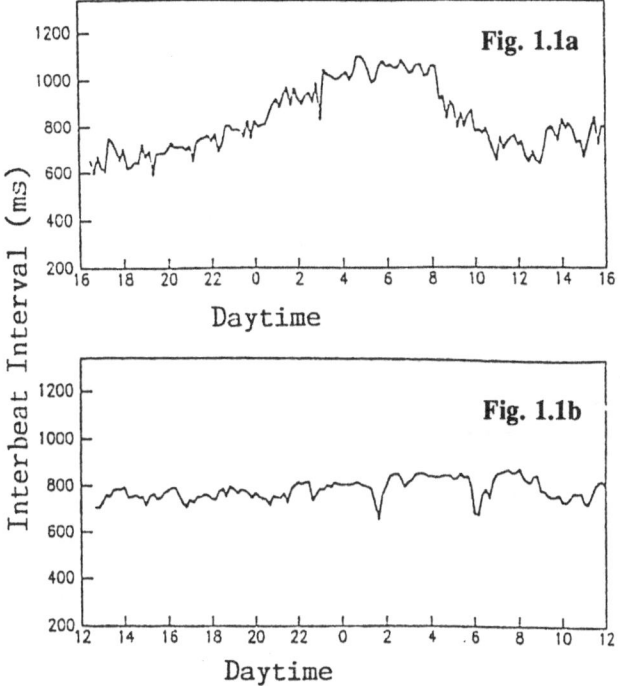

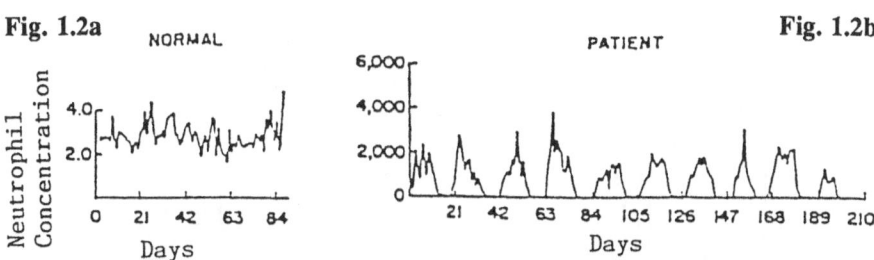

Fig. 1. Empirical data of chaotic type in health and disease; **Fig. 1.1a.** Record of interbeat intervals of heart beat from a healthy female subject over one day; **Fig. 1.1b.** Same kind of record as in Fig. 1.1a, however from a female patient after acute myocardial infarction. (Data kindly provided by H. Bettermann & P. van Leeuwen, University Hospital Witten/Herdecke.); **Fig. 1.2a.** Concentration of neutrophils (a type of white blood cells) of a healthy person, measured over 484 days; 1.2b Neutropil concentration of a patient suffering from cyclic neutropenia (after Guerry et al. 1973); **Fig. 1.3a.** Concentration of parathormone (PTH) in the blood of a healthy person. Measurements were taken every 2 minutes over 9h. **Fig. 1.3b.** PTH-concentration in a case of hyperparathyroidism. Measurements every 2 minutes over 5.5h. Data kindly provided by Prof. R.D. Hesch and Dr. H.M. Harms, Medical High School Hannover; **Fig. 1.4.** Seizure diaries of a patient with epilepsy. Hour of occurrence of seizures are indicated by a dot (after Griffiths and Fox, 1938); **Fig. 1.5.** Manic and depressive phases of a patient with manic-depressive illness (after T.A. Wehr & F.K. Goodwin 1979)

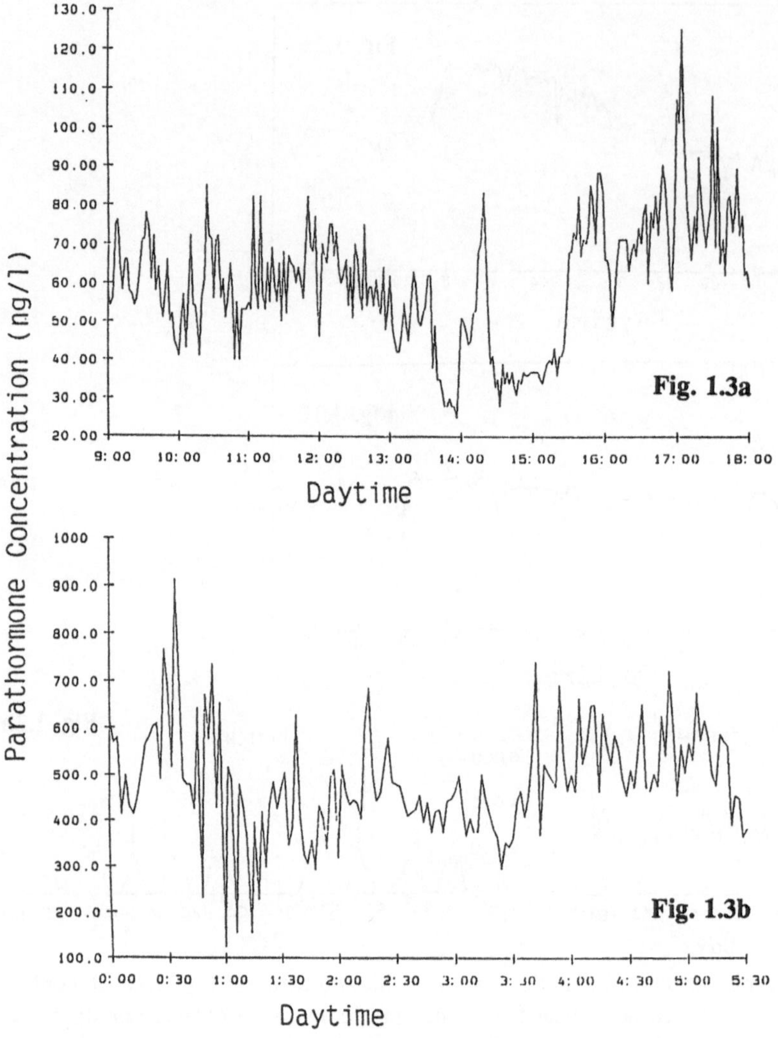

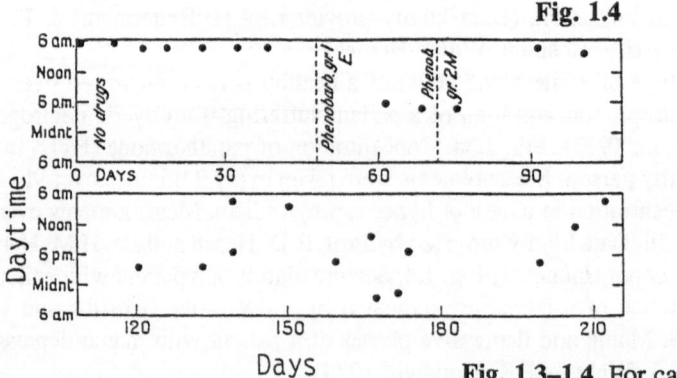

Fig. 1.3a

Fig. 1.3b

Fig. 1.4

Fig. 1.3–1.4. For caption see p. 57

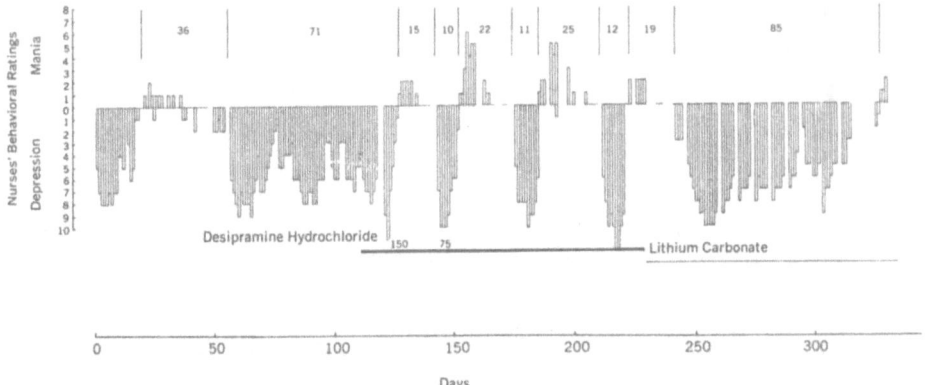

Fig. 1.5. For caption see p. 57

of a patient suffering from cyclic neutropenia. Whereas Fig. 1.2a gives the impression of rather irregular fluctuations around a high average level, Fig. 1.2b contains a basic periodicity which is somehow mixed with irregularity.

Another example of highly aperiodic behavior in the healthy state is the endocrine system. Fig. 1.3a displays a 24h record of the concentration of PTH ("parathyroid stimulating hormone") which is produced by the pituary. Similar chaoticity can be observed in the adrenal and thyroidal systems. The idea that there can be different types of chaos under healthy and pathological conditions is supported by comparison of Fig. 1.3a with Fig. 1.3b where the PTH-pattern of a person with hyperparathyroidism is shown. As with the heart rhythm, the variability is reduced and the total average is shifted. These observations were made by the group of R.-D. Hesch at the Medical High School, Hannover, who measured the hormone concentration every 10 min or even every 2 min (Brabant et al., 1987; Hesch, 1989; Harms et al., 1989).

A case of manic-depressive illness is portrayed in Fig. 1.5. Without treatment manic and depressive phases alternate four or more times yearly, phases are very different in length such that it makes no sense to speak of a periodic process. Under treatment with the antidepressants desipramine and lithium the time pattern significantly changes. In the first case, periodic behavior with a period of about 30 days is observed. In this patient lithium had a negative effect by leading to prolonged intervals of depression.

Another impressive example of how drugs can change the temporal development is given by Fig. 1.4 which shows seizure diaries of a patient with epilepsy. Whereas without drugs seizures always occur at about six o'clock in the morning, under medication seizures are distributed rather randomly throughout the day.

Later on (Sect. 4.2) we will consider a model that may be of importance in understanding some of the processes leading to the onset of epileptic seizures. In an attempt to understand the complex sequence of events leading to the onset of petit mal and grand mal epileptic seizures, neurophysiologists and neurologists have often employed the penicillin-induced epileptic discharge model. In this

59

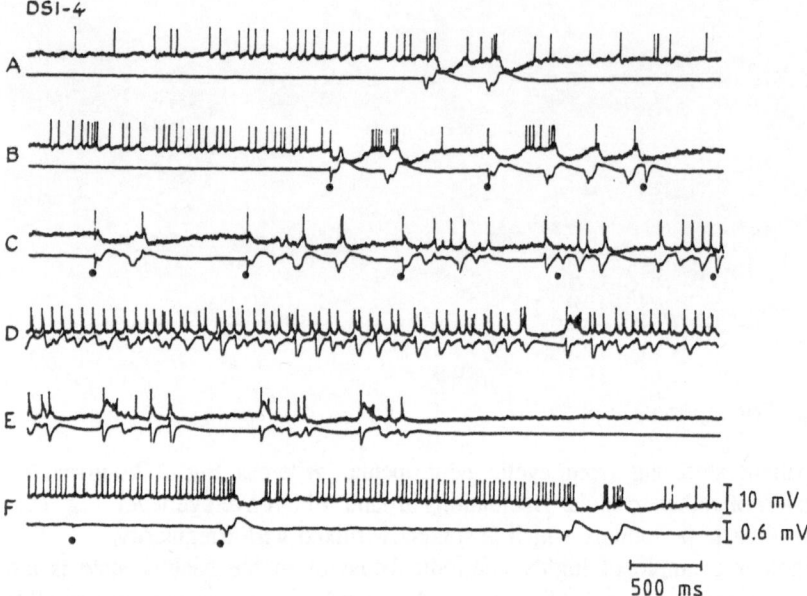

Fig. 2. Electrophysiological recordings from a neuron in the pericruciate cortex of a cat. At certain points in time (indicated by dots) the brain tissue has been stimulated by different doses of penicillin, leading to a variety of different firing patterns of pyramidal cells (after Prince 1969)

preparation, the topical application of penicillin to various cortical structures leads to a discharge pattern in cortical neurons quite similar to those observed in naturally occurring epileptic seizures. There patterns are generally characterized by gradual shift from low frequency burst-like firing patterns to one in which there is continuous and sustained high frequency irregular neural firing. Fig. 2 illustrates the type of firing patterns observed in a hippocampal neuron following the application of penicillin.

More examples could be given from Parkinson's disease (see e.g. Kraus et al., 1987), pupil movements (see e.g. Milton & Longtin, 1990; Milton at al., 1990), apneic respiratory rhythms (see e.g. Specht & Fruhman, 1972; Mackey & Glass, 1977), ventricular fibrillation (see e.g. Lindsay & Budkin, 1975) etc. A complete list is beyond the scope of this contribution.

3. What is Chaos?

The modern concept of *deterministic* chaos, and it is this kind of chaos we are speaking about, is very different from what is classically called a random process. In a random process a subsequent event has either total or partial independence from a previous event. In deterministic chaos causality is completely preserved in the sense that the present state of a system completely determines all future states.

In principle it is possible to give some laws or rules which determine the transition from the present state to a state a little later in the future. These rules are generally describable in terms of mathematical equations.

The state of a system at a certain point t of time is normally given by values $x_1(t)$, $x_2(t)$, $x_3(t)$, $x_4(t)$, etc. of the quantities or variables x_1, x_2, x_3, x_4, etc., which characterize the system and which can change in time.

There are essentially two types of equations for the transition rules, one operating with a discrete time, and the other in which time is a continuous quantity. With discrete time there is a fixed step Δ of time and the process is only considered at times $0 = 0 \cdot \Delta$, $\Delta = 1 \cdot \Delta$, 2Δ, 3Δ, 4Δ, etc. Let us call $x_i(n)$ the value of the quantity x_i at time $n \cdot \Delta$. A transition rule states which value $x_i(n+1)$ will occur at time $(n+1)\Delta$ if at time $n\Delta$ the values $x_1(n)$, $x_2(n)$, etc., are given. This means that for each variable x_1 there is a function (or mapping in the mathematical sense) f_1 such that

$$x_i(n+1) = f_i(x_1(n), x_2(n), ..., x_k(n)), \qquad i=1, 2, ..., k. \qquad (1)$$

Here it is assumed that there are altogether k variables. Mathematically, (1) is called a system of k *difference equations*. If the time discretization step Δ is taken to be sufficiently small then every real system can in principle be described by a system of difference equations (1), but only up to a certain preciseness depending on the size of Δ. When time is considered to be continuous, then there is no unique point in time following the point t. Therefore the description of the rules of transition from a state $x_1(t)$, $x_2(t)$, ..., $x_k(t)$ to future states has to be given in terms of *differential equations*

$$dx_i(t)/dt = f_i(x_1(t), x_2(t), ..., x_k(t)), \qquad i=1, 2, ..., k. \qquad (2)$$

Here the function f_i tells how the rate of change $dx_i(t)/dt$ of quantity x_i at time t depends on the state $x_1(t)$, $x_2(t)$, ..., $x_k(t)$ at time t. The theory of differential equations (see e.g. Verhulst, 1990) shows (we do not go into detail here) that given some initial condition $x_1(0)$, $x_2(0)$, ..., $x_k(0)$ there is a uniquely determined time course $x_1(t)$, $x_2(t)$, ..., $x_k(t)$ for all t≥0, which obeys (2) for all t≥0 and which starts with the given initial condition. Note that different initial conditions lead to different time developments of the system variables. (The theory requires some technical assumptions about the functions f_i, which we do not specify here and which are not restrictive to any applications).

The systems (1) and (2) of equations make precise what is meant by a *deterministic* system : Once an initial condition $x_1(0)$, $x_2(0)$, ..., $x_k(0)$ is fixed the future of the k variables x_1, x_2, ..., x_k is uniquely determined for all t≥0 (of course the time point 0 can be chosen arbitrarily).

We now try to characterize the concept of a *chaotic attractor* (sometimes also called *strange attractor* or *fractal attractor*). To this end the notions of a *state space* and *trajectory* are very helpful. For each point t in time one considers $\{x_1(t), x_2(t), ..., x_k(t)\}$ a point in the k-dimensional Euclidean space $\mathbf{R}^k$. For k=1, k=2, k=3

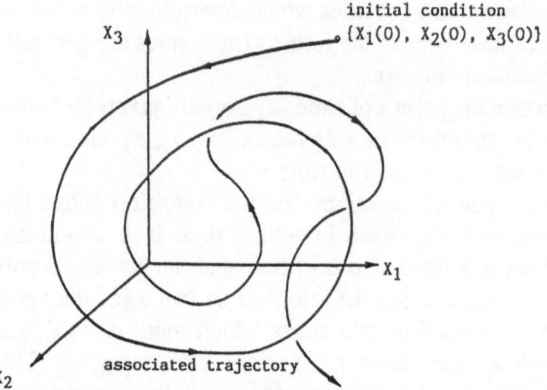

initial condition
$\{x_1(0),\ x_2(0),\ x_3(0)\}$

Fig. 3. Scheme of a three-dimensional state space. An example of an initial condition and of the associated trajectory is shown. The reader has to imagine a point moving in time along the trajectory in the direction of the arrows

this space coincides with what is normally called a line, a plane, and a volume respectively. Fig. 3 illustrates the case k=3. Each point in the space $\mathbb{R}^k$ is specified by special values of the k coordinates $x_1, x_2, ..., x_k$. Since the state variables are just the coordinates of this space, it is called the state space. Each point in this space represents a special state of the system.

In particular the initial condition $\{x_1(0), x_2(0), ..., x_k(0)\}$ is itself one point in state space, cf. Fig. 3. The whole set of points $\{x_1(t), x_2(t), ..., x_k(t)\}$ for all $t \geq 0$ which describes the development of the system, starting from a given special initial condition, is called the *trajectory* or *orbit associated with the initial condition*. Such a trajectory can be determined either as a solution of the system (2) of differential equations or by measuring the variables of some real system. Note, however, that by the second method, generally only a finite number of points of the trajectory can be obtained; some interpolation procedure has to be used in order to arrive at a continuously connected curve.

The concepts of state space and trajectory allow one to characterize different types of a system's behavior. Before we give such a classification, a fairly general property of trajectories of either real systems or of mathematical systems of type (2) which describe or model real world systems should be mentioned. Such systems generally possess only a finite amount of energy. As a consequence none of the variables $x_i(t)$ can grow indefinitely as $t \to \infty$, and thus the trajectory $\{x_1(t), x_2(t), ..., x_k(t)\}$ for all $t \geq 0$ remains in a bounded area of state space.

Under this fairly general condition, the course of a trajectory and thus the development of the system, starting from some initial condition $\{x_1(0), ..., x_k(0)\}$, can be classified roughly and exhaustively into four categories (for illustration see Fig. 4):

i) As $t \to \infty$ the trajectory converges toward a single point, which is a *steady state* of the system (Fig. 4a).

62

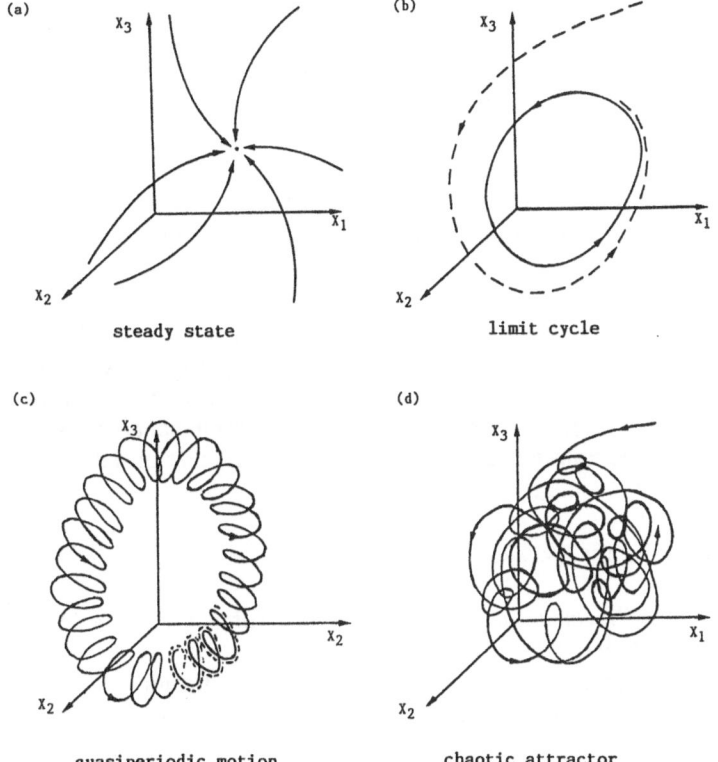

(a) x_3 x_1 x_2

steady state

(b) x_3 x_1 x_2

limit cycle

(c) x_3 x_2 x_2

quasiperiodic motion

(d) x_3 x_1 x_2

chaotic attractor

Fig. 4. The four possible characteristics of development in state space: (a) Motion towards a stable steady state. (b) Motion towards a stable periodic orbit (=limit cycle). (c) Quasiperiodic motion: there is never an exact repetition of events; however, neighboring trajectories remain neighbors. (d) Motion in a chaotic attractor, which is characterized by sensitive dependence on initial conditions

ii) As $t \to \infty$ the trajectory converges toward a closed loop, which is a *periodic orbit* or *limit cycle* of the system (Fig. 4b).

iii) The trajectory describes a quasiperiodic motion, which can be understood as a superposition of several periodic orbits (Fig. 4c).

iv) None of the previous three cases holds.

In this last case, which is by no means exceptional but forms the broadest class, the trajectory is called *chaotic*. There is no universally accepted positive definition of a chaotic trajectory. However, we have the following situation:

The trajectory stays in a finite, bounded domain of state space, it is infinitely long, but it does not converge toward any fairly simple object in the classical sense. "Classical" means that the object can be attributed a dimension which is an entire number (0, 1, 2, etc.). A steady state, e.g., is a point, having dimension 0. A limit cycle has dimension 1. In fact, it has been a substantial discovery that the trajectory can fill a region in state space which must be attributed a so-called *fractal dimension*, i.e. a dimension which is represented by a noninteger number, e.g. 3.7 or $\sqrt{7}$.

Presupposing at the moment, that a reasonable definition of a fractal dimension can be given (we will take up this subject again in Sect. 5), we say that a trajectory is chaotic if it fills in state space a set of fractal dimension. It is possible, by some mathematical reasoning (see e.g. Schuster, 1989), to conclude fairly generally from this property another important property, which is often used as an alternative definition of deterministic chaos. It is called *sensitive dependence on initial conditions*. This concept breaks with the traditional idea that similar causes have similar effects. On the contrary, it may happen that for any two initial conditions $\{x_1(0), x_2(0), ..., x_k(0)\}$ and $\{y_1(0), y_2(0), ..., y_k(0)\}$ which are arbitrarily close to each other (but not identical), i.e. for which $|x_i(0)-y_i(0)| < \varepsilon$ with some arbitrarily pregiven small, but positive number ε, the associated trajectories $\{x_1(t), ..., x_k(t)\}$ and $\{y_1(t), ..., y_k(t)\}$, $t \geq 0$, diverge from each other such that after a while (which depends on ε) the trajectories are far apart from each other in state space. The latter means that for some $t > 0$

$$|x_i(t) - y_i(t)| > \delta \quad \text{for} \quad i=1, 2, ..., k \tag{3}$$

with a positive number δ *which does not depend on the value of* ε, i.e. δ can be an arbitrarily large multiple of ε.

We are now able to define the concept of a chaotic attractor (or strange attractor). A *chaotic attractor* is a bounded subset A of the state space $\mathbf{R}^k$ with the following properties:

i) there is at least one trajectory coming arbitrarily close to all points of A;

ii) each pair of trajectories within the attractor shows sensitive dependence on initial conditions, i.e. (3) holds with a positive number δ which is independent of the initial conditions;

iii) there is a neighborhood of A such that all trajectories starting in this neighborhood approach A as $t \rightarrow \infty$.

Sensitive dependence on initial conditions in a bounded domain, which is never left by the trajectories, implies a certain mixing of the trajectories and gives trajectories their chaotic, apparently irregular appearance. It can be shown that such systems have properties which in earlier times were believed to be typical only for random processes and probabilistic systems (Lasota & Mackey, 1985). For example, the power spectrum of a component $x_1(t)$, $t \geq 0$, as a function of time can be broad-band and continuous (without peaks), or the amplitudes of the fluctuations in time can obey a continuous probability distribution.

A chaotic attractor, defined by the three properties above generally has a fractal dimension. Thus, the two concepts of chaos given above are, in all practical respects, identical. A detailed discussion of deterministic chaos can be found in Schuster (1989).

4. Sources of Chaos

In this chapter we shall discuss some of the known "mechanisms" or organizational principles which can generate chaos. In this young field, no nearly complete catalogue on the origin of chaos is available and it is to be supposed that none will ever exist. There are strong indications that infinitely many different types of chaos are possible. In the following we describe two basic mechanisms which have shown wide applicability in biology and medicine.

4.1 Circular Organization, Mixed Feedback, and Time Delays

By the very nature of life as an autocatalytic, self-generating process (an der Heiden et al., 1985; an der Heiden, 1991c), a predominant feature of biological organization is *circularity* or *feedback*. The basic scheme is outlined in Fig. 5. Via several intermediary steps a quantity (or time-dependent variable) $x_1 = x_1(t)$ feeds back onto itself: $x_1 \to x_2 \to x_3 \to x \dots x_{n-1} \to x_n \to x_1$. Each of the intermediary steps can be modulated by external influences e_i, as indicated in Fig. 5.

Circularity can be observed at all levels of biological organization. At the molecular level, a famous example is regulation of transcription of DNA, which via several steps leads to proteins and finally some product which gives a signal to the enzymes controlling DNA-transcription, cf. Fig. 6a. Another molecular example is the cAMP-cycle in *Dicteostelium* where extracellular cAMP binds to a membrane receptor; see Fig. 6b. The cAMP receptor complex activates adenylase cyclase which results in the production of intracellular cAMP which, by permeating the membrane, changes the concentration of extracellular cAMP (Goldbeter, 1990, p.145).

At the cellular level, an example is hematapoiesis: Red blood cells (and also other components of the blood) give a signal (erythropoietin) to the stem cells in

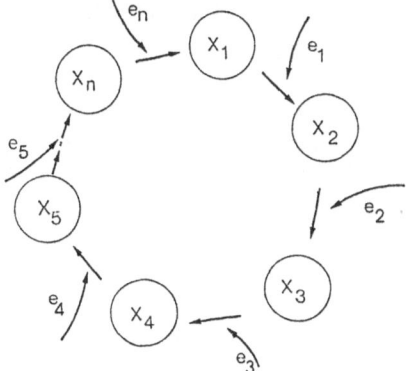

Fig. 5. Basic scheme of circular organization: A certain number (n) of subsystems with variables x_1, x_2, x_3, etc. are concatenated in a circular order such that the last one (x_n) feeds back on the first one (x_1). External influences on the cycle are indicated by curved arrows and external variables e_1, e_2, ..., e_n

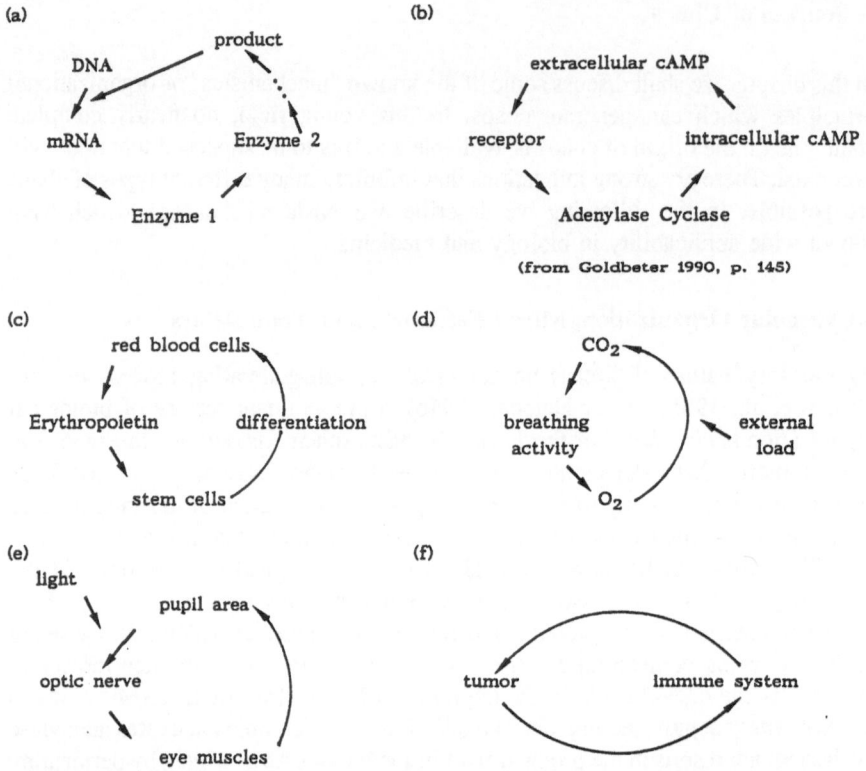

Fig. 6. Examples of circular organization. (a) Metabolic regulation: Signals of end products feed back to modulate transcription. (b) Feedback loop of intra- and extracellular cAMP (Goldbeter, 1990, p.145). (c) Circular control of hematopoiesis (Mackey, 1978). (d) Circular control of O_2 and CO_2 concentrations in the blood. (e) The pupil-light-reflex. (f) Circular dependence of immune system and tumor: The tumor stimulates activity of the immune system which influences growth and development of the tumor (an der Heiden, 1991b)

the bone marrow to enhance or reduce the production of red blood cells; cf. Fig. 6c (Mackey, 1978). An example with nerve cells will be given in Sect. 4.2.

At the level of organs, there is a circular relationship between the breathing activity of the lungs and the CO_2-content of the blood, see Fig. 6d (Mackey & Glass, 1977). Another example is the pupil-light reflex: the diameter of the pupil controls light intensity on the retina, which influences across the optic nerve and other nervous organs muscles controlling the pupil area (Milton &Longtin, 1990). Another interesting example is the mutual interaction between a tumor and the immune system. It would not be difficult to find examples even at the organismal and behavioral level and at the level of species.

If one tries to describe circular organization formally one of the simpler approaches is to start with k time-dependent quantities $x_i = x_1(t)$, i = 1, 2, ..., k. The assumption of circular organization results in a set of k differential equations

$$dx_1(t)/dt = f_1(x_k, x_1, e_1)$$

$$dx_i(t)/dt = f_i(x_{i-1}, x_i, e_i) \qquad \text{for } i=2, 3, \dots k. \tag{4}$$

The meaning of these equations is that the momentary change dx_i/dt at time t of quantity x_i is a function f_i of the preceding quantity x_{i-1} (which in case of x_1 is x_k), the quantity x_i itself and some external influence e_i.

In many applications the influence functions f_i can be decomposed into two terms

$$f_i(x_{i-1}, x_i, e_i) = p_i(x_{i-1}, x_i, e_i) - d_i(x_{i-1}, x_i, e_i). \tag{5}$$

The first term p_i describes factors involved in the production, activation, or magnification of x_i, whereas the second term addresses destructive, inhibitory, or diminishing factors (an der Heiden & Mackey, 1982).

A well known example of such a structure is the Goodwin model for enzyme regulation (cf. Fig. 6a) which has the following special form of (4):

$$dx_1(t)/dt = f(x_k(t)) - a_1 x_1(t)$$

$$dx_i(t)/dt = b_i x_{i-1}(t) - a_i x_i(t) \qquad \text{for } i = 2, 3, \dots, k. \tag{6}$$

Here all equations with the exception of the first one contain only linear terms with constant coefficients a_i and b_i (which in this context are all positive). Thus, the destruction terms simply model linear decay or death rate. The first equation, originally describing the rate of mRNA synthesis, contains a nonlinear feedback function f. In earlier investigations it has been assumed that f is either monotonically decreasing, as illustrated in Fig. 7a, corresponding to supression or negative feedback, or it is monotonically increasing, as illustrated in Fig. 7b corresponding to induction or positive feedback. However there are good biological reasons (Mackey & Glass, 1977; an der Heiden, 1979) why a non-monotonic feedback function of the type illustrated in Fig. 7c is often realistic.

One reason for this latter shape is that none of the quantities x_i should become too large in order to avoid biological disorder. Therefore for large values of x_k the

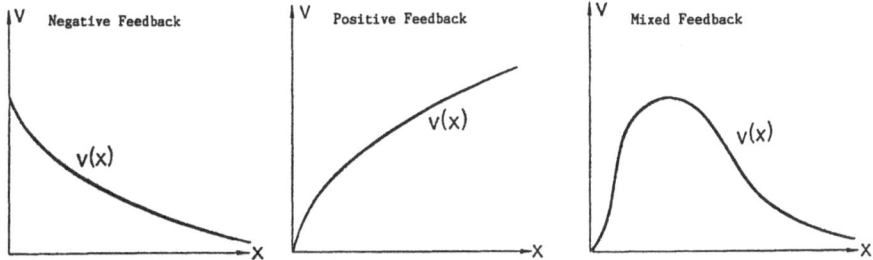

Fig. 7. Three types of feedback functions: (a) negative feedback, (b) positive feedback, (c) mixed feedback

value $f(x_k)$ should be small. On the other hand, if the quantity x_k is functionally important for the organism a too small value of x_k would mean that the system has run out of the "physiological range", leading to a low production rate $f(x_k)$ for small values of x_k.

In case of negative feedback (Fig. 7a) it has been shown (Hastings et al., 1977; Mallet-Paret & Smith, 1990) that system (6) can display just two types of behavior. Either for arbitrary initial conditions the solution $x_1(t)$, $x_2(t)$, ..., $x_k(t)$ approaches temporally constant values, i.e. in state space a steady state is approached as $t \rightarrow \infty$ (point attractor). Or the solution trajectory moves into a periodic orbit, meaning that each of the quantities $x_i(t)$ oscillates in time in such a way that ultimately $x_i(t)$ approaches a periodic function of time (limit cycle). No chaos is possible in this case (Mallet-Paret & Smith, 1990).

In case of positive feedback (Fig. 7b) generally some steady state is approached by the trajectories (Tyson & Othmer, 1977). However, there may exist more than one steady state. Thus, the system has the possibility of reaching one of several equilibria ("multistability"). The initial condition $\{x_1(0), x_2(0), ..., x_k(0)\}$ determines which of the steady states is actually approached.

The situation is much more complicated in the case of mixed feedback (Fig. 7c). It has been demonstrated (an der Heiden, 1978, 1979) that besides steady states and periodic solutions there also exist chaotic solutions to system (5). Which of these three types of behavior actually occurs essentially depends on the parameters (a_i, b_i) of the system and on the steepness of the feedback function f. Fig. 8 shows some of the periodic and chaotic solutions (only the component $x_1(t)$ is drawn as a function of time). The periodic solutions may be rather complicated (with more than one maximum within one period). As the parameter a_1 is varied

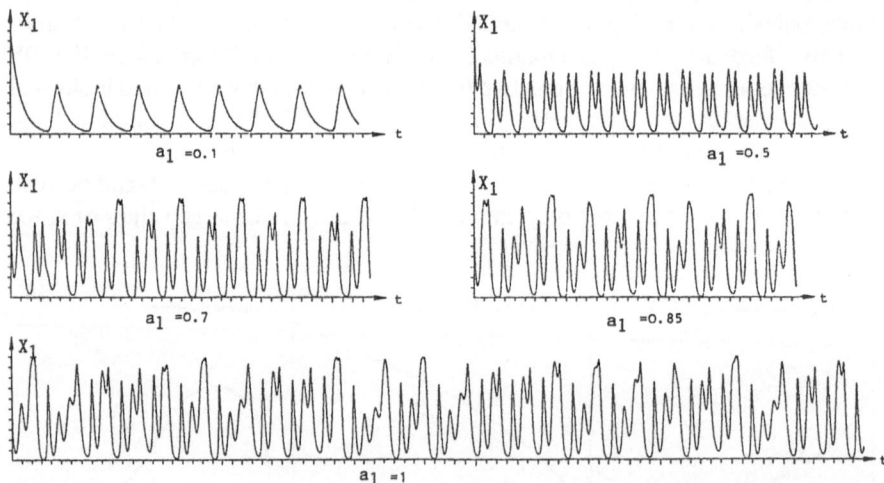

Fig. 8. Periodic and chaotic computer solutions to the feedback system (5) with mixed feedback. Only the parameter a_1 is varied, values of a_1 are given below the plots of x_1 versus t. Lack of smoothness in the graphs is due to the plotter resolution. After an der Heiden, 1979

(all other parameters being held constant) a transition from periodic to chaotic behavior takes place, apparently via a series of so-called period-doubling bifurcations. This is one typical way of transition from non-chaotic to chaotic behavior, sometimes called "Feigenbaum sequence" after M.J. Feigenbaum.

These investigations show that feedback, used in many technical systems to stabilize a steady state and later on realized as a source of periodic oscillations, can also be a source of chaotic time patterns.

This feature becomes even more salient if one additionally takes time delays into account. With system (5) the number k, which measures the length of the feedback chain, must be rather large (the minimum value is k=3) in order to obtain chaos. It was discovered by M.C. Mackey and L. Glass (1977) that a single equation is sufficient if one combines mixed feedback with a delay. Their equation is

$$dx(t)/dt = f(x(t-\tau)) - ax(t).$$ (7)

The essential difference between a system of type (6) and a system of type (7) is that in (6) the rate of change $dx_i(t)/dt$ at time t depends on the state $\{x_1(t), x_2(t), ..., x_k(t)\}$ at the same time t. In (7) the rate of change $dx(t)/dt$ at time t depends on the state $x(t-\tau)$ at a time $t-\tau$ in the past, with a positive constant time delay τ. Thus the states in the future not only depend on the present state (because of the term $-ax(t)$) but also on the history of the system some time τ ago.

In fact, (7) can be viewed as a simplification of the system (6): Since the second up to the k-th equation in (6) are linear they can be formally integrated and hence x_k can be expressed as a functional of x_1:

$$x_k(t) = \int_{-\infty}^{t} x_1(t') \, g(t-t') \, dt'$$ (8)

with a function g which is illustrated in Fig. 9 (an der Heiden, 1979). If one chooses τ simply to be that time where the function g has its maximum, then equation (7), with the identification $x=x_1$, is in a certain sense an approximation of system (6). The delay τ is roughly the time needed for a signal to travel once through the feedback cycle of Fig. 5. Mackey and Glass (1977) demonstrated that, under the assumption that f is a suitable mixed feedback function, the delay

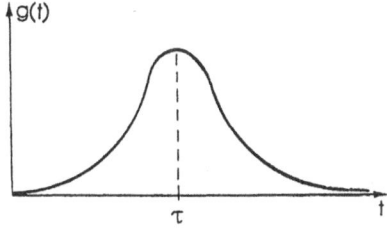

Fig. 9. Kernel function g to explain the formal connection between the Goodwin model and the Mackey-Glass differential delay equation, see text for explanation

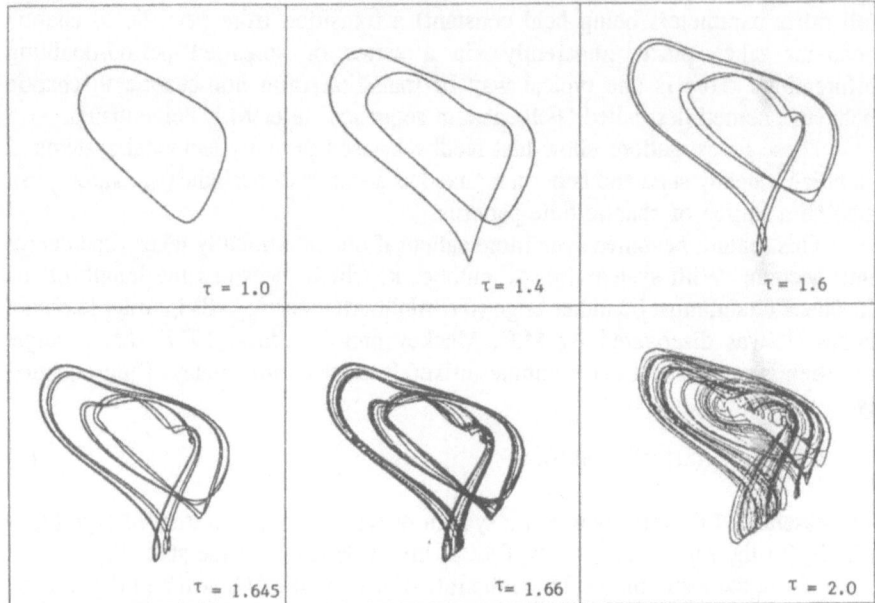

Fig. 10. Phase portraits of solutions to the differential delay equation (7) with a mixed feedback function f given by equation (9). Horizontal axes always represent $x(t)$, vertical axes $x(t-\tau)$. Therefore, each trajectory is to be considered as an ensemble of points $(x(t), x(t-\tau))$ with t varying over some interval of time. The different trajectories (a)-(f) are obtained for different values of the delay parameter τ (all other parameters have been held constant). (a)-(d) show increasingly complex periodic orbits, which originate through a series of period-doubling bifurcations. (e) and (f) represent trajectories within a chaotic attractor

differential equation (7) exhibits both complicated periodic solutions with the phenomenon of period doubling bifurcations, and chaotic solutions. Fig. 10 displays projections of trajectories onto a plane where one axis represents $x(t)$ and the other $x(t-\tau)$. Figs. 10a-d show periodic orbits with increasing complexity as the delay constant τ is increased. The function f chosen for the calculations is (after Mackey & Glass, 1977)

$$f(x) = \frac{2x}{1+x^{10}} - x \ . \tag{9}$$

Figs. 10e,f show projections of a chaotic attractor apparently present in this equation. The computer calculations leading to these pictures strongly suggest the existence of such an attractor, but a mathematical proof is hardly ever available. However, for certain special functions f a proof has been given; see an der Heiden & Walther (1983), an der Heiden & Mackey (1982).

4.2 Application to Penicillin-Induced Epileptic Seizures

Equation (7) has found many applications in biology, and even in economics. Applications range from models for the production of red blood cells (Wazewska-Czyzewska, 1984), for the origin of a variety of haematological diseases including aplastic anaemia and periodic haematopoeisis (Mackey, 1987; Mackey & Dörmer, 1981), periodic excitations of neurons (Coleman & Renninger, 1976), the pupil light reflex (Milton & Longtin, 1990; Milton et al., 1990), population dynamics of whales (May, 1980), respiratory cycle (Mackey & Glass, 1977), to psychiatric disorders like schizophrenia and panic attacks (King et al., 1983) and to blowflies (Nisbet & Gurney, 1976).

Here we mention an investigation (an der Heiden et al., 1981; Mackey & an der Heiden, 1984) concerning neurology, which may serve as a mathematical and neurophysiological model of epileptic seizures, artificially elicited by topical application of penicillin to brain tissues.

Feedback via inhibitory interneurons is present in many if not in all brain areas. A particularly well-documented example is the hippocampus CA3-area: the mossy fibre-CA3 pyramidal cell-basket cell complex. The CA3 pyramidal cells receive excitatory presynaptic input from the mossy fibres which form the main input to the hippocampus. The pyramidal cells excite basket cells which in turn inhibit the pyramidal cells, see Fig. 11.

The mathematical model (for details see Mackey & an der Heiden, 1984) is briefly summarized as follows. In simple models of neurons it is often assumed that deviations x(t) of the membrane potential from its resting state follow first order kinetics:

$$m \, dx(t)/dt = - x(t) \tag{10}$$

with a membrane time constant m. In case of the pyramidal cells the deviations originate from the excitatory input e(t) of the mossy fibres and from the inhibitory input i(t) of the basket cells leading, by inclusion into equation (10), to

$$m \, dx(t)/dt = e(t) - i(t) - x(t). \tag{11}$$

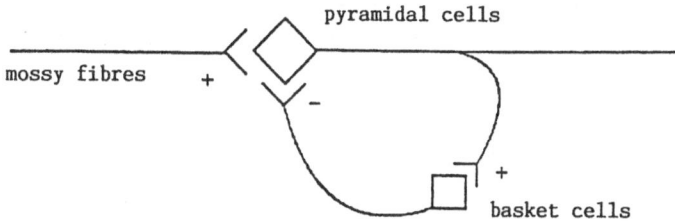

Fig. 11. Recurrent inhibitory circuit in the hippocampus: Pyramidal cells excite basket cells which inhibit the pyramidal cells. The pyramidal cells receive excitatory input from mossy fibres

71

The transformation of the membrane potential into the instantaneous firing rate of the pyramidal cells is modelled by

$$F(x(t)) = \begin{cases} 0 & \text{if} \quad x(t) \leq T \\ ax(t) & \text{if} \quad x(t) > T \end{cases} \tag{12}$$

where T denotes the threshold potential for the generation of an action potential, and a is a constant related to the rate of change of firing frequency with respect to potential changes above threshold.

In the model, the inhibitory potential $i(t)$ of the pyramidal cell is assumed to be proportional both to the amount C of inhibitory transmitter released from the basket cells and to the fraction $G(C)$ of receptors at the postsynaptic site available to be activated by that transmitter:

$$i(t) = N \, V_m \, G(C)C \; . \tag{13}$$

Here N is the total number of receptors at the postsynaptic site, V_m the inhibitory potential resulting from activation of a single receptor.

The fraction $G(C)$ of receptors which is free from inhibitory transmitter (in this case GABA) can be determined from quasi-equilibrium receptor-transmitter kinetics (Mackey & an der Heiden, 1984). It is described by the expression

$$G(C) = \frac{K}{K + C^n} \tag{14}$$

where K is the equilibrium constant for the transmitter-receptor reaction and n the number of molecules of transmitter required to activate one receptor.

The amount $C(t)$ of transmitter released from the basket cells is assumed to be proportional to their firing frequency $F^\sim$

$$C(t) = b1 \, F^\sim(t) \; . \tag{15}$$

In order to avoid a too complicated model it is finally assumed that $F^\sim(t)$ is simply proportional to the firing frequency $F(x(t-\tau))$ of the pyramidal cells. Here the delay τ is due to the time required to transmit a signal from the soma of the pyramidal cell back to the inhibitory synaptic connection of the basket cell onto the soma of the pyramidal cell.

Taking (13-16) together we arrive at

$$i(t) = bNV_m F(x(t-\tau)) \frac{K}{K + [F(x(t-\tau))]^n} \; . \tag{17}$$

Inserting (17) into (11) results in the *complete model*

$$m\frac{dx}{dt}(t) = e(t) - bNV_m F(x(t-\tau)) \frac{K}{K + [F(x(t-\tau))]^n} - x(t) \; . \tag{18}$$

Equation (18) essentially coincides with the type (7) of equations. The second term (equal to (17)) represents the mixed feedback since the function f(x) = KF(x)/(K + (F(x))^n) is not monotonic for n>1. The experimental estimate of n is 2 to 4.

Penicillin binds competetively to the GABA receptor, and indeed the topical application of penicillin is a common experimental tool for the generation of epileptic-like behavior in cortical cells. In the context of the model (18) this means that application of penicillin is equivalent to reducing the number R of GABA receptors on the CA3 cells.

Fig. 12 shows eight time traces of the model membrane potential x(t) for eight different values of N. This number of receptors per cell ranges in steps of 200 from N=1900 at the top of Fig. 12 to N=500 at the bottom panel. In each of these

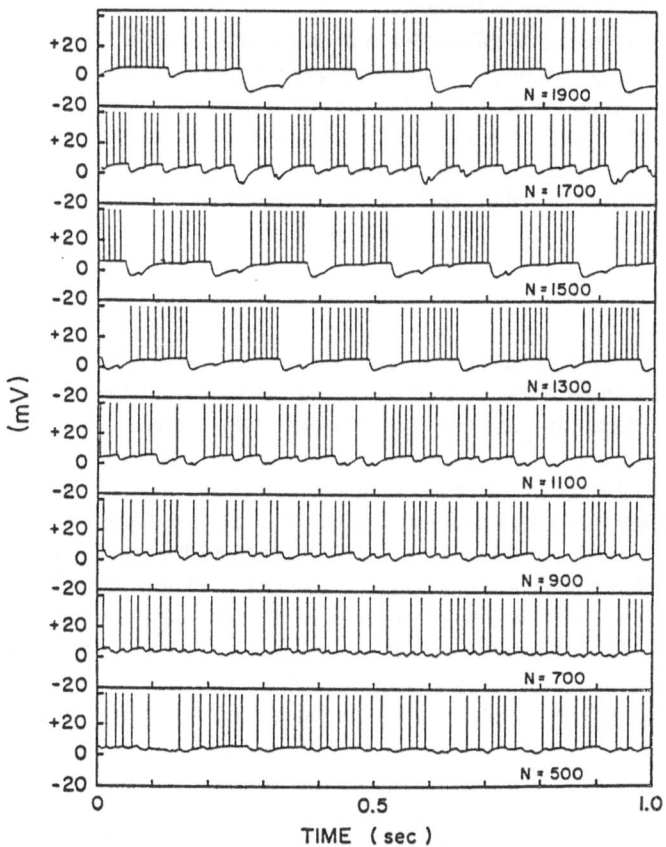

Fig. 12. Model predicted behavior of the feedback circuit in hippocampus as a function of GABA receptor density. Each horizontal panel represents one second of pyramidal cell membrane potential (continuous curve) and superimposed nerve impulses. Panels differ in the assumed number N of inhibitory GABA receptors on the soma of the pyramidal cells. From top to bottom this number decreases from N = 1900 to N = 500 in steps of 200. After Mackey & an der Heiden (1984)

solutions of (18) the excitatory input e(t) from the mossy fibres has been held temporally constant at the same value e=1.6 mV. Therefore, all the dynamics visible in these plots results from the internal delayed mixed feedback of this neuronal circuit. As an aid to understanding, superimposed vertical lines corresponding to the occurrence of action potentials are included.

For very high values of N, the neuronal circuit produces a constant but low firing frequency (not shown in Fig. 12). As N decreases the firing frequency undergoes a transition to periodic firing of CA3 cells with different kinds of bursting behavior (upper panels of Fig. 12) and then to progressively more complicated asychronous firing patterns. At very low densities the model predicts that the CA3 cells will fire in a uniform fashion at a relatively high frequency. This sequence of events essentially mimics, despite many differences in the details, those observed in real hippocampus tissues.

In addition to the relevance of this model to understanding penicillin-induced behavior, interesting notions are raised concerning variability in neuronal activity. This variability is commonly attributed to stochastic sources. Yet, in this completely deterministic model for recurrent inhibition there are parameters values associated with highly irregular, "stochastic-like" neuronal firing. These observations raise the possibility that much of the variation observed in nervous electrical activity is not a reflection of complicated neuronal feedback effects.

4.3 Coupled Oscillators

An *autonomous oscillator* is a system which is able to generate oscillations in time of one or several of its variables without being driven by any external oscillator. For example, the systems discussed in the previous two sections are such oscillators. Each organism is a large ensemble of autonomous oscillators. These oscillators operate on the molecular and cellular levels, and on the levels of network of cells (e.g. immune and nervous systems) and of organs. Since all parts of an organism are more or less interconnected and influence each other it is quite natural to assume that the organism is a system of coupled autonomous oscillators.

There are numerous examples of coupling between biological oscillators. Large populations of coupled cellular oscillators are present in the respiratory centre, in the sinoatrial node, in the gut and in locomotory rhythms. Interactions between oscillators at the level of organs have been observed with the breathing rhythm and locomotory rhythm, and between the breathing rhythm and heart rate. Periodic sympathic activity modulates the rhythms of many organs like the lung, the gut, and also the endocrine system. The brain generates a variety of rhythms, e.g. the circadian rhythm in the suprachiasmatic nucleus, endogenous rhythms in the hypothalamus, probably also rhythms of mood, mania and depression, Parkinsonian rhythms, etc. Nearly all of these rhythms have some influence on many other of the body rhythms. Research in all of these areas is still very much in its infancy, both experimentally and theoretically.

Rhythms of the organism can also be influenced by rhythms external to the body. For example, the circadian oscillator is entrained by the light/dark cycle of

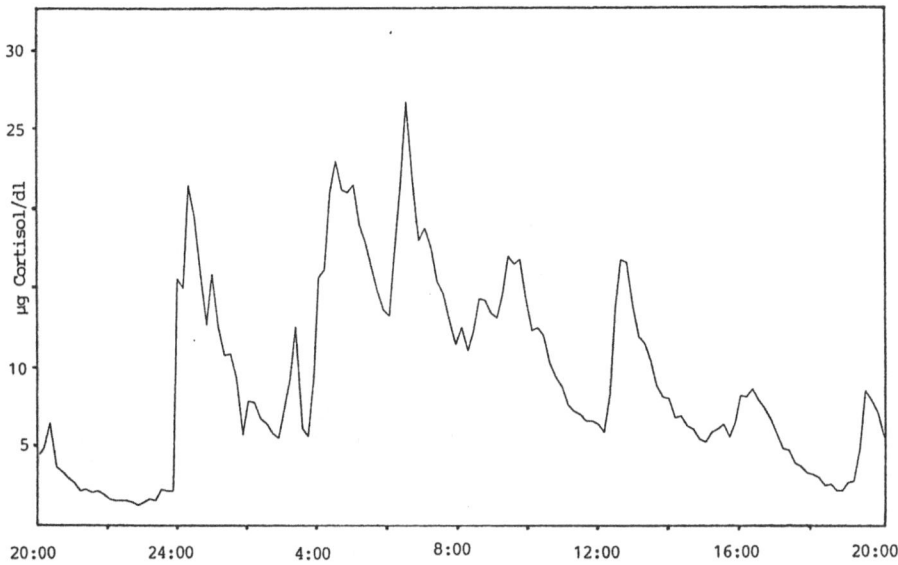

Fig. 13. Concentrations of cortisol in the blood over one day. Measurements have been taken every ten minutes. Data kindly provided by Dr. T.H. Schürmeyer, Medical High School Hannover

the earth. Important with respect to diseases is the fact that many drugs are applied in a periodic fashion. There are still by far too few investigations on the effect of different time patterns of drug administration (however, see Lemmer, 1984). A dramatic example of the coupling between an external and an internal oscillator are artificial pacemakers both for the heart and for respiration (Glass, 1987).

Coupled oscillators can exhibit extremely complicated behavior and we are far from having understood all phenomena. Severe problems arise even in simple physical systems like a pair of coupled pendula. In many technical systems intentional and unintentional coupling of oscillatory components often creates great difficulties. It is impossible to give here, or in any book, a complete overview of all the effects of coupling two or more oscillators. As a sample of references we mention Othmer (1986), Rensing & Jaeger (1985), Rensing et al. (1987), and Glass (1987).

Instead of studying the problem of interacting oscillators in full generality, which is impossible, we now turn to a specific example in more detail.

4.4 Periodic, Quasiperiodic and Chaotic Rhythms in the Endocrine System Elucidated on the Basis of Coupled Oscillators

Most of the hormones in the body are produced and secreted not at a nearly constant rate ("tonically") but in a pulsatile, dramatically oscillatory time pattern. Figs. 1.3a,b showed examples from the parathyroid gland. Fig. 13 represents a "typical" time behavior over one day of cortisol concentrations in the blood of a healthy person.

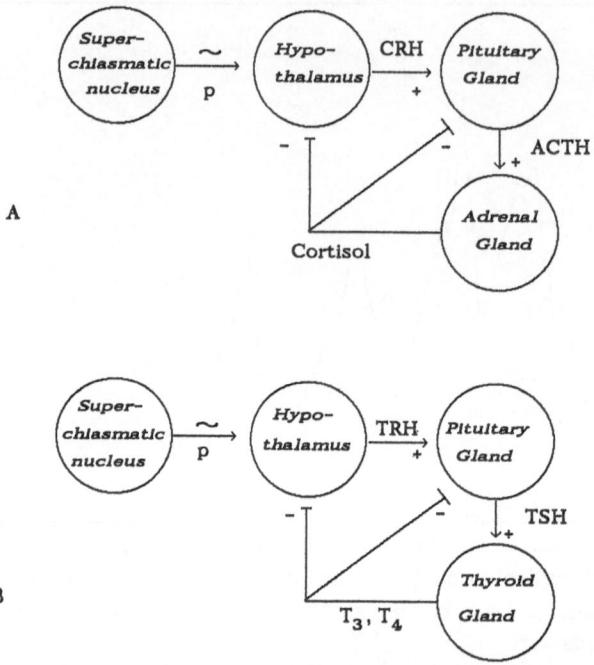

Fig. 14. Scheme of the adrenal (A) and the thyroidal (B) system. The suprachiasmatic nucleus is a small brain area supposed to produce a circadian rhythm (~) which is mediated to the hypothalamus. The hypothalamus is able to activate cells of the pituitary gland either by release of CRH (in case of the adrenal system) or of TRH (in case of the thyroid). The pituitary stimulates the adrenal gland and the thyroid gland by ACTH and TSH, respectively. Inhibitory influences are mediated from these two glands both to the hypothalamus and the pituitary by the hormones cortisol and $T_{3,4}$, respectively

It is now well known (Knobil, 1981; Li & Goldbeter, 1989; Hesch, 1989) that fluctuations in the level of hormone concentrations are necessary for a functional stimulation of the target organ or target cells. Constant application of a hormone leads to adaptation and desensitization.

Besides the functional meaning of the pulsatile secretion, an important and difficult problem is the origin of the apparently very irregular, aperiodic rhythms of hormone activity.

My central approach (an der Heiden, 1991) to this problem is the idea that at least part of the complexity of these hormonal time patterns results from the interaction of several oscillators.

Fig. 14a shows a schematic representation of the cortisol system made up of the axis hypothalamus-pituitary gland-adrenal gland. The hormone CRH released from the hypothalamus activates the pituitary gland to produce ACTH which stimulates the adrenal gland to produce cortisol. Cortisol itself is able to inhibit both the hypothalamus and the pituitary (Brabant et al., 1987).

In a first simplified model for this feedback loop we took the hypothalamus and the pituitary together as a single unit. Together they produce ACTH, the concentration of which is denoted by x(t). The production rate dx(t)/dt of ACTH is described by the delay-differential equation

$$dx(t)/dt = (c + p(t)) \, f(y(t - \tau_1)) - ax(t). \qquad (19)$$

Here y(t) denotes the time-dependent concentration of cortisol which inhibits the production of ACTH. Consequently f(y) is a monotonic decreasing function of y, e.g.

$$f(y) = \frac{1}{k+y^n}, \qquad (20)$$

with constant parameters k, n for effectiveness and cooperativity in the hormone-receptor interaction kinetics. The delay constant τ_1 measures the time needed to transport the signal from the adrenal to the pituitary plus the time from binding of the cortisol receptor until release of ACTH is effectively reduced (possibly across some second messenger and transcription system).

The constant c in equation (19) represents the strength of ACTH production if no inhibition by cortisol were present. The constant production rate c is modulated not only by this inhibition but also by a periodically varying influence p(t) from a circadian oscillator in the brain. This oscillator is probably located in the superchiasmatic nucleus. The function p(t) can alternatively be thought of as describing a mixture of other rhythms mediated from the brain to the endocrine system. A third interpretation could be that p(t) describes some rhythm produced inside the hypothalamus or the pituitary.

The equation for the rate of change of cortisol concentration is given by

$$dy(t)/dt = g(x(t - \tau_2)) - by(t) . \qquad (21)$$

Just like in equation (19) b is a constant factor in the linear decay rate. The function g(x) describes the strength of activation of the adrenal gland by ACTH, and hence depends in a monotonic increasing fashion on x, e.g.

$$g(x) = d \cdot \frac{x^m}{K+x^m} \qquad (22)$$

with positive constants d, K, m.

Equations (19) and (21) together completely describe the model of the cortisol system presented here. They can also be considered as a model for the thyroidal system which has a similar structure as illustrated by Fig. 14b. Similar models have been proposed by other authors, e.g. Danziger & Elmergreen, 1956; Rössler et al., 1979; Smith, 1980; Murray, 1989. However, they did not incorporate the periodic (or aperiodic) stimulation factor p(t) in equation (19), thus they did not consider the coupling of oscillators in the hormone system.

Without this periodic stimulation term the feedback system (19) and (21) is able to produce autonomous periodic rhythms p(t), i.e. this system is an oscillator in itself. Fig. 15a shows one of the periodic solutions generated under this condition (p(t)=0 for all t). Note that in this case we have a negative feedback system, since the function f is monotonically decreasing and the function g is monotonically increasing, resulting altogether in a negative feedback loop. The theory (Mallet-Paret & Smith, 1990) says that in this case periodic but not chaotic behavior is to be expected.

If the brain stimulates the endocrine system periodically (i.e. p(t) is a periodic function of t) we have the situation of coupled oscillators (only the simplest case where the coupling is only in one direction).

The behavior of the system (19) and (21) can be very different depending on the amplitude and frequency of the periodic forcing term p(t). Figs. 15b,c show solutions x(t), y(t) together with the periodic function p(t). In all of these figures the parameters a, b, c, d, τ, k, K, m, n have the same values. In calculating these solutions from the system (19) and (21) we have chosen for simplicity

$$p(t) = A \sin \omega t . \tag{23}$$

In Figs. 15b,c the amplitude A is equal to 1.5. They only differ in the value of the frequency ω. In Fig. 15b ω is equal to 0.01 and we observe a periodic development of x(t) and of p(t). However, within a single (smallest) period four maxima appear. This period is exactly as long as the forcing period $2\pi/\omega$. Counting the numbers of maxima per period of p(t) we obtain here a relationship of 1:4 between p(t) and x(t). Such phenomena are known as phase locking between two oscillators.

There are many more possibilities of phase-locking relationships, and we cannot discuss them here in any detail. However, there are also situations where phase locking does not occur, and apparently aperiodic and even chaotic time behavior evolve. Fig. 15c shows the case $\omega=0.145$. No periodicity in x(t) or y(t) could be detected. Projections of these solution trajectories (x(t), y(t)) into the phase plane lead to pictures like those of Fig. 16. Some of these plots show quasiperiodic trajectories, others chaotic trajectories, situations sometimes not easily to be distinguished. Fig. 16c ($\omega=0.27$) represents the orbit of a complicated periodic solution.

It is very difficult to obtain a complete overview of all possible types of behavior of the only seemingly simple system (19) and (21). Moreover, we do not think it to be a very realistic model of any of the existent hormone systems. Nevertheless, we believe that we are on the way to understanding endocrine systems as systems of oscillators coupled either unidirectionally or in feedback loops. The periodic term p(t) in equation (19) need not necessarily be interpreted as a periodic forcing of the hormone system by brain structures. An alternative view would be that a gland, e.g. the pituitary, produces by itself the rhythm p(t). This means that ACTH is produced periodically, and the release of ACTH is modulated by the feedback loop pituitary $\rightarrow$ adrenal gland $\rightarrow$ pituitary. It is easily

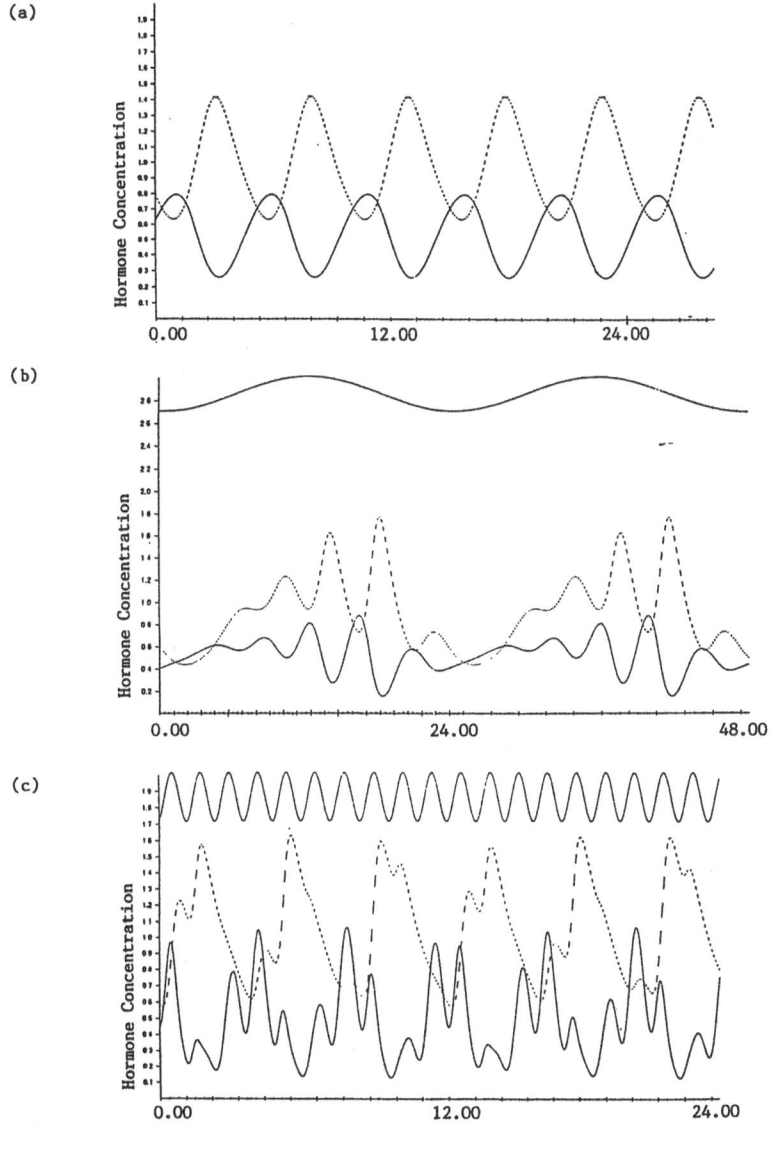

(a)

(b)

(c)

Daytime

Fig. 15a-c. Solutions of the hormone model given by the equations (19) and (21). Top curve gives the oscillation of the forcing oscillator p(t). Lower full curve represents x(t), dotted curve y(t); **Fig. 15a:** simple period behavior occurring under the assumption that the suprachiasmatic nucleus does not stimulate the hypothalamus (i.e. p(t)=0 in equation (19)); **Fig. 15b:** More complicated behavior occurring under periodic stimulation of the hormone system by the brain with period of about one day, compare Fig. 13; **Fig. 15c:** Apparently aperiodic solutions occurring for high frequency stimulation of the feedback circuit

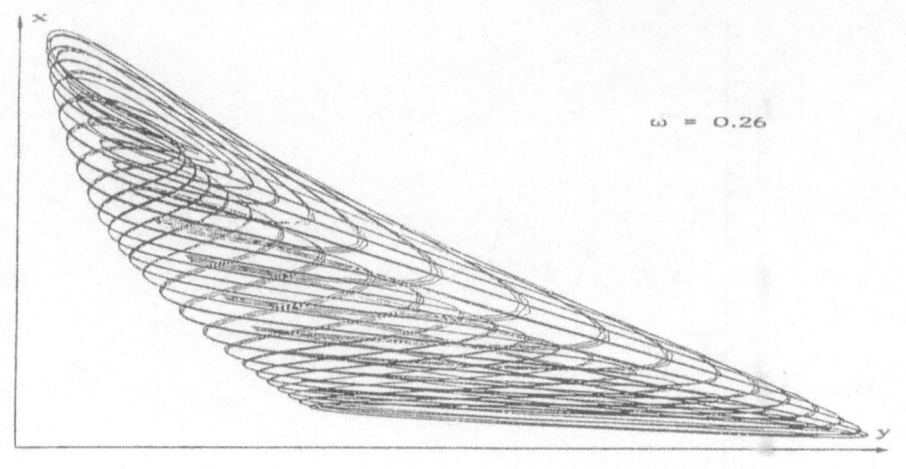

ω = 0.26

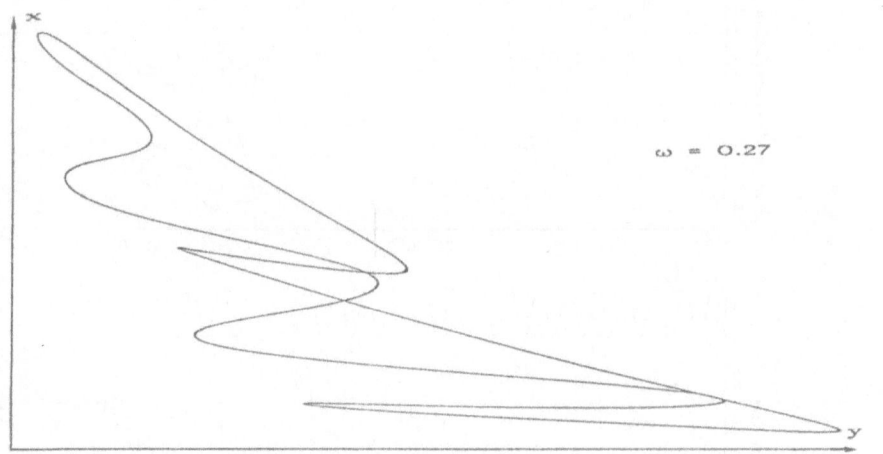

ω = 0.27

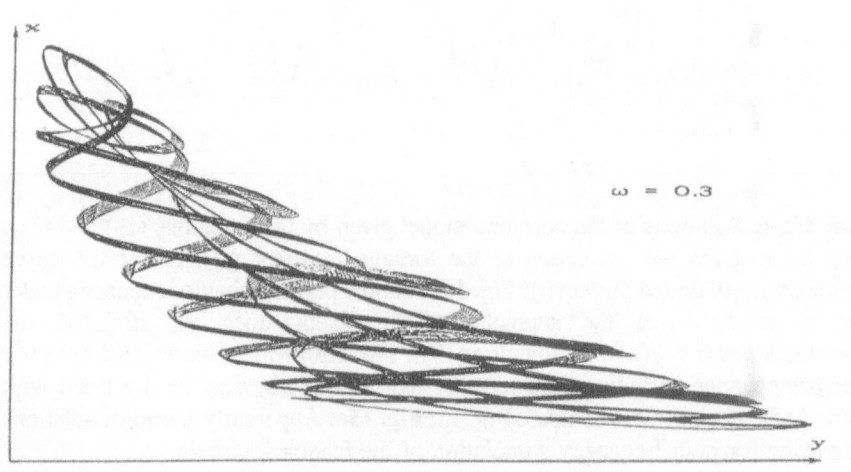

ω = 0.3

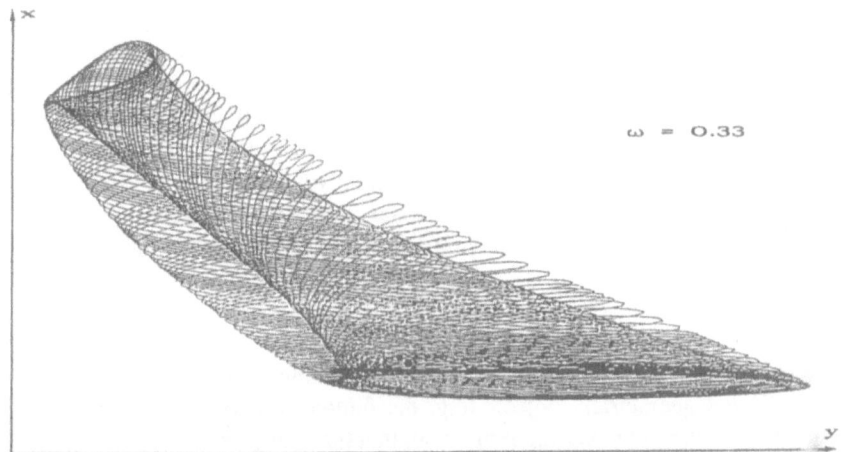

Fig. 16. Trajectories (x(t), y(t)) of the hormone model equations (19) and (21) plotted in the (x,y)-plane. Different solutions (a-d) are obtained by choosing different frequencies ω of the driving oscillator (which is either to be located in the brain or in one of the hormone glands). Complexity of behavior is apparently different for different values of ω

imaginable that the adrenal gland too is itself an oscillator coupled to the pituitary oscillator. This would lead to an even higher level of complexity and chaoticity, which comes nearer to reality, but we shall not pursue this any further here.

5. Detection and Quantification of Chaos in Empirical Systems with Applications to the Heartbeat

In the last ten years great efforts have been made towards the development and application of methods for the detection of deterministic chaos and to distinguish it from pure random noise. This problem and corresponding results are not reviewed in depth here, nor can a complete list of references of this broad and rather sophisticated topic be given. The interested reader is referred to e.g. Schuster (1989), West (1990), Mayer-Kress (1986), Farmer et al. (1983). There are now a number of methods available, known under names like "fractal dimension", "Lyapunov exponents", "correlation dimension", "entropy".

Let us concentrate on the dimension of a chaotic attractor. A chaotic attractor is a certain bounded subset of the state space: Infinitely long trajectories fill a certain bounded domain with a very complicated structure, see e.g. Fig. 16. In order to get an idea of this complexity and somehow to measure it the notion of fractal dimension has become very fruitful. Ordinary geometrical objects have a dimension which is expressed by an integer number: A line of finite length is one dimensional, rectangles are two-dimensional, and cubes are three-dimensional. More generally in k-dimensional space, where k is now some arbitrary positive integer, one can speak of k-dimensional cubes ("k-cubes"), which are generalizations of 3-dimensional cubes. (A 2-dimensional cube is a square, and a 1-dimensional cube is a bounded interval.) An arbitrary k-dimensional object (not necessarily cubes) can be measured by covering it with small k-dimensional cubes and counting their number. The smaller the cubes the more precise the measurement. An important property of the notion "dimension" is that the number N of k-cubes needed to cover an m-dimensional object (m≤k) depends on the edge length l of the k-cubes such that N is proportional to l^{-m}, at least in the limit l→0:

$$N(l) \propto l^{-m} . \tag{24}$$

As already discovered early in this century by the mathematician Hausdorff there are complicated geometrical objects (e.g. the famous Cantor set) for which, by applying this method of covering, it turns out that their dimension is not an integer, but a nonintegral number like 2.312 or log2/log3. Following the mathematician Mandelbrot (1987) such objects are now called fractals. They have a fractal dimension, sometimes also called "Hausdorff dimension". There are other notions of dimension in the literature like "correlation dimension", "generalized dimension"; however, they are all related to this basic idea of Hausdorff.

In order to measure the dimension of a chaotic attractor, this attractor must be specified somehow. With real systems in nature the attractor, if present, is generally unknown, since not all variables can be measured. However, according to a theorem of Ruelle and Takens the following is possible. If at least one variable, say x(t), of a deterministic system can be measured over a sufficiently long time at sufficiently dense time intervals, then from the resulting time series an artificial state space and an artificial trajectory in this state space can be constructed with the following property: If the underlying real system is deterministic and has a chaotic attractor then the dimension of this real attractor is equal to the dimension of the artificial chaotic attractor which is approximately represented by the trajectory constructed from the time series. For details of this method the reader is again referred to the literature given above.

In the research laboratory in our university hospital my colleagues Henrik Bettermann and Peter van Leeuwen (1991) evaluated the 24-hour records of electrocardiograms of normal and ill persons. In fact, they determined a certain form of the correlation dimension. Data were the temporal distances of consecutive heart beats (R-R intervals). They verified that a sequence of five hundred beats is sufficient to provide a reliable dimension number. Covering one day by 140 time windows of length 10 min, they calculated the dimension in each of these windows. Moreover, they calculated the average R-R interval length for these windows. In Fig. 17a the 144 values of these averages are plotted as a function of time for a healthy person. There is a clear circadian rhythm in this plot which is modulated by a rather irregular pulsatile pattern. For the same heart data Fig. 17b shows the 144 calculated values of the correlation dimension. There is a considerable variation of dimension between four and nine. Frequency of variation of the dimension appears comparable to that of the average R-R interval. There

(a)

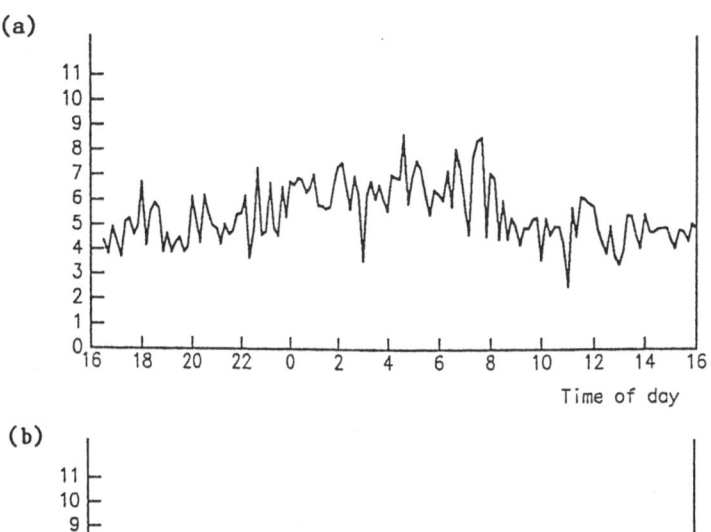

Time of day

(b)

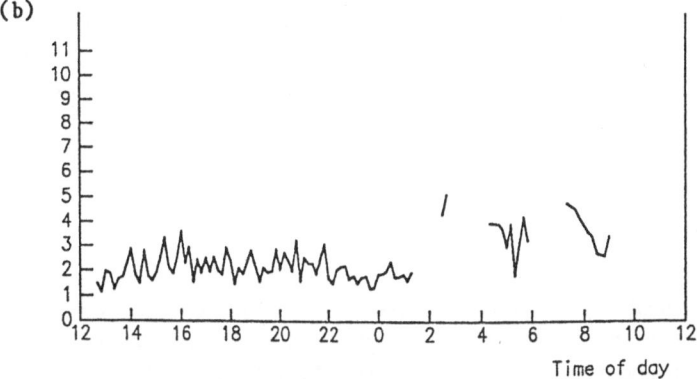

Time of day

Fig. 17. Time varying fractal dimension of the heart beat; **Fig. 17a:** for the healthy subject of Fig. 1.1a, **Fig. 17b:** for the patient of Fig. 1.1b. Horizontal axes represent daytime. Vertical axes represent fractal dimension (for definition see text) of R-R intervals calculated for 144 successive time windows, each of length 10 min. In Fig. 17b for some times after 1 a.m. the dimension values could not be determined reliably and a distinction between deterministic chaos and random noise was not possible. Calculations performed and kindly made available by H. Bettermann and P. van Leeuwen, University of Witten/Herdecke

seems to be a trend in that higher heart beat frequency is to some degree correlated with lower dimensionality. These data suggest that the heart may be in different functional states which correspond to attractors of different dimension. Of course, such a hypothesis is in need of much more empirical and theoretical support by future research.

Babloyantz & Destexhe (1988) found correlational dimensions ranging from 3.5 to 5.2. They used 4-minute windows with 6×10^4 data points from a continuous record of the ECG. Like Bettermann and van Leeuwen, Babloyantz and Destexhe also evaluated the correlation dimension of R-R intervals and arrived at values around 5.9 with about 1000 intervals in the series. The different numbers emerging from different investigations must lead to a critical use of the method, which must still be theoretically and practically improved. However, there are other indicators that the normal heart beat is not strictly periodic. For example, Kobayashi & Musha (1982) and Goldberger et al. (1985) noted the 1/f spectrum observed in interbeat variations in healthy subjects. Altogether, borrowing the words of B.J. West (1990, p. 222) "there is no way that the conventional wisdom of the ECG of periodic oscillations can be maintained in light of these new results".

Finally, Fig. 17b shows the same kind of data as in Fig. 17a, however this time from a patient during stationary treatment shortly after a heart attack. Whereas the averages both of the R-R intervals and of the dimensions across a whole day are not considerably different from those of the healthy person, the variability of these two quantities is much smaller with the patient. These findings are in line with the observation emphasized by Goldberger and West that the heart activity of young, healthy people may be much more irregular than that of older people or in case of some heart diseases.

For other applications of these methods for detecting empirical chaos in biology and medicine the reader is referred to the books edited by Mayer-Kress (1986), Rensing, an der Heiden and Mackey (1987), and to the monograph of B. West (1990).

Acknowledgement. I thank Prof. R.D. Hesch, Dr. H.M. Harms, and Dr. T.H. Schürmeyer, Medical High School Hannover, for kindly providing the data on hormone concentrations. I also thank Henrik Bettermann at my institute for supplying me with the data and plots of the heart beat dynamics. Finally, I thank Ralf Scholl at my institute for his help in producing Figs. 15 and 16. This work has been supported in part by the Deutsche Forschungsgemeinschaft, Grant He 1653/1-1.

References

BABLOYANTZ, A. & DESTEXHE, A. 1988. Is the Normal Heart a Periodic Oscillator? Biol. Cybern. **58**, 203-211.

VON BERTALANFFY, L., BEIER, W. & LAUE, R. 1977. Biophysik des Fließgleichgewichts. Vieweg, Braunschweig.

BETTERMANN, H. & VAN LEEUWEN, P. 1991. Dimensional Analysis of RR Dynamics in 24 Hour Electrocardiograms. Acta Biotheoretica (submitted).

BRABANT, G., BRABANT, A., RANFT, U., OCRAN, K., KÖHRLE, J., HESCH, R.D. & VON ZUR MÜHLEN, A. 1987. Circadian and Pulsatile Thyrotropin Secretion in Euthyroid Man Under the Influence of Thyroid Hormone and Glucocorticoid Administration. J. Clin. Endocrin. Metab., **65**, 83-88.

COLEMAN, B.D. & RENNINGER, G.H. 1976. Theory of the Response of the Limulus Retina to Periodic Excitation. J. Math. Biol., **3**, 103-120.

DANZINGER, L. & ELMERGREEN, G.L. 1956. Bull. Math. Biophys., **18**, 113.

FARMER, J.D., OTT, E. & YORKE, J.A. 1983. The Dimension of Chaotic Attractors. Physica, **7D**, 153-180.

GLASS, L. 1987. Coupled Oscillators in Health and Disease. In: Rensing, L., an der Heiden, U. & Mackey, M.C. (Eds.): Temporal Disorder in Human Oscillatory Systems. Berlin: Springer-Verlag, 8-14.

GOLDBERGER, A.L. 1987. Nonlinear Dynamics, Fractals, Cardiac Physiology and Sudden Death. In: Rensing, L., an der Heiden, U. & Mackey, M.C. (Eds.): Temporal Disorder in Human Oscillatory Systems. Berlin: Springer-Verlag, 118-125.

GOLDBERGER, A.L., BHARGAVA, V., WEST, B.J. & MANDELL, A.J. 1985. On a Mechanism of Cardiac Electrical Stability: The Fractal Hypothesis. Biophys. J., **48**, 525-528.

GOLDBETER, A. 1990. Rhythmes et Chaos. Paris: Masson.

GOODWIN, B.C. 1965. Oscillatory Behaviour in Enzymatic Control Processes. Advances in Enzyme Regulation, **3**, 425-438.

GRIFFITHS, G.M. & FOX, T.A. 1938. Lancet ii: 409-415.

GUERRY, D., DALE, M., OMINE, M., PERRY, S. & WOLFF, S.M. 1973. Periodic Hematopoiesis in Human Cyclic Neutropenia. J. Clin. Invest., **52**, 3220-3230.

HARMS, H.M., KAPTAINA, U., KÜLPMANN, W.R., BRABANT, G. & HESCH, R.-D. 1989. Pulse Amplitude and Frequency Modulation of Parathyroid Hormone in Plasma. J. Clin. Endocrinol. Metabol., **69**, 843.

HASTINGS, S., TYSON, J. & WEBSTER, D. 1977. Existence of Periodic Solutions for Negative Feedback Cellular Control Systems. J. Diff. Equ., **25**, 39-64.

AN DER HEIDEN, U. 1978. Chaotisches Verhalten in zellulären Kontroll-prozessen. In: L. Rensing & G.Roth (Eds.): Zelluläre Kommunikations- und Kontrollmechanismen. Bremen: Universitätsverlag, 34-45.

AN DER HEIDEN, U. 1979. Delays in Physiological Systems. J. Math. Biol., **8**, 345-364.

AN DER HEIDEN, U. 1991. Nonlinear Delay Equations in Neuro- and Endocrinology. In: J. Demongeot (Ed.): Mathematics Applied to Biology and Medicine. Grenoble University Press, 35-41.

AN DER HEIDEN, U. 1991b. A Basic Mathematical Model of the Immune Dynamics (to appear).

AN DER HEIDEN, U. 1991c. Der Organismus als selbstherstellendes dynamisches System. In: Zänker, K. (Ed.): Kommunikationsnetzwerke in unserem Körper. Heidelberg: Spektrum Akad. Verlag.

AN DER HEIDEN, U. & MACKEY, M.C. 1982. The Dynamics of Production and Destruction: Analytic Insight into Complex Behaviour. J. Math. Biol., **16**, 75-101.

AN DER HEIDEN, U., MACKEY, M.C. & WALTHER, H.O. 1981. Complex Oscillations in a Simple Deterministic Neural Network. In: Hoppenstaedt, F. (Ed.): Mathematical Aspects of Physiology. Providence, R.I.: Amer. Math. Soc., 355-360.

AN DER HEIDEN, U., ROTH, G. & SCHWEGLER, H. 1985. Die Organisation der Organismen: Selbstherstellung und Selbsterhaltung. Funktionelle Biol. Med., **5**, 330-346.

AN DER HEIDEN, U. & WALTHER, H.O. 1983. Existence of Chaos in Control Systems with Delayed Feedback. J. Differential Equations, **47**, 273-295.

HESCH, R.-D. (Ed.) 1989. Endocrinology. München: Urban & Schwarzenberg.

KING, R., BARCHAS, J.D. & HUBERMANN, B. 1983. Theoretical Psychopathology: An Application of Dynamical Systems Theory to Human Behavior. In: Basar, E., Flohr, H. & Haken, H. (Eds.): Synergetics of the Brain. Berlin: Springer Verlag, 352-364.

KNOBIL, E. 1981. Patterns of Hormonal Signals and Hormone Action. N. Engl. J. Med., **305**, 1582-1583.

KOBAYASHI, M. & MUSHA, T. 1982. IEEE Trans. Biomed. Eng., **29**, 456-475.

KRAUS, P.H., BITTNER, H.R., KLOTZ, P. & PRZUNTEK, H. 1987. Investigation of Agonistic/Antagonistic Movement in Parkinson's Disease from an Ergodic Point of View. In: Rensing, L., an der Heiden, U. & Mackey, M.C. (Eds.): Temporal Disorder in Human Oscillatory Systems. Berlin: Springer Verlag, 110-115.

LASOTA, A. & MACKEY, M.C. 1985. Probabilistic Properties of Deterministic Systems. Cambridge: Cambridge University Press.

LEMMER, B. 1984. Chronopharmakologie. Stuttgart: Wissenschaftl. Verlagsgesellschaft.

LI, Y.-X. & GOLDBETER, A. 1989. Frequency Specifity in Intercellular Communication: The Influence of Patterns of Periodic Signalling on Target Cell Responsiveness. Biophys. J., **55**, 125-145.

LINDSAY, A. E. & BUDKIN, A. 1975. The Cardiac Arrythmias. Chicago: Yearb. Med. Publ. (2nd edition).

MACKEY, M.C. 1978. A Unified Hypothesis for the Origin of Aplastic Anemia and Periodic Hematopoiesis. Blood, **51**, 941-956.

MACKEY, M.C. & GLASS, L. 1977. Oscillation and Chaos in Physiological Control Systems. Science, **197**, 287-289.

MACKEY, M. C. & DÖRMER, P. 1981. Enigmatic Hematopoiesis. In: Rothenberg M. (Ed.): Biomathematics and Cell Kinetics. Amsterdam: Elsevier/North Holland Biomedical Press, 87-103.

MACKEY, M.C. & AN DER HEIDEN, U. 1984. The Dynamics of Recurrent Inhibition. J. Math. Biol., **19**, 211-225.

MALLET-PARET, J. & SMITH, H. L. 1991. The Poincaré-Bendixson Theorem for Monotone Cyclic Feedback Systems. J. Dynamics and Different. Equats. (to appear).

MANDELBROT, B.B. 1987. The Fractal Geometry of Nature. Basel: Birkhäuser Verlag.

MAY, R. 1980. Mathematical Models in Whaling and Fisheries Management. Lectures on Mathematics in the Life Sciences, **13**, 1-64.

MAYER-KRESS, G. (Ed.) 1986. Dimensions and Entropies in Chaotic Systems. Berlin: Springer Verlag.

MAYER-KRESS, G. YATES, F.E., BENTON L., KEIDEL, M., TIRSCH, W., PÖPPL, S.J. & GEIST, K. 1988. Dimensional Analysis of Nonlinear Oscillations in Brain, Heart and Muscle. Math. Biosci., **90**, 155-182.

MILTON, J. & LONGTIN, A. 1990. Clamping the Pupil Light Reflex with External Feedback: Evaluation of Constriction and Dilation from Cycling Measurements. Vision Res., **30**, 515-525.

MILTON, J., AN DER HEIDEN, U., LONGTIN, A. & MACKEY, M.C. 1990. Complex Dynamics and Noise in Simple Neural Networks with Delayed Mixed Feedback. Biomed. Biochim. Acta, **49**, 697-707.

MONOD, J., WYMAN, J. & CHANGEUX, J.-P. 1965. On the Nature of Allosteric Transitions: A Plausible Model. J. Mol. Biol., **12**, 88-118.

MURRAY, J. D. 1989. Mathematical Biology. Berlin: Springer-Verlag.

NISBET R. & GURNEY W.S.C. 1976. Nature, **263**, 319.

OTHMER, H.G. (Ed.) 1986. Nonlinear Oscillations in Biology and Chemistry. Lect. Notes Biomath., **66**. Berlin: Springer Verlag.

PRINCE, D.A. 1969. Microelectrode Studies of Penicillin foci. In: Jasper, H.H, Ward, & Pope, A. (Eds.): "Basic Mechanisms of the Epilepsies". Boston: Little & Brown, 320-328.

RENSING, L. & JAEGER, N.I. (Eds.) 1985. Temporal Order. Berlin: Springer Verlag.

RENSING, L., AN DER HEIDEN, U. & MACKEY, M.C. (Eds.) 1987. Temporal Disorder in Human Oscillatory Systems. Berlin: Springer Verlag.

RÖSSLER, R., GÖTZ, F. & RÖSSLER, O.E. 1979. Chaos in Endocrinology. Biophys. J., **25**, 216a.

SCHUSTER, H.G. 1989. Deterministic Chaos. Weinheim: VCH Publishers.

SMITH, W. R. 1980. Hypothalamic Regulation of Pituitary Secretion of Luteinizing Hormone. Bull. Math. Biol., **42**, 57-78.

SPECHT, H & FRUHMAN, 1972. Bull. Physiol. Pathol. Respir., **8**, 1075.

STARK, L. 1977. Neurological Oscillations: Formulations of Mathematical Control Models and Applications to Clinical Syndromes. In: Rensing, L., an der Heiden, U. & Mackey, M.C. (Eds.): Temporal Disorder in Human Oscillatory Systems. Berlin: Springer-Verlag.

TYSON, J.J. & OTHMER, H.G. 1977. The Dynamics of Feedback Control Circuits in Biochemical Pathways. Progr. Theor. Biol., **5**.

WAZEWSKA-CZYZEWSKA, M. 1984. Erythrokinetics. Foreign Scientific Publications, National Center for Scientific, Technical and Economic Information.

VERHULST, F. 1990. Nonlinear Differential Equations and Dynamical Systems. Berlin: Springer Verlag.

WEHR, T.A. & GOODWIN, F.K. 1979. Arch. Gen. Psychiatry, **36**, 555-559.

WEST, B. J. 1990. Fractal Physiology and Chaos in Medicine. Singapore: World Scientific.

Chaos and Disorder

Karl W. Kratky

With 1 Figure

Abstract. This article deals with the question of whether nature is dominated by chance or necessity. If one tries to answer this question, one has to distinguish between a basic level and the world as we see it. This is also connected with the distinction between objectivity and subjectivity. It turns out that there is a whole spectrum of complex phenomena ranging from order to disorder. Deterministic chaos is treated in some detail. It is shown that it has more to do with self-organization than with disorder. Finally, the possible relevance for clinical psychology is taken into account.

1. Introduction

1.1 Law Versus Randomness

The development of the sciences is characterized by the search for laws which explain or at least describe the world around us. This has been very successful in the physical sciences, especially in mechanics. As a consequence, the idea of the universe being a large clock became quite popular. However, some kind of randomness could not be removed when describing nature properly. As an example, the weather forecast shall be mentioned. Should a large cloud be a more appropriate model for the universe (Popper)? After all, it may be that the "clouds" are in fact "clocks" (or the other way round).

The question of law or randomness has a different answer depending on the level of description. As an example, we consider the toss of a coin. Such a process is a typical case for a random process. However, if we toss the coin many times, it turns out that head and tail appear equally often. There seems to be a law behind this reproducible result.

Now we look at a single coin toss. This is a mechanical process, at least when we neglect the influence of the air. If we knew exactly the initial position, velocity, and rotation of the coin at the beginning of the toss, we could predict the outcome (Laplace). So, where is there randomness ... only in our mind? However, if we try to describe such a toss properly, we have to bear in mind that classical mechanics is not "really" true. Quantum mechanics has to be taken into account. Then, uncertainties and probabilities come in. At the lowest level, there are random quantum fluctuations.

The above example shows that necessity (regularity) may emerge from chance (randomness) and vice versa if we compare different levels of description. A

88

philosophical consideration of law, probability, and the other terms mentioned above can be found in Armstrong (1985).

By the way, there are still problems with the interpretation of quantum mechanics (see, e.g., Selleri 1990) and its actual application to macroscopic processes. Thus, we stick to classical considerations throughout the article. When quantum physics is relevant for our arguments, this will be stated explicitly.

1.2 Correlations Versus Causality

In experiments, we can only find correlations between measured quantities, and no causal relation. The latter is constructed by our mind, but there are always pitfalls. If one sits on the shore and looks at two small boats rocking in parallel, one may argue that one boat (which one?) beats time and forces the other one to obey the rhythm. If we make an "experiment" and move one boat with our hands, the synchronicity stops, and we find that such a dependence does not exist. Finally, the water waves (as a common cause for the movement of both boats) save our causal thinking.

The above example is very simple, but similar considerations are helpful in many fields. For instance, illness is connected with several interrelated symptoms. If one tries to cure a sick person, this usually means to act on one or more symptoms. Via the relation to the other symptoms, this should lead to health. However, this treatment (which is sometimes confused with a causal procedure) may yield counterintuitive results.

A simple example is fever as a "symptom of illness". Just suppressing fever without any further treatment is silly because usually the illness then becomes worse. In fact, fever is a sign of healing by the body itself. Thus, the correlation is the other way round!

Another example: A very high level of a specific substance in the urine may be an indication of two opposite facts:
i) There is too much of that substance in the whole body, and thus in the urine.
ii) Too much of that substance is secreted from the body to the urine. Thus, not enough of the substance will remain in the body.

Counterintuitive behavior often occurs in feedback loops, as first studied by cybernetics. Even if the single processes building up the loop can be described by causal laws, the total system may behave in a surprising way. At this stage, one may argue that this is only a problem of subjectivity, which does not contradict our idea of "true" causality and determinism. We will come back to that question several times.

1.3 A Simple Model for Self-Organization

In this section, we will treat the case where law and randomness are not opponents, but allies. Consider the following simple model for the behavior of a system: A ball rolls in a model landscape containing several basins. Depending on the starting point, the ball will come to rest at the lowest point of one of the basins. These minima represent, e.g., stable patterns of animal behavior. The lowest basin contains the absolute minimum, i.e., the most stable and most probable behavior. The landscape represents the internal rules; the location of the starting point of the ball is due to the influence of the environment. There seems to be a simple relationship between cause (starting point) and effect (final location of the ball). However, this is not the case when the landscape is rugged. Moreover, the landscape symbolizes an internal rule which is not accessible to someone outside. We can influence an animal, but the possible patterns of its behavior are already there.

Now we assume that one of the minima corresponds to the desired behavior. What happens if the ball comes to rest in the "wrong" basin? There are two ways to correct that. First, one may change the location intentionally so that the ball will fall into the desired basin. Second, one just shakes the ball. If this is done vigorously enough, the ball will leave the basin and hopefully fall into the desired one. A similar case occurs in mathematics when the absolute minimum of a given function has to be determined. In general, the methods only yield one of the (relative) minima as a result. Thus, an additional random process is included in order to solve the problem.

In medicine, one may think of different states of health, the absolute minimum of the landscape meaning good health. In situations where this model is valid, it is sufficient to "disturb" a sick person sufficiently. Then the healing ability of the person will do the rest. The unspecific random action is just a trigger. For obvious reasons, the whole process is often called self-organization.

The most interesting point is that randomness helps the absolute minimum to appear. Expressions like "order from noise" (von Foerster) or "order through fluctuation" (Prigogine) refer to similar cases which seem paradoxical at first sight. Collaboration of deterministic and stochastic principles is productive not only in mathematics, but also, for instance, in biology. In a simple picture, mutations occur randomly, whereas selection is deterministic. Both together are very efficient to promote evolution.

Synergetics is an important (meta)theory of self-organization, cf. Haken (1983), Ebeling & Ulbricht (1986), and Haken (1988). Another route was chosen by Prigogine. His school emphasizes dissipative systems and complexity as central terms, cf. Nicolis & Prigogine (1989). Various approaches to self-organization are treated in Yates (1987), Dalenoort (1989), Kratky & Wallner (1990), and Niedersen & Pohlmann (1990). In Sect. 4.1, we will come back to the concept of self-organization.

2. Equilibrium in Isolated Systems

2.1 Microscopic and Macroscopic Determinism

In order to study questions of causality and randomness further, we will concentrate on complex behavior occurring in physics. To begin with, we consider a fluid in a box, the total system being isolated from the environment. We start with a macroscopic view of the fluid. We can define density, pressure, or temperature via (equilibrium) thermodynamics. The system is macroscopically at rest, no fluctuations are observed. It is easy to predict the future since equilibrium is timeless. Now we "zoom" into the fluid. At a mesoscopic level, where we consider millions of molecules, we observe fluctuations of density, pressure, and temperature. These are the "local" equilibrium fluctuations which cancel out on the macroscopic scale. Now, we see that the equilibrium is not static but dynamic.

In a last step, we zoom into the microscopic level, where we see molecules fly about and collide. These single processes can be described by mechanics. Thermodynamic quantities like density, pressure, or temperature have lost their natural meaning at this level, and the term "equilibrium" makes no sense, either. Instead, we consider the positions and velocities of the molecules. Solving the equations of motion makes it possible to predict the future – at least in principle. In practice, there are several problems even if we assume that we know the correct equations of motion:

i) To solve these equations, we have to know the initial state. However, there is always a (small) uncertainty, our knowledge is always restricted.

ii) When we solve the equations of motion on a computer, we use numerical methods which contain approximations. Moreover, only a finite number of digits can be handled by a computer.

iii) Using the zoom, we only see a small section of the fluid, which is not isolated from the remainder. Even if we could calculate the future of the few molecules inside, molecules enter the section from outside in time intervals which are not predictible.

Thus, it does not help much that the equations of motion are deterministic. If we zoom back to the mesoscopic level, it is no longer possible to solve the millions of equations. We have to work with probabilities and fluctuations. Unexpected influences occur not only from the outside (the mesoscopic section also contains an open system), but also from the inside: Now, the influences from the microscopic level also seems to be stochastic. Back at the macroscopic level, we may even ignore the existence of molecules. The fluctuations from inside are smeared out, and the outside is just the isolating box. At this level, the system is deterministic.

The behavior of a macroscopic fluid can be described by phenomenological equations, relating, e.g., density, pressure, and temperature. If one is interested in explaining the relations from first principles, one has to connect the microscopic (mechanical) and the macroscopic (thermodynamical) levels. This is formally done with the help of statistical physics. However, we will not go into further details.

2.2 Weak and Strong Causality

Now, the connection between determinism and predictability will be investigated. It seems clear that a deterministic system is also a predictable one. The considerations of Sect. 2.1, however, suggest that the relation is not so simple. This has to do with the distinction between weak and strong causality. The determinism shown by the equations of motion means "same initial conditions – same future".

More general, "same cause – same effect". Tacitly, this is often changed to "similar cause – similar effect". These two versions correspond to the weak and strong causalities, respectively. It is the second one which guarantees predictability and, e.g., the reproducible outcome of an experiment. One can never exactly repeat an experiment, not in psychology, and not even in physics. The assumption of strong causality makes it possible to mix up the determinism of two levels: that of the rules or basic equations of a system and that of its behavior. This seems so self-evident that the cases where this does not work are considered as puzzles or even as paradoxes.

Before we deal with two of these puzzles, deterministic chaos and irreversibility, we come back to Sect. 2.1. There, (iii) seemed to be the greatest problem for predictability. However, if only weak and not strong causality is valid, (i) and (ii) are equally problematic. At first sight, (i) is a subjective argument. In fact, in classical mechanics it is possible to reduce any uncertainty arbitrarily. But even this may not be sufficient if only weak causality holds. One would need infinite accuracy to be safe. Correspondingly, an approximate solution of the equations of motion, argument (ii), is then crucial.

2.3 Chaos and Irreversibility

What about the actual situation described in Sect. 2.1? If the molecules are considered to be spherical balls interacting via collision, it turns out that all kinds of (small) uncertainties result in the lack of long-term predictability, even if the number of particles is small. For instance, if the initial state of the molecules is not known exactly, it can be shown that uncertainty increases exponentially from collision to collision, so that after a few collisions the trajectories are totally different for initially very close states. This microscopic chaos, however, has no influence on the fluid when taking the macroscopic level into account. In Sect. 3, we will come across macroscopic chaos as well. To stress the fact that this chaos is based on deterministic equations, it is often called deterministic chaos. As scientific books introducing to this field, Bergé et al. (1986) and Schuster (1988) shall be mentioned. Several books of the Springer Series on Synergetics also deal with chaos, e.g., Haken (1984), Mayer-Kress (1986), and Haake (1991).

The puzzle of irreversibility is based on the discrepancy between the time-symmetric basic physical equations and the obvious arrow of time we find in our world. There are may suggestions for a solution, cf. Prigogine (1980), Kratky (1989), and Zeh (1989). One of the most popular arguments is that the reversible equations refer to the microscopic (mechanical) scale, whereas irreversibility occurs on the macroscopic (thermodynamical) scale. Thus, something may happen when bridging the two levels. However, this is not necessarily the important point. Computer experiments investigating systems with just a few particles show that irreversible thermodynamics yields the correct results (Kratky & Hoover, 1987; Posch et al., 1990).

To produce chaos, a few particles are sufficient, too (see above). Maybe an analogous statement is true for synergetics as well? Usually it is assumed that synergetics is valid if a large number of elements are working together. Especially for cognitive processes and behavior, the number of elements is not easy to determine. So it would be interesting to know if this number is really crucial for all applications.

Back to irreversibility: A hundred years ago, Boltzmann investigated the problem of irreversibility in detail. He showed that his time-symmetric Boltzmann equation (valid for a dilute gas) resulted in an irreversible solution. However, to get a solution at all, he had to incorporate the assumption of "molecular chaos", a chaos in another sense. This assumption means that two colliding particles "forget" each other after the collision, thus introducing an element of irreversibility. Strangely enough, this "error" yields correct results, but the situation is not very satisfactory. It reveals another possibility for the discrepancy between basic laws and the (calculated) behavior of a system: an approximation when analytically deriving a solution from the basic equation(s).

One may argue that quantum mechanics would change the situation considerably: it includes uncertainty relations which may replace the assumption of molecular chaos. This would mean, however, throwing together part of mechanics (the trajectories of molecules) and quantum mechanics (uncertainty). An appropriate procedure would be to start from quantum mechanics alone. However, the basic Schrödinger equation is time-symmetric in the probabilities. To get irreversibility, one has to take the measurement process into account. Then, one is in the midst of the problems of interpretation of quantum mechanics (Selleri 1990). So we would avoid one problem only to find another one. In both cases, irreversibility comes from "outside".

If basic laws and the behavior of the corresponding system fall apart, the terminology is delicate. Usually, "irreversibility" refers to the behavior, but "deterministic" to the basic laws. Thus, one has to be careful when interpreting such terms.

2.4 Equilibrium as an "Attractor"

To end this section, we return to the fluid in the box. If the initial state is a special one (e.g., all molecules being in the left half of the box), after a short time the fluid will relax to equilibrium, the density being constant over the volume. Equilibrium seems to be an attractor. This face of irreversibility has to do with the overwhelming number of equilibrium states, an argument already due to Boltzmann. In the above example, the "unnatural" initial conditions are responsible for the irreversible process.

If one waits long enough, isolated systems are in equilibrium. In the following, we will consider open systems. As for isolated systems, we are not interested in transient states, but in the long-term behavior of the system.

3. Pattern Formation in Open Systems

3.1 Bénard Convection

As an example of an open system, we consider a modification of the fluid in the box. Now, it may exchange heat with its environment. The height H of the box is much smaller than its length and width. On the upper (lower) side, the box has the constant temperature T_- (T_+), $T_- < T_+$. The almost identical temperatures are induced by the environment. The fluid reacts with a linear decrease of T from below (h=0) to above (h=H):

$$T = T_+ - (h/H)\, \delta T , \qquad\qquad 0 \le h \le H , \qquad\qquad (1a)$$
$$\delta T \equiv T_+ - T_- . \qquad\qquad\qquad\qquad (1b)$$

There is a heat flux through the system from the lower to the upper side. The system is no longer in equilibrium, but still macroscopically at rest (steady state). The influence of the environment just results in boundary conditions (fixed temperatures T_+, T_-). Thus, in this simple model, the system is as deterministic as if it were isolated.

If δT is slowly increased by changing the environment, the heat flux also increases, the system remaining at rest. At a critical temperature difference, however, a horizontal arrangement of macroscopic parallel rolls appears (Bénard convection). Adjacent rolls rotate in opposite directions. Thus, a distinct spatial structure appears which does not vary with time. Therefore, this is again a steady state.

If δT is increased further, the rolls begin to oscillate at another critical temperature difference. The details depend on the experimental conditions. At still higher δT, deterministic chaos appears: the rolls change their direction of rotation in a sequence that is not periodic and thus not predictable. However, during the oscillation and even in the chaotic state, the coherence of different rolls (spatial order) is not destroyed.

The Bénard case can be appoximately treated via the Lorenz equations. They model the experimental findings quite well, only the oscillation of the rolls does not come out.

3.2 The Lorenz Equations

Lorenz (1963) approached the problem from the standpoint of weather forecasting. For some reason or other, sometimes two horizontal layers of different constant temperature develop in the air. If the temperature difference is high enough, then large rolls appear between the layers. This can be seen if there are clouds.

Lorenz simplified the microscopic equations and their mesoscopic formulation. Finally, he came to the following three coupled differential equations for the macroscopic variables X, Y, and Z as a function of time t:

$$dX/dt = -\sigma X + \sigma Y , \qquad\qquad (2a)$$
$$dY/dt = -XZ + rX - Y , \qquad\qquad (2b)$$
$$dZ/dt = XY - bZ . \qquad\qquad (2c)$$

Strictly speaking, he got three ordinary differential equations of first order. The change of the variables in time (left-hand side) is determined by the variables (right-hand side). Their meaning is the following: If we pick out a specific roll, X is the velocity of the rotation. The sign depends on the direction of rotation: clockwise or counter-clockwise. Y characterizes the difference between temperatures in the ascending and descending parts of the roll. Z specifies the overall deviation of the vertical temperature profile from linearity (1a). σ and b are parameters referring to the fluid itself. As usual, we assume $\sigma=10$ and $b=8/3$ (see Bergé et al. 1986, p. 308). The remaining external or control parameter r couples the system to the environment, r being proportional to the temperature difference δT.

The behavior of the system, i.e., the solution of the three equations, may be characterized by a trajectory in the three-dimensional space spanned by coordinates X, Y, and Z. At $t=0$, the trajectory starts at a given initial point. This so-called phase space should not be confused with ordinary space.

At first sight, the Lorenz equations are astonishingly simple, the whole open system being expressed by three differential equations only. In the following, we will consider them as "basic laws". The reason for the surprising complexity of behavior lies in two facts. First, the equations are coupled, each variable influencing all others directly or indirectly. Second, nonlinear terms occur: –XZ in (2b) and XY in (2c). Usually, simplification means linearization. However, retaining nonlinear terms is crucial in the present case, because the following conditions have to be fulfilled for chaotic (nonperiodic) solutions to appear: there are at least three coupled differential equations which contain at least one nonlinear term. Lorenz did not know this, he found chaos to his surprise.

Other possible solutions of coupled differential equations are steady states (fixed points) and periodic (oscillatory) solutions. In these two cases, the minimum number of equations needed is one and two, respectively. Nonlinearities are not necessary then.

In general, nonlinear differential equations cannot be solved analytically (exactly with paper and pencil), but only numerically (approximately by computer). Special solutions may, however, be found quickly. For the Lorenz equations, a steady state is given by

$$dX/dt = dY/dt = dZ/dt = 0 . \qquad\qquad (3)$$

This reduces the differential equations (2a-c) to normal algebraic equations, the solutions being fixed points of (2a-c). The first solution is the trivial one:

$$X_1 = Y_1 = Z_1 = 0 . \qquad\qquad (4)$$

Furthermore, a pair of fixed points exist:

$$X_{2,3} = Y_{2,3} = \pm [b(r-1)]^{1/2}, \quad Z_{2,3} = r-1 . \tag{5}$$

There are no more solutions for a steady state.

3.3 Solutions for Different Values of r

Stability analysis reveals the regions where the steady-state solutions are (locally) stable. This means that after a small disturbance or perturbation the system returns to the steady state which thus serves as an attractor.

i) $0 \leq r \leq 1$:
At small temperature differences, $r \leq 1$, the trivial steady state, is stable, cf. (4). Thus, the corresponding fixed point is an attractor. The stability is even global since the Lorenz equations have no other solution for $r \leq 1$. Starting from any initial state, the long-term behavior is given by the trivial steady state. The system is macroscopically at rest there. In phase space, this solution is characterized by a point at the origin. By the way, the special case $r=0$ means that there is no temperature difference and thus no heat flux through the system. This corresponds to equilibrium.

ii) $1 < r \leq 24.06$:
Instead of (4), the fixed points (5) are now stable. This means that convection rolls appear. The two solutions correspond to the clockwise or counter-clockwise rotation of a given roll. As in (i), the system is macroscopically predictable. However, one now has to know which of the two steady states is valid at a given time. Then, one can predict that the corresponding rotation of the roll will be found in the future, too.

Incidentally, the trivial steady state (4) is still a solution of the Lorenz equations, but it is no longer stable (repellor instead of attractor). If we start exactly at this fixed point, we remain there. If we start close to it, the state will be attracted by one of the two stable fixed points. The whole phase space is now divided into three parts: the unstable fixed point and the basins of the two attractive fixed points. Any kind of noise suppressed in the macroscopic formulation of the Lorenz equations will destabilize the trivial steady state, but we cannot predict into which basin the trajectory will move. Here the purely macroscopic view fails, only weak causality being valid.

The case $r=1$ is especially interesting. There, steady state (4) bifurcates into the pair (5). One cannot predict which of the two possibilities is realized when increasing r slightly above 1. Small influences from "inside" (mesoscopic fluctuations) and "outside" (temperatures T_+, T_- not being strictly constant) are then crucial.

Two comments to round off the above considerations: First, each of the two stable solutions (convection rolls) breaks the spatial symmetry of the system represented by the Lorenz equations. Only both solutions together retain the symmetry. Second, the sensitivity to small influences at and close to the trivial steady state can be interpreted in a positive way: With a small effort, one can give the roll the desired direction of rotation, e.g., by stirring the fluid slightly.

iii) 24.06 < r < 24.74 :

For higher temperature differences the convection rolls begin, in reality, to oscillate. This means that the corresponding point in phase space circles around one of the two steady states (5). As already mentioned, such a periodic attractor (limit cycle) does not occur for the Lorenz equations. However, something else happens: In addition to the stable fixed points (5), there appears another attractor. This chaotic or strange attractor being the only stable solution in region (iv), it will be discussed below. In the region now under investigation, the portion of the basin of the strange attractor increases from 0% to 100% of phase space for r increasing from 24.06 to 24.74. We do not notice it when slowly increasing from r=24.06. The convective rolls remain (locally) stable, their rotation velocity slowly increasing due to (5). Only if we start from initial conditions away from these fixed points may the trajectory become chaotic.

iv) 24.74 ≤ r ≤ 31.1 :

The strange attractor (Lorenz attractor) is globally stable in this region. The corresponding chaotic movement can be described in the following way (Fig. 1): Several times, the point in phase space circles around one of the two "centers of gravity" (the steady-state solutions which are unstable now). This movement corresponds to a roll oscillating around its mean rotation velocity. Every time the point comes closer to the other center of gravity, it may be attracted by that one. If this is the case, the roll changes the direction of rotation. These jumps backwards and forwards occur in a nonperiodic manner. One may bet when the next change will come as if it were a game of chance. In both cases, lack of strong causality is the reason for the breakdown of predictability. For instance, slightly different initially conditions, which cannot be controlled in a real experiment, determine whether the rolls will change their direction of rotation at a given time.

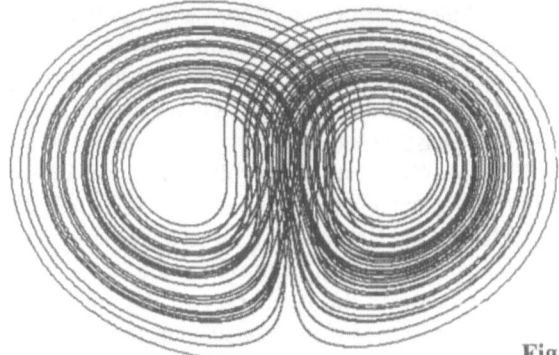

Fig. 1. Lorenz attractor at r=28.0

This may be compared with the instability of the trivial steady state in (ii), including the bifurcation at r=1. However, it is much more severe now. Instead of one point, the whole phase space now shows weak causality.

v)   r > 30.1 :
Here the solutions are very complex, with alternating chaotic and periodic regions. Interestingly enough, chaos disappears at all for r>214.4. Above, only periodic solutions exist.

4. Another Approach to Chaos

4.1 Chaos as a Face of Self-Organization

Now we consider the sequence of solutions of the Lorenz equations from the standpoint of self-organization. To do this, we start with the bifurcation point. If r is increased beyond 1, the rolls emerge spontaneously. The appearance of coherent structures is usually connected with the term "self-organization". This term is a bit misleading (as is the expression "chaos"). It does not mean that something appears only by itself. The external influence via the temperature difference is necessary. However, the control parameter r is not simply the "cause" for the formation of the rolls. It is more appropriate to speak of a trigger for the spontaneous pattern formation the system is capable of. The (hidden) potential to form rolls is unfolded by the choice of an appropriate r, the implicit is made explicit. The stimulus is unspecific in the sense that the temperatures at the upper and lower sides of the box are spatially constant. Nevertheless, there is a specific response: the structured rolls. This coherent structure is due to a collective movement of the molecules, which is remarkable since there are only short-range interaction forces between the particles.

The emergence of oscillating structures in complex systems is usually also subsumed under self-organization. The chaotic case, however, is often contrasted with self-organization. Unlike the order and predictability inherent in steady or oscillating structures, chaos is rather classed with disorder. The choice of the word "chaos" has also to do with that. Sometimes, it is even confused with randomness.

Here, another point of view will be taken. Self-organization will be used in a broader sense, chaos then being but a face of self-organization. For the chaotic solution of the Lorenz equations, not "everything is possible". On the contrary, the strange attractor has zero volume in phase space. In this respect, there is no difference to fixed points or limit cycles. The convection rolls exist even in the chaotic case, and they oscillate in a typical manner. The actual future of the direction of rotation is not predictable, but the strange attractor as a whole reveals a delicate kind of high order in phase space. The single trajectories (as a function of time) are very sensitive to perturbations and the inaccuracy of the computer. This is not the case, however, for the limiting attractor considered as a timeless structure in phase space.

In order to underline that shift of viewpoint, this contribution is called "chaos and disorder" instead of the usual "chaos and order". The use of the word "chaos" itself also has interesting consequences. In spite or because of frequent misunderstanding of the term, it has generally become quite popular: "It's nice that even science has found chaos". In this statement, deterministic chaos is usually confused with the meaning of chaos in everyday life. Furthermore, chaos is often misused as a synonym for the new development in science studying complex dynamical systems (deterministic chaos being only part of it), cf. the popular books of Gleick (1987) and of Briggs & Peat (1989). This may be compared with the investigation of "true" chaos and the evolution of this term. See, e.g., Cvitanovic (1989) and Hao (1989). Both books contain a large reprint section where the most important steps of chaos research can be looked up. By the way, Lorenz (1963) used the term "nonperiodic". The choice of the word chaos and its different use in science and by the public has caused much confusion. It would be a good time to change the term, but the expression is too "attractive".

4.2 Different Meanings of Self-Organization

As already indicated, not only chaos, but also self-organization can be looked at in different ways. If self-organization is interpreted as the autonomous reaction to a stimulus, then also the "trivial" steady state (Sect. 3.3 i) close to equilibrium shows a simple kind of self-organization. Due to the temperature difference stemming from outside, a linear temperature profile is developed inside. This is also an overall macroscopic reaction which is not easily traced back to the molecules. Bearing this in mind, the structures appearing at higher temperature differences are not as exceptional as they are often considered to be.

There is a further important point concerning self-organization: namely, the role of boundary conditions. If macroscopic structures (including chaos) appear, they are co-determined by the boundaries. In experiments, these are often man-made (e.g. the box of a fluid). In nature, however, the emergence of structures and of their boundaries show a kind of co-evolution, be it meteorological realizations of Bénard convection or the development and growth of a biological cell, the boundary being the membrane. The structure is not just determined by a given boundary, but is also constructing its own boundary. This is a marked type of self-organization.

From these considerations, it is not far to autopoiesis. There is an ongoing discussion as to whether this variant of self-organization fits into the whole concept or not. In normal self-organization, the aspect of open systems is stressed. In autopoiesis, the aspect of boundaries to the environment is more important. Autopoiesis has come from biology (Maturana & Varela, 1980). Now, a lively discussion is taking place in psychiatry and family therapy. Self-organization in general and autopoiesis in particular is treated in Kratky & Wallner (1990) as well as in Tschacher (1990).

4.3 The Merits of Chaos – also for Clinical Psychology

Having considered various aspects of self-organization, we come back to chaos. If one looks at it in detail, it reveals its merits. The argument (at the end of Sect. 3.3 ii) concerning the trivial steady state holds for chaos as well. Due to the large effects of small perturbations, it is very difficult to predict the future of a chaotic state if one sits and looks at it from outside (as physicists usually do), but it is especially easy to regulate the future of a chaotic system if one persistently interacts and disturbs the system. A juggler or a tight-rope walker are good examples for that view. These artists handle shaky situations with small corrections (and only small corrections!).

It is not surprising that nature uses chaotic situations and transitions for the (self-) regulation of living systems (Degn et al., 1987; Skarda & Freeman, 1987; Glass & Mackey, 1988). Neither the EEG nor the cardiac rhythm consist of strictly periodic oscillations, there is always a chaotic portion. Thus, the merits of chaos should also be taken seriously by and used in science and technology.

What can we learn from all this for clinical psychology? First, the existence of chaos has an exonerative function: Even in physics, there are processes which are not predictable and reproducible in the usual sense. Second, behind the seemingly unpredictable behavior of a single person or a group there may be strict rules. This is in accordance with a typical (pattern of) behavior; however, the moment of change from one to the next cannot be foreseen. To stabilize a behavior, a small but continuous interaction is likely to be most efficient. The merits of chaos are utilized then. Alternatively, one may guide a person to the basins of attraction (fixed points or limit cycles) which come close to the desired behavior. Then, after triggering the process no interaction is necessary any more in this ideal case of self-organization. Much effort, however, is needed if it turns out later that the basin of attraction should be changed again.

Thus, the phenomena of chaos and self-organization can provide a framework which may help very much in clinical psychology, even if some of the results of complex dynamical systems can only be used in a qualitative way.

References

ARMSTRONG, D.M. 1985. What is a Law in Nature? Cambridge: Cambridge University Press.

BERGÉ, P., POMEAU, Y. & VIDAL, C. 1986. Order within Chaos. Towards a Deterministic Approach to Turbulence. New York: Wiley.

BRIGGS, J. & PEAT, F.D. 1989. Turbulent Mirror. An Illustrated Guide to Chaos Theory and the Science of Wholeness. New York: Harper and Row.

CVITANOVIC, P. 1989. Universality in Chaos. Bristol: Adam Hilger.

DALENOORT, S.J. (Ed.) 1989. The Paradigm of Self-Organization. London: Gordon and Breach.

DEGN, H., HOLDEN, A.V. & OLSEN, L.F. (Eds.) 1987. Chaos in Biological Systems. New York: Plenum.

EBELING, W. & ULBRICHT, H. (Eds.) 1986. Selforganization by Nonlinear Irreversible Processes. Berlin: Springer.

GLASS, L. & MACKEY, M.C. 1988. From Clocks to Chaos. The Rhythms of Life. Princeton, N.J.: Princeton University Press.

GLEICK, J. 1987. Chaos: Making a New Science. New York: Viking.

HAAKE, F. 1991. Quantum Signatures of Chaos. Berlin: Springer.

HAKEN, H. 1983. Synergetics. An Introduction. Berlin: Springer.

HAKEN, H. (Ed.) 1984. Chaos and Order in Nature. Berlin: Springer.

HAKEN, H. 1988. Information and Self-Organization. A Macroscopic Approach to Complex Systems. Berlin: Springer.

HAO, B.-L. 1989. Chaos. Singapore: World Scientific.

KRATKY, K.W. & HOOVER, W.G. 1987. Hard Sphere Heat Conductivity via Nonequilibrium Molecular Dynamics. J. Statist. Phys. **48**, 873-900.

KRATKY, K.W. 1989. Entropie, Irreversibilität und Strukturbildung. In: Kratky, K.W. & Bonet, E.M. (Eds.), Systemtheorie und Reduktionismus. Wien: Österreichische Staatsdruckerei, 147-182.

KRATKY, K.W. & WALLNER, F. (Eds.) 1990. Grundprinzipien der Selbstorganisation. Darmstadt: Wissenschaftliche Buchgesellschaft.

LORENZ, E.N. 1963. Deterministic Nonperiodic Flow. J. Atmos. Sci. **20**, 130-141.

MATURANA, H.R. & VARELA, F.J. 1980. Autopoiesis and Cognition: The Realization of the Living. Boston: Reidel.

MAYER-KRESS, G. (Ed.) 1986. Dimensions and Entropies in Chaotic Systems. Quantification of Complex Behavior. Berlin: Springer.

NICOLIS, G. & PRIGOGINE, I. 1989. Exploring Complexity. An Introduction. New York: Freeman.

NIEDERSEN, U. & POHLMANN, L. (Eds.) 1990: Selbstorganisation und Determination. Berlin: Duncker & Humblot.

POSCH, H.A., HOOVER, W.G. & HOLIAN, B.L. 1990. Time-Reversible Motion and Macroscopic Irreversibility. Ber. Bunsenges. Phys. Chem. **94**, 250-256.

PRIGOGINE, I. 1980. From Being to Becoming – Time and Complexity in the Physical Sciences. San Francisco: Freeman.

SCHUSTER, H.G. 1986. Deterministic Chaos. Weinheim: VCH.

SELLERI, F. 1990. Quantum Paradoxes and Physical Reality. Dordrecht: Kluwer.

SKARDA, C.A. & FREEMAN, W.J. 1987. How Brains Make Chaos in Order to Make Sense of the World. Behav. & Brain Sc. **10**, 161-195.

TSCHACHER, W. 1990. Interaktion in selbstorganisierten Systemen. Grundlegung eines dynamisch-synergetischen Forschungsprogramms in der Psychologie. Heidelberg: Asanger.

YATES, F.E. (Ed.) 1987. Self-Organizing Systems. The Emergence of Order. New York: Plenum.

ZEH, H.D. 1989. The Physical Basis of the Direction of Time. Berlin: Springer.

Instability and Cognitive Order Formation: Self-Organization Principles, Psychological Experiments, and Psychotherapeutic Interventions

Peter Kruse, Michael Stadler, Boris Pavleković, and Vladimir Gheorghiu

With 6 Figures

Abstract. The transfusion of basic principles of self-organization theory to cognitive processes is a necessary and promising task of general and applied psychology. For historical reasons and because of practical evidence self-organization concepts have already gained considerable acceptance in psychotherapy. But the attempt to link the basic theoretical principles with cognitive phenomena and with concrete therapeutic interventions is still in its infancy. The interventions often seem to be more a product of the therapists intuition than a consistent deduction, and in experimental psychology only few examples of phenomena analysed in terms of self-organization theory can be mentioned.

In self-organization theory the dimension of stability and instability of system behavior is of central importance. Systemic instability is a critical agent of order formation. In the phase of instability the system becomes sensitive to minute influences in the sense of little cause having great effect. For psychological processes it can be hypothesized that any cognitive organization or reorganization is guided by phases of internal instability and that the impact of external fluctuations varies with the degree of this instability. As a concrete example, a close connection between suggestibility and cognitive instability can be postulated.

In this report experiments are described to prove these assumptions and some theoretical and practical consequences of the empirical results are discussed. The idea of a synergetic approach to psychotherapy is outlined. The special therapeutic relevance of hypnosis and ritual behavior is briefly explained.

1. Introduction

Usually meta-theoretical concepts are transferred into practice with a considerable time delay and after a variety of intermediate scientific stages. The theoretical framework of cybernetics (see Wiener, 1948), the principle of systemic self-stabilization by negative feedback and the balancing of acquired and present state, for example, was adopted by psychology decades after the first formulations (e.g. Miller et al., 1960) and again several years passed before its useful application in psychotherapy (e.g. Mahoney, 1974). Concerning self-organization theory this typical development seems to be significantly altered. Basic aspects of the self-organization perspective were elaborated in parallel as a meta-theoretical view of complex system behavior and as a concrete psychotherapeutic approach to mental problems. An elaborated psychological theory of cognition – explicitly functioning

Springer Series in Synergetics, Vol. 58   ·   **Self-Organization and Clinical Psychology**
Editors: W. Tschacher, G. Schiepek, and E.J. Brunner   © Springer-Verlag Berlin Heidelberg 1992

as the normally expected link between meta-theory and practice – is largely lacking.

Concerning the systemic aspects the meta-theoretical view is connected with the work of Eigen, von Foerster, Haken, and Prigogine. The epistemological and cognitive consequences were first formulated by von Glasersfeld, Maturana, and Varela. The corresponding psychotherapeutic approach can be subsumed under the concept of systemic therapy (e.g. Reiter et al., 1988; Simon, 1988). Here names like Bateson, Selvini-Palazzoli, and Watzlawick have to be mentioned as initiators.

Several possible explanations can be suggested for this asymmetric development. One possible reason is that in a way the idea of autonomous order formation and cognitive reality construction is less counter-intuitive for a practically experienced therapist than for an empirically oriented psychologist who is trained to establish distinct relationships of cause and effect and who has a predominantly information processing view of cognition. A second reason is already implicitly mentioned in this argument. In the scientific history of psychology, holistic concepts became less and less important during recent decades whereas analytical concepts which try to model cognitive phenomena as the result of hierarchically ordered elements or elementary functions are generally accepted. But the necessity to deal with the natural complexity of psychological phenomena and the severe problems analytical concepts have in this respect now seem to be changing the situation again. In the course of this change the old psychological concept of gestalt theory is reassessed.

Gestalt theory can be understood as a legitimate precursor of self-organization theory and the central ideas and methodological procedures developed in this context may help to elaborate the missing link of a psychological view of cognitive self-organization (Stadler and Kruse, 1990a,b; Haken, 1991). There are astonishing correspondences between the terminology and the argumentation of gestalt theory and modern self-organization concepts. In particular the basic assumptions of Wolfgang Köhler are nearly identical to modern approaches (see Kruse, 1988a). Köhler was only limited by the physical knowledge of his time. Following the tenets of linear thermodynamics the general systemic tendency towards final equilibrium was the only principle of self-organization he could use in his modelling (see Köhler, 1938a). But Köhler was fully aware that this limitation was rather unsuitable for application to biological systems (Köhler, 1955). Throughout the whole history of gestalt theory the autonomy of order formation in cognition was the explicit starting point of conceptualization and experimental work. Therefore gestalt theory offers a lot of well-described phenomena and stimulating ideas for the elaboration of a self-organization theory of cognition (see Stadler and Kruse, 1990c).

2. Stability and Instability

In gestalt theory and in concepts of self-organization cognitive order formation is not seen as the more or less trivial result of an ordered environment or of an organizing will. Perception is not understood as information processing by the use of properly tuned filter mechanisms and processing algorithms, or as the direct input of relevant cues. Behavior is not understood as the product of a hierarchy of feedback loops governed by will and environmental conditions. The cognitive system is only modelled as an open system with regard to energy and is a closed system with respect to information. The environment functions as the unspecific initial condition of an autonomous process of order formation. The brain is modelled as a system which produces global states of order out of the free inner dynamics of its own elementary units. "That these processes, occurring in the ... nervous system, should be passive copies of stimulus patterns is certainly an idea which can no longer be seriously held" (Köhler, 1938b, p. 92). This view necessarily ends up in a variety of fundamental reorientations of theoretical questioning and empirical research (see Stadler and Kruse, 1990c).

The central aspect to be discussed in the following is the aspect of stability and instability in cognitive order formation. The stability of our world of phenomena is the necessary basis of action and usually it is seen as self-evident. It was one major contribution of gestalt theory to bring this view into question. In a variety of experiments gestalt psychologists demonstrated that the experienced order and stability of our world is a product of the cognitive system itself and not a derivative of the given situation. What we see, hear, or feel depends on the activity of cognitive principles of order formation which themselves are usually not experienced. Gestalt principles decide what is perceived. Perception is not a representation of a given stimulus situation. Stability is only the result and instability is the basic characteristic of cognition (Kruse, 1988; Kruse and Stadler 1990). "When we look at ... a pattern our first impression might be that a pattern is something static. However, when thinking somewhat more about patterns we realize that patterns are intimately connected with processes" (Haken, 1979, p. 2).

stable state A   stable state B
faces   vase

Fig. 1. Rubin's "vase-faces" figure. Spontaneous reversals of figure and ground

For gestalt theory to search for intra-systemic principles of order formation is the explicit goal in perception as well as in other cognitive functions like remembering, thinking or in motor behavior (Vogt et al., 1988; Stadler, Vogt, et al., 1991; Stadler, Richter et al., 1991).

Consequently gestalt psychologists were especially interested in phenomena of transition from one stable order to another, and in phenomena of emergence of macroscopic order. Mostly the conscious realisation of these phenomena is limited to rare exeptions which need a lot of training or special conditions to be produced. In perception our experience is predominated by the rapidity, ease, and stability of actual genetic order formation and the efficiency of the process prevents a questioning attitude. Phenomena like perceptual multistability, where the spontaneous transition from one stable state to another can be perceived directly (Fig. 1), seem to be exceptional and curious. But in view of the theoretical perspective of autonomous order formation these rare phenomena of experienced cognitive instability are no longer curious exceptions but paradigmatic tools of research. Situations of transition or emergence open the possibility to investigate the intrinsic dynamics underlying the process of cognitive order formation (see Köhler, 1940; Haken and Stadler, 1990). For gestalt theory as well as for self-organization theory the task is to look for instabilities and to use these instabilities to analyse the process of cognitive order formation (Kruse et al., 1991).

In self-organization theory intra-systemic instability is of major importance for the process of autonomous order formation. To change from one stable state to another or to approach a stable state, the system has to pass through or to be in a phase of instability. During instability very subtle and far-reaching interactions between the elements of the system occur which enable the spontaneous order formation. Haken (1977) describes the dynamic behavior of a self-organizing system by the overdamped movement of a particle in a potential field. This mathematical description can be visualized by a ball moving in a landscape. If the system is in a stable state the ball is in one valley and its own movements are not strong enough to leave it. A transition from one stable state to another (*phase transition*) is only possible when the shape of the landscape changes or when the movement of the ball is increased up to a critical degree (*critical fluctuations*). When the ball moves from one valley to another it has to pass a point of complete instability at the top of hill. This situation is called *symmetry breaking* because it has to be decided in which direction the ball will move, in which direction the symmetry of possible stable states will be broken. In the situation of symmetry breaking the behavior of the system is open to very subtle influences. Even a force of otherwise neglegible influence is able to determine which stable state the system will reach. In the phase of maximal instability little causes have great effects (see Haken, 1977; Prigogine, 1980). Therefore it can be predicted for the process of autonomous order formation that any organization and reorganization is dependent on the amount of intra-systemic instability and that during instability the reactivity of the system is critically enhanced.

This synergetic metaphor of system behavior comes very close to Lewin's (Lewin, 1936) topological psychology and has already been introduced more or less explicitly in psychotherapeutic modelling (see e.g. Schneider, 1983; Brunner, 1988). Concerning cognitive processes a variety of concrete assumptions can be derived from the metaphor leading to important theoretical consequences and to interesting empirical hypotheses (see Kruse, 1988; Kruse and Stadler, 1990). As already mentioned, a close connection is to be expected between the amount of intra-systemic instability, the readiness to change to another stable state of order, and the reactivity to subtle internal or external influences. The higher the intra-systemic instability the higher is the readiness for alternative order formation and the higher is the sensitivity of the system (see also von der Malsburg, 1984). Since the early days of psychology this expectation has been the basis for a variety of different empirical investigations. For perceptual research it resulted in the methodological idea of using ambiguous or multistable patterns to analyse the processes of perceptual order formation (Kruse et al., 1986). Recently Ramachandran and Anstis declared this strategy to be "the basic strategy for studying perception" (Ramachandran and Anstis, 1985, p. 142).

Another field of research in which instabilities have been of major interest right from the beginning is the investigation of *top down influences* in cognition. Due to every-day experience and experimental research under special conditions, a thought, an expectation, or an imagination are able to alter perception or to influence basic physiological processes. This is the case in e.g. psychosomatic phenomena like the placebo effect, or hypnotic analgesia, and generally in the case of suggestive influence. Psychologists such as Sherif (Sherif, 1935) and Coffin (Coffin, 1941) introduced the idea that cognitive instability is a necessary condition or at least an opportunity for suggestive influences to change our phenomenological

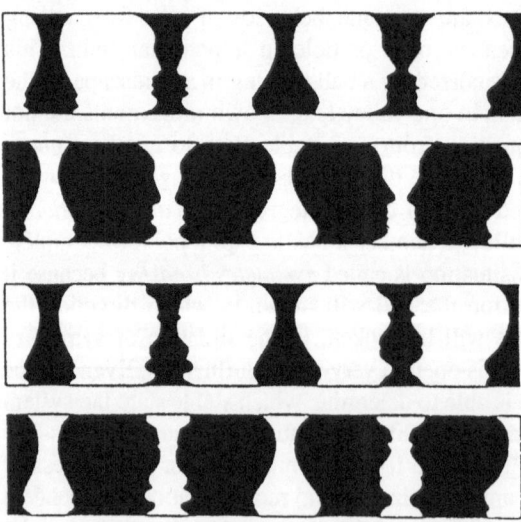

Fig. 2. Top-down influence to perception. In the variation of Rubin's "vase-faces" figure (Kruse, 1986) meaning dominates the basic gestalt factor of symmetry. In the upper two patterns usually the faces and in the lower two patterns usually the vases are seen

world. As an empirical realization Sherif performed his experiments on suggestion by using the instable perceptual situation caused by the autokinetic effect. He was able to demonstrate that a suggestion in the form of a verbally created expectation is able to determine what is perceived. The same can be shown by using multistable patterns. Perceptual multistability can be compared to the above-described situation of symmetry breaking. When confronted with a multistable pattern the system has to decide which stable state it will reach. A suggestive cue or an association can induce one perceptual alternative to be preferred (Fig. 2). In the case of instability, a thought is not "only" a thought but also a major determinant of the state of order of the cognitive system.

In the following, experiments are presented to illustrate and analyse the connection between intra-systemic instability, suggestion, and suggestibility in more detail. On basis of the results and after integrating other empirical data, some conclusions concerning the relationship to the process of cognitive order formation are derived. Finally, the general relevance of the argumentation for the field of psychotherapy and some practical consequences for psychotherapeutic interventions are discussed.

3. Multistability, Suggestion, and Suggestibility

In spite of the self-evident stability of perception there are some phenomena known where perceptual instability can be perceived directly or where the system spontaneously changes between different stable states (see Kruse, Stadler, et al., 1991). In particular, perceptual multistability has been found or demonstrated for the whole range of cognitive complexity from the basic process of perceptual object formation up to the ambiguity of meaning. Starting with the first systematic investigations in the laboratory of Wundt (Wundt, 1898) at the end of the 19th century, multistability has been more frequently subject to, or part of, psychological experimentation than any other perceptual phenomenon. During these investigations it has often been reported that there are significant *interpersonal differences* in the frequency of the spontaneous changes between the perceptual alternatives. Some people display quite low values of the parameter of rate of apparent change and others quite high. Depending on the stimulus conditions this ratio may be about one to twenty. The rate of apparent change (RAC) typical for a person shows a high degree of re-test reliability (Bergum and Bergum, 1980). In view of the outlined self-organization perspective, it can be concluded that the amount of intra-systemic instability varies from person to person. Some systems obviously operate closer to instability points than others.

As already mentioned, systems which operate close to instability points are more sensitive and therefore more able to adapt to changing conditions but less reliable than systems in stability. Basically the same idea was already formulated by Pavlov in his theory of types of higher nervous activity (Pavlov, 1955). In his typology individual nervous systems are mainly classified concerning the dimensions of strength and mobility. This classification was connected to intra-systemic instability by Nebelytsyn who concluded that "...in the simplest, the pure case – the most stable system is the strongest and at the same time the least sensitive, and the least stable system is the weakest and the most sensitive" (Nebelytsyn, 1964, p. 405). In the synergetic metaphor of system behavior, the differences in the amount of intra-systemic instability can be explained by differences in internally produced fluctuations.

On the psychological level these ideas suggest a strong relationship between habitual personal differences in intra-systemic instability, adaptive mental functions like cognitive flexibility, and suggestibility. Using multistable patterns two different measures of the amount of intra-systemic instability are possible. One measure is the individual rate of apparent change. Another possible parameter is the stability of one perceptual alternative when the multistable pattern is biased more and more in favor of one of the other alternatives. This measurement of the tendency to persist in one stable state is called hysteresis. People who have a high rate of apparent change tend to be low in hysteresis because they change easily between different stable states. Concerning suggestibility this results in the following hypotheses:

– The rate of apparent change (RAC) in observing multistable patterns correlates positively with suggestibility. (Additionally positive correlations are to be expected for other adaptive functions like cognitive flexibility).

– The amount of hysteresis when observing gradually biased multistable patterns correlates negatively with suggestibility.

3.1 Experiments

For the actual experiments multistable apparent motion (AM) patterns were used because these patterns show a phenomenologically impressive change and because they can be very easily biased in favor of one perceptual alternative (see Kruse et al., 1991). Additionally these patterns are more reliable concerning their perceptual effects than static multistable patterns (see Kruse, 1988). Two different multistable AM patterns were used (Figs. 3 and 4): the stroboscopic alternative motion (SAM) and the circular apparent motion (CAM). In a first experiment the RAC and the hysteresis was measured on basis of the SAM pattern. To measure the RAC of different persons the SAM pattern was presented for one minute. The task of the subjects (Ss) was to indicate by help of push buttons which perceptual alternative they perceived and when their perception changed during each trial. By this the rate of apparent change for any person was measured five times. The actual RAC value of each person was calculated as the mean of these five trials. To measure the hysteresis of the Ss first the SAM was presented with a strong bias in favor of the vertical motion alternative. Then the vertical distance of the inducing lights was gradually increased (Fig. 5). The task of the subjects was to push a button when the vertical changed into the horizontal motion alternative. As a quantification of the hysteresis effect the critical vertical distance was again measured five times and the mean was calculated.

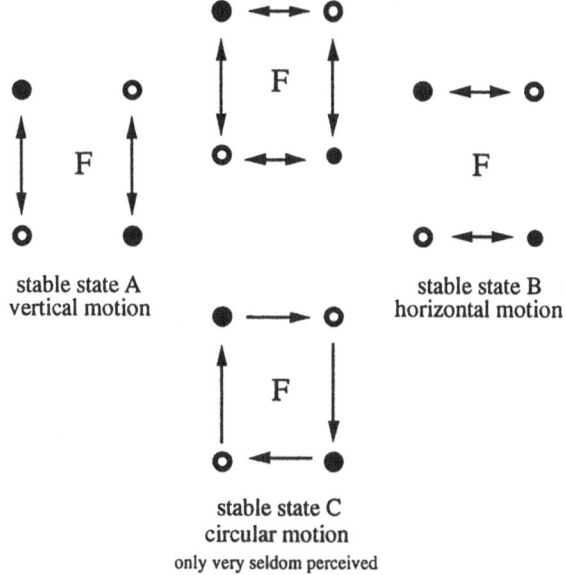

stable state A
vertical motion

stable state B
horizontal motion

stable state C
circular motion
only very seldom perceived

depending on the frequency of flashing lights sometimes no motion can be seen

(succession or simultaneity)

F = fixation point

Fig. 3. Stroboscopic alternative motion (SAB). Four points are positioned at the corners of an upright rectangle. When one pair of points in diagonal corners is presented simultaneously and in alternation with the other pair, alternation between different stable states of apparent motion appears

The suggestibility of the Ss was measured using the Sensory Suggestibility Test (SST) constructed by Gheorghiu (see e.g. Gheorghiu and Reyher, 1982). In this test Ss are confronted with a number of situations which seem to allow measurement of threshold perception. In tactile, auditory, and visual items real stimulation alternates with situations where stimulation is only simulated. To enhance the plausibility of the procedure and to prevent a loss of confidence the Ss are warned of the possibility that sometimes no objective stimulation is applied. The score of the test depends on how often a subject reports a perception in absence of objective stimulation.

It was expected that the RAC correlates positively with the score of the SST and that the hysteresis correlates negatively with the score of the SST. The RAC and the score of the SST was measured for 85 Ss and the hysteresis was measured for a sub-group of 45 Ss. The results confirm the hypothesis. There is a significant positive correlation between the rate of apparent change and the score of the SST ($n=85$, $r=.25$, $p<.05$). The hysteresis shows a significant negative correlation to the score of the SST ($n=45$, $r=.42$, $p<.01$). Taking only extreme groups – Ss with a high rate of apparent change (high rate persons HRP) and Ss with a low rate of apparent change (low rate persons LRP) and correspondingly Ss low and Ss high

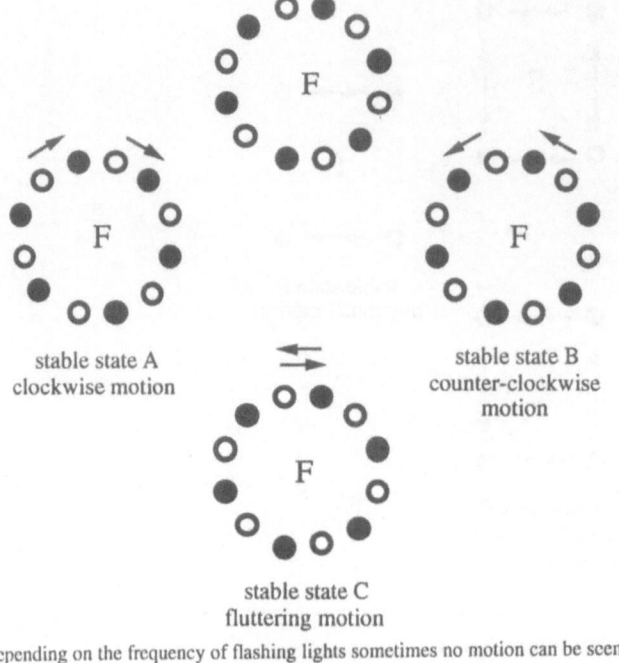

stable state A
clockwise motion

stable state B
counter-clockwise
motion

stable state C
fluttering motion

depending on the frequency of flashing lights sometimes no motion can be seen

(succession or simultaneity)

F = fixation point

Fig. 4. Circular apparent motion (CAM). An even number of points are positioned in a circle. When all points are presented simultaneously which are not directly side by side and alternated with the other points, reversion between different stable states of apparent motion appears

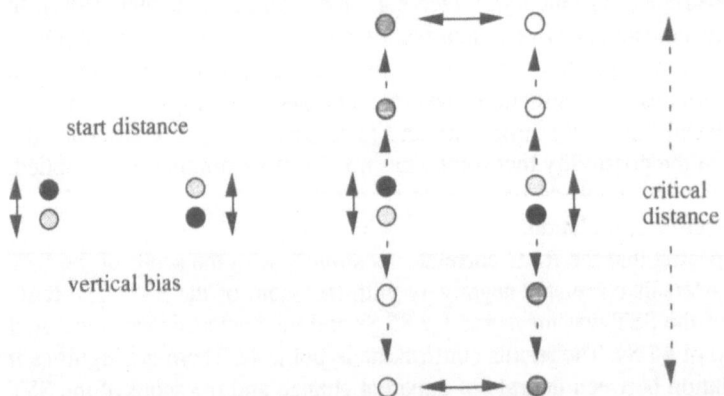

Fig. 5. Hysteresis. A simple way to measure the tendency to persist in one stable state is the critical distance where the vertical motion alternative of the SAM changes perceptually to the horizontal motion when the vertical distance is gradually increased starting from the situation of strong vertical bias

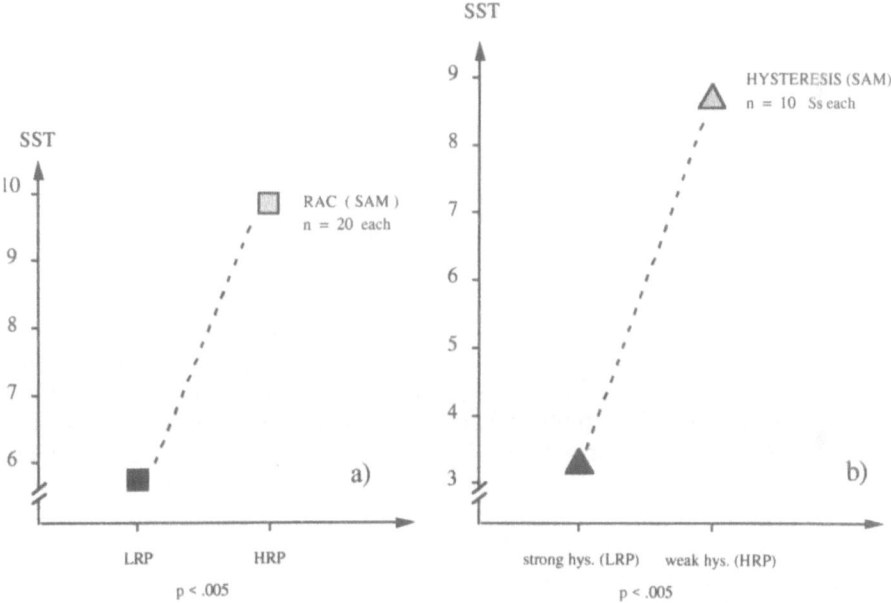

Fig. 6. Extreme group differences. (a) The difference in sensory suggestibility (SST score) between low rate persons (LRP) and high rate persons (HRP). (b) The difference in sensory suggestibility (SST score) between persons with a strong hysteresis effect and persons with a weak hysteresis effect (see text for details)

in hysteresis – a highly significant difference between the mean scores of the SST can be shown (Fig. 6). People who are high in rate of apparent change and low in amount of hysteresis tend to be highly suggestible and people who are low in rate of apparent change and high in hysteresis tend to be low in their readiness to respond to suggestive influences.

Another experiment was designed to demonstrate that suggestive influences are able to alter the perception of multistable apparent motion patterns. In this experiment the SAM and the circular apparent motion (CAM) were used. Again the task of the Ss was to indicate by help of push buttons which perceptual alternative they perceived and when their perception changed. The SAM and the CAM was presented three times for one minute and the relative durations of the different perceptual alternatives were calculated. The Ss were divided into two groups balanced concerning the parameter of RAC. To balance the groups the RAC of every subject was measured in a pre-test. In the control group only the SAM and the CAM were presented to the Ss. In the experimental group during presentation of the multistable patterns verbal suggestions were given. For the SAM the vertical motion alternative and for the CAM the counter-clockwise motion alternative was supported by verbal suggestions. The mean of the duration of the motion alternatives chosen for verbal support was significantly higher with (experimental group) than without (control group) suggestions (t-test, p<.01). In case of symmetry breaking perception is obviously very sensitive to suggestive cues. Even introducing the idea of a special motion direction indirectly, e.g. by using little arrows instead of point-lights in the CAM, is able to enhance the indicated perceptual alternative (Kruse et al., 1991).

3.2 Discussion

As predicted by self-organization theory, there is a close connection between intra-systemic instability, sensitivity and cognitive order formation. Given that the measured parameters of multistable patterns are accepted as indicators of the amount of intra-systemic instability, one can conclude that the reactivity to subtle cues and the readiness to create perceptual states of order covaries with the amount of intra-systemic instability. Obviously individual cognitive systems differ concerning the basic cognitive strategy they prefer. Some systems tend to lean more strongly towards stability and are very reliable concerning their inner states but less sensitive and less able to adapt. And some systems are more often and longer in instable states and therefore less reliable but more sensitive and able to adapt. Seeking psychological and physiological functions which are responsible for, or which correlate with these individual differences is an interesting approach to understanding suggestibility. The idea that cognitive instability is a central aspect of suggestive processes is a stimulating viewpoint for suggestion research (see Gheorghiu and Kruse, 1991).

The results of these experiments are additionally confirmed by more indirect correlations reported in other empirical findings (Table 1). A positive relationship has been shown for cognitive flexibility, originality in thinking, ability to imagine, and rate of apparent change (Gräser, 1977; Klintman, 1984). Comparable correlations were found for these functions of cognitive order formation and suggestibility in the context of hypnosis (Crawford, 1989). Suggestibility and rate of apparent change are reduced with age (Feingold, 1982; Heath and Orbach, 1963).

Besides these interpersonal differences, it is of special interest to inquire about *intrapersonal variations* in intra-systemic instability. If the connection between instability and cognitive order formation can be further substantiated, to understand the conditions under which the system is in a more stable or unstable state is theoretically and practically extremely relevant from a psychological point of view. Reliability and sensitivity as well as the readiness to create or change stable states

Table 1. Indirect correlative relations between perceptual multistability and suggestibility

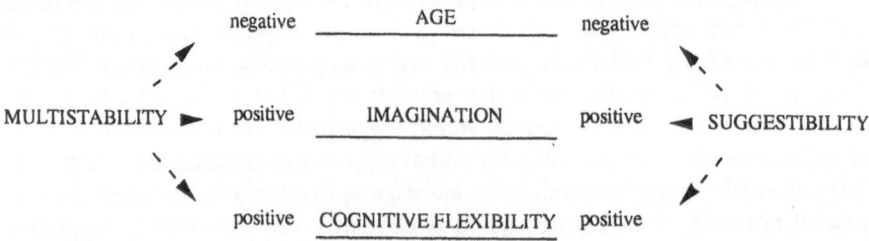

of order are central constructs for the description of processes of cognitive order formation in perception, imagination, remembering, thinking, learning, and so on. In a first experiment we found some evidence that the rate of apparent change and therefore the intra-systemic instability is increased after hypnotic induction. This would explain the enhanced readiness for cognitive construction and reconstruction of individual reality which can be attributed to hypnosis (see Kruse, 1989; Kruse and Gheorghiu, 1990). The higher intra-systemic instability associated with hypnosis is indirectly supported by the previously mentioned work of Crawford (1989) and more directly confirmed by experimental data which show that the variability of EEG alpha amplitude is higher in hypnosis (DePascalis, 1990). The self-organization perspective of cognitive order formation opens a variety of relevant and concrete empirical questions to be answered by future research.

4. Cognitive Instability and Psychotherapy

Applying the basic principles of self-organization theory to psychotherapeutic processes leads to a variety of concrete assumptions which define the role of the therapist in a way which is in some aspects different from more traditional or common therapeutic belief systems. In a synergetic approach to psychotherapy the process of change induced by therapeutic interventions is understood as a process of alternative order formation or alternative reality construction. Symptoms are not malfunctions but states of order which emerge on the basis of certain rules and under certain conditions. The therapist has the task to stimulate the system – a person, a family, a group – to create new stable states of order. This can be done e.g. by making inherent rules, implicit tendencies, and structures of the actual order formation transparent to the system e.g. by means of iterative enfolding (circular questioning etc.). The therapeutic contract is limited to movement. Change can appear in form of a phase transition, suddenly and spontaneously. The complexity of the symptom to be changed need not be identical to the complexity of its explanations. Complex symptoms may be changed by simple interventions. Symptoms which have had a long evolutionary development may be changed in a revolutionary manner. The appearance and the detailed elementary structure of the new state of order the system will reach can not be anticipated and – in this sense – the therapist is not and cannot be responsible for it (see Grawe, 1987). In a way the competence of order formation is left to the system and the therapeutic competence and responsibility has to be defined in a more process-oriented manner. One aspect which is of importance for this process-oriented definition of therapeutic competence and responsibility is the discussed dimension of intra-systemic stability and instability.

If the psychotherapeutic process is understood as autonomous formation of alternative states of order, phases of intra-systemic instability are a necessary condition of change. One central competence of the therapist is to recognize the state of stability or instability of the system and one goal of psychotherapeutic interventions may be to influence the system with respect to this dimension.

Following the line of argumentation of the theoretical and empirical part of the article, instability is necessary for change and for adaptation to new conditions, but stability is necessary as a reliable basis of action. Instability is part of the psychotherapeutic process but not the goal. During instability the therapist has to be very careful and intentional concerning verbal interpretations and nonverbal cues because of the enhanced suggestibility and the enhanced readiness for alternative order formation. Interpersonal differences in the amount of habitual intra-systemic instability have to be taken into account. The therapist should not try to force the system to change. He should wait till the system is ready for change or he should try to enhance this readiness. As mentioned above, hypnotic induction may be one efficient therapeutic intervention to enhance the intra-systemic instability of a person and the well-defined expectations induced by ritual behavior sequences can be used to stabilize or destabilize the reality constructions.

In the view of a synergetic approach to psychotherapy it is of special interest to develop relevant criteria for the amount of stability or instability of a person or a group of persons. Additionally, to create useful therapeutic interventions it is a necessary task to evaluate in more detail under what conditions systems tend to stabilize or destabilize. Especially for the understanding of psychosomatic interactions, this analysis seems to be promising and necessary (Kruse and Stadler, 1990).

Acknowledgement. The authors wish to express their appreciation to the members of the "Norddeutsches Institut für Kurzzeittherapie (Nik)", W. Eberling, H. Dreesen, and M. Vogt for stimulating discussions.

References

BERGUM, J.E. & BERGUM, B.O. 1980. Reliability of reversal rates as a measure of perceptual stability. Perceptual and Motor Skills, **50**, 1038.

BRUNNER, E.J. 1988. Pioniere systemischen Denkens. In: Reiter, L., Brunner, E.J. & Reiter-Theil, S. (Eds.): Von der Familientherapie zur systemischen Perspektive. Berlin: Springer, 273-284.

COFFIN, T.E. 1941. Some conditions of suggestion and suggestibility: A study of certain attitudinal and situational factors influencing the process of suggestion. Psychological Monographs, **4**, 53.

CRAWFORD, H.J. 1989. Cognitive and physiological flexibility: multiple pathways to hypnotic responsiveness. In: Gheorghiu, V.A., Netter, P., Eysenck, H.J. & Rosenthal, R. (Eds.): Suggestion and Suggestibility. Berlin: Springer, 155-167.

DEPASCALIS, W. 1990. Sensory suggestibility and self-organization. 5th European Congress of Hypnosis in Psychotherapy and Psychosomatic Medicine. University of Constance, FRG.

FEINGOLD, E. 1982. Untersuchungen zur sensorischen Suggestibilitt sowie zum Zusammenhang zwischen sensorischer Suggestibilität und der Placebo-Ansprechbarkeit im Schmerzbereich. Mainz: Doctoral Dissertation.

GHEORGHIU, V.A. & KRUSE, P. 1991. The psychology of suggestion: An integrative perspective. In: Schumaker, J.F. (Ed.): Human Suggestibility. Advances in Theory, Research, and Application. New York: Routledge.

GHEORGHIU, V.A. & REYHER, J. 1982. The effect of different types of influence of an "indirect-direct" form of a scale of sensory suggestibility. American Journal of Clinical Hypnosis, **24**, 191-199.

GRÄSER, H. 1977. Spontane Reversionsprozesse in der Figuralwahrnehmung. Trier: Doctoral Dissertation.

GRAWE, K. 1987. Psychotherapie als Entwicklungsstimulation von Schemata. Ein Prozeß mit nicht vorhersehbarem Ausgang. In: Caspar, F. (Ed.) Problemanalyse in der Psychotherapie. Tübingen: DGVT, 72-87.

HAKEN, H. 1977. Synergetics. An Introduction. Berlin: Springer.

HAKEN, H. 1979. Pattern formation and pattern recognition – An attempt at a synthesis. In: Haken, H. (Ed): Pattern Formation by Dynamic Systems and Pattern Recognition. Berlin: Springer.

HAKEN, H. 1991. Synergetic Computers and Cognition. Berlin: Springer.

HAKEN, H. & STADLER, M. (Eds.) 1990. Synergetics of Cognition. Berlin: Springer.

HEATH, H.A. & ORBACH, J. 1963. Reversibility of the Necker cube: IV. Responses of elderly people. Perceptual and Motor Skills, **17**, 625-626.

KLINTMAN, H. 1984. Original thinking and ambiguous figure reversal rates. Bulletin of the Psychonomic Society, **22**, 129-131.

KÖHLER, W. 1938a. Some gestalt problems. In: Ellis, W.D. (Ed.): A Source Book of Gestalt Psychology. London: Routledge & Kegan Paul.

KÖHLER, W. 1938b. The Place of Value in a World of Facts. New York: Liveright.

KÖHLER, W. 1940. Dynamics in Psychology. New York: Liveright.

KÖHLER, W. 1955. Direction of processes in living systems. Scientific Monthly, **80**, 29-32.

KRUSE, P. 1986. Wie unabhängig ist das Wahrnehmungsobjekt vom Prozeß der Identifikation. Gestalt Theory, **8**, 141-143.

KRUSE, P. 1988. Stabilität, Instabilität, Multistabilität. Selbstorganisation und Selbstreferentialität in kognitiven Systemen. Delfin, **11**, 35-57.

KRUSE, P. 1989. Some suggestions about suggestion and hypnosis. In: Gheorghiu, V.A., Netter, P., Eysenck, H.J. & Rosenthal, R. (Eds.): Suggestion and Suggestibility. Berlin: Springer, 91-98.

KRUSE, P. & GHEORGHIU, V.A. 1990. Self-organization theory and radical constructivism: A new concept for understanding hypnosis, suggestion, and suggestibility. 5th European Congress of Hypnosis in Psychotherapy and Psychosomatic Medicine. University of Constance, FRG.

KRUSE, P. & STADLER, M. 1990. Stability and instability in cognitive systems: multi-stability, suggestion and psychosomatic interaction. In: Haken, H. & Stadler, M. (Eds.): Synergetics of Cognition. Berlin: Springer, 201-215.

KRUSE, P., STADLER, M. & WEHNER, T. 1986. Direction and frequency specific processing in the perception of long-range apparent movement. Vision Research, **26**, 327-335.

KRUSE, P., ROTH, G. & STADLER, M. 1988. Self-organization and Köhler's psychophysical field theory. German Journal of Psychology, **12**, 334-336.

KRUSE, P., STADLER, M. & STRÜBER, D. 1991. Psychological modification and synergetic modelling of perceptual oscillations. In: Haken, H. & Koepchen, H.P. (Eds.): Synergetics of Rhythm in Biological Systems. Berlin: Springer.

LEWIN, K. 1936. Principles of Topological Psychology. New York: McGraw-Hill.

MAHONEY, M.J. 1974. Cognition and Behavior Modification. Cambridge, Mass.: Ballinger.

MILLER, G.A., GALANTER, E. & PRIBRAM, K.H. 1960. Plans and the Structure of Behavior. New York: Henry Holt.

NEBELYTSYN, V.D. 1964. An investigation of the connection between sensitivity and strength of the nervous system. In: Gray, J.A. (Ed): Pavlov's Typology. Oxford: Pergamon Press, 402-445.

PAVLOV, I.P. 1955. Selected Works. Moscow: Foreign Language Publishing House.

PRIGOGINE, I. 1980. From Being to Becoming. San Francisco: Freeman.

RAMACHANDRAN, V.S. & ANSTIS, S.M. 1985. Perceptual organization in multistable apparent motion. Perception, **14**, 135-143.

REITER, L., BRUNNER, E.J. & REITER-THEIL, S. (Eds.): 1988. Von der Familientherapie zur systemischen Perspektive. Berlin: Springer.

SCHNEIDER, H. 1983. Auf dem Wege zu einem neuen Verständnis des psychotherapeutischen Prozesses. Bern: Huber.

SHERIF, M. 1935. A study of some social factors in perception. Archives of Psychology, **187**, 60.

SIMON, F.B. 1988. Unterschiede die Unterschiede machen. Berlin: Springer.

STADLER, M. & KRUSE, P. 1990a. Cognitive systems as self-organizing systems. In: Krohn, W. & Küppers, G. (Eds.): Self-organization. Portrait of a Scientific Revolution. Dordrecht: Kluwer, 181-193.

STADLER, M. & KRUSE, P. 1990b. Theory of Gestalt and self-organization. In: Heylighen, F., Rosseel, E. & Demeyere, F. (Eds.): Self-Steering and Cognition in Complex Systems. Towards a New Cybernetics. New York: Gordon and Breach, 142-159.

STADLER, M. & KRUSE, P. 1990c. The self-organization perspective in cognition research: historical remarks and new experimental approaches. In: Haken, H. & Stadler, M. (Eds.): Synergetics of Cognition. Berlin: Springer, 32-52.

STADLER, M., RICHTER, P. H., PFAFF, S. & KRUSE, P. 1991. Attractors and perceptual field dynamics of homogeneous stimulus areas. Psychological Research, **53**.

STADLER, M., VOGT, S. & KRUSE, P. 1991. Synchronization of rhythm in motor actions. In: Haken, H. & Koepchen, H.P. (Eds.): Synergetics of Rhythm in Biological Systems. Berlin: Springer.

VOGT, S., STADLER, M. & KRUSE, P. 1988. Self-organization aspects in the temporal formation of movement Gestalts. Human Movement Science, 7, 365-406.

VON DER MALSBURG, C. 1984. Algorithms, brain, and organization. In: Demongeot, J., Goles, E. & Tchuente, M. (Eds.): Colloque sur le compartement dynamique des réseaux d'automates. New York: Academic Press.

WIENER, N. 1948. Cybernetics or Control and Communication in the Animal and the Machine. New York: MIT Press.

WUNDT, W. 1898. Zur Theorie der räumlichen Gesichtsempfindungen. Philosophische Studien, 14, 1-118.

Simulation and Empirically Based Models
of Self-Organizing Processes

Simulation of Psychological Processes:
Basic Issues and an Illustration
Within the Etiology of a Depressive Disorder

Harald Schaub and Günter Schiepek

With 18 Figures

Abstract. This contribution is subdivided into two parts. The first section surveys some of the common approaches used to simulate psychological processes by means of models implemented on the computer. (a) Parallel distributed processing and connectionism, (b) the concept of production systems, (c) semantic and propositional networks, (d) differences and differential equation sets and (e) cellular automata will be presented. In the second part, the concept of production systems will be used to simulate a concrete example of the development of a depressive disorder. The recursive nesting of seven relevant variables serves as the basis for this simulation. The variables were obtained by the detailed analysis of an individual case study. The simulation, formulated in 46 *productions* (IF-THEN rules) underwent extensive model testing.

1. Survey of Simulation Concepts Commonly Used in Psychology

Some of the variants and problems of computer simulations in psychology will be discussed in some detail prior to a description of the simulation of a depression etiology.

Aided by the possibilities which the modelling and displaying of theories by computer programs offer, the theorist can evaluate whether the ideas and hypotheses underlying his model plausibly describe the phenomenon he is dealing with. It has to be emphasized, however, that simulation programs on a whole do not represent theories. They include two aspects:

- theoretical assumptions and
- assumptions with exclusively computer-technical relevance.

A computer program contains those theoretical assumptions, data and structures necessary to adequately represent and compute the theory to be simulated. If the simulation can be brought to "run", this is a proof of the consistency of the underlying model. However, the question immediately arises as to whether the model is also valid and whether the simulated assumptions and the simulation's behavior are in agreement with reality. Simulation models necessarily include assumptions about variables or processes that are not directly observable; this makes validity testing difficult. We will return to this problem later.

Springer Series in Synergetics, Vol. 58   Self-Organization and Clinical Psychology
Editors: W. Tschacher, G. Schiepek, and E.J. Brunner   © Springer-Verlag Berlin Heidelberg 1992

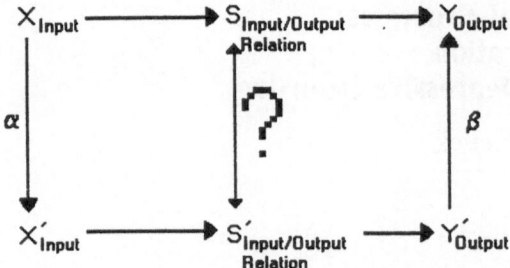

Fig. 1. Relations between a real system S and a simulated systems S'. α and β are the simulation relations (Zinterhof 1987)

Two information processing systems are involved in modelling psychological processes: the human psyche (a bio-psycho-social system) and the simulation of this psyche implemented on the computer. The relation between these two systems can be described as the relation between two triplets (Zinterhof 1987), (X,Y,S) and (X',Y',S'). X stands for the input into the real system's psyche, X' is the input into the simulation. Y and Y' stand for the respective output; S and S' represent the input-output relations.

If the internal structure of a system is not observable, as is generally the case for the human psyche, the connection between the two triplets can only be established by the input and output conditions (α and β in Fig. 1). The adequacy of the model is then estimated using α and β. Thence one can derive the relation between S and S'. Numerous ways exist to transfer the hypotheses held about S to S'. These different options are theoretical frameworks into which the hypotheses are converted, to then be formulated as computer models. A model can then be formulated as a "production system", a "connectionistic net", a "semantic network", a "cellular automaton", a "differential equation set" and so forth. Various approaches are used to describe psychological phenomena. Generally, all the mentioned approaches can be reduced to elementary mathematical functions (this derives from the necessity of computer implementation). Thus the choice of one of these theoretical frames is generally governed more by pragmatic than by theoretical aspects.

1.1 Parallel Distributed Processing, Connectionism and Neural Nets

The concepts of connectionism and neural nets have been widely discussed in recent years. McCulloch & Pitts in 1943 already showed that a network of neuron-like units is able to process information. Hebb (1949) developed ideas about the processes of learning in neural nets. Lashley (1950) generated concepts about the representation of knowledge in memory. He introduced the idea of distributed representation in neural nets. The "Perceptrons" of Rosenblatt (1959, 1962) were the first implementations of all these ideas. They became famous, because they were discussed by Minsky & Papert (1969), whose critique was fatal, since by a mathematical analysis of the devices they could prove that information could not be processed in the depicted fashion.

The difficulties faced by traditional AI in constructing intelligent systems and the problems in modelling psychological processes in the cognitive sciences again drew "theoretical neurology" into the center of interest. Having tried to adjust models of psychological processes to the serial architecture of von Neumann computers for quite some time, new programs and psychological models were oriented toward the parallel and distributed network mechanism of the human brain. Grossberg (1976,1982) and Marr & Poggio (1976) presented work on parallel and distributed information processing (PDP). The terms "PDP" (Rumelhart & McClelland 1986) and "connectionism" subsume all the efforts in the area. Despite the common name, a single connectionism does not exist. All of the approaches do however have certain concepts in common. The functional mechanism is based on the neural structure of the human brain, either in the sence that physiology offers notions for connectionistic systems or vice versa, that connectionistic systems offer possible ideas as to how human information processing might look.

The underlying, micro-level structures in a neural net are the processing units, idealized neurons, which connect to other units by way of idealized axons. The information processing activity of a single unit is generally very simple. The unit has a certain state of activation, and it receives activity from the units it is connected with. From the incoming information and depending on its own activation, the unit by way of its activation rule calculates an output value. This value is then sent to the units it impinges on. Changes in a neural net can be implemented by changes in the structure of the net, for example via a change of the unit number, changes in the number of connections or in the connection strength, or changes in the learning rule of the net. To retain the analogy with the human brain, generally only the connection strength is varied. Fig. 2 shows a simple neural net. The basic assumptions of connectionistic systems (Strohschneider & Schaub 1991) result in a number of characteristics which are of interest from the psychological viewpoint, and which we will discuss in the following.

The pre-eminent characteristic of neural nets is their error tolerance. The operating mechanism of connectionistic nets in the case of a faulty or incomplete input was called "graceful degradation" by Rumelhart & McClelland (1986). The system even for degraded inputs (up to a certain degree) will produce the answer expected for a correct and complete input. The system does not break down, but instead produces an output corresponding to the respective input. This error tolerance results from the network features and is not explicitly implemented (and to implement it would be an extremely arduous undertaking). For a correct input a connectionistic net works like a classical AI-system, it acts like a production system, even though it does not contain any explicitly stored rules. Spontaneous generalization as well as abstraction also result from the net structure and do not have to be implemented. If a neural net is confronted with a number of inputs which all are variations of a basic type, the system produces a

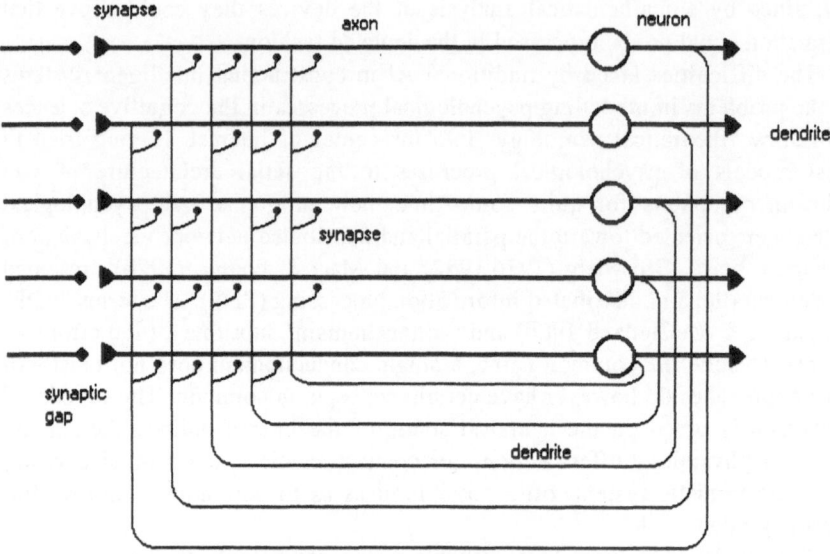

Fig. 2. Neural network (Rumelhart, McClelland 1986)

pattern of connectivity representing those features common to all of the inputs. Thus, an underlying general type is produced which corresponds to the input patterns leaving out their individual variations.

The representation of knowledge in psychological modelling is of central interest. In classical memory models of the type of semantic networks, every node is semantically interpretable, relations between the nodes are also semantically interpreted. Connectionistic systems store information in quite another way. Single units of a neural net are not semantically interpretable. They are just carriers of activity. Semantically interpretable knowledge is stored in the pattern of activity all over the net. Every unit therefore contributes with its information to the whole pattern and therefore to the overall meaning.

Tied to knowledge representation of course is the question of memory processes. Here it has to be mentioned that for neural nets the usual distinction between the structure and the processes working on the structure is no longer sensible. The different basic concepts of neural nets can immediately be transformed into memory processes. The activation and propagation rules of the system are also the inference rules which the system follows in processing the input. The rules which determine the connection strength between units are the learning and the memory processes which determine changes in the system. These aspects of cognitive processes are implicit parts of connectionistic nets.

Of course, connectionistic models also face considerable problems. The main problem is the arbitrariness in specifying important net parameters, which generally not can follow any psychological plausibility. Propagation rules, thresholds, learning- and decay-rates simply have no psychological equivalents. This leads to a "TWITIT philosophy" (Lehnert, 1987: tweak it till it thinks). For every input-output combination countless connectionistic architectures are conceivable. This is a problem all modelling approaches have to face, but in the case of semantic networks or production systems it is possible to transfer elements of the model to other theories and thus test their plausibility, whereas connectionistic nets can only be considered as a whole.

1.2 Production Systems

Production systems, in contrast to connectionistic models, formulate theories on a high level of abstraction. For this purpose a number of assumptions have to be made, in particlar, the assumption that psychological processes are representable by a process of executing rules.

The term "production system" is used in various ways: first of all it characterizes a formal language to describe algorithms. Secondly, it is used as a psychological concept for the architecture of the human cognitive machine, and, thirdly, the term refers to program systems or programming languages (as for example in Ohlsen & Langley 1986).

Production systems were first introduced in the 1940s by Post (1943) as models to describe algorithms. The adaption of production systems in psychology was basically brought about by the work of Newell (1973). ELIZA (Weizenbaum 1966) was a program demonstrating that by assuming a few rather simple rules a quasi-natural dialogue between man and machine is possible. The ACT model (adaptive control of thought) of Anderson (1983) can be considered as the most elaborate and well-known production system serving psychological theory formulation.

Production systems are useful in psychological theory formulation since it is possible to construe rather complex models of human information processing activity with them. For this purpose it is "only" necessary to formulate the knowledge relevant for the problem space in the form of IF-THEN rules. Thus a set of different rules is obtained. If used for psychological facts these rules are never really exclusive in their conditional part. This means that under a certain condition or in a certain situation more then one of them can be applied. Therefore preparations have to be made for cases in which two or more rules apply, and for cases where not a single one of them can be applied.

For the rules to be correctly adapted to the surrounding situation, data are required about the world and of course about the actual situation. This means that the system, besides a number of rules, also needs a lot of data. Finally the application of the rules and the incorporation of the data have to be coordinated; knowledge is therefore necessary exclusively for the purpose of coordination.

We have now mentioned the essential parts of production systems. These are:

- the *working memory*, which contains the system's declarative knowledge (knowledge about what to do)
- the *production memory*, which offers the procedural knowledge (knowledge

how to do). The conditional part of the rules is compared to the working memory, if this comparison renders a "true" result to the action part of the rule, the operation is executed.

- the *interpreter* contains the system's control-knowledge on the application of declarative and procedural knowledge. The interpreter serves three functions:
  - analysis of the conditional parts of a production (pattern matching) and determining the truth value for the conditional part
  - selection of one or more of the productions from the set of applicable productions (conflict resolution)
  - execution of the operations laid down in the action part of the production. These productions can be directed to the system itself as well as to the outside world.

  The operating mechanism of the interpreter is called the "recognize-act" cycle.

A number of psychologists consider production systems to be the most elaborate way of simulating psychological processes. In their opinion the programming language to construct a production system model should be at least as advanced as PROLOG. Much better is a "real" production-system programming language like PRISM (Ohlsen & Langley 1986) or GRAPES (Sauers & Farrell 1982). This means that a language should be chosen which in itself offers as many production system constructs as possible (pattern matching etc).

Production systems offer a number of advantages. The possibility to easily model a variety of psychological phenomena, for example Schmalhofer & Polson's (1986) solution to the missionary-cannibal problem. From the assumptions made about the problem space a set of rules can be almost directly derived which can be executed within a production system. This, on the other hand, makes use of the fact that a production system, in contrast to a neural net, in its inner structure remains psychologically understandable. This offers the opportunity to test the inner structure not against reality but against other theoretical concepts. The simplicity in formulating productions on the other hand might result in almost arbitrary assumptions. Productions of a different "resolution level" can exist side by side. If an unexpected system behavior occurs, a new production can immediately be created and so forth. Production systems are a powerful tool for creating models. The formulation of a theory in the form of a production system is, however, very abstract and very far from the physiological level of implementation. The gap between neural nets and production systems is immense. By deciding to model with production systems one has decided for a relatively small scope of data and especially processes that can actually be modeled. Psychological processes are described by the processes of "pattern matching" and "conflict resolution"; other information processing activities are not presupposed.

1.3 Semantic and Propositional Networks

Another means to model mental structures and processes is via semantic and propositional nets. The basis of propositional nets is predicate calculus. A predicate is a relation between arguments. These arguments might be either constants or variables. A predicate with included arguments is a proposition. In the proposition "subject 49 watered field 1" the relation "watered" exists between subject 49 and field 1. Therefore in this case the arguments are constants. Generally, in the literature, propositions are written with preceding predicate, for example in the form: 'water(subject_49, field_1)' or '(water, subject_49, field_1)'. A proposition is either true or false; this means that two truth values exist, one of which the proposition actually has.

Predicate logic offers tools to describe facts by propositions. By way of these tools propositions can be connected. The truth value of a statement which was generated by way of connecting elements can be read from a truth table. P and q are both propositions, t and f are truth values, 'and', 'or', 'implies', 'not' are the possible connectors. The truth value of a statement therefore derives from the truth values of the single statements dependent on the type of their connector.

Table 1. Truth table

p q (and p q)	p q (or p q)	p q (implies p q)	p (not p)
t t t	t t t	t t t	t f
t f f	t f t	t f f	f t
f t f	f t t	f t t	
f f f	f f f	f f t	

Quantors and functions are further possibilities to process propositions. 'For all' and 'if exists' are quantors; examples are "for all x p is true" or "at least one x exists for which p is true". Further and more sophisticated inference rules beyond the truth table also exist, such a the "Modus ponens" or "Modus tollens".

- Modus ponens: if p implies q and p is given then q is also given. An example for this is: "If it rains, the street gets wet. It is raining, therefore the street also gets wet."
- Modus tollens: p implies q and q is not given, then p is also not given. Such as: "If it rains the street gets wet. The street is not wet, therefore it also did not rain."

Predicate calculus is employed to depict cognitive processes. Lists of propositions are in this case the data structure to which processes and inference rules are applied. What, for example, would the following sentence look like in predicate calculus?

1) "Freda buys two rolls and a bottle of milk in a small corner-shop because she is hungry. (After Anderson 1980, quoted by Wender 1988, p. 64)
   First of all the sentence has to be dissected into underlying propositions.
2) Freda buys rolls and milk in a shop

3) the number of rolls was two
4) the shop was small
5) Freda was hungry
6) (5) includes the reason for (2)

These sentences are then transformed into propositional syntax.

7) buy Freda milk and roll shop past tense
8) number roll 2
9) size shop small
10) state Freda hungry
11) reason (10) (7)

This example immediately shows two things: First of all the representation of more complex knowledge by long lists of apparently unconnected statements is confusing and counterintuitive. Secondly this kind of information representation is typical for computers, which can process information of the presented form very well.

Because of the lack of clarity such proposition lists are often represented graphically thus forming semantic networks. Semantic networks have the form of directed marked graphs. The graphs consist of nodes and edges. The nodes represent the propositions and the arguments of the propositions. The edges represent the relations between the nodes (Fig. 3).

Semantic networks are graphical representations of propositional lists. Except for the differences in clarity and the possibility of implementation on the computer no fundamental difference between the two concepts exists. Semantic nets, however, are often extended far beyond the frame of predicate calculus and therefore allow the theorist numerous degrees of freedom. Thus nodes for example might be restricted to facts only, or facts and propositions might be

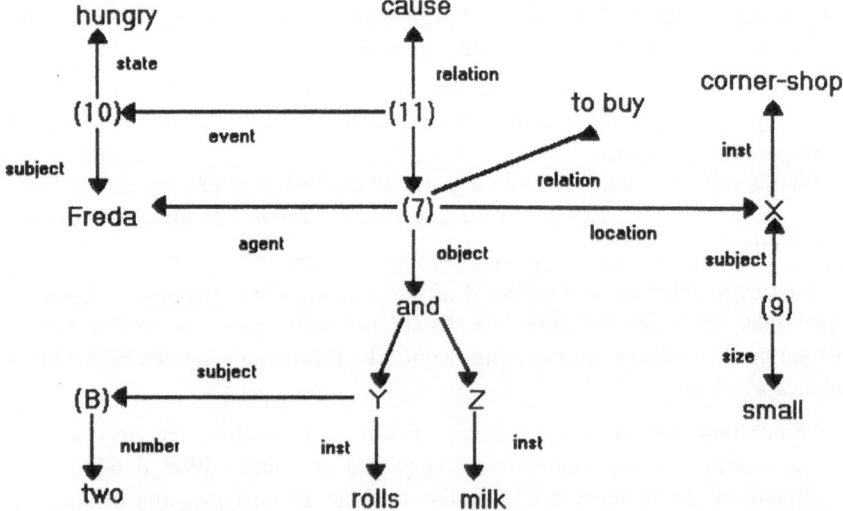

Fig. 3. Semantic network

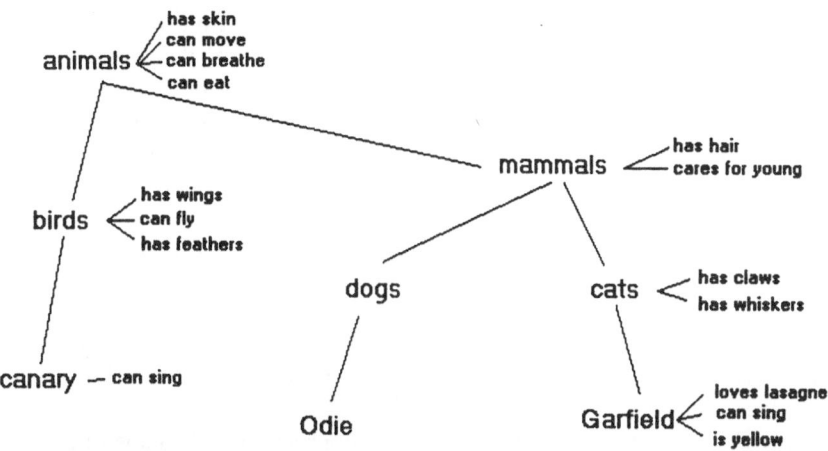

Fig. 4. Semantic network (Collins, Quillian 1972)

admitted, different edges might also be defined. One of the earliest approaches in this area is the theory by Collins and Quillian (Quillian 1966, 1969; Collins & Quillian 1972; Collins & Loftus 1975). In this theory the aspect of hierarchy which can be build up within semantic networks is especially stressed.

The authors distinguish two kinds of nodes and two kinds of edges. There are object nodes (animal, cat, Garfield) and there are feature nodes (has skin, claws and is yellow) (Fig. 4). Collins and Quillian distinguish between edges defining class membership (Garfield is a cat) and edges which connect facts and features (Garfield is yellow). Features which are assigned to a generic term automatically also apply to all subsumable concepts, without having to be named there explicitly. This makes economical storage possible; exceptions require extra mention if necessary. Thus ostriches belong to the class of birds even though they cannot fly. Smith et al. (1974) introduced the concept of feature lists. According to this concept every object has a list of defined features which identify it as a member of a certain class and a list of characteristic features which are typical for such objects but do not necessarily have to be given. Thus all birds lay eggs (defined feature) and most but not all of them are able to fly (characteristic feature).

Tulving (1972) introduced the distinction between a semantic memory (containing abstract, rule-based knowledge) and an episodic memory containing individual events. The knowledge that birds lay eggs can be ascribed to the semantic memory, the statement that Henry is the budgerigar of my grandma is episodic knowledge.

Semantic networks are often combined with other approaches. Generally the assumptions about the structure of data storage are of the type of semantic networks on which certain processes, as for example production systems work. The diverse possibilities of semantic networks do however also have their disadvantages.

"A difficulty with semantic network theories results from their enormous flexibility. This flexibility stems from the fact that for this approach no restriction exists on the number and the type of the introduced nodes and relations. Even more flexibility results if varying strength and activity is assigned to the nodes and edges. If different processes are then also allowed to work on the net, the resulting theory is so flexible that it cannot be tested any more." (Wender 1988, p. 71)

We have already met this argument in relation to network approaches. A further problem of semantic nets is the fact that no excess exists to the external meaning of the objects stored in the net. Nodes in the net only receive internal meaning, i.e. only references to other nodes in the net. The nets therefore are disjoined from reality. The node "dog" does not have any connection to the really existing animal.

Semantic networks make plausible assumptions about the storage of data in human memory. Knowledge is stored in terms (Hoffmann 1986) which can have various relations to each other (Klix 1984). The question of how pictorial information is stored was for example discussed by Tulving, who failed however to find an answer to it. Not many assumptions are made about the processing of information; thus the combination of semantic networks and production systems has been implemented in numerous theories (Anderson 1983).

1.4 Simulations Based on Difference and Differential Equation Sets

Difference equations formalize the change of discrete variables over time, that is, from one time-step to the next ($\Delta x = x_{t+1} - x_t$). Differential equations on the other hand are chosen to describe continuous variables and their change over time. The discrete or continuous change of a variable in these models is formalized by a mathematical function of the variable itself or other variables which are part of the system at time t or an earlier time t-τ (delay term); for example:

$$\dot{x} = f(x_t, y_{t-\tau}, z_t)$$
$$\dot{y} = f(x_{t-\tau}, y_t)$$
$$\dot{z} = f(x_t, y_t).$$

For difference and differential equations *variables* and *parameters* are distinguished. The equation model describes changes in the variables and, if no stochastic processes are involved, determines them. Parameters are the constants of the equations. Equation systems are abstract and formal models of dynamic systems. As such, they cannot be claimed to validly model real psychological phenomena, whereas for example connectionistic models follow the real functioning of neural nets. To simulate psycho-social processes (as for example in the frame of a structural model; refer to Schiepek, Schoppek & Tretter, in this volume) it first has to be determined which variables are connected by way of autocatalytical or crosscatalytical, inhibitory or excitatory relations, as well as which parameters should be included into the system. Generally, plausible system dynamics only result for a certain range of parameter tunings. Here again the TWITIT philosophy applies.

Non-linear difference and differential equations are of importance in the area of self-organisation and chaos. Here variable changes over time do not follow a linear function, but result from non-linear processes of two or more variables (or one variable with itself). In the area of deterministic chaos, computer-based numerical solutions of non-linear equation systems are often necessary because analytical solutions do not exist (Haken 1983; Seifritz 1987; Schuster 1988).

It is noteworthy that even though difference and differential equation models not only in the classical natural sciences but also in biology, medicine and economics belong to the fundamental methodological tools, in clinical psychology they are still exceptional. Clinical psychology still basically leans methodologically toward the general linear model of univariate and multivariate inference statistics.

1.5 Cellular Automata

Cellular automata are discrete, dynamical systems which, by employing simple local rules (on the micro-level), are able to produce complex patterns on the macro-level. Hayes (1988) names the following four features of cellular automata:

1) The automaton consists of a space, subdivided into cells. Normally these cells are ordered as a grid, variations of the cell forms, honeycombs for example, or variations of the dimensions (for example three-dimensional cell structures) are possible.
2) Each cell has a determined neighborhood from which it can be activated and thus changed. In rectangular formations the von Neumann neighborhood (only those four neighbors sharing edges with the cell are possible) and the Moore neighborhood (neighbors are also those four cells sharing the corners with the cell) are customary.
3) A cell can have different states; in the simplest case it is set in a binary fashion.
4) The state of a cell is determined by the constellation of the states of the neighboring cells. In "the game of life", a cellular automaton by Conway, a cell can be either dead or alive (binary). A Moore neighborhood is implemented. If a cell is alive it stays alive if two or three other cells are also alive in its neighborhood. A dead cell becomes alive during the next iteration if at least three neighboring cells are alive. In all other cases it dies or remains dead. This results in a complex process of emergence and decay of structures. Depending on the starting condition, static or oscillatory structures develop (Eigen & Winkler 1975).

Cellular automata are suitable for simulating the reciprocal activation or inhibition processes between sub-systems of a system (for example, the signal exchange or the path of infection in a cell culture). A typical example is the wave pattern of the chemotactical activity in the cellular layers of slime moulds. The cells of the moulds are able to send cAMP in pulses into their surrounding and to strengthen such impulses. This leads to a collective emission of chemical pulses, which spread from a center in the form of concentrated waves and produce a concentration gradient of the cAMP.

In psychology cellular automata are seldom applied. Tschacher (1990) proposed to simulate the nearness/distance regulation in groups on this basis (Tschacher et al. in this volume). For dynamical systems theory it is of importance that in cellular automata phenomena of symmetry breaking, of order development, of fractality and chaos might occur (Wolfram 1984; Toffoli & Margolus 1987; Hayes 1988).

The approaches presented here of course do not represent all the possibilities to theoretically capture cognitive processes. There are further theories, for example the script approach (Schank & Abelson 1977), the frame approach (Minsky 1975) or scheme approach (Bartlett 1932) as well as theories of human cognitive processes which fit into one or the other theoretical frame (Anderson 1983).

2. A Simulation of Depression Development

2.1 The Model

A production system was chosen for the purpose of the present simulation because this approach seemed most appropriate to transfer the existing model of a depression etiology into a computer program. We wanted to save the psychological plausibility of the single aspects of the model as far as possible. In the present case this was most easily achieved by transforming the linear parts of the originally ideographic model (for the method of ideographic system modelling see Schiepek 1986) into a system of IF-THEN rules. We did not use a formal production system language for this purpose; instead we implemented the model in PASCAL. We have some reservations about production system languages because choice and control processes cannot be directly modelled in these languages. A "classical" programming language is more open in this respect because all the control and choice processes have to be programmed anew.

The production part of the model consists of a number of productions, all having the same structure. They have a conditional "if" part and an action part "then" and sometimes an additional action part "else". The action part is executed if the condition is true, the optional action part "else" is executed if the condition is false.

The control structures of our model are intentionally kept simple. The "interpreter" part chooses all those productions for which the conditional part is true at a given time. The action parts of the productions are then executed in the proper production sequence. If for example at a certain time the conditional part of production 34 is true, in other words the conditions formulated for this production correspond to the data in memory, this production is executed. Afterwards, for production 35 it is checked whether the conditional part is true. If this is the case, the action part of this production is executed as well. The next step is checking the next production and so forth. This procedure has the effect that productions which are checked later in time already have a different surrounding to the ones checked earlier. Production 2 works with the memory already changed by production 1, production 3 again uses a memory changed by 1 and 2.

The simulation is based on detailed biographical and casuistical material, gained from and with a young adult at the psychological therapy center of the University of Bamberg within a number of therapeutical sessions. The client was a 24-year-old farmer who had always lived (except for the time of his military service) on the farm of his parents. His parents expected him to take over the farm. The client felt that too much was expected from him by all the duties he had on the farm. When he came to our therapy center he brooded about alternative professional careers with increasing intensity; however, he felt unable to realize them. As a whole, he felt an inner vacuum, fatigue, fear and thought that he could not take a lot.

For research purposes the simulation was based on a system model which was constructed in a very detailed fashion. All accessible information was taken into account (interviews, questionnaires on the life history as well as on problems). Following the construction, the number of variables in the system was however reduced again. Seven variables were chosen:

1) the *demands* made by the parents and by the duties on the farm
2) *realization* of these demands in the sense of adjusted functioning
3) development of certain *schemes* (about the self as well as negative emotional schemes) as special factors of vulnerability
4) *fear of failure* concerning the amount and difficulty of work on the farm as well as in his independent coping
5) *coping efforts* to reduce and cope with the fear of failure
6) *motivation* toward a success-oriented occupation with different tasks, going hand in hand with own initiative
7) *depression*, of the status of illness, concerning the experiences and behavior patterns occurring as well as the self-description of the client and his description by his social surrounding.

According to a dependency-effectiveness analysis (see Dörner et al 1983,p. 399) and based on the considerations of further cybernetic criteria of the original system model, these seven variables proved to be very influential. They, so to speak, explained most of the variance. Further variables either covaried with

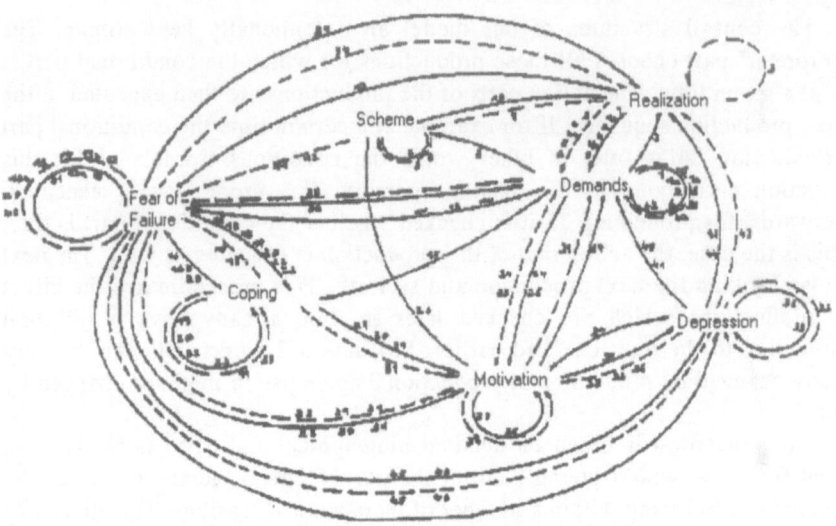

Fig. 5. The net structure between the variables incorporated in the system. The figures indicate the numbers of the productions (IF-THEN rules)

those mentioned in the sense of a dependency or it was found, that a further "fine tuning" of the original model by means of one or more of these variables did not produce important consequences for the system dynamics. The family dynamics for example were neglected as they did not actually push the system dynamics into another direction, and complexity reduction was absolutely essential. This decision was supported by findings (Reiter 1988) which cast doubt on a direct influence of the family structure on the development of depression. Generally, we tried to choose those variables which in the clinical literature are most often mentioned as relevant to etiology. Fig. 5 shows the recursive nesting of the seven variables.

The simulation runs over 4000 time-steps. Thus for every variable thus a time series is generated of a length that could not be gained empirically. One day can be considered as a plausible real time reference for one time-step. Every step in this time series therefore characterizes the hypothetically measured state of a variable on one day. The total development covers a time period of about 11 years, which corresponds to the depicted biography of our client. We are considering the period between ages 14 and 24. Corresponding to that the DSM-III (Wittchen et al., 1989 p. 232) defines the common onset of dysthymic dysfunctioning for early adulthood. The variable "depression" in the simulation increases after about two thirds of the time, in other words a comparatively long development beforehand seems essential.

Every variable at every instant shows a mean and an actual value. The mean represents the respective level of the variable around which the actual value of the variable oscillates. Depending on the production rules the means of the different variables undergo changes which can even follow each other in rapid succession. On the whole the level of the actual variable state serves to stabilize the respective developments.

The productions incorporated in the program determine the variable values at time t, given the value at time t-1, or sometimes dependent on t at a much earlier point in time or long-term developments of other variables or variable constellations. Every variable receives a random value prior to each new iteration. This value can deviate from its value at time t-1 by 1, -1 or 0. This random process represents the ever-present spontaneous fluctuations, comparable to thermodynamic molecular movements, which cannot be attributed to the mechanisms implemented in the model (Bischof 1985, p. 427).

Due to the limited space here we cannot specify all 46 productions determining the interconnections between the seven variables. A detailed description can be found in Schiepek & Schaub (1991). Here three productions will be described in more detail: Production 10 for example describes the psychological fact that not only the concrete occurrence of demands can lead to efforts towards their realization, but that certain inner motives can also produce this effect. The activation of some inner schemes and the respective action plans ("show your willingness, be of use, try to read other persons wishes form their eyes") might be sufficient to initiate demand oriented actions. The rule is: "if variable "scheme" increases by more than two scale points (SP) from t-1 to t, then variable "realization" should also increase by two SP".

Production 15 takes into account the fact that "demands" and even only a minor degree of "fear of failure" can motivate considerable coping efforts. If the degree of "fear of failure" is below a threshold value of 10 and if both "demands" and "fear of failure" increase by one SP from t-1 to t then the variable "coping" should also be increased by one SP. Rule 38 describes the fact, that anxiety not only serves as a signal in case the self is threatened, but can also be a defensive strategy to self-presentation. An increase in "fear of failure" by at least 2 SP from t-1 to t under certain conditions reduces the actual "demands" from relevant peers for the same interval by 2. For this to happen a mean "fear of failure" of above 10 is required as well as a mean in "demands" of above 20. The frequency of anxiety increases necessary for the reduction in "demands" is defined by the relation of the actual value of "demands" (numerator) to "fear of failure" (denominator). If the amount of "demands" is relatively high, production 38 is generally activated less often; if denominator and numerator are closer together it is activated more often. All this is based on the assumption that a high level of anxiety serves the communicative effectiveness of the emotional self-presentation well.

The total simulation model is constructed in a way to best depict both the individual biography of the client as well as the actual knowledge of psychology in the area, applying 46 productions (hypotheses) for this purpose. Psychology, as well as its neighboring disciplines, therefore constitutes the sources from where hypotheses and diagnoses can be taken to construe an ideographic system model for a single case. The rules are taken for example from results from the area of clinical depression research (for a survey refer to Lewinsohn et al. 1985;

Alloy et al. 1985; Reiter 1988; Hautzinger & de Jong 1990), from learning theory (for example Lewinsohn 1974), from schema theory (for example Grawe et al. 1987), coping research, research concerning self-presentation and impression management, social psychology as well as family research. Of course an even finer level of resolution might have been chosen (as for example by taking biochemical, physiological or more detailed psychological descriptions into account). This was avoided for reasons of clarity.

The rates of change of one or two scale points up or down per iteration, are 'compared to the total range of variable (from 0 to 45)' small and plausible but arbitrary. It is therefore worth mentioning that, despite this arbitrariness on the level of these single rules (micro-level), a psychologically plausible dynamical pattern of the system dynamics is produced on the macro-level. The total system dynamics also shows behavior that was not implemented on the level of the individual rules, a phenomenon, which is generally called emergence (see sect. 2.2 on model testing).

The starting values of the variables were set as follows: demands 3, realization 1, all others 0. Figs. 6-8 show the simulated developments of the variables. Fig. 9 shows a part of the development of the variable "fear of failure" in a (t, t-1) diagram.

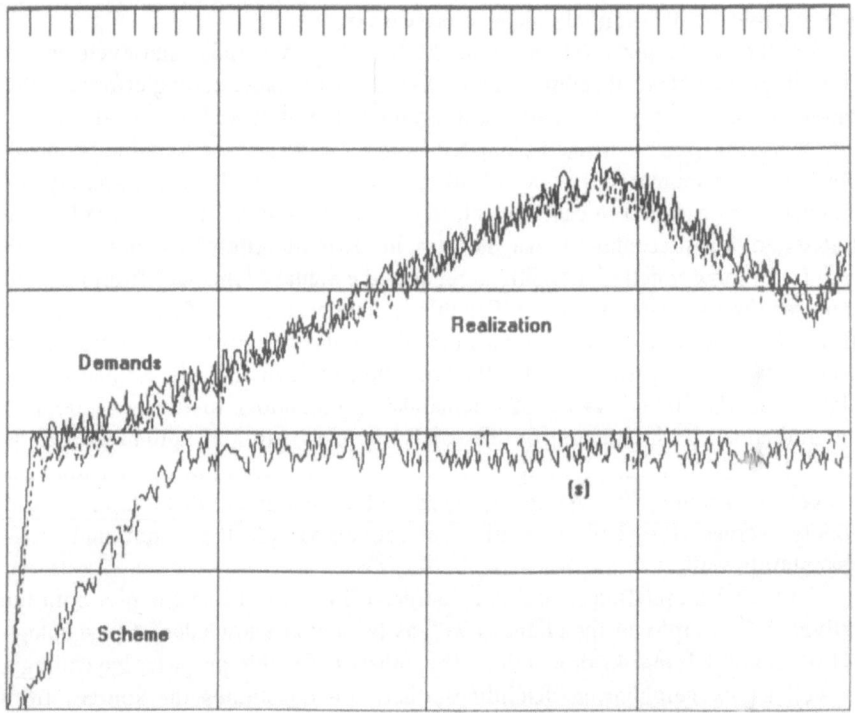

Fig. 6. Developments of the variables "demands", "realization" and "scheme" over all 4000 time-steps. (In this as in the following charts:x-axis 0 to 4000, y-axis 0 to 50)

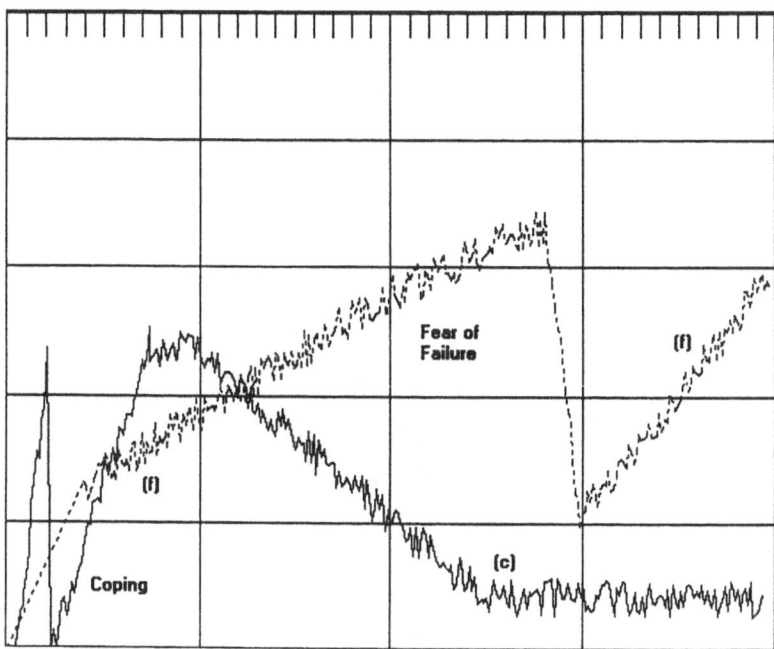

Fig. 7. Developments of the variables "fear of failure", and "coping" over all 4000 time-steps

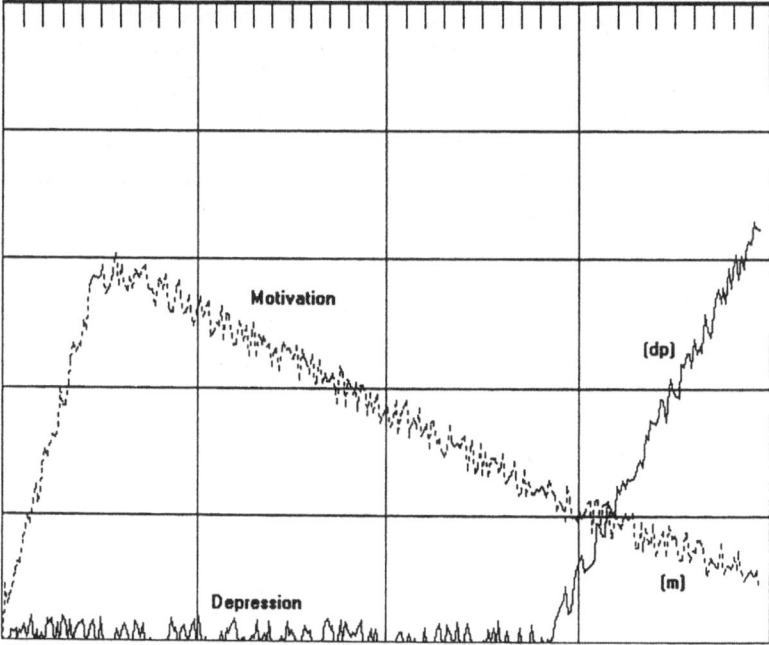

Fig. 8. Developments of the variables "motivation", and "depression" over all 4000 time-steps

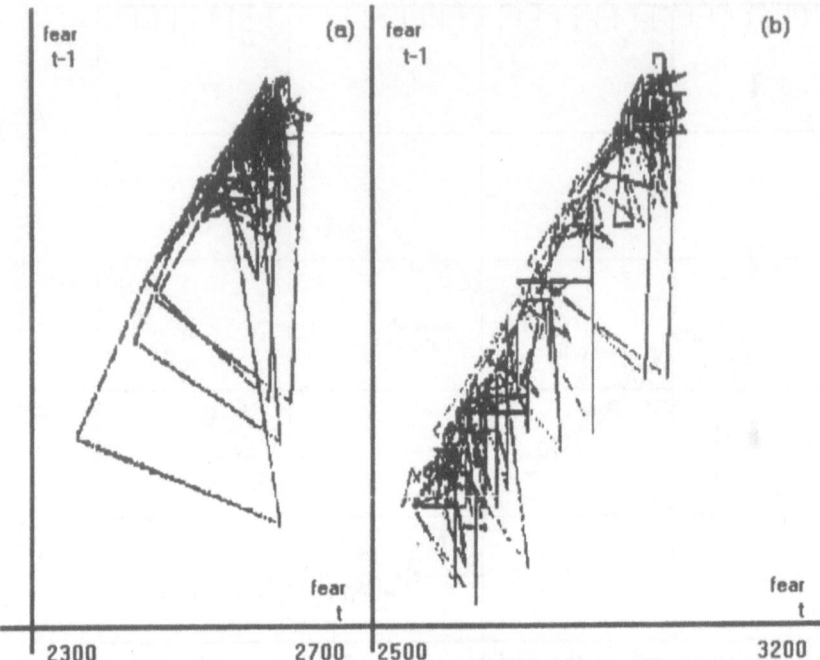

Fig. 9. (a) Development of the variable "fear of failure" for time-step 2300 to 2700 in a (t,t-1) diagram. The shape does not close into a limit cycle. The development therefore does not represent a returning periodicity, but seems like chaotic behavior. What makes this process more complicated is the shift in the variable level. (b) With the onset of depression around time-step 2700 "fear of failure"(v) topples over into the reach of another attractor (from the top at the right to the bottom at the left), which however proves to be unstable. The edged form of the "curves" is produced because the simulation program produces discrete, not continuous time series

2.2 Model-Testing

Attention has already been drawn to the problems of validating computer simulations. Even if empirical data series exist, this does not in itself render a convincing validation criterion. This is even more difficult, if, as in our case, no empirically gained time series of the variables included in the model are available. This forces the researcher to check the model for inner inconsistency and for the plausibility of its behavior, especially after interventions (for example changes of variables or rules) have been made.

The unmanipulated developments can already be considered as quite plausible. The variables "demands" and "realization" show a sharp and quick increase, which corresponds to their positive feedback in this phases. Later their development is more attenuated. With the onset of depression "demands" and "realization" considerably decrease; however, this is not a persistent state. "Fear

138

of failure" shows a similar development. Its level increases negatively accelerated, till by onset of the depression a significant fear reduction is possible, which then, however, gives way to a longer-term increase.

The negative feedback between "fear" and "coping" is noteworthy: in phases of relative stability it almost leads toward a limit-cycle behavior of "fear of failure" (Fig. 9). However this is not a matter of recurrent periodicity which would have to generate a complicated but closed loop in the (t,t-1) diagram. The behavior in this diagram rather resembles a chaotic attractor.

Like the variable "coping" the variable "motivation" first increases and then slowly falls again. With the onset of depression, further downward motivational deviations occur. After reaching its maximum value of 20, the variable "scheme" deviates only slightly from this maximum, which can be interpreted as a constant disposition. The variable "depression" first of all adopts an escalating, after that however a negatively accelerated development. The onset of the depression marks a phase transition, a new order of the system dynamics, for which the depression might serve as an order parameter. It should be noted, that the change in system dynamics is not generated by external input.

Let us now consider the consequences of well-defined interventions in the system dynamics. As the dynamics in the beginning was determined by the positive feedback processes between the variables "demands" and "realization", the question arises of what would happen if "demands" were initially kept low. In answering this question we observe a critical instability between the "demand" means of 10 and 11. If the mean for "demand" is kept at a level of 0 to 10 over 500 time steps, no depression occurs (Fig. 10). This phenomenon is mediated by the variable "motivation" which under these conditions does not reach sufficiently high values to make a decrease of 15 scale points possible. This would, however, be a prerequisite for a depression to occur. If, however, the mean for demands is kept at 11 scale points over 500 time steps, a depression occurs with all the known consequences for the development of "demands" and "realization" (Fig. 11).

Furthermore, the developments in Figs. 10 and 11 illustrated the special depression concept underlying this simulation. It is not the absolute degree of stress caused by demands that play the most significant role, and neither it is the low degree of motivation (for example: low demands, minor interest). All this would only lead to a reluctance to develop initiative and would leave the motivation for one's own performance to external demands. The relative *reduction* of motivation is what is really of importance: loss of happiness, demands which are impossible to fulfil, insidiously growing resignation, diminishing of hope.

The question thus arises of whether depression could be prevented if the demands stemming from parents and peers could be reduced (for example on the advice of the family physician) as it might well be concluded that these demands "caused" the problematic development. To test this, the mean for demands was reduced by 50% of its value at t=1000 for time-steps 1000 to 2000 (Fig. 12). The simulation then shows that "demands" are not in fact the critical variable for the

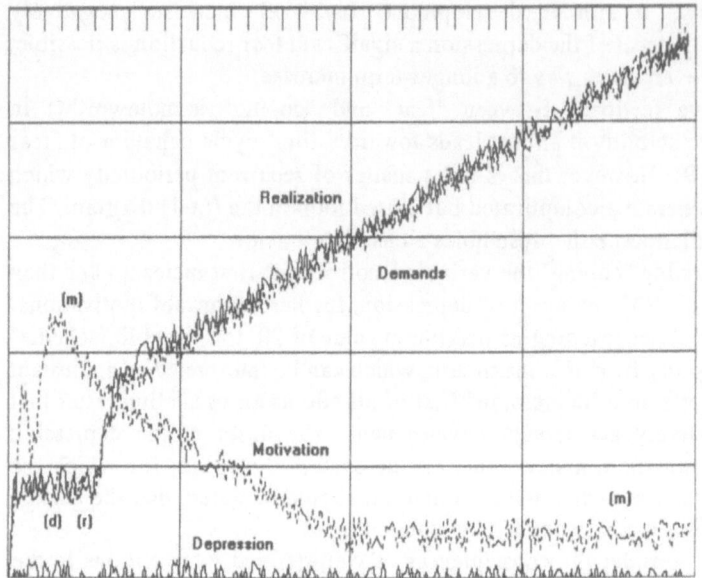

Fig. 10. Developments of the variables "demands", "realization", "motivation" and "depression". The mean for "demands" was kept at value 10 for the first 500 time-steps

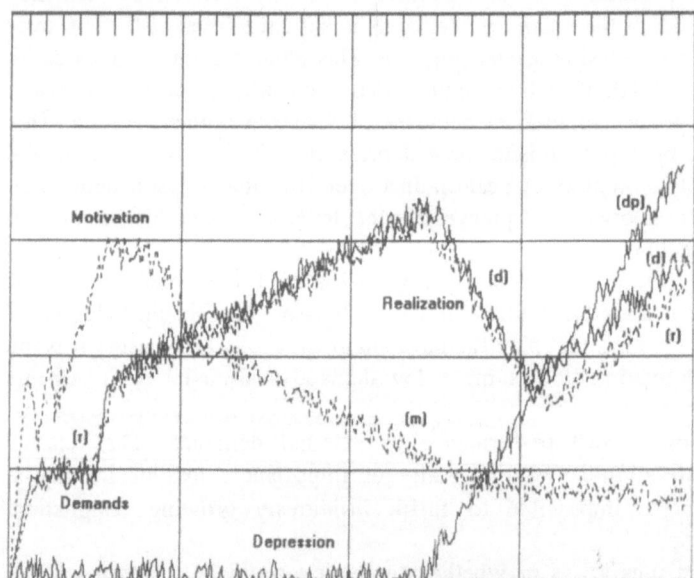

Fig. 11. Developments of the variables "demands", "realization", "motivation" and "depression". The mean for "demands" was kept at a value of 11 for the first 500 time-steps. In comparison to Fig. 10 the consequences of this minor starting difference for the further system development become apparent (critical instability)

140

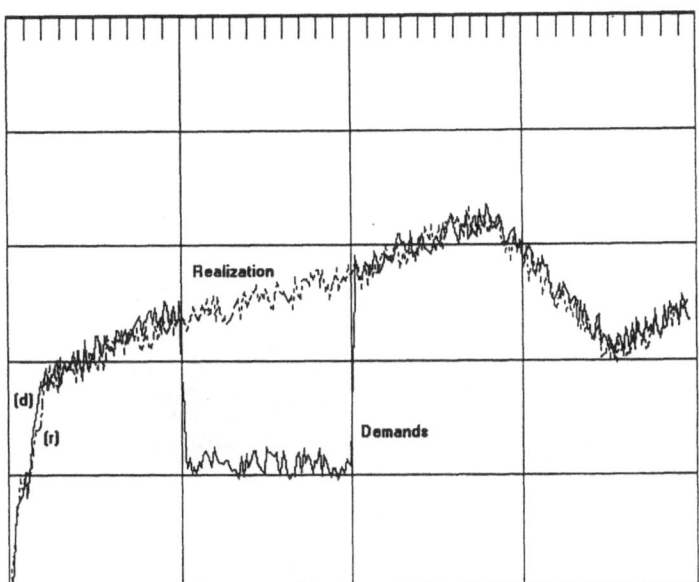

Fig. 12. The mean for "demands" for time-steps 1000 to 2000 was fixed at half of the value it had at time-step 1000

development of the depression since the total development of the system is not much changed by the decrease of demands, at least within this period of time. The willingness for realization seems to have dissociated from the real demands and instead have been internalized. What is prohibited within time-steps 1000 to 2000 is merely an increase in this internalized need to fulfil demands. The intervention cannot, however, prohibit the onset of the depression.

The effect achieved by decreasing the mean of the variable "scheme" to 3 at a certain point is also comparatively minor (Fig. 13). Within a time span of 800 steps it grows back to its original maximum of 20. If schemes are considered to be a central organizing principle of the self, this system behavior is not satisfactory. More effect would have been expected. It should be taken into account however, that the living conditions as well as the learning mechanisms and psychological experiences did not change, and this would indeed make a relapse into old and well-established behavior mechanisms and schemes seem very likely. It can also be expected that the change of a numerical value is not apt to actually effect structural changes. This is because psychological organization principles should not happen at the surface but in the depth of the personality, which means that for the purpose of the simulation the rules would probably have to be changed. A preliminary change of the mean does not actually mean that alternative schemes are developed (consider the therapeutical postulate that it is not the reduction of problems which helps but rather the construction of alternatives).

A drastic interference with the level of "fear of failure" also shows that an exclusive value change does not necessarily lead to a major change in

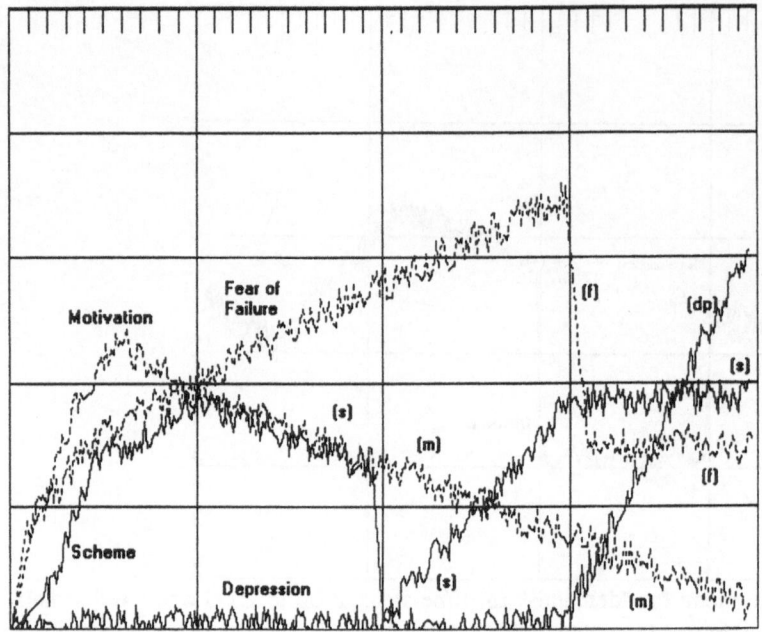

Fig. 13. The effect of decreasing the mean of the variable "scheme" to a value of 3 at time-step 2000

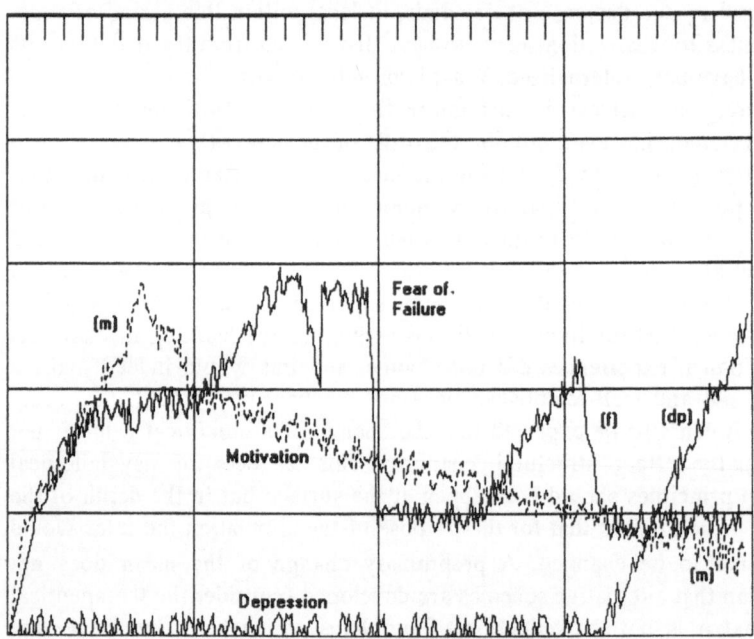

Fig. 14. The mean of the variable "fear of failure" for time-steps 2000 to 2500 was reduced by half of its level at time-step 2000. Here the effect on the variables "fear of failure", "motivation" and "depression"

development characteristics (Fig. 14). This variable was reduced by half of its previously observed value for the time-steps 2000 to 2500. Afterwards it was again left to itself. It immediately increased again. The depression development could only be postponed by this interference but not prevented. This surprised us at first, since removing fear is supposed to allow for a totally different psychical and interactional dynamics. However, it was plausible at second sight. Without a fundamental change in the living conditions, a sudden reduction in the experience of fear should not really occur. (Refer for example to conditioning models which rather plead for a further manifestation of fear over time, Mowrer 1960). The effect of our intervention therefore comes closest to the application of psychopharmaca: fear is reduced, but everything else remains the same. Coping is not enhanced (Fig. 15) and motivational problems do not get solved either. The continued decrease in motivation and the onset of depression are only postponed (Fig. 14).

It is a well-established finding in coping research, that depressive persons orient in their coping strategies toward avoidance and consequently do not carry those coping efforts out (Reicherts 1988). If it were possible, by the persons own initiative or by the aid of therapy, to maintain the periodically effective coping efforts over longer time spans, a variety of further consequences for the development could be expected. A knowledge of the rules implemented into the system allows the assumption that both actual values as well as the means of "fear of failure" would considerably decrease, which would also lead to a

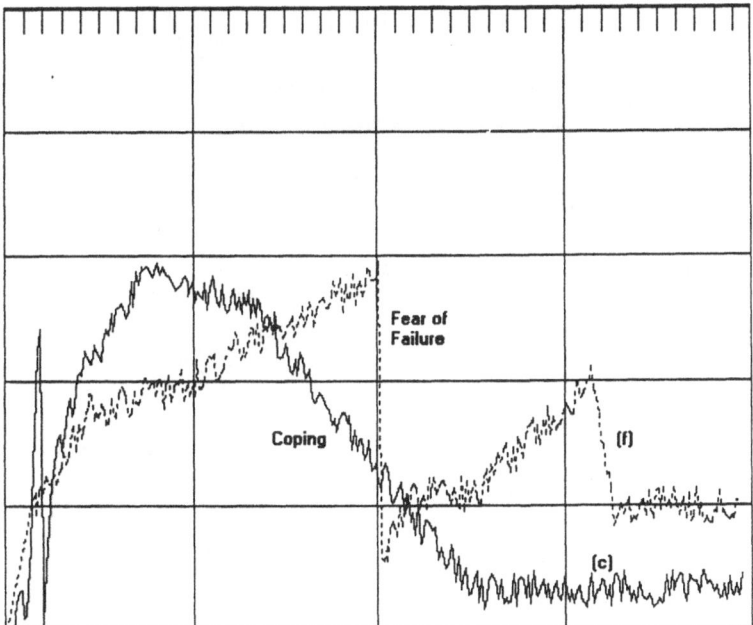

Fig. 15. As Fig. 14. In addition to "fear of failure" the variable "coping" is also plotted

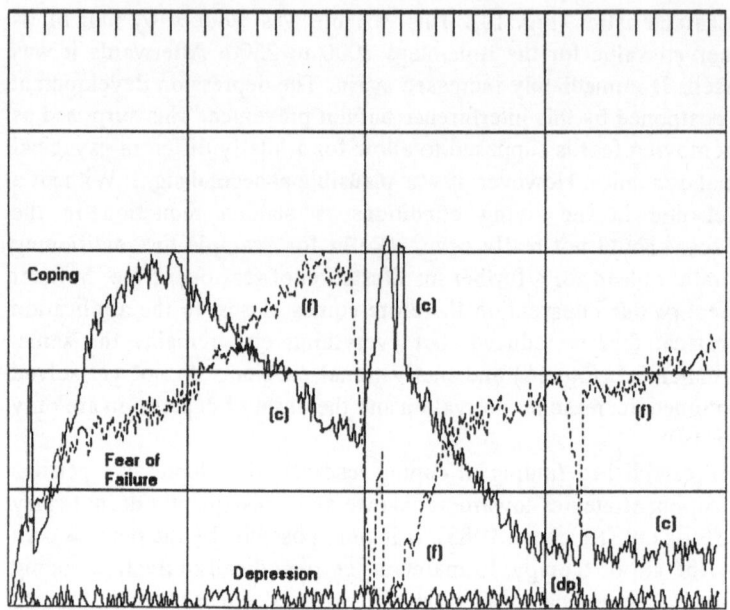

Fig. 16. The effect of repeated and continuous coping efforts, distributed over 8 phases of 20 time-steps each between time-steps 2000 and 2230. The variables "fear of failure", "coping" and "depression" are depicted

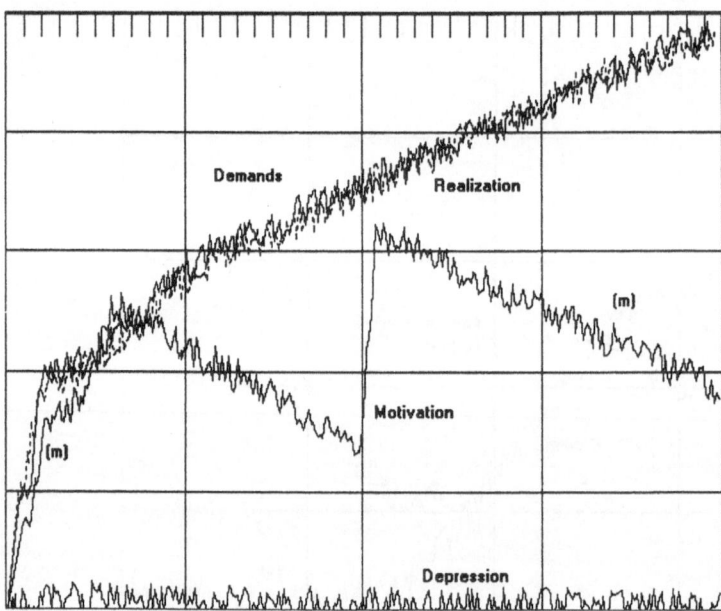

Fig. 17. As Fig. 16, but for the variables "motivation", "demands", "realization" and "depression"

considerable increase in the mean of "coping" (coping disposition). "Motivation" should also increase. Model testing is carried out by establishing an extra rule which is effective from time-step 2000 onward. For 8 phases (duration of each phase 20 time-steps), each separated by 10 steps, it asks for coping approaches to be intensified (positive feedback of "coping" on itself). The simulated prevention experiment confirms the expectations (Figs. 16 and 17). It is remarkable however, that "fear of failure", even though it is considerably reduced in the manner described, always increases again. "Motivation" also drops again after a spontaneous increase. The onset of "depression" is prohibited however. The whole system proves to be very stable, since even though fluctuations are considerable, lasting new trends cannot be established (see Figs. 12-15). The variables "demands" and "realization" particular are not much influenced by the "coping" episodes (Fig. 17). To speak in clinical terms, the system develops a constant vulnerability which makes continuous coping strategies necessary.

The system shows an interesting behavior when the previously described coping intensifying rule is activated at time-step t=1000 and another rule, influencing the mean of "demands", is additionally implemented. This other rule dictates that a momentary reduction of the actual "demand" value should also lead to a reduction in the "demand" mean. The time-step at which the production is again "allowed to fire" is defined by a quotient between the "demand" mean times 10 (numerator) and the "motivation" mean. Thus, if the "demand" mean is

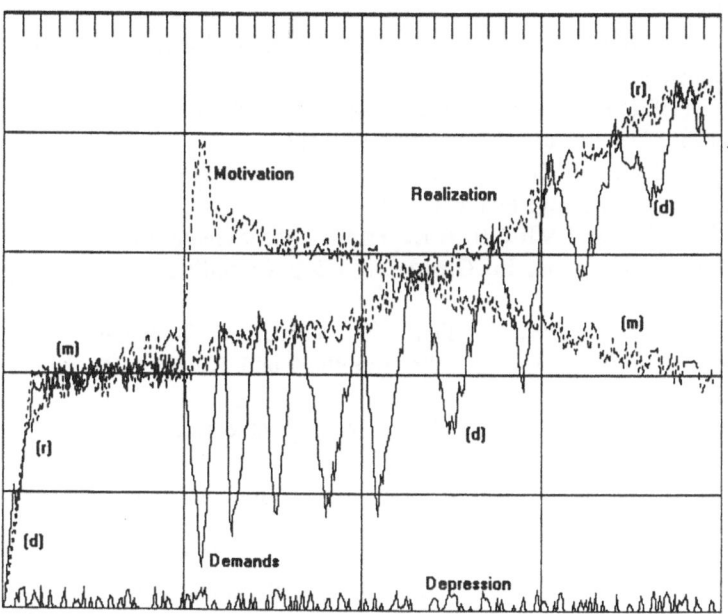

Fig. 18. The effect of coping efforts repeated eight times over phases of 20 time-steps starteing at time-step 1000. At the same time an additional rule was installed, influencing the mean of the variable "demand". Depicted are the variables: "demands", "realization", "motivation" and "depression".

high, and the "motivation" mean is low, it takes a considerable time until the "demand" mean will slowly start to fall. If, however, denominator > numerator, this will happen much more quickly. Except for the coping effects already familiar from Figs. 16 and 17 on "fear of failure", "motivation" and "coping" (with the additional fact that after these effects the system dynamics relax to the attractor of the uninfluenced system behavior) the variable "demands" produces oscillating behavior, whereby its total level increases and the amplitude of oscillation slowly decreases (Fig. 18). This decrease can be explained by the fact that the mean for "motivation" slowly decreases, whereas the mean for "demands" slowly increases, whereby a reduction of the "demand" mean through the extra rule becomes more and more difficult. The reason for the cyclical behavior of "demands" remains obscure however. This demonstrates that complex, dynamical systems hold some surprises even for their creators.

References

ALLOY LB , CLEMENTS C, KOLDEN G (1985) The Cognitive Diathesis-Stress Theories of Depression: Therapeutic Implications. In: Reiss S, Bootzin RR (Eds): Theoretical Issues in Behavior Therapy. Academic Press, New York,pp 379-410

ANDERSON JR (1980) Cognitive Psychology and its Implications . Freeman, San Francisco

ANDERSON JR (1983) The Architecture of Cognition. Harvard University Press, Cambridge

BARTLETT FC (1932) Remembering. Cambridge University Press, London

BISCHOF N (1985) Das Rätsel Ödipus. Piper, München

COLLINS AM, LOFTUS EF (1975) A Spreading-activation Theory of Semantic Processing. Psychological Review 82, 407-428

COLLINS AM, QUILLIAN MR (1972) Experiments on Semantic Memory and Language Comprehension. In: Gregg LW (Ed): Cognition and Learning. Wiley, London

DÖRNER D, KREUZIG HW, REITHER F, STÄUDEL T (Hrsg) (1983) Lohhausen. Vom Umgang mit Unbestimmtheit und Komplexität. Huber, Bern

EIGEN M, WINKLER R (1975) Das Spiel. Naturgesetze steuern den Zufall. Piper, München

GRAWE K (1987) Psychotherapie als Entwicklungsstimulation von Schemata. Ein Prozeß mit nicht vorhersehbarem Ausgang. In: Caspar F (Hrsg): Problemanalyse in der Psychotherapie. Forum 13, dgvt, Tübingen: 72-87

GROSSBERG S (1976) Adaptive Pattern Classification and Universal Recording: Part 1 Parallel Development and Coding of Neural Feature Detectors. Biological Cybernetics 23, 121-134

GROSSBERG S (1982) Studies of Mind and Brain. Reidel, München

HAKEN H (1983[3]) Synergetics. An Introduction. Springer, Berlin, Heidelberg

HAUTZINGER M, DE JONG-MEYER R (1990) Depressionen. In: Reinecker H (Hrsg) Lehrbuch der Klinischen Psychologie Hogrefe, München, 126-165

HAYES B (1988) Zelluläre Automaten. Spektrum der Wissenschaft, Sonderband Computer II, S 60-67

HEBB D O (1949) The Organization of Behavior. Wiley, New York

HOFFMANN J (1986) Die Welt der Begriffe. Deutscher Verlag der Wissenschaften, Berlin

KLIX F (1984) Über Erkennungsprozesse im menschlichen Gedächtnis. Zeitschrift für Psychologie 192, 18-46

LASHLEY K (1950) In Search of the Engram. In: Hillman JR (Ed) Society of Experimental Biological Symposium No 4. Cambrige University Press, London, 478-505

LEHNERT WG (1987) Possible Implications of Connectionism. Procedures of TINLAP-3, Las Cruces, pp 78-83

LEWINSOHN PM (1974) A Behavioral Approach to Depression. In: Friedman RI, Katz MM (Eds) The Psychology of Depression: Contemporary Theory and Research. Wiley, Washington DC, pp 157-178

LEWINSOHN PM, HOBERMAN HM , TERI L, HAUTZINGER M (1985) An integrative Theory of Depression. In: Reiss S, Bootzin RR (Eds): Theoretical Issues in Behavior Therapy. Academic Press, New York, pp 331-359

MARR D, POGGIO T (1976) Cooperative Computation of Stereo Disparity. Science 194, pp 283-287

MCCULLOCH WS, PITTS W (1943) A logical Calculus of the Ideas Immanent in Nervous Activity. Bulletin Mathematics Biophysics 5, pp 115-133

MINSKY M (1975) A Framework for Representing Knowledge. In: Winston PH (Ed), The Psychology of Computer Vision. McGraw-Hill, New York

MINSKY M, PAPERT S (1969) Perceptrons. MIT Press, Cambridge

MOWRER OH (1960) Learing Theory and Behavior. Wiley, New York

NEWELL A (1973) Production Systems: Models of Control Structure. In: Chase WG (Ed), Visual Information Processing. Academic Press, New York, pp 463-526

OHLSEN S, LANGLEY P (1986) PRISM: Tutorial and Manual Technical Report. University of California Department of Information & Computer Science, Irvine

POST EL (1943) Formal Reductions of the General Combinatorial Decision Problem. American Journal of Mathematics 65, pp 197-215

QUILLIAN MR (1966) Semantic Memory. Bolt, Beranek and Newman, Cambridge

QUILLIAN MR (1969) The Teachable Language Comprehender. Communications of the Association for Computing Machinery, 12, pp 459-476

REICHERTS M (1988) Diagnostik der Belastungsverarbeitung. Huber, Bern

REITER L (1988) Auf der Suche nach einer systemischen Sicht depressiver Störungen. In: Reiter L, Brunner EJ, Reiter-Theil S (Hrsg): Von der Familientherapie zur systemischen Perspektive. Springer, Berlin

ROSENBLATT F (1959) Two Theorems of Statistical Separability in the Perceptron. In: Mechanisation of Thought Processes. Proceedings of a Symposium held at the National Physical Laboratory, Nov 1958, HM Stationary Office, London

ROSENBLATT F (1962) Principles of Neurodynamics. Spartan, New York

RUMELHART DE, MCCLELLAND JL, PDP Research Group (1986) Parallel Distributed Processing: Explorations in the Microstructure of Cognition. Vol 1 Foundations. MIT Press, Cambridge

SAUERS R, FARRELL R (1982) GRAPES User's Manual. Carnegie-Mellon University Press, Pittsburgh

SCHANK R C, ABELSON R (1977) Scripts, Plans, Goals, and Understanding. Erlbaum, Hillsdale

SCHIEPEK G (1986) Systemische Diagnostik in der klinischen Psychologie. Beltz, Weinheim

SCHIEPEK G, SCHAUB H (1991) Als die Theorien laufen lernten. Ein Simulationsmodell zur Depressionsentwicklung. In: Schiepek G: System-theorie der klinischen Psychologie. Vieweg, Braunschweig, S 221-305

SCHMALHOFER F, PLOSON P G (1986) A Production System Model for Human Problem Solving. Psychological Research 48, pp 133-122

SCHUSTER H G (1988) Deterministic Chaos. VCH, Weinheim

SEIFRITZ W (1987) Wacshtum, Rückkopplung ubd Chaos. Hauser, München

SMITH EE , SHOBEN EJ, RIPS LJ (1974) Structure and Process in Semantic Memory A Featural Model for Semantic Decisions. Psychological Review 81, pp 214-241

STROHSCHNEIDER S, SCHAUB H (1991) Konnektionismus und kognitive Psychologie. Systeme, 2, S 132-152

TOFFOLI T, Margolus N (1987) Cellular Automata Machines. A New Environment for Modelling. MIT Press, Cambridge

TSCHACHER W (1990) Interaktion in selbstorganisierten Systemen Asanger, Heidelberg

TULVING E (1972) Episodic and Semantic Memory. In: Tulving E, Donaldson W (Eds). Organization of Memory. Academic Press, Cambridge, pp 381-403

WEIZENBAUM J (1966) ELIZA - A Computer Program for the Study of Natural Language Communication between Man and Machine. Communications of the ACM 9, pp 36-45

WENDER FW (1988) Semantische Netze als Bestandteil gedächtnis-psychologischer Theorien. In: Mandl H, Spada H (Hrsg), Wissens-psychologie. Psychologische Verlags-Union, München, S 55-73

WITTCHEN U U, SASS H, ZAUDIG M, KOCHLER K (1989) Diagnostisches und Statistisches Manual psychischer Störungen (DSM-III-R) der APA. Deutsche Übersetzung der dritten Auflage. Beltz, Weinheim

WOLFRAM S (1984) Universality and Complexity in Cellular Automata. Physica 10d; S 1-35

ZINTERHOF P (1987) Mathematische Systemtheorie. In: Roth E (Hrsg). Sozialwissenschaftliche Methoden. Oldenbourg, München, S 105-121

Simulating Clinical Processes by Population Dynamics

Jürgen Kriz

With 4 Figures

1. Introduction

Over the last few decades new concepts have emerged in clinical psychology designed to understand psychopathological and psychotherapeutic processes. These concepts of the so-called "systems approach" represent a valuable change of perspective: mental illnesses that were considered "individual" in other approaches (whether they were caused by inner conflicts, by learning, by "faulty thinking", or whatever) are analyzed in terms of their status and their function for the communicative structures in a social system – especially in the family.

What appears as "illness" from a certain perspective can be seen from another vantage point as the "ability" of the system (and the persons in it) to adapt to all the given circumstances. As a consequence, the underlying perspective of this understanding of pathology and of therapeutic intervention focuses on the patterns of the interactions (or "communications" in a wide sense, which also reflects expectations, interpretations, definitions of "reality", etc. by the family members). A "pathological" family thus regulates itself through patterns of communication that reflect the type of symptom. However, the focus on family interactions in no way denies that the persons who have come together (usually the spouses) have brought certain experiences and habits into this relationship, which have an effect on their setting up certain rules. Moreover, communication, something interpersonal, a mark of the system "family", is directly related to intrapersonal aspects (e.g. biochemical and neuronal processes, muscular and general physical constitutions, experiences and the ways of storing them, self-esteem, etc.). By the same logic, it would be difficult to overestimate the value of larger social system's processes and their rules for the interaction patterns of a family (e.g. the economic situation, social values and norms, etc.). However, attention must be drawn in the dynamic-systemic circularity to the fact that patterns of processes in systems have only been developed relative to the patterns of processes in the meta-systems (the "environment"), and now both perpetuate the pattern of interactions as well as being perpetuated themselves by these interactions. Of course, most theorists agree that these patterns are the results of self-organized processes. For example, on the level of interactions in the family, there are no external orders or rules specified such that each person has to act in a well-defined way to fulfil these, but the persons act together by some kind of mutual understanding, each one acting so as to fulfil the rules of a game. By the same logic, thought, cognition, processes of the central nervous system and, on the other hand, society and several sub-systems

150

Springer Series in Synergetics, Vol. 58   **Self-Organization and Clinical Psychology**
Editors: W. Tschacher, G. Schiepek, and E.J. Brunner   © Springer-Verlag Berlin Heidelberg 1992

of the society can be understood as self-organized processes. This interpretation is supported by a large body of research.

In summary, in systemic clinical psychology the focus is on the system "family" or, more precisely, on communication patterns, CP. A "pathological" structure, CP_p, is then understood as a specific self-organized pattern including the "symptoms". The transition from CP to CP_p is due to internal or external changes – e.g. changes in processes of a higher or lower system levels – and is one specific adaptation to all given circumstances. Lack of flexibility keeps the system in the status CP_p. By the same logic, the goal of therapy is to "perturb" (Maturana & Varela, 1980) the system in order to enable the system to make a transition from CP_p to CP' (hoping that CP' might show fewer symptoms than CP_p – otherwise therapy has to continue).

2. Population Dynamics and Communication

In order to throw more light on this issue, the aim of this investigation is to examine the potential application of synergetics to such communication processes. From a theoretical point of view, this would seem most reasonable: In contrast to other conceptions of self-organization (e.g. the concept "autopoiesis" by Maturana & Varela, 1980, or Luhmann, 1984), the "synergetics" of Haken (1983) gives a profound answer to the question of how we can conceptualize the emergence of self-organized patterns and which principles govern the change from this order to a new pattern. Moreover, Haken's synergetics stresses the relationship of systems and subsystems by order parameters and control parameters.

Haken himself and quite a number of contributors to his conferences on synergetics have shown the profound and amazing analogies between completely different systems when they pass through an instability to a new pattern. However, in order to apply this fruitful concept to the description of clinical processes one main difficulty must be overcome: According to Haken's definition, "synergetics deals with systems composed of many parts which can form spatial, temporal or functional structures on macroscopic scales". As a consequence, there must be a minimum number of components in order to allow for collective behavior. The question to be answered therefore becomes: "How can we describe family interactions in terms of such multicomponent systems?"

In order to answer this question, it is necessary to consider the meaning of the term "communication pattern" or even of the term "family". More specifically, it must be borne in mind that the systems approach – and especially terms like "process", "phase transition", etc. – reflect the notion of a flowing universe. Moreover, "reality", from the perspective of the flow of time, may be viewed as a unique, constantly changing, complex interdependent process.

In contrast, "reality", observed and described in the context of everyday life as well as in (social) science, is constituted according to certain categories. This is done, firstly, by cutting the time-dimension into delimited pieces. The results of this procedure are called "situations". However, no two situations (or, more precisely, two space-time constellations) are equal to each other in all possible aspects which could be taken to describe these situations completely. On the other hand, a successful typology of possible experience, of whatever kind, cannot be based on an unlimited sequence of singular situations because each theory can only consider a limited number of parameters. Therefore, complexity is strongly limited by creating categories which simply take into account the "similarities" of different situations and, correspondingly, of individual experiences. Only in this way are communications about experiences and observations in the past as well as prognoses about the future possible. As a consequence, reality that appears phenomenologically in an endless diversity of "situations" has to be divided and structured into relevant equivalence classes.

By the same logic, terms like "house", "child", "communication", "advice", "question", "conflict" are categories on different levels and with differing generalities which not only unite different "elements" (in the sense of objects or delimited phenomena) but also (at least in most cases) map processes onto "things" and "elements" (or, more abstractly, onto "entities"). The important question "How can we ensure that the meaningful structuring of phenomena into "equal" (within an equivalence class) and "unequal" (between classes) can be sufficient to make meaningful and relevant statements?", is a matter for empirical investigation and depends on the theory and on the questions concerning a particular study. From a theoretical point of view, no general answer exists; a solution can only be arrived at through answering questions such as: Which phenomena fall into the class which constitutes the characteristic features? For example, in considering the relationship between the "actions" and "thoughts" of a person (in the frame of "interaction" in a particular family), the question is, in essence, whether there is a relationship between a particular set of phenomenological equivalence classes – categories of "actions" – and, on the other hand, another set of phenomenological equivalence classes – categories of "thoughts" – which are relevant to the realm the researcher is interested in.

Let us take a very simple example: In order to investigate the "communication patterns" of a family (of, say, four members), we must cut the time-dimension into pieces and define what a communication could be. Moreover, let us take an average time of 10 s (which is rather long) for one communication. Then at least 10,000 to 20,000 communications a day are created by the system – or some millions per year. In order to make any statement about this family, we have to find classes of phenomenological equivalence (empirically, with respect to this particular family, and/or theoretically, with respect to a particular theory of systemic interaction, or of clinical processes, or whatever). Moreover, even if we reduce drastically the classes of communications which we take into account, we have to deal with a highly complex interwoven dynamical system. Therefore, classes of communications – such as "asking", "crying", "begging", "placating" – define populations of phenomena. The same is true with respect to even more abstract categories, e.g. "conflict", "enmeshment" etc.

Given this discussion, it would seem reasonable to suggest that the meaning of "family" is this complex interwoven dynamical system of communications, and,

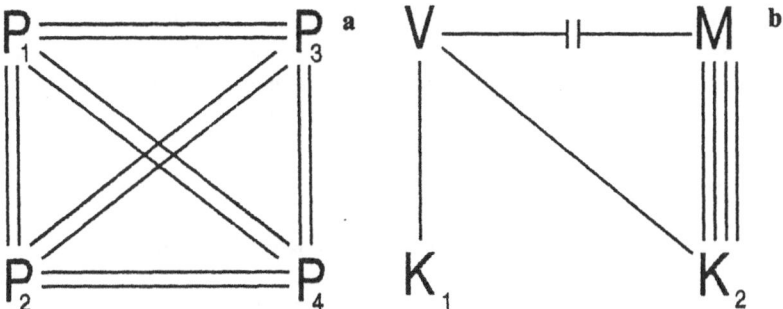

Fig. 1. (a): Communication pattern of the "standard average family"; (b): Example of the interaction pattern of a "real" family

moreover, that the meaning of particular "communication patterns" refers to multicomponent systems of (categories of) communications. By the same logic, the same is true with respect to "thoughts" and, moreover, with respect to "processes in the body".

How, then, can we distinguish between different structures of communications (i.e. different "types of families")? In order to answer this question, let us take four persons realizing all possible communications (in an observer/researcher defined category-set) between them in an equal manner. This might be called the "standard average family, SAF" which is shown schematically in Fig. 1a. Any other pattern of interactions – for example, the one shown in terms of a so-called family map in Fig. 1b – actually represents a "real" family. In this family map (see Minuchin & Fishman, 1982, for details) the symbol $-\|-$, for example, represents a "conflict". However, if an observer identifies a "conflict", he has to refer to (tacitly or explicitly defined) categories of communications, say "C", (operationalizations of "conflict"), which occur much more often than do the same categories in SAF. Alternatively, this aspect of the pattern may be described as increasing the "population" of communications which belong by definition to C, while decreasing other "populations" of communications (because it seems reasonable to suggest that the total number of communications in a given time is somewhat limited). Expressed in this alternative way, it is of some interest to understand the emergence and transition of a particular communication pattern in terms of competition (and, by the same logic, of symbiosis etc.) of communcation-populations (which are defined by names like "asking", "crying" etc.).

From the above discussion, it is clear that it may be useful to adopt population dynamics in order to describe the emergence and transition of communication patterns.

3. Methodological Remarks

In population dynamics major phenomena, e.g. the increase of a population, competition and coexistence, symbiosis etc., are described by dynamical laws of the type $n(t+1) = f(n(t))$, where n is the number of "species" of a given population.

In the case of a single population (one kind of species) the growth rate $R=(n(t+1)-n(t))/n(t)$ is proportional to n with the rate factor g. Due to depletion of the ("food")-resources, there should be an additional density-dependent term βn or "death"-coefficient. The resulting equation is of the type

$$n(t+1) = f(n(t)) = (1+g - \beta n(t)) \, n(t) \, .$$

We may write $\Delta n = n(t+1) - n(t)$; then the equation takes the form

$$\Delta n = (g - \beta n)n \, . \tag{1}$$

Among the earliest studies in population dynamics were those made by Verhulst, who investigated population growth models and formulated the "law" of equation (1) in 1845.

Iterative processes are very important. Peitgen and Richter (1986) pointed out that "processes of this kind are found in all exact sciences. Indeed, the description of natural phenomena by differential equations, introduced by Sir Isaac Newton and Gottfried W. Leibniz some 300 years ago, is based on a feedback principle... It is not essential whether the process is discrete – that is, takes place in steps – or continuous. Physicists like to think in terms of infinitesimal time-steps: *natura non facit saltus*. Biologists, on the other hand, often prefer to look at the changes from year to year or from generation to generation. Obviously, both views are possible, and the circumstances stipulate which description is appropriate" (p. 5).

In this sense, the physicist Haken, the founder of synergetics, prefers the continuous view and its expression by differential equations. The growth rate in the above case is dn/dt, and the dynamics is described by the equation

$$dn/dt = (g - \beta n)n \, . \tag{2}$$

Haken (1973) pointed out that this equation "holds equally well for a single mode laser or for an autocatalytic reaction" (so that n can be interpreted as photons or molecules respectively). Moreover, it should be noticed that n is the order parameter of that system in terms of synergetics. In contrast to many examples stemming from the natural sciences (laser, ferromagnet, etc.), where equations for order parameters can be derived from microscopic theories (for example, the connection of the photon number n and the electric field strength E of the laser light on the basis of a detailed microscopic theory for E), we can obtain adequate equations in clinical psychology, at least at the present time, only by plausibility arguments. Moreover, the descriptions have only qualitative character and enable only a phenomenological discussion of the possible states and pattern transitions of the communication system "family" (including symptoms).

Therefore, a serious question may arise: does it make sense to use, at this stage, mathematics instead of natural language to describe these processes? There are at least some reasons, I believe, to answer this question in the affirmative: Even the "simple case" of the growth of one population in (1) gives a lot of evidence that our brain without any mathematical tool is unable to understand, and, furthermore, that natural language is rather inappropriate to describe nonlinear systemic processes. For example, the famous Mandelbrot process (which is mathematically equivalent to the Verhulst process – see Peitgen and Richter, 1986) and its fractal structures remained largely unknown, even to most mathematicans, although the famous French mathematicans Julia and Fatou stated a lot about these processes. But even to mathematicans it was too difficult to understand fully the meaning of their statements and to communicate the subtle ideas without the tool of modern computer simulation and computer graphics. Moreover, one may ask if a description of a dynamical process would "really" work and sufficiently explain the observed phenomena – but using just words there is no evidence. In contrast, a precise formulation of the relationships and a computer model may lead to new and/or more questions about the principles that govern the phenomena. Our brain is not able to compute highly complex verbal information in this precise manner without any "help" – errors might enter into a study as a result of ignoring some interaction effects or making mistakes in reasoning. A mathematical description and a computer simulation based on this description may be understood as such a "help" in understanding and reasoning (see Kriz (1988) for more details of this argumentation).

However, before we go into more detailed considerations of communication patterns, one more methodological remark seems to be necessary: In the following, we use the discrete notion on the basis of (1). We are going to investigate the distribution of n_A, n_B,... in given time intervals which are understood to be macroscopic patterns of communications. From the above discussion, it is clear that the number of communications, n_A, only makes sense in terms of time intervals. For our present purpose, we need not be concerned with the differences in the notions of (1) and (2). However, empirical research is being initiated and this problem will arise when we start to make operational definitions of the communication categories which we want to observe. Moreover, even for our phenomenological considerations the difference between (1) and (2) can become important. Under some conditions, special phenomena emerge for (1) which are absent for (2): When the growth-parameter g in (1) increases slightly beyond the value 2, the iterative process of n will not reach only one "attractor" (i.e. one fixed value) but will show a steady oscillation between two values. When the growth-parameter is increased beyond 2.45 more complicated behavior of (1) occurs. The oscillations involve 4, then 8, then 16 different values, and so on. Finally, at g=2.570, the process is no longer periodic at all, and chaos sets in: the process is still deterministic but cannot be predicted other than by letting it develop because it never turns into an ordered sequence of events.

It is not the aim here to discuss such interesting phenomena. Moreover, for the present purpose the growth dynamics of "communication-populations" can be described by differential equations of the type (2) as well, because in the discrete case of the type (1) we limit g to a maximum of about 2 (at least, I presently have no interpretation as to what the above "over-adaptation" could phenomenologically mean in the case of a communication system). Consequently, as we take care that no g (or when we later combine some g-terms, the total g) exceeds about 2, there is no loss of generality compared to the "equivalent" differential equations.

4. Some Aspects of Modeling

In order to discuss a concrete example, consider the simplest case of a communication pattern in a family based on two persons (1,2) and two different types or categories of communications (a,b). In practice, for empirical research, a and b have to be adequately operationalized, of course. We denote the number of communications in a given time interval (e.g. one day) by A and B. If all communications are independent of each other we may write four equations of the type (1) and, after some days, A_1, A_2, B_1, and B_2 will have values which depend only on the g/β-terms. Of course, this is not a realistic example; it is essential for a family that the communication of the members is related to (or dependent on) each other. Then, two basically different cases may occur: symbiosis and competition:

   a) **Symbiosis.** Let us assume that the growth rate for A_1 depends on the presence of B_2 and vice versa (this is called "symbiosis" in population dynamics). For A_1 and B_2 we may write the following equations

$$\Delta A_1 = (g_1 B_2 - \beta_1 A_1)A_1 + \text{fluctuations,} \tag{3}$$

$$\Delta B_2 = (g_2 A_1 - \beta_2 B_2)B_2 + \text{fluctuations.} \tag{4}$$

The term "fluctuation", typical for such synergetic equations, describes the spontaneous creation of communication of the particular type; the meaning in this context will be discussed later.

   b) **Competition.** In addition to symbiosis, competition plays an important role for the dynamics of the communication populations. Each person has limited resources of energy and, therefore, of possible communications in a given time. As a consequence, A_i and B_i stem from the same pool of resources. For that reason, the coefficient β depends not only on one type of communication but on all the communications of a person i. From population dynamics we know that for two populations with

$$\Delta n_1 = [g_1 - \beta_1 (n_1 + n_2)]n_1 \quad \text{and} \tag{5}$$

$$\Delta n_2 = [g_2 - \beta_2 (n_1 + n_2)]n_2 \tag{6}$$

only one population (that with the large value of g) survives (except for the unrealistic case with $g_1/\beta_1 = g_2/\beta_2$).

   c) **Modeling the Influences.** In (3) the quantity B_2 and the coefficient g_1 are combined to a function of the type

$$\hat{g} = gB \tag{7}$$

which postulates a proportional influence. This seems to be rather unrealistic. We prefer a weaker influence, given by the linear function

$$\hat{g} = g' + gB .\tag{8}$$

We can then modify (3) to read

$$\Delta A_1 = (g_1' + g_1 B_2 - \beta_1 A_1)A_1 + \text{fluctuations}\tag{9}$$

and (4) in an equivalent manner. Equations of this type have been discussed in the context of "symbiosis" in Haken (1983). However, these considerations naturally lead us to the problem of how the influence on g may be modeled. Of course, not only functions like (7) or (8) are possible, but also other types. For example, in order to keep the g limited (which was discussed above) we may use the function

$$\hat{g} = g' + g \ [B/(B+1)]\tag{10}$$

because for $0<B<\infty$ the $B/(B+1)$ are limited to the interval $(0,1)$. If we replace $B/(B+1)$ by $[B/(B+1)]^n$ (with $n>0$) we will find that g becomes more and more "sensitive" to large B.

In summary, these considerations may be understood as tools used to model the interactions between different categories of communications. After these preliminaries we can now write the equations for some simple models and study, for example, reasonable cases of bifurcation in the behavior of such a model. As I have presented such a bifurcation model in some detail elsewhere (Kriz, 1990), we may now switch to a substantially more important model as a basis for further discussion.

5. Triangulation: An Illustrative Example

A concept which plays an important role in the context of systemic clinical psychology is the so-called "triangulation". Triangulation is defined as the extension of the conflictive relationship between two persons (in most cases the parents) to include a third person (in most cases a child), who conceals and/or neutralizes the conflict. The conflict is covered up if the child develops "a problem" (e.g. psychosomatic symptoms) because the parents, whose conflicts are concealed and detoured via such triangulation, can and must concentrate on the child's problem together.

In terms of a family map the core of the concept "triangulation" is represented in the form of Fig. 2a. Of course, there might be more children in the family and some other important communicative relationships between the family members. Moreover, triangulation is discussed in verbal terms in the clinical literature. For our purpose we focus on the core of the concept and quantify the assumptions in order to give an example of a simple model for triangulation. Let us assume the

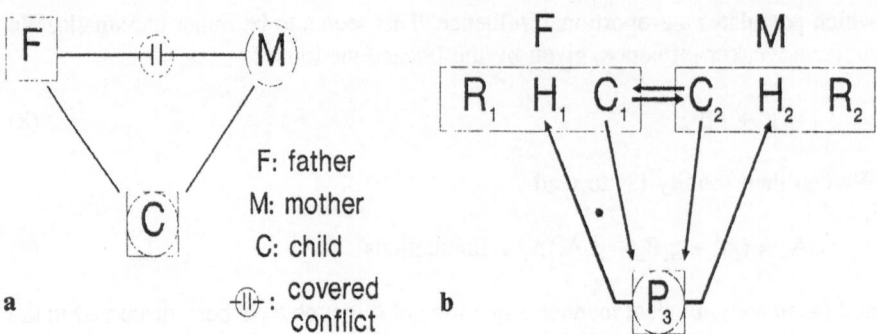

Fig. 2. (a): "Family map" representation of triangulation; (b): Diagram of the modeled influences

following "populations" of communication behavior (when we conduct empirical research, all categories will have to be well defined; moreover, a time interval has to be defined in which the events can be counted):

C: "Conflict behavior" (which must be noticeable to the child)
H: "Helping behavior" (or the parents' "reaction" to P below)
R: "Other behavior" – this is a "remainder" category, including in particular all the "normal" behavior.
P: "Problem behavior" – the "symptoms" of the child.

The assumptions of the influence are shown in Fig. 2b.

According to our considerations above we can establish the following equations for the model:

$$\Delta C_1 = [g'_{11} + g_{11}C_2 - \beta_{11}(C_1 + H_1 + R_1)]C_1 + \text{fluctuations},$$
$$\Delta H_1 = [g'_{12} + g_{12}P_3 - \beta_{12}(C_1 + H_1 + R_1)]H_1 + \text{fluctuations},$$
$$\Delta R_1 = [g'_{13} + g_{13} - \beta_{13}(C_1 + H_1 + R_1)]R_1 + \text{fluctuations},$$
$$\Delta C_2 = [g'_{21} + g_{21}C_2 - \beta_{21}(C_2 + H_2 + R_2)]C_2 + \text{fluctuations},$$
$$\Delta H_2 = [g'_{22} + g_{12}P_3 - \beta_{22}(C_2 + H_2 + R_2)]H_2 + \text{fluctuations},$$
$$\Delta R_2 = [g'_{23} + g_{23} - \beta_{23}(C_2 + H_2 + R_2)]R_3 + \text{fluctuations},$$
$$\Delta P_3 = [g'_3 + g_3 f(C_1 C_2) - \beta_3 P_3]P_3 + \text{fluctuations}.$$

The equations are written in a form that makes it easy to see the influences. The f in the equation for P_3 is due to the fact that we used a function of the product $C_1 C_2$ of the type (9) in order to limit the total g – as we discussed above. In order to make very few assumptions about the coefficients we set all g'_{ij} to 0.5, all β_{ij} to 0.006 and all g_{ij} to 0.005 except $g_{13}=g_{23}=0.2$ in order to give a higher rate to the "normal" behavior.

When we set the fluctuations to very small values, say, 0.5, and start with a "normal" family – that is, the C, H and P have low values, say, about 5 to 10, and R is high, say, about 50 or more – the C, H, and P are repressed to leave only the contribution of their fluctuations, while R increases until the "resources" of behavior get depleted. The model is then stable (except for slight variation due to the fluctuations). If we add values of, say, 10 or 20 or 30 to C in one single

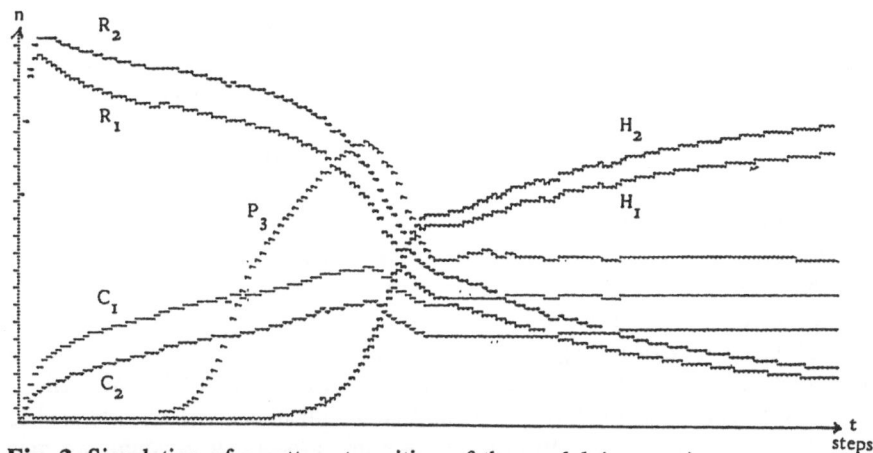

Fig. 3. Simulation of a pattern transition of the model (see text)

iteration of the running model, after a while, the model settles down again to the stable values. The same will happen if we increase the fluctuation very slightly. However, if we add values to C in some more iterations and/or increase the fluctuation a little more the system eventually will start a process which leads to a totally new pattern – for example, R rather low and P, H and C fairly high: Fig. 3 shows a simulation of a pattern transition due to the fact that we set the fluctuations of C_1 to 4 and of C_2 to 2. After an instability, the system becomes stable again with a new pattern of communications, in which "normal" behavior has nearly died out but conflict- and symptom-behavior have stabilized at a medium level while the helping-behavior is rather high. (Parents would argue that this is "due to the symptoms". But the systems therapist would say, conversely, that the symptoms are also "due to the H-behavior", and that both, symptoms and H-behavior, are "due to C-behavior", etc. It is simply a self-organizing pattern).

6. Towards a Multilevel Concept of Communications Synergetics

It can be seen from the above example that the occurrence of a new structure can be initiated properly if we take fluctuations into account. As a consequence, it may be useful to consider the purpose of the fluctuations and, moreover, discuss the question of what fluctuation means in this context and how it can be introduced into the model.

Indeed, at the beginning of this article we argued for a conceptualization of the family communication system as an open system. Given the level of communications, the family members not only communicate with each other but also with other people. These communications may be so relevant that the family system becomes seriously affected – for example, if the father is so overworked and tense that he takes out all his aggression on his family.

In the above model, there are at least two reasonable possibilities to take this fact into account:

a) The fluctuations, represented by a random variable, may increase the mean and/or variance. (As we deal with sums in a given time interval and the fluctuations are assumed to be of short time effect the mean is significantly more relevant than the variance). Therefore, in the equations for the C we may add a term F which reflects an increased occurrence of "conflict potential" due to the stress at the father's place of work. Mathematically, this F represents a shifted mean in our model. This is exactly what has been done in Fig. 3.

b) Alternatively, we may introduce a third g-term, say g^*, in the equations for C, which represents the influence of this external "stress" on the conflict-behavior. When we let this parameter g^* increase then the influence of the spontaneous fluctuation on C grows bigger and bigger. This is related to the well-known phenomena of "critical slowing down" and "critical fluctuations" in synergetics.

However, as stated before, the family communication process is related to intrapersonal processes, too. In order to take this aspect into account, we may introduce, additionally, another term, say g', which represents the influence of these "internal" processes (i.e. the interrelationships to sub-systems).

In summary, these considerations lead to the conclusion that it seems reasonable to decompose the g-terms into three components. The increasing or decreasing g^* in the discussion above might be seen as the influence from a meta-system on the family (e.g. society – or, better, parts of society). Thus, g^* may itself be seen as a variable of this meta-system with a higher constancy in time. By the same logic, the g'-terms refer to inner processes of the individuals, that is, they refer to a lower system level. Expressing these ideas symbolically, we obtain for a class of communication behavior CB

$$\Delta CB_{ij} = (\overset{\downarrow}{g'_{ij}} + \overset{\rightarrow}{g_{ij} CB_{kl}} + \overset{\uparrow}{g^*_{ij}} - \beta_{ij} \Sigma_j^n CB_{ij}) CB_{ij} + \text{fluctuations},$$

where $\downarrow$ represents a connection to a lower, $\uparrow$ to a higher, and $\rightarrow$ to the same level in the system hierarchy. At least for communications, it is very important that these connections are taken into account. They might allow one to model the relationships between processes in the family and in "society" on the one hand, and between processes in the family and in the individual on the other hand.

It may be noted that in the above discussion, the increasing and decreasing g-terms can be seen as a slight variation of the values. However, it is nowhere claimed that this has to be the case. Specifically, from research into biochemical processes in the body we know that these can be periodic as well as chaotic (see, for example, Markus & Hess, 1984). Even if the change of a parameter may be described by a single Verhulst-equation, we are in for surprises. For example, the above process (1) will show for g=2.73817 (and n_0=10) the behavior shown in Fig. 4, where over long periods of time the system seems to be stable (with an oscillation of period 5), but with windows of chaotic behavior. It should be noted that this g=2.73817 is in the midst of the chaotic region of the period doubling scenario of the Verhulst process, but near to the "window" in which the attractor again consists of distinct points. Moreover, for g=2.738172372 the process seems

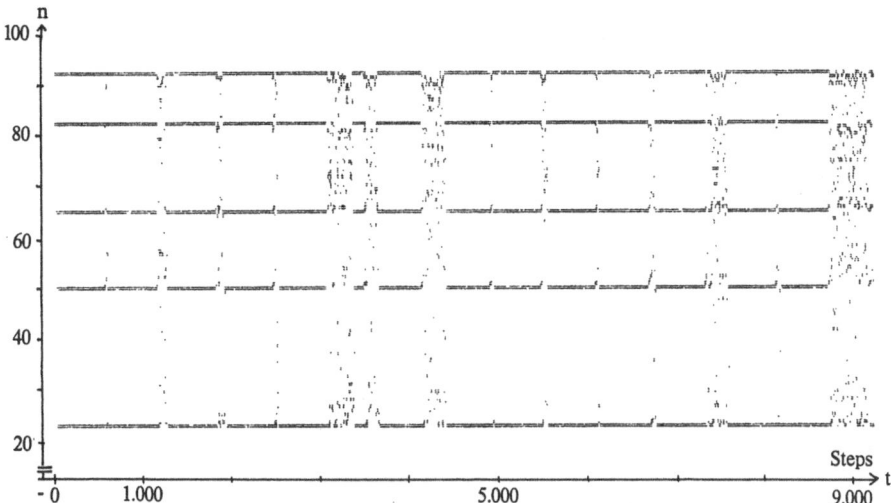

Fig. 4. Time series of (1) – stability with chaotic windows

to approach a stable oscillation of period 5 for some 10,000 (!) iterations, but then shows chaotic behavior for some hundred iterations, and further seems to be stable again with the same oscillation for again some 10,000 iterations.

It is evident that very interesting effects would occur if the behavior of a (micro level-)system, as shown in Fig. 4, were to modify the g-term of an other (i.e. macro level-) system.

It was not our aim to discuss all the possible effects in detail here. Rather, we wanted to give a rough impression of some ideas concerning the application of synergetics and population dynamics to clinical processes. The discussion has, at the present stage, only qualitative and phenomenological character. But future work will attempt to give exact operational definitions to some "species" of communications in the light of clinical theories, and then to collect empirical data of processes (documented on video tapes) which can be observed in systemic clinical psychology.

References

HAKEN, H. 1973. Synergetics. Cooperative Phenomena in Multi-Component Systems. Stuttgart: Teubner.

HAKEN, H. 1983. Synergetics. An Introduction. Berlin: Springer.

KRIZ, J. 1985. Grundkonzepte der Psychotherapie. München: Urban & Schwarzenberg.

KRIZ, J. 1988. Facts and Artefacts in Social Science. An Epistemological and Methodological Analysis of Social Science Research Techniques. New York: McGraw-Hill.

KRIZ, J. 1990. Synergetics in Clinical Psychology. In: H. Haken & M. Stadler (Eds.): Synergetics of Cognition. Berlin: Springer, 393-404.

LUHMANN, N. 1984. Soziale Systeme. Grundriβ einer allgemeinen Theorie. Frankfurt: Suhrkamp.

MARKUS, M. & HESS, B. 1984. Transitions between Oscillator Models in a Glycolytic Model System. Proc. Natl. Acad. Sci. USA, 4394-4398.

MATURANA, H.R. & VARELA F.J. 1980. Autopoiesis and Cognition. The Realization of the Living. Dordrecht: Reidel.

MINUCHIN, S. & FISHMAN, B.J. 1982. Family Therapy Techniques. Cambridge, Ma.: Harvard University Press.

PEITGEN, H.-O. & RICHTER, P.H. 1986. The Beauty of Fractals. Berlin: Springer.

Synergetics in Psychiatry –
Simulation of Evolutionary Patterns of Schizophrenia
on the Basis of Nonlinear Difference Equations

Günter Schiepek, Wolfgang Schoppek, and Felix Tretter

With 17 Figures

This article is an attempt to conceptualize schizophrenia as a dynamical disease (Mackey and an der Heiden). In accordance with synergetic models, different evolutionary patterns of schizophrenia are regarded as representing coherent dynamic patterns evolving out of a complex biopsychosocial system. Findings indicate the significance of cognitive disorders, stress, withdrawal, expressed emotion and delusions for the development of schizophrenia, and the nonlinear interactions of these variables (constructs) have been taken as the basis for a computer simulation of the evolution of schizophrenia. The model allows the generation of different evolutionary patterns according to the parameter values chosen.

1. Introduction

The search for the one, ultimate cause responsible for the pathogenesis of schizophrenia has been in vain. The investigations following this avenue of research have arrived at a dead end. Thus, linear models of the etiology of schizophrenia have been practically abandoned and even their multifactorial varieties no longer guide research. The number of causal factors hypothesized does not change the fruitlessness of linear thinking — unproductive not only because of justifiable doubts regarding the uniformity of the schizophrenic syndrome or the processes involved (Sarbin & Mancuso 1982; Strauss 1989; E. Bleuler mentioned the "group of schizophrenias" as early as 1911). Also, decades of research have shown that the different processes investigated are recursively interrelated to a very high degree when the life pattern of schizophrenic individuals is taken into consideration (Böker & Brenner 1989). "There is a need for models which focus on the interactions between biological and psychosocial, intrapsychic and interpersonal, cognitive and affective aspects, as well as the positive and negative psychological symptoms of schizophrenia" (Ciompi 1986a, p.49)

Recent meta-theoretical models aim to conceptualize the influence of several factors and their interactions in the pathogenesis of schizophrenia. These include:

- the diathesis-stress model (schizophrenic disorders evolve on the basis of a genetic disposition (diathesis) under the influence of stress, cf. Gottesmann & Shields 1976);

Springer Series in Synergetics, Vol. 58   Self-Organization and Clinical Psychology
Editors: W. Tschacher, G. Schiepek, and E.J. Brunner   © Springer-Verlag Berlin Heidelberg 1992

- the model of interactive evolution (the evolution of schizophrenia and other mental disorders depends on their complex and continuous interaction with favorable and unfavorable environmental conditions, cf. Strauss & Carpenter 1981);
- the concept of neural plasticity (neurons tend to react to repeated stimuli by functional and anatomical changes which lead to a progressive facilitation of those central nervous pathways which are most often and most intensively used. Due to this neural plasticity, interactive effects of unspecific genetic and environmental influences may lead to a specific vulnerability to schizophrenia, cf. Haracz 1984).
- The best-known models may be the vulnerability concept of schizophrenia put forward by Zubin & Spring (1977) or Nuechterlein & Dawson (1984) and
- the biological-psychosocial evolutionary model of schizophrenia by Ciompi (1982; 1986a,b; 1989).

There are several other concepts, such as the stimulus-window model, the concept of basic disorders, and different information-processing theories. All of the above-mentioned approaches are designed to integrate various findings and micro-theoretical assumptions to varying degrees. Within the vulnerability concept of schizophrenia, for example, a number of different specific, etiological assumptions are integrated into its submodels. Thus, it should not be considered an independent etiological model (Zubin 1989).

An extensive review of factors thought to interact and contribute to the evolution of schizophrenic episodes has been undertaken by Nuechterlein (1987; cf. also Nuechterlein & Zaucha 1990; see Fig.1).

The biological-psychosocial model of Ciompi (1989) assumes three phases in the long-term course of schizophrenia: phase I: premorbid evolution; phase II: psychotic decompensation; phase III: long-term evolution. Feedback-loops are assumed to exist between the decisive processes in each phase (Fig. 2). Thus, the further course of the illness is influenced by processes which have occurred before: for example, vulnerability to the outbreak of acute psychosis differs, depending on whether an acute episode has already occurred or not. According to Dauwalder (1988), different subprocesses are constitutive for the different phases. The fact that these sub-processes correspond to well-known and widely discussed theoretical approaches and empirical findings once more emphasizes the integrative power of the model proposed by Ciompi.

Apart from its heuristic, integrative and didactic values, such a wide-range synopsis also poses certain problems:

a) The problem of the "precise identification and operationalization of the relevant biological and psychosocial variables or markers" (Ciompi 1989, p.19).
b) The task of the precise description of the interactions and feedback processes between the constructs, variables or subsystems.
c) The difficulties of a dynamic approach going beyond the mere postulation of phases. Phases assumed a priori, such as "acute psychosis", "intermediary non-

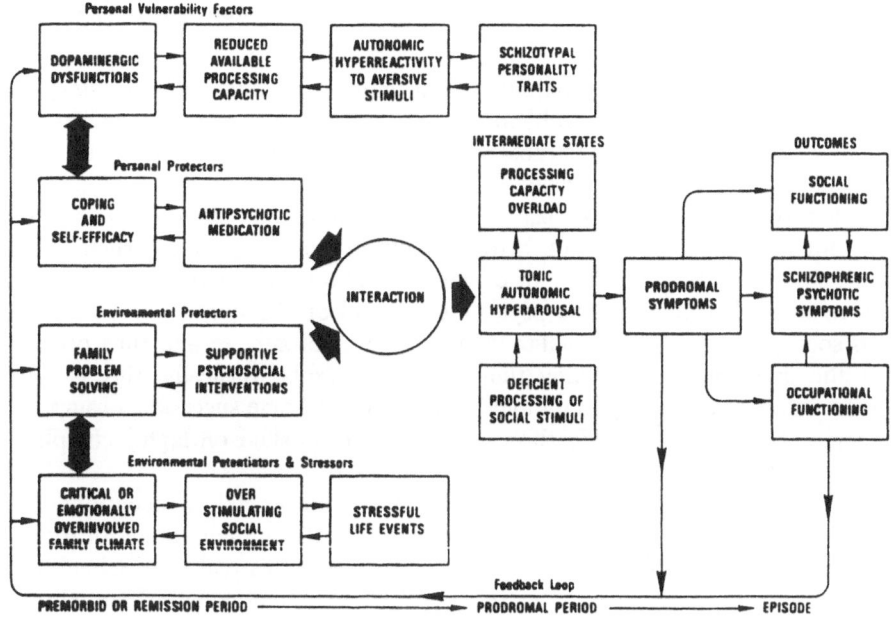

Fig. 1: A heuristic conceptual framework for possible factors in the development of schizophrenic episodes (from Nuechterlein 1987, p 81).

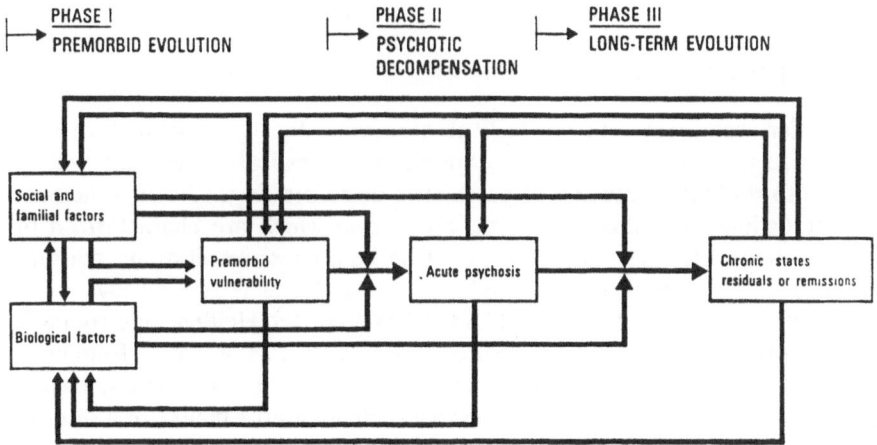

Fig. 2: A biological-psychosocial model of the long-term evolution of schizophrenia in three phases (Ciompi, 1989, p 19).

psychotic state", or "chronification", are certainly plausible at a phenomeno-logical level. However, they must be explained and — even more demanding — a prognosis should be possible, if theory is to be applied to actual cases. They do not possess explanatory power in themselves and therefore should not be regarded as given by the model builder. Rather, they should be generated by the systems model on the basis of recursively interrelated processing and thereby be explained.

Problem (a) is of a methodological nature and will not be treated in depth here. However, the problem of valid and reliable variables indicating changes in the pertinent constructs must be solved for a valid confirmation of dynamic models, and especially of computer simulations. A solution to problem (c) has as its prerequisite at least an attempted solution of problem (b), for a mere "drawing of arrows between boxes" (or constructs and subprocesses) is not sufficient, of course. The usual procedure is to describe the relations between various psychological, and also between psychological and social, processes in a qualitative way — i.e. verbally — and thus obtain specific assumptions regarding the course of events. Basically, this is the present state of theorizing in the field of psychology. Some successful examples (Ciompi 1982; Bischof 1985) show that light can be shed on highly complex problems in this way.

If we wish to generate concrete dynamics with the help of such models, however, it will be necessary to attempt a simulation and quantification. Several possibilities present themselves for achieving this end (see Schaub & Schiepek, this volume). The "construction of heuristic mathematical models of dynamics" (Ciompi 1989, p. 19) is one of these. This is the goal of the present contribution, a contribution to the solution of problems (b) and (c) thereby also being made.

2. Schizophrenia as a Dynamical Disease

If one seriously considers the task of generating the dynamic pattern of a process (e.g. an illness) by an appropriate mathematical model, one will be obliged to study the dynamics of psychiatric disorders. In medicine, a number of dysfunctions of biological systems are known which are characterized by specific temporal patterns, such as Cheyne-Stokes respiration, periodic hematopoiesis, epileptic seizures and cardiac arrythmias (so-called "dynamical diseases", see Mackey & an der Heiden 1982; an der Heiden, this volume). Under normal conditions, physiological variables such as breathing frequency, blood cell concentration, neuronal activity, or heart beat frequency, display either a stationary or fairly regular periodic activity. In the course of a disease, however, the behavior of these variables may undergo a sequence of distinct and characteristic oscillatory patterns. However, the transition from periodicity to irregularity and chaos cannot always be considered equivalent to the transition from a healthy state to a pathological one, as has recently been shown by research on the endocrinology of osteoporosis (Hesch 1990; Gerok 1989). Measurements of the calcium and parathormone levels at two-minute intervals indicate that it is precisely in healthy persons that the concentration of parathormone is subject to chaotic fluctuations with irregular microfluctuations being superposed by irregular macropulses, whereas such fluctuations are almost completely absent in individuals suffering from osteoporosis. Clearly, normal bone formation depends on irregular fluctuations of the parathormone signal, whereas a decline in irregular fluctuations — that is, a rigid order, with the parathormone concentration kept constantly at the same level — is characteristic for osteoporosis. In accordance with Gerok's approach (1989), the conclusion can be drawn that a healthy state encompasses dynamics of both order *and* chaos, whereas illness on the other hand means either "rigid" order or "irregular" disorder.

This leads to the question of whether such dynamic approaches can also be productively employed in psychiatry, and applied not only to the level of functioning of neurobiological systems (see Babloyantz & Destexhe 1987 for the nonlinear analysis of EEG recordings; Mayer-Kress & Holzfuss 1987) or to the time-scale of circadian rhythms (Pflug 1987), but also to the level of the long-term evolution of biopsychosocial systems. "Indeed, due to the scalar invariance inherent to chaotic dynamic systems in general, one can well expect a dynamic approach to schizophrenia to be justified at the "social time scale" of long-term course (years to decades), as well as at the "psychological time scale" of a single exacerbation of illness (days to months) or the "biological time scale" of a single psychotic episode (milliseconds to hours)" (Schmid 1991). A better understanding of the long-term course of illnesses provides the basis for gaining insight into the nature of psychiatric disorders. The significance of long-term rhythms is obvious with regard to manic-depressive psychosis or unipolar cyclic depression (Wehr & Goodwin 1979), making their dynamic nature evident ("dynamical diseases", see Fig. 3).

Increasing attention has also been paid to the temporal patterns found in schizophrenia. Some time ago, the field was still dominated by the notion of an unstructured and progressive course and an unfavorable prognosis, based

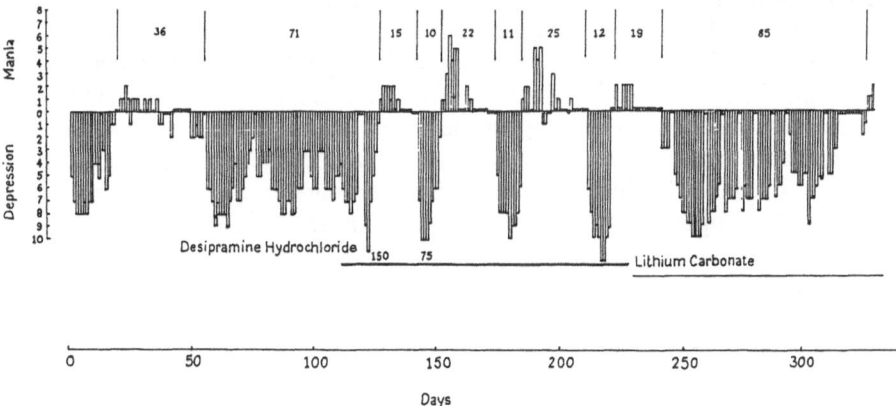

Fig. 3: Daily mood ratings of a bipolar patient with rapid cycling between mania and depression. Vertical lines above ratings indicate where switches into mania or depression occurred; numbers above ratings indicate days elapsed between switches in either direction (Wehr & Goodwin 1979).

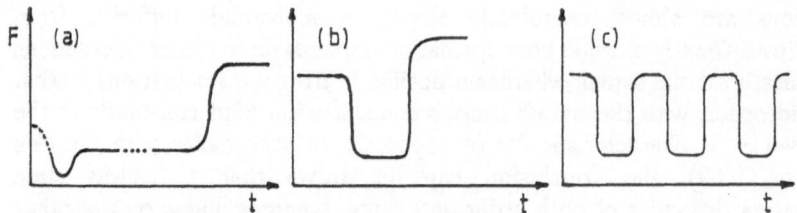

Fig. 4:(a) Woodshedding; (b) The lowturning point; (c) Oscillating levels of function (F = level of psychosocial functioning)

upon Kraepelin's description of dementia praecox; today, however, a number of distinct clinical patterns may be described. Strauss (1989, pp. 23 ff.) describes three longitudinal patterns:

a) *Woodshedding*. Patients who have experienced a psychotic episode afterwards remain on a relatively low plateau of functioning characterized by apathy and withdrawal for an extended period of time. The woodshedding pattern "appears to reflect a pathologic withdrawal — a negative symptom; but it may actually be a mechanism to prevent recurrence of positive symptoms" (autoprotective strategies, see Böker 1986; Gross 1986) "by providing respite of the demands of interaction with the environment" (Strauss 1989, p.23). "Patients then begin to improve and may reach a higher level of functioning than they had ever previously accomplished"(Strauss 1989, p.23) (Fig.4a).

b) *The low turning point*. "The second longitudinal pattern is one of organization followed by disorganization followed by reorganization with a somewhat new psychological structure" (Strauss 1989, p.23) (Fig.4b). In terms of self-organisation, this could be considered a two-step order-to-order transition (or phase transition), since a schizophrenic episode represents a highly structured pattern (states of mind and interactional patterns, see Horowitz 1987; "lifestyle scenario with strong attractor characteristics", see Schiepek, Fricke & Kaimer, this volume), although it may be experienced and described on a phenomenological level as a state of disorganization. A behavior which began as coping becomes more and more maladaptive. "Disorganization of a massive type ensues" (a schizophrenic state of order) "which in spite of its "illness" characteristics appears to open the way to a new approach to coping as part of the reorganization" (Strauss, 1989, p.25).

c) *Oscillating levels of function*. "The third pattern in the course of the disorder involves oscillation around a mean level of functioning" (Strauss, 1989, p. 25), i.e. oscillation between functioning extremely well on the one hand and psychotic decompensation on the other (Fig.4c).

Strauss (1989) emphasizes that all three patterns have one characteristic in common: discontinuity in the evolution of patient functioning over time. Such qualitative shifts in the behavior of a system, usually determined by a change in one or more parameters (control parameter(s)), are termed phase transitions. With the help of nonlinear systems theory, phase transitions between different states of order of psychological functioning can be not only analyzed but also generated or simulated (see Haken 1983; 1990a). Models accordingly suggest themselves which allow a nonlinear analysis of dynamic patterns. Within the framework of the theory of dynamic systems, attempts to this end are made by the concept of dynamical diseases, which focuses particularly on the analysis of bifurcations.

Of special theoretical and therapeutic interest is the fact that, despite dramatic changes in the behavior of functional variables, the underlying control system may still be intact, with the exception of a single (or a few) parameter(s) being out of its (their) physiological ("healthy") range. This has already been shown to be the case in physiological and biochemical systems (Mackey & an der Heiden 1982; an der Heiden & Mackey 1987; Glass 1987; for a comprehensive survey see Rensing et al. 1987). It seems worthwhile to attempt an application of this promising model to schizophrenic phenomena. This could in fact be very rewarding, as various distinct longitudinal patterns characterized by discontinuous phase transitions between psychotic episodes, healthy functioning and chronic states have been detected in the long-term course of schizophrenia (cf. the findings of Bleuler 1972; Ciompi & Müller 1976; Huber et al. 1979; Harding 1984). Investigations have shown that in the long run one can expect about 25% complete remissions, about 40-50% moderate residual states and only about 35% cases of permanent, severe, mental invalidity. There can be no doubt that an improved care system and more differentiated therapeutic and rehabilitative facilities contribute to these results. The different patterns of the long-term evolution of schizophrenia as found by Ciompi and Müller (1976), together with their frequencies, are shown in Fig.5 (compare Ciompi 1989). These patterns are the result of various combinations of slow vs. acute onset, acute episodes vs. progressive deterioration, and remission vs. chronic end-state. These patterns are the empirical basis, which will be explained (explanandum) by the simulation model presented below.

3. Development of a Structural Model

The etiology of schizophrenia has been the subject of research in various quite disparate disciplines: phenomenological and experimental psychopathology, psychology, physiology, biochemistry, neurobiology, communications and family research, psychoanalysis, sociology and ethnology, to name only a few. Each of these disciplines focuses on different aspects of the highly complex biopsychosocial system of "schizophrenia", so that it is hardly possible any longer to maintain an overall perspective. Accordingly, the selection of variables and parameters given below does not claim completeness. At the same time, however, it forms the basis for a structural model which does not postulate the status of a theory. The variables have been selected on the basis of reviews of psychiatric and psychological research from recent years (e.g.

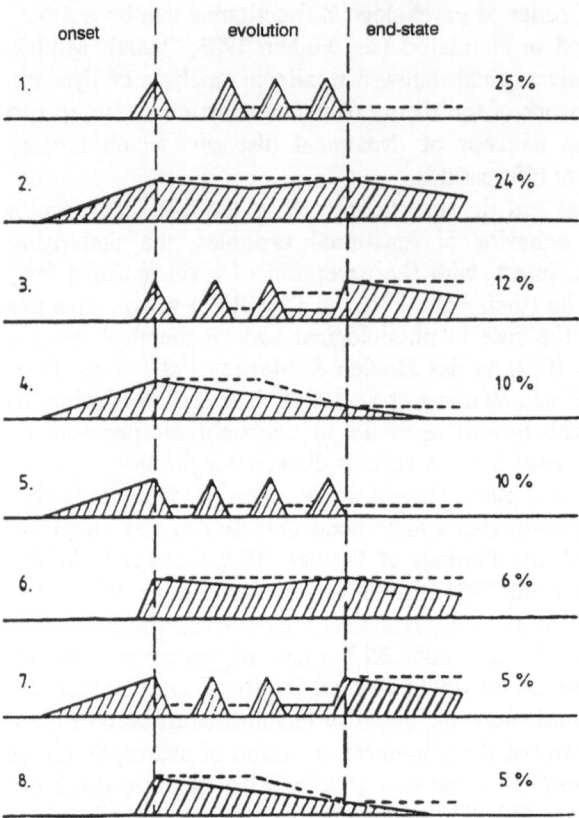

Fig. 5: Long-term evolution of schizophrenia (Ciompi & Müller 1976; Fig. from Ciompi 1989, p. 16).

Böker & Brenner 1986; 1989), as well as the presentation by Ciompi (1989). Each of the constructs included could be regarded as a system of its own and analyzed at a still finer level of resolution (see also the concept of the microlevel approach as a virtual horizon of the high resolution of a system in psychological synergetics, Schiepek & Tschacher, this volume). However, this will not be pursued here.

3.1 Variables

The model interrelates five different variables or constructs:

1) *Cognitive disorders (c)*. These correspond to those cognitive dysfunctions which have been termed first rank symptoms by Kurt Schneider (thought disorders, such as broadcasting of thought, thought hearing, thought blocking) or disturbances of association as defined by E.Bleuler. "Basic symptoms" or "uncharacteristic basic disorders", as described by Süllwold, Huber and others, should also be included (cf. Süllwold 1986; compare also the items of the

Frankfurt Complaints Questionnaire (FBF), Süllwold 1983). These include difficulties in maintaining focused attention, loss of coherent thinking, and perceptual disorders experienced as visual or acoustic overtaxation (filter defect).

2) *Stress (s)*. Any kind of overtaxation, whether perceptual, cognitive, performance-related or interpersonal, may be experienced as stressful. Diffuse, highly emotional or overly critical styles of interaction have been shown to support relapse. This is also true for sensory overstimulation and overtaxation at work. Stress not only leads to endosecretory and physiological modifications, but also affects the central nervous system. These modifications can become more or less fixed in the functioning of biochemical and physiological systems and also in the cerebral substratum.

3) *Withdrawal (w)*. This construct refers to phenomena of covert (emotional) and overt (spatial) withdrawal, e.g. blunting of affect, inapproachability, social isolation, increased involvement in intrapsychic affairs, or even actually shutting oneself away in a room. Extreme cases include catalysis and catatonia. The variable referred to covers the area of autism as defined by E.Beuler, who subdivides the psychopathological phenomena of schizophrenia into autism, affective disorders and disturbances of association. With regard to the recent differentiation of positive and negative psychological symptoms, it clearly belongs to the latter.

4) *Expressed Emotion (e)*. There have been a great number of studies investigating the stress-inducing and relapse-supporting effects of a family atmosphere characterized by criticism and emotional over-involvement. In contrast to previous concepts of family-dynamics, such as the "double-bind" hypothesis, the concept of pseudo-community, mystification, etc. (for a survey see Bateson et al. 1984), all theoretically more ambitious but lacking empirical confirmation, the concept of expressed emotion (EE) has considerably stimulated empirical family research. Various measuring scales have been developed (e.g. the Camberwell Family Interview), different aspects of the construct worked out, temporal patterns differentiated. The measurement of physiological correlates of the EE construct and an assessment of their relationship with interactional and coping styles of the patients, as well as the devising of treatment methods or, rather, management strategies (family education regarding the nature of the disorder, communication and problem-solving training, limitation of face-to-face contact), have also resulted from the intense research in this field.

5) *Delusion (d)*. In addition to systematized delusions, this construct also comprises phenomena such as delusional ideas, percepts and moods, as well as misidentifications or illusions. In a broad sense, it corresponds to the positive symptoms of schizophrenia.

3.2 Parameters

The values introduced into equations as constants are termed parameters. Therefore, constructs which are invariable or subject to very slow modifications over the course of time (as opposed to the temporal dynamics of variables) take on the function of parameters in the equations proposed below. Their purpose is to mediate in specific ways the interactions of the variables described above.

1) *Diffuseness of affective-cognitive schemata* (σ). The construct of affective-cognitive frames of reference (or schemata) lies at the core of Ciompi's conceptualization of the "Affektlogik" (Ciompi 1982 (in English 1988); 1986a, b). Programs for thinking, feeling and behaving are internalized as a condensation of the entire experience of an individual: external dynamics become internal structure (Ciompi 1986a, pp.51 ff.). Due to the diffuseness and contradictions experienced in interpersonal encounters with close relatives, the affective-cognitive systems of reference which mediate emotional experience and interpersonal behavior seem to be extremely diffuse and unclear. Yet one must not assume linear causal effects. In the first place, the vulnerable person participates in the specific dynamics of communication (within the family) from the very beginning. Secondly, biological factors, such as a genetically rooted tendency toward synaptical over-reacting, also play a part in contributing to the impairment of schemata development. Thus Ciompi (1986a) argues that quite specific defects in the relevant affective-cognitive systems of reference (schemata) generate a vulnerable premorbid terrain for psychotic decompensation. Interestingly, the diffuseness of these systems of reference evidently corresponds to certain phenomena which can be discovered in the neuronal substratum and are explained by the concept of "neural plasticity" (Haracz 1984;1985). As with all learning processes, the evolution of schemata necessitates a biological basis, consisting of functional and anatomical changes in the neuronal system (e.g. electrophysiological potentation and synaptic-dendritic sprouting).

2) *Dopamine and serotonin metabolism* (δ). An important mediator in the evolution of schizophrenia seems to be the metabolism of neurotransmitters such as dopamine, serotonin and norepinephrin. Findings indicate an over-activity of the dopaminergic and serotonergic receptors, evidence for this including the blocking of dopamine receptors by neuroleptics. Recently developed drugs (e.g. pipamperone, setoperone, ritanserin, risperidone) affect the serotonin metabolism in an antagonistic manner (Gelders 1989). Dopamine metabolism is activated by amphetamine-like drugs, which can provoke psychoses. Modifications in the plasma level of the dopamine metabolite homovanillin acid parallel the intensity of psychotic symptoms, and increased levels of serotonin have been found in chronic schizophrenics. A recent finding is the increased number of dopamine receptors in schizophrenics post-

mortem. Also, observations have shown that dopamine metabolism is greatly affected by stress (Haracz 1984;1985; Trulson & Preussler 1984). On the basis of these findings, dopamine may be assumed to be an important mediator between environmental stimuli and the cerebral substratum, thus gaining theoretical significance particularly with respect to concepts emphasizing the relationship between emotional reactions on the one hand and cognitive performances on the other (Ciompi 1982). If the cerebral areas relating to the dopaminergic system are examined, several parts are seen to relate the old, basal parts of the brain through the limbic system to the cortex, and particularly to the frontal lobe. Also, serotonin receptors have been shown to be concentrated in the mesolimbic system, the frontal lobe and the striatum.

3) *Social incompetence* (κ). As in the case of σ and δ, the high values of the parameter social competence κ (indicating a high degree of incompetence) suggest an increased vulnerability to schizophrenia. Prior to the onset of acute psychosis, vulnerable individuals seem to be characterized by deficiencies regarding their social skills. This is especially true in cases of poor premorbid adjustment. Social skills can hardly be developed to a sufficient degree within stressful family environments and contradictory communication structures accompanied by diffuse cognitive schemata for social perception and interpersonal interaction. Therefore, various treatment programs (such as the Integrated Psychological Treatment Program (IPT), Brenner et al. or the social skills training by Liberman) are directed at improving the social perceptual and interactional skills of schizophrenic patients.

Apart from their contributing to pathogenesis, impaired abilities in establishing social relations represent a key characteristic of acute and chronic schizophrenic disorders, and may also be considered an artefact of hospitalization. Social competence may serve as an auto-protective strategy in coping with stress and stressful social interactions (high expressed emotions) and thus represent a "protective shield" which becomes "pervious" under certain conditions. An investigation by Käsermann and Altorfer (1989) found that a patient's physiological reactions (e.g. pulse rate) differed depending on whether he succeeded in reacting appropriately to invasive and stressful communication on the part of his parents, or whether he reacted in an evasive manner.

4) β_c 5) β_s 6) β_w 7) β_e 8) β_d: Within the mixed feedback model proposed here, the extent of negative feedback is mediated by means of these five β-parameters: that is, increased values for β indicate stronger damping effects on the pertinent variable. Similar parameters are employed within models of population dynamics (cf. Krebs 1985; Odum 1983; Kriz 1990) to determine the effects of limited resources (resulting from competition or high population density, for example) on the decrease of population density. With regard to the dynamics of schizophrenia, high values of β_c, β_w and β_d, referring to cognitive disorders, withdrawal and delusions respectively, could be

interpreted as an antipsychotic tendency on the way to recovery, whereas high values of β_s and β_e, referring to stress and high expressed emotion respectively, would indicate a natural relaxing tendency.

9) *Genetic risk (γ)*. Our model does not provide for specific and direct influence by this parameter. Rather, it is assumed that certain genetic factors constitute the background for the vulnerability expressed by the parameters σ, δ and κ to evolve (see sect. 6c). However, genetic factors alone are not sufficient to transform the vulnerability into clinically overt schizophrenia. As was convincingly shown by a recent Finnish adoptive-family study of schizophrenia (Tienari et al. 1989), this requires a specific interaction with a disturbed rearing family environment.

3.3 Modelling

In this section we will elaborate the hypotheses by means of which the variables described in section 3.1 are related to one another. As in sections 3.1 and 3.2, a vast number of references in support of these hypotheses could be quoted; however, we will refrain from doing so for reasons of space. The following description of the interactions between the variables can be compared directly with the difference equations (1) through (5) given in section 4, and also with Fig.6.

1) It is proposed that an increase of *cognitive disorders (c)* depends linearly on the prevailing intensity of cognitive disorders, and is mediated by the

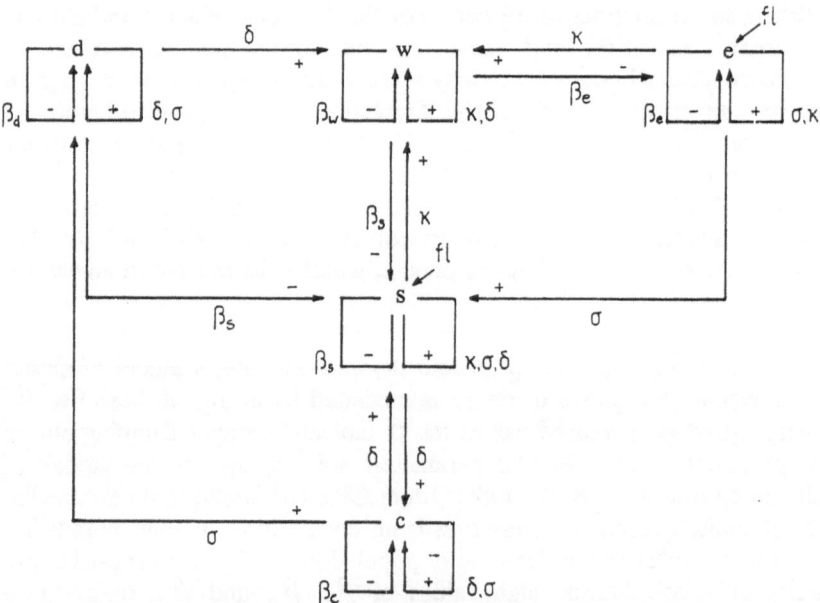

Fig. 6: The structural model as a basis for simulation.

parameter σ. This is to say that the higher the degree of disturbance of affective-cognitive schemata for the perception and processing of internal (intrapsychic) and external (environmental) events, the smaller the possibility that the patient will comprehend and cope with perceived anomalies of cognitive functioning, and the greater the chance for an intensification and escalation of the cognitive disorders. In addition, cognitive disorders interact nonlinearly with stress. On the one hand, they evoke anxiety and insecurity and are thus stress-inducing; on the other hand, they are activated by stressful events. The interaction is mediated by the dopamine and serotonin metabolism (parameter δ). This is in agreement with findings indicating that medication reducing serotonin and dopamine activity has a positive effect on cognitive disorders and may also reduce the impact of stress on cognitive disorders. Finally, the model includes a nonlinear negative feedback loop with respect to cognitive disorders (parameter β_c; see sect. 3.2).

2) An increase of *stress (s)* depends linearly on the degree of prevailing stress. The Parameter κ mediates this linear dependency. Stress leads to more stress if social skills of regulating interpersonal closeness, coping with difficult situations and stress management are only poorly developed. A nonlinear decline of stress also depends on the current degree of stress. Thus, there is a large decline in stress (e.g. through exhaustion or habituation), outweighing further increase, for very high degrees of stress. Furthermore, the decline of stress is regulated by the interactive effects of stress and withdrawal, as well as stress and delusions. Intrapsychic or actual withdrawal means avoiding stress, whereas delusions reduce stress induced by experiences such as loss of coherent thinking, perceptual overtaxation or the dissolution of ego boundaries (represented by the variable of cognitive disorders). It is exactly this nonlinear interaction between stress and cognitive disorders which has a stress-inducing effect mediated by the serotonergic and dopaminergic activity (δ). Additionally, social interactions of the high expressed emotion type are stress-inducing, the impact being greater on a patient already suffering from a high degree of stress (nonlinear interaction). This effect grows more intense with decreased abilities to comprehend, and cope with, stressful interpersonal experiences. Therefore, the parameter σ (diffuseness of affective-cognitive schemata) is formally employed for mediating the stress-inducing effects of the interaction $e \cdot s$.

Since the occurrence of stressful events is determined not only by the processes described above but also by random events in the patient's environment which are neither predictable nor possible to control, random fluctuations (fl) generating stress must also be included in the model.

3) The intensity of a patient's *withdrawal (w)* is considered to be linearly dependent on the current degree of withdrawal, since autistic encapsulation and spatial isolation are thought to have a self-reinforcing effect which becomes increasingly marked with reduced social skills, allowing the person to

keep or get in contact with others (parameter κ). Furthermore, there is a nonlinear interaction between withdrawal and stress, especially for low levels of social and stress-management skills, resulting in withdrawal as an auto-protective effort in the case of stressful events, be they intrapsychic or environmentally caused. The experience of highly emotional and overly critical social interactions (e) leads to basically analogous results. Furthermore, delusions interact with preexisting withdrawal to effect further withdrawal. After all, delusions are defined as a process of "wrapping oneself up" in an idiosyncratic world of one's own. The parameter δ (serotonergic and dopaminergic activity) was chosen as the mediator, since positive symptoms such as delusions and hallucinations can be effectively influenced by neuroleptic medication. The parameter β_w determines the impact of withdrawal on future withdrawal by quadratic negative feedback.

4) In our model the variable e (*expressed emotion*) not only represents the "actual" behavior of the patient's relatives but also, more importantly, describes the patient's subjective experience of EE attitudes. The change of e over the course of time is considered a linear function of e in two different ways. It is mediated by the parameter σ, on the one hand, and the parameter κ, on the other, for difficulties in coping with highly emotional and overly critical interpersonal encounters are the result both of diffuse schemata and of poorly developed social skills. Additionally, patients' socially incompetent reactions to EE attitudes contribute to further increases in criticism and highly emotional statements on the part of the family members. The nonlinear decline of e (mediated by β_e) is regulated by the interaction between e and w, as withdrawal means interrupting face-to-face encounters and escaping emotionally from stressful communications, and by means of a quadratic term, e^2. The model allows for random fluctuations (fl) of e, since this variable also represents relatives' behavior as influenced by external conditions.

5) The temporal changes in the degree of *delusion* (d) are, first of all, considered to be a linear function of the preexisting degree of delusion; this represents the tendency of delusional processes to spread out, affecting an ever-increasing part of ideation. This growth function is related to neurochemical processes of the dopamine and serotonin metabolism via the parameter δ. Intrapsychic processes resulting in delusion are represented by the nonlinear interaction between cognitive disorders, which may occasion delusional ideas by their very occurrence — at least when consciously experienced as such — and the tendency for delusional interpretations currently existing. The parameter σ is suggested as mediator of this kind of interaction, as the inability to perceive and react appropriately to complex or ambiguous intrapsychic but also social stimuli represents the basis for strange interpretations and paranoidal ideas. Finally, the equation governing the dynamic processes of delusions contains a quadratic feedback term, relating the degree of delusions to the parameter β_d.

3.4 Production and Protection

Following more sophisticated descriptions of the dynamic processes of schizophrenia, it becomes evident that some symptoms may be regarded as efforts at coping with primary symptoms or previously unsuccessful coping mechanisms (see also the concept of auto-protective strategies by Böker 1986; Gross 1986). This is apparently true with regard to withdrawal following perceptual overtaxation, stressful requirements and interpersonal encounters. Yet, going beyond this approach, "...the notions of schizophrenia put forth by Hughlings Jackson (1958), Bleuler (1950), and Freud (1959) consider even such symptomatic phenomena as delusions and hallucinations to be coping mechanisms as opposed to primary illness mechanisms such as disorders in formal thought" (Strauss 1989, p.20; for extensive coverage see Böker & Brenner 1983; Böker 1986). It can be readily understood that it is easier to cope with a coherent pattern of interpretations, although it cannot be shared with other persons, than an inconsistent, uncontrollable and disintegrated world of subjective experience causing great anxiety. Following this line of argument, delusions may be regarded as self-healing or auto-protective strategies. In accordance with a theory of auto-protective strategies (Böker & Brenner 1983; Brenner 1986), the variables included in our model can be categorized as those contributing to the *production* of schizophrenia, such as cognitive disorders, stress, expressed emotions, and those serving self-healing or *protective* efforts, such as withdrawal and delusions (Fig.7).

This view corresponds exactly to that put forward within the theory of dynamical diseases. Within this approach, the rate of change of a certain substance (here a variable) with time, $x(t)$, is considered as given by the difference between the production rate, p, and the destruction (or loss) rate, d:

$$\frac{dx}{dt} = p - d.$$

We would like to emphasize that, in synergetic terms, the parameters introduced in section 3.2 can be considered the control parameters of a complex biopsychosocial system, which is described on the macroscopic level by the variables c, s, w, e and d (compare Haken 1988). It is even more reasonable to regard the parameters σ, δ and κ as control parameters, as they refer to three of the most important fields of the therapeutic treatment of schizophrenia, namely those of offering the patient clear and unambiguous communication, and training relatives to get acquainted with these principles, together with further treatment programs promoting the development of appropriate cognitive schemata aimed at improving the management of social situations (σ), neuroleptic medication (δ) and social skills training (κ) (Fig.7).

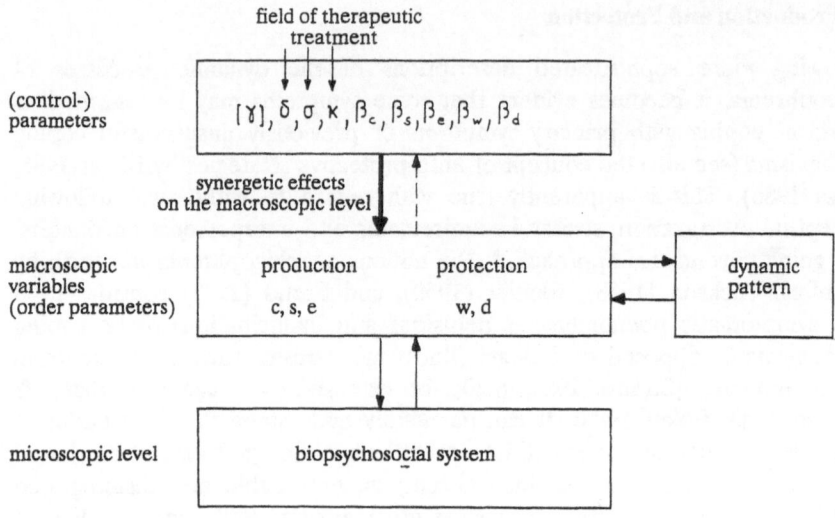

Fig. 7: Synergetic model of schizophrenia.

4. Transformation of the Structural Model into a Set of Equations

There have been several attempts to date to create a mathematical model of the dynamic patterns of the evolution of schizophrenia. For example, Cronin (1977) describes the periodic patterns of psychomotor symptoms and biochemical processes occurring in periodic catatonic schizophrenia by means of a set of differential equations. The theory of dynamic systems has been applied in an analysis of dysfunctions within the feedback regulation of the dopamine metabolism by King and Barchas (1983). Elkaim et al. (1980) have attempted an analysis of the intrafamiliar patterns of interactions in terms of symmetry-breaking bifurcations.

Nonlinear differential or difference equations are especially suitable for a modelling of the dynamics of complex systems because of their ability to generate asymptotically stable, simple or complex periodic or even chaotic patterns, depending on the choice of parameters (Mackey & an der Heiden 1982; Seifritz 1987; Schuster 1988).

Another reason lies in the fact that it is possible to transform graphically represented structural models into sets of equations. This will be demonstrated using the Lotka-Volterra model, which describes in an idealized way the interactions of a prey population (x) and a predator population (y):

$$\dot{x} = \alpha x - \beta xy$$
$$\dot{y} = -\gamma y + \delta xy$$

where the parameters α, β, γ and δ represent positive constants.

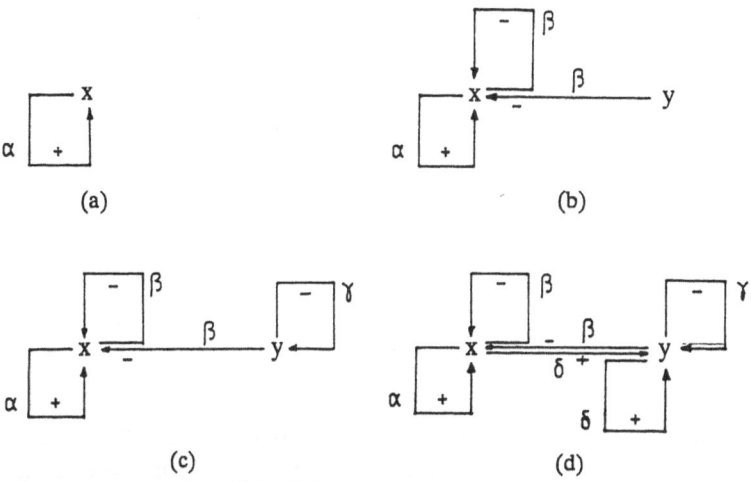

Fig. 8: Structural model of the predator (y) —prey (x) interaction according to the Lotka-Volterra equations.

(Note: The parameters contained in this example are not identical to those introduced in section 3.2.)

The first of the equations given implies that there would be unlimited growth of the prey population (the differential equation $\dot{x} = \alpha x$ corresponds to the experimental growth function $x(t) = e^{\alpha t}$ if predators did not exist ($y = 0$) (Fig.8a). Also, a decline of the prey population ($x < 0$) can take place only if both prey and predators are present (Fig.8b). The second equation implies that the predators would become extinct ($y(t) = e^{-\gamma t}$) without prey ($x = 0$)(Fig.8c). The predator population will start growing after the prey population has grown to some threshold density. Consequent increased predation will again reduce the prey population (Fig.8d) (compare May 1980b; Seifritz 1987). This results in the well-known oscillatory pattern of predator-prey interaction.

The analogy between the structural model of a dynamic system and the corresponding equations clearly shows that a fundamental precondition for mathematical formalization lies in the systemic analysis of the interactions between the relevant system variables. A detailed description of the development of structural models of biopsychosocial systems is given in Schiepek (1986; 1991), where it is referred to as "idiographic systems modelling". The combination of reductionism and holism characteristic of systems approaches becomes obvious in this process (see Odum 1983 with regard to ecology).

The Verhulst equation ("logistic map") provides the basis for the computer simulation of the dynamics of schizophrenia:

$$\dot{x} = \alpha x(1 - x/K)$$

where α is the growth rate (often called r instead of α), x is the population size and K is the maximum value of x. This equation, known in ecology as the logistic function of population growth (taking the general form of a sigmoid curve when numbers are plotted against time), has a major place in theoretical models of population growth (cf. May 1980a; Krebs 1985, chap. 12; see also Seifritz 1987, chap. 2, for mathematical details, especially with respect to chaotic systems).

Rearranging the above equation yields

$$\dot{x} = (\alpha - \beta x)x$$

where the growth rate α is equal to the difference between the birth rate γ and the death rate δ, the latter assumed to be constant for a certain species within its specific niche ($\alpha = \gamma - \delta$). The coefficient $\beta = 1/K$ denotes the limitations of food and space, indicating that with growing population density there will be a decrease in population size exceeding that of the natural death rate ($-\beta x^2$). The population will remain stable when $x = \alpha/\beta$, largely irrespective of the initial population size x_0 (but with $x_0 > 0$).

The term αx can be understood to represent an autocatalytic process of the system in question, denoting the exponential rate function $x(t) = e^{\alpha t}$. Autocatalytic processes are considered a fundamental mechanism effective in many kinds of self-organizing dynamical systems (cf. Nicolis & Prigogine 1987).

The size or density of a population (variable x) takes the role of an order parameter governing the system dynamics. On the microscopic level, this corrresponds to the birth and death of a large number of individuals. Analogously, the macroscopic variables in our theoretical model of schizophrenia take the role of order parameters of a dynamic pattern, one constituted through a highly complex biopsychosocial system on the microscopic level (Fig.7).

Another analogy is given by the fact that the change in the population size x is a function of the production and destruction rates (Haken 1990, chap. 19; compare the above statements with respect to the concept of Dynamical Disease). When $\alpha = \gamma - \delta$,

$$\begin{aligned} \dot{x} &= (\gamma - \delta)x - \beta x^2 \\ &= \gamma x - (\delta + \beta x)x \\ &= p - d. \end{aligned}$$

This suggests that the production rate p equals the population size multiplied by the natural birth rate, while the destruction rate d equals the natural death rate multiplied by the population density together with the resource limiting coefficient multiplied by the square of the population density.

However, it must be pointed out that theoretical models of population dynamics have been chosen as an aid in the simulation of psychosocial processes on account of their presenting readily observable analogies. We do not mean to imply that — apart from formal aspects — the growth of animal populations bears any resemblance to personal suffering and interpersonal conflicts. Besides, this approach does not imply a direct transference of ecological models to the realm of psychiatry. Rather, the parallels that have been pointed out demonstrate how the theory of dynamic systems may be applied within a wide range of quite different disciplines to model complex dynamic phenomena. In this sense, the synergetic approach is equally applicable to ecology, population dynamics and psychiatry (e.g. serving the understanding of etiological or therapeutic processes) as well as to a number of other disciplines (cf. Haken 1983; 1990; Kriz 1990 and in this volume; Schiepek, Fricke & Kaimer, this volume).

A further advantage of models of population dynamics may be seen in the fact that they allow a prediction of the outcomes of different types of interactions between variables (animal populations), such as cooperation (facilitation) or interference. If one variable (x) exerts a facilitating influence on another variable (y) (crosscatalytic processes), this may be described as symbiotic cooperation between these two variables (or populations):

$$\dot{x} = (\alpha_x + \alpha'_x y - \beta_x x)x$$
$$\dot{y} = (\alpha_y + \alpha'_y x - \beta_y y)y$$

where the parameters α'_x and α'_y mediate the crosscatalytic or symbiotic impact of y on x and of x on y respectively (cf. Haken 1990, p.311; Kriz 1990).

The impact of two variables interfering with each other can be conceived of as being analogous to competition: that is, two populations simultaneously competing for the same resource. The resource limiting terms of both variables include values x and y to give

$$\dot{x} = (\alpha_x - \beta_x(x+y))x$$
$$\dot{y} = (\alpha_y - \beta_y(x+y))y$$

In the following we would like to turn to an examination of discrete steps of change, as opposed to the continuous processes of change considered up to now. Thus, the change of a variable x at time $t+1$ becomes a function of x at time t, which gives $x_{t+1}=f(x_t)$. The term $\dot{x} = dx/dt$ thus becomes $\Delta x = x_{t+1} - x_t$, where Δ denotes the difference operator. (The index n is often given instead of the index t; this denotes, analogously, discrete points in time, successive generations of a population.) With respect to the simulation of schizophrenia, one week has been chosen as a reasonable time interval $(t+1)-t$, where Δc, Δs, Δw, Δe and Δd represent the change in the level of the corresponding variables within the space of a single week. (For more detailed information on discrete maps and difference equations, see Goldberg 1968; Haken 1990a; 1983, chap. 12.6; Seifritz 1987, chap. 2; Martienssen 1989.)

The above considerations present the basis which allows the transformation of the structural systems model of schizophrenia described in section 3.3 into a set of nonlinear difference equations (abbreviations correspond to those introduced in sect. 3):

$$\Delta c = \sigma c + \delta sc - \beta_c c^2 \tag{1}$$

$$\Delta s = \kappa s + \sigma es + \delta cs - \beta_s(w + d + s)s + fl \tag{2}$$

$$\Delta w = \kappa w + \kappa sw + \kappa ew + \delta dw - \beta_w w^2 \tag{3}$$

$$\Delta e = \sigma e + \kappa e - \beta_e(w + e)e + fl \tag{4}$$

$$\Delta d = \delta d + \sigma cd - \beta_d d^2 . \tag{5}$$

These equations combine the effects of symbiosis and competition between the variables, designating a model of mixed feedback in accordance with Fig.6 (cf. an der Heiden & Mackey 1987). Equations (2) and (4) provide for random fluctuations, since it must be assumed that stress and expressed emotion are also activated by external environmental influences.

5. System Dynamics

Depending on the parameter values chosen, each single equation describes a logistic rate function which takes the form of a sigmoid curve with a more or less steep slope representing the growth or the decline of the pertinent population (variable) (Fig.9). The y-coordinates may be interpreted roughly as higher or lower values; precise quantitative interpretations are not possible. This applies to the following figures as well.

As is shown by the well-known Feigenbaum route (Fig.10), the system behavior converges towards a fixed point for low values of the control parameter. If these values are increased, periodic oscillations will result, with two fixed points occurring for each parameter value beyond the first bifurcation. Beyond a critical value of the control parameters, the system behavior is driven into the chaotic regime following several successive period doublings.

However, in the present simulation the system behavior of each single equation converges asymptopically towards a fixed point in accordance with the parameter values chosen, which corresponds to points "to the left of" the first bifurcation of the Feigenbaum route. Thus, the overall behavior of the system results not from the intrinsic dynamics of each single equation, but

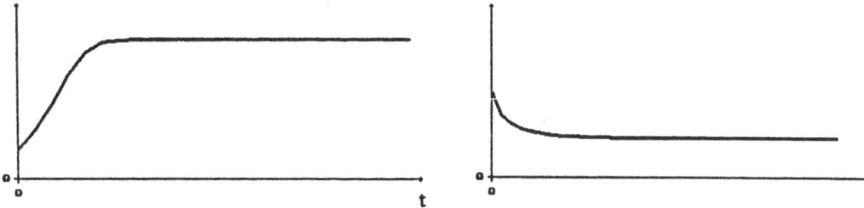

Fig. 9: Evolution of variable c according to equation (1): asymptotic growth or decline depending on the parameter chosen.

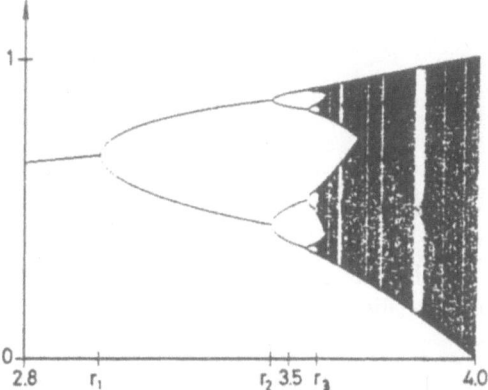

Fig. 10: The "Feigenbaum route" according to the logistic map
$x_{n+1} = rx_n (1 - x_n)$; $0 < r \leq 4$; $n = 0,1,2,...$ For control parameters $2,8 \leq r \leq 4$ the figure shows the quantities of iterates x_n (attractors) produced. Up to r_1: fixed point attractor; r_1 to r_2: simple periodic attractor; $r_2 - r_3$: period doubling; $r_3 - 4$: chaotic attractors. Standard range of ordinate values is [0,1] (Martienssen 1989, p 84).

rather (a) from their interactions — that is, from the interconnectedness of the complete system — and (b) from the external impacts (input) via random fluctuations.

The system behavior resulting from the interactions of all five equations (still without random fluctuations) depicts one single schizophrenic episode (Fig.11).

Initial conditions are represented by an increased value of the variable c (cognitive disorders), while the remaining variables begin at relatively low levels. The level of cognitive disorders decreases continuously, yet its initial value is sufficiently high to activate stress, withdrawal and delusion. The autonomous growth of e is regulated through the negative feedback of e on itself and increased withdrawal. Future elaborations of this model should take into consideration the fact that highly emotional and critical reactions on the part of a patient's relatives, such as worry and feelings of insecurity, may be caused not only by e, but also by the patient's delusional and autistic behavior.

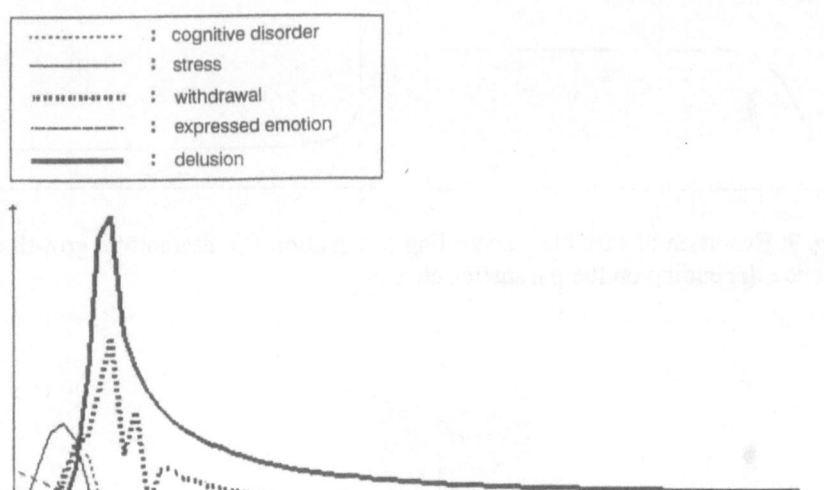

....	: cognitive disorder
———————	: stress
⊪⊪⊪⊪⊪⊪⊪⊪⊪⊪	: withdrawal
‒‒‒‒‒‒‒‒‒	: expressed emotion
▬▬▬▬▬▬▬	: delusion

Fig. 11: Evolutionary patterns of variables c, s, w, e and d according to equations (1) to (5); without random fluctuations.

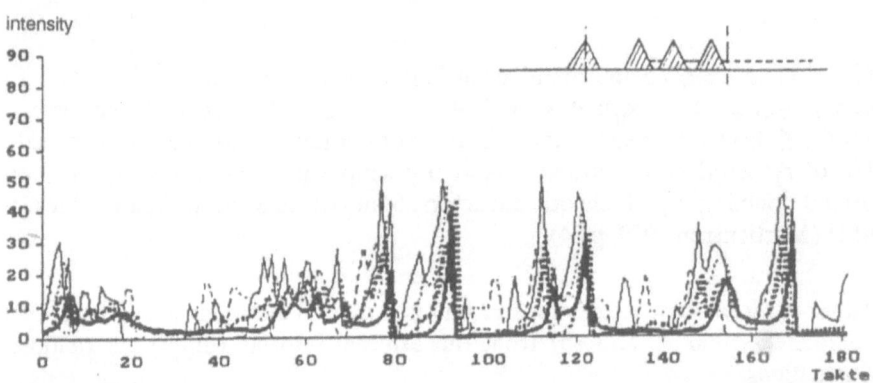

Fig. 12: Episodic evolutionary pattern with prodromal symptoms, acute onset. The parameter values are : $\sigma = 0.055$; $\delta = 0.041$; $\kappa = 0.034$; $\beta_c = 0.050$; $\beta_s = 0.030$; $\beta_w = 0.100$; $\beta_e = 0.009$; $\beta_d = 0.080$.

Stress ceases almost completely for high values of the autoprotective efforts of delusion and withdrawal. This is accompanied by a decrease in the values of d and w, as these types of coping strategies no longer seem necessary. The reasons for specific system behavior may be taken from the structural model presented in Fig.6.

The following figures (figs. 12-16) represent examples of the different system behavior resulting when random fluctuations are effective with respect to "stress" and "expressed emotion".

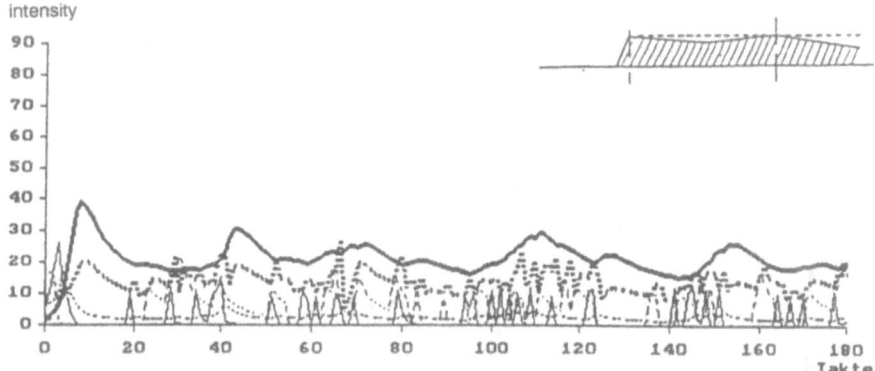

Fig. 13: Acute onset, chronic evolution, chronic end state. Same parameter values as in Fig. 12, except for: $\beta_w = 0.080$ and $\beta_d = 0.009$

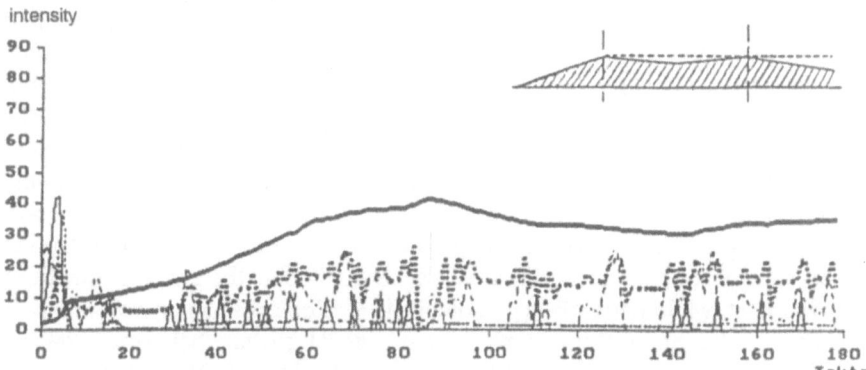

Fig. 14: Slow onset, chronic evolution. Same parameter values as in Fig. 12, except for: σ(only in equation (5)) = 0.015 and $\beta_d = 0.002$.

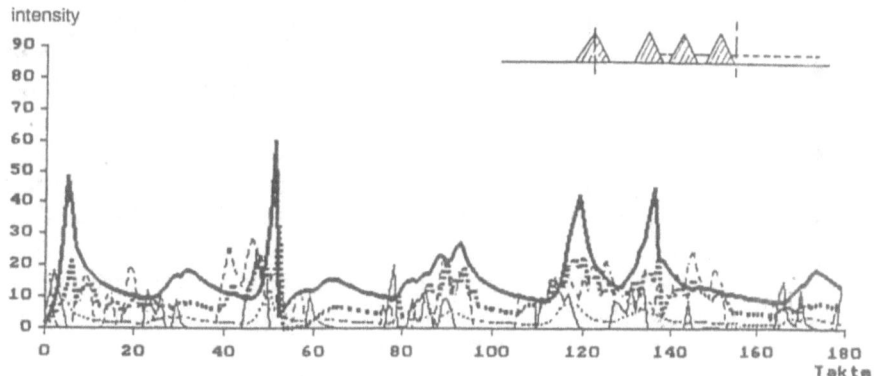

Fig. 15: Episodic evolution; acute onset without complete remission of symptoms between episodes. Same parameter values as in Fig. 12, except for: σ (only in equation (5)) = 0.130 and $\beta_d = 0.028$.

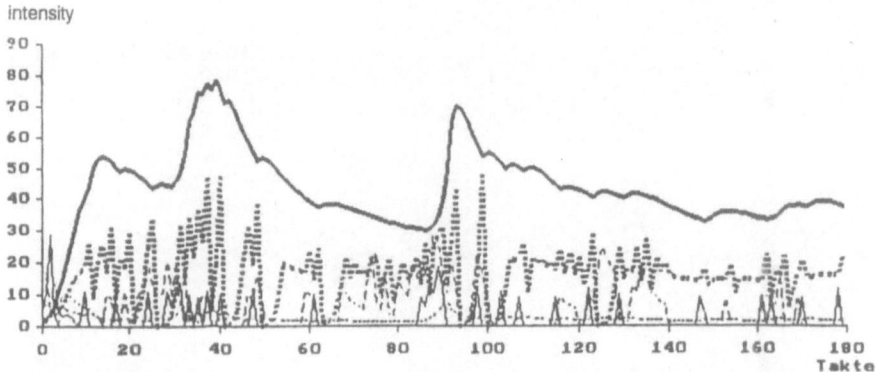

intensity

Fig. 16: Nontypical evolution. Same parameter values as in Fig. 12, except for: $\beta_d = 0.004$

The activation of schizophrenic processes by random fluctuations and their "dying out" without such external influence — that is, their tending towards latency (Fig.11) — correspond exactly to the assumptions made by the diathesis-stress model of schizophrenia. This model maintains that the physiological and intrapsychic systems and those pertaining to the regulation of interpersonal interaction are not pathological per se. Rather, they are very sensitive and vulnerable in their response to external irritations and demands, tending to overreact in particular ways. This view, in turn, corresponds to that of the concept of dynamical diseases, which holds that the systems' functions remain intact, while particular (pathological) — albeit principally reversible — changes in the parameters result in patterns of system behavior differing qualitatively from those of "healthy" system behavior (cf. Mackey & an der Heiden 1982; Hesch 1990).

This very phenomenon is demonstrated by the different simulations presented in Figs. 12-16, for each single simulation was generated by means of the same mechanisms contained in the structural model (Fig.6) and the set of equations. They differ with respect to only a few parameter values, which are stated for each figure. Moreover, the law of parsimony of scientific explanations (Occam's razor) has been very closely observed.

One might argue that it is merely the coincidence of two marked random fluctuations which causes the peaks (or schizophrenic episodes) such as those shown in Fig. 12. However, this objection is overruled by the fact that such an episodic pattern did not occur in every simulation and, moreover, that the patterns produced vary systematically in accordance with the parameter values chosen, largely independent of the flickering of external noise. This would certainly be different for parameter values within the chaotic regime, where there is a sensitive dependence on initial conditions.

The astonishing similarity between the evolutionary patterns obtained via simulation and those found by empirical research into the long-term evolution of schizophrenia can be regarded as the essential result of this simulation (see Fig.5). The patterns obtained empirically are shown again for those cases most obviously exhibiting this similarity (Figs. 12-15).

6. Prospects for Future Development

It was the purpose of the present simulation to investigate a possible means of modelling by means of nonlinear equations. This seems to be a promising approach for future work. New perspectives have been opened up, yet certain questions must also be raised.

a) First of all, the model should be improved with respect to not only its content but also its mathematical foundations. Further examinations are needed to explore the model's consistency with empirical findings regarding the evolution of schizophrenic processes. Formal considerations arise with regard to the question of whether delay terms (t-τ) should be introduced into the model. After all, it does not seem reasonable that the interactive effects of the variables should take place instantaneously or even within a single time interval (t to $t+1$). Furthermore, a study should be carried out to ascertain whether some of the numerous interactional terms contained in the equations could be adequately replaced by simple linear terms. Such interactional terms stem from the basic structure of equations applied in population dynamics, which in turn stem from the Verhulst equation extended to include terms representing symbiotic and competitive mechanisms. In addition, the fundamental question arises of whether the use of theoretical models of population dynamics for simulations of psychosocial phenomena implies an argumentation based upon inadequate analogies or even means a needlessly restrictive commitment to just one possible approach. At any rate, the theory of dynamic systems offers many more alternatives in guiding the formulation of nonlinear equations.

b) It is mainly due to the variation of the β-parameter values that each simulation generates a different evolutionary pattern; less influence is produced by variations of σ, δ and κ, which represent constructs of greater theoretical significance. (An exception is the variation of σ in equation (5), generating the patterns shown in Figs. 14 and 15.) In this connection, the model is being tested at present to ascertain whether different values of σ, δ and κ, selected on the basis of theoretical considerations, yield patterns in accordance with the theoretical concepts of schizophrenia. Accordingly, it is possible to simulate the effects of therapeutic treatment aimed at the modification of these parameters (see sect. 2.4).

c) The parameters σ, δ and κ may be considered to indicate temporally relatively stable states of intrapsychic, psychosocial and biochemical systems. However, their relatively slow rate of change, as opposed to that of the

187

variables, does not mean that they are invariable. It is precisely the development of cognitive-affective schemata (in German: majorisierende Äquilibration, Affektlogik) and social skills, together with the evolution of the neuro-peptide system, upon which current etiological theories focus. Thus, a separate model should describe the evolutionary dynamics of the parameters, taking into account their interactions with the dynamics of the variables. The implication here is that we are concerned with coupled dynamic systems (cf. Glass 1987; an der Heiden, this volume), with the only constant factor being the genetic disposition, γ.

The vulnerability of the neurotransmitter system may be interpreted as a consequence of certain genetic features regulating the processes of protein synthesis and responsible for the interactive effects of the agonistic and antagonistic properties of neurotransmitters as well as for receptor characteristics ($\gamma \delta$), i.e. internal processes of the system, together with their interaction with the development of intrapsychic structures ($\sigma \delta$).

A specific constitutional disposition, as well as prolonged periods of stress, s, and inconsistent interpersonal experience, e, the effects of which become apparent only after some delay (t-τ), are responsible for the development of diffuse intrapsychic schemata. Furthermore, findings concerning neuronal plasticity (Haracz 1984;1985) indicate an interaction between the intrapsychic systems and the biochemistry of neurotransmitters ($\delta \sigma$)(see sect. 3.2(2)).

Social incompetence (κ) results from the interaction between κ and the diffuseness of intrapsychic schemata guiding perception and behavior within social contexts σ, as well as from prolonged periods of stress and impairment due to cognitive disorders.

These hypotheses lead to the formulation of the following difference equations (6) to (9):

$$\gamma = \text{const.} \tag{6}$$

$$\Delta \delta = \gamma \delta + \sigma \delta \quad - \beta_\delta \, \delta^2 \tag{7}$$

$$\Delta \sigma = \gamma \sigma + \delta \sigma + \rho_s \bar{s}_{t-\tau} + \rho_e \bar{e}_{t-\tau} - \beta_\sigma \sigma^2 \tag{8}$$

$$\Delta \kappa = \sigma \kappa + \rho_s \bar{s}_{t-\tau} + \rho_c \bar{c}_{t-\tau} - \beta_\kappa \kappa^2 \tag{9}$$

where $\bar{s}$, $\bar{e}$ and $\bar{c}$ denote the mean values of the corresponding variables obtained within a certain time interval τ following the current point in time t, ρ_s, ρ_e and ρ_c represent damping constants, and $\beta_\delta, \beta_\sigma$ and β_κ denote feedback parameters. It is undoubtedly rather unusual in mathematics to treat the parameters of one set of equations as variables in the next. However, findings are available indicating a differentiation of variables into slowly and quickly varying variables in oscillating systems (cf. Tyson 1976, p.54ff.). Haken (1990b, p.11) also proposes possible interactions between order and control

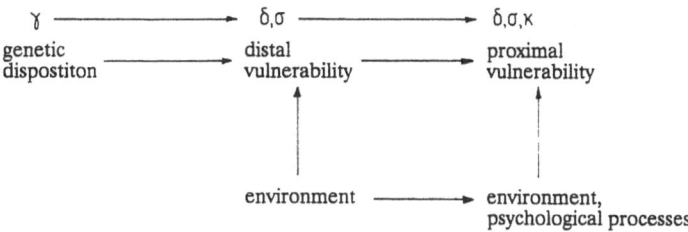

Fig. 17: The vulnerability model interpreted as evolutionary model of relevant parameters.

parameters: "More generally we may expect an interplay between control parameters and order parameters where for instance an order parameter at one hierarchical level acts as control parameter at another, etc."

Establishing a system of parameter dynamics means adhering even more closely to the frames of reference given by the metatheoretical approaches to schizophrenia described in section 1. For specific vulnerabilities to evolve, these models hypothesize (a) specific interactions of a biopsychological system with itself, (b) specific interactions with the system's social and biological environment, and finally, (c) specific interactions with illness and health care processes (Fig.17).

Furthermore, it is possible for the parameter values to enter a regime within which they cause the set of equations (1)-(5) to generate chaotic behavior (on the right-hand side of r in Fig.10). Whether they remain within this chaotic region, or begin to oscillate between asymptotically stable, periodic or chaotic regimes on their own, all these phenomena open up new perspectives for an analysis of schizophrenia as a dynamical disease by means of the framework provided by the theory of dynamic systems and chaos theory.

d) Simulations are judged as to their internal consistency, plausibility and theoretical productivity. Also, they must be compared with data empirically obtained. Thus, in addition to the comparison with general evolutionary patterns (see Fig.5), we intend to utilize genuine time series of psychiatric ratings as an empirical basis. This will be achieved through cooperation with the team of Prof. Ciompi and Prof. Dauwalder in Bern, Switzerland. However, being able to measure the parameters used is an important precondition, one that must be fulfilled before single case simulation can become possible.

e) If the repeated single case measurement of relevant parameters were to be achieved, this could offer the opportunity of recognizing evolutionary trends towards the manifestation of clinical symptoms at an earlier date than is currently possible by means of prodromal symptoms. The basic idea of such a "trend estimation" is that of a separatrix, separating two attractor states: that of a "harmless" basin of attraction and that of a "psychotic" basin of attraction. Measured values of parameters which constitute such types of separatrices within the phase space could indicate how far the system has approached the separatrix prior to entering the regime of the "schizophrenia attractor".

Acknowledgement. We would like to thank Dipl.-Psych. Martina Weisensel (Staatl. gepr. Übersetzerin) for the translation of the present contribution.

References

Babloynantz A, Destexhe A (1987) Strange Attractors in the Human Cortex. In: Rensing L, an der Heiden U, Mackey MC (Eds) Temporal Disorder in Human Oscillatory Systems. Springer Series in Synergetics, Vol.36. Springer, Berlin, pp 57-68

Bateson G et al (1984) Schizophrenie und Familie. Suhrkamp, Frankfurt am Main

Bischof N (1985) Das Rätsel Ödipus. Piper, München

Bleuler E (1911) Dementia praecox oder die Gruppe der Schizophrenien. Deuticke, Leipzig

Bleuler M (1972) Die schizophrenen Geistesstörungen im Lichte langjähriger Kranken- und Familiengeschichten. Thieme, Stuttgart

Böker W (1986) Zur Selbsthilfe Schizophrener: Problemanalyse und eigene empirische Untersuchungen. In: Böker W, Brenner HD (Hrsg) Bewältigung der Schizophrenie. Huber, Bern, S 176-188

Böker W, Brenner HD (1983) Selbstheilungsversuche Schizophrener: psychopathologische Befunde und Folgerungen für Forschung und Therapie. Nervenarzt 54: 578-589

Böker W, Brenner HD (Hrsg) (1986) Bewältigung der Schizophrenie. Huber, Bern

Böker W, Brenner HD (Hrsg) (1989) Schizophrenie als systemische Störung. Huber, Bern

Ciompi L (1982) Affektlogik. Klett-Cotta, Stuttgart

Ciompi L (1986a) Auf dem Weg zu einem kohärenten multidimensionalen Krankheits- und Therapieverständnis der Schizophrenie: Konvergierende neue Konzepte. In: Böker W, Brenner HD (Hrsg) Bewältigung der Schizophrenie. Huber, Bern, S 47-61

Ciompi L (1986b) Zur Integration von Fühlen und Denken im Licht der "Affektlogik". Die Psyche als Teil eines autopoietischen Systems. In: Kisker KP et al (Hrsg) Psychiatrie der Gegenwart Bd 1. Springer, Berlin, S 373-410

Ciompi L (1989) Zur Dynamik komplexer biologisch-psychosozialer Systeme: Vier fundamentale Mediatoren in der Langzeitentwicklung der Schizophrenie. In: Böker W, Brenner HD (Hrsg) Schizophrenie als systemische Störung. Huber, Bern, S 27-38

Ciompi L, Müller C (1976) Lebensweg und Alter der Schizophrenen. Eine katamnestische Langzeitstudie. Springer, Berlin

Cronin J (1977) Mathematical Aspects of Periodic Catatonic Schizophrenia. Bull Math Biol 39: 187-197

Dauwalder JP (1988) A Comprehensive View on Affect and Logic: Some Implications for Treatment and Prevention of Schizophrenia. Psychopathology 21: 95-110

Elkaim M, Goldbeter A, Goldbeter E (1980) Analyse des transactions de comportement dans un système familial en termes de bifurcation. Cahiers Critiques de Thérapie Familiale et de Pratique des Réseaux 3: 18-34

Gelders YG (1989) Thymosthenische Wirkstoffe, ein neuartiger Ansatz in der Behandlung der Schizophrenie. In: Böker W, Brenner HD (Hrsg) Schizophrenie als systemische Störung. Huber, Bern, S 65-74

Gerok W (1989) Ordnung und Chaos als Elemente von Gesundheit und Krankheit. In: Gerok W (Hrsg) Ordnung und Chaos in der unbelebten und belebten Natur. Wissenschaftliche Verlagsgesellschaft, Stuttgart, S 19-41

Glass L (1987) Coupled Oscillators in Health and Disease. In: Rensing L, an der Heiden U, Mackey MC (Eds) Temporal Disorders in Human Oscillatory Systems. Springer Series in Synergetics. Vol.36. Springer, Berlin, pp 8-14

Goldberg S (1986) Differenzengleichungen und ihre Anwendung in Wirtschaftswissenschaften, Psychologie und Soziologie. Oldenbourg, München

Gottesman II, Shields J (1976) A Critical Review of recent Adoption, Twin and Genetic Perspectives. Schizophrenia Bulletin 2: 360-398

Gross G (1986) Basissymptome und Coping Behavior bei Schizophrenen. In: Böker W, Brenner HD (Hrsg) Bewältigung der Schizophrenie. Huber, Bern, S 132-141

Haken H (1983) Synergetics. An Introduction. Springer Series in Synergetics, Vol.1. Springer, Berlin (3.Aufl.)

Haken H (1988) Information and Self-Organization: A Macroscopic Approach to Complex Systems. Springer Series in Synergetics, Vol. 40. Springer, Berlin

Haken H (1990a) Synergetik. Eine Einführung. Springer, Berlin

Haken H (1990b) Synergetics as a Tool for the Conceptualization and Mathematization of Cognition and Behavior - How Far Can We Go? In: Haken H, Stadler M (Eds) Synergetics in Cognition. Springer Series in Synergetics, Vol.45. Springer, Berlin, pp 2-31

Haracz JL (1984) A Neural Plasticity Hypothesis of Schizophrenia. Neuroscience and Biobehavioral Reviews 8: 55-71

Haracz JL (1985) Neural Plasticity in Schizophrenia. Schizophrenia Bulletin 11 191-229

Harding CM (1984) Long-term Outcome Functioning of Subjects rediagnosed as meeting the DSM III Criteria for Schizophrenia. Doctoral Dissertation, University of Vermont

an der Heiden U, Mackey MC (1987) Mixed Feedback; A Paradigm for Regular and Irregular Oscillations. In: Rensing L, an der Heiden U, Mackey MC (Eds) Temporal Disorder in Human Oscillatory Systems. Springer Series in Synergetics, Vol.36. Springer, Berlin, pp 30-46

Hesch RD (1990) Sein im deterministischen Chaos: Ansätze einer neuen Systemtheorie der Medizin. Vortragsmanuskript, Hannover

Horowitz MJ (1987) States of Mind. Plenum Press, New York

Huber G, Gross G, Schüttler R (1979) Schizophrenie. Eine Verlaufs- und sozialpsychiatrische Studie. Springer, Berlin

Käsermann ML, Altorfer A (1989) Physiologische Korrelate unterschiedlich belastender Situationen in Familiengesprächen. In: Böker W, Brenner HD (Hrsg) Schizophrnie als systemische Störung. Huber, Bern, S 300-314

King R, Barchas JD (1983) An Application of Dynamic Systems Theory to Human Behavior. In: Basar E, Flohr H, Haken H, Mandell AJ (Eds) Synergetics of the Brain. Springer Series in Synergetics, Vol.23. Springer, Berlin

Krebs CJ (1985) Ecology. Harper & Row, New York

Kriz J (1990) Synergetics in Clincal Psychology. In: Haken H, Stadler M (Eds) Synergetics of Cognition. Springer Series in Synergetics, Vol 45. Springer, Berlin, pp 393-404

Mackey MC, an der Heiden U (1982) Dynamical Diseases and Bifurcations: Understanding Functional Disorders in Physiological Systems. Funkt Biol Med 1: 156-164

Martienssen W (1989) Gesetz und Zufall in der Natur. In: Gerok W (Hrsg) Ordnung und Chaos in der belebten und unbelebten Natur, Wissenschaftliche Verlagsgesellschaft, Stuttgart, S 77-99

May RM (1980a) Populationsmodelle für eine Art. In: May RM (Hrsg) Theoretische Ökologie. Verlag Chemie, Weinheim, S 5-23

May RM (1980b) Modelle für zwei interagierende Populationen. In: May RM (Hrsg) Theoretische Ökologie. Verlag Chemie, Weinheim, S 47-65

Mayer-Kress G, Holzfuss J (1987) Analysis of the Human Electroencephalogram with Methods from Nonlinear Dynamics. In: Rensing L, an der Heiden U, Mackey MC (Eds) Temporal Disorder in Human Oscillatory Systems. Springer Series in Synergetics, Vol.36. Springer, Berlin, pp 57-68

Nicolis G, Prigogine I (1987) Die Erforschung des Komplexen, Piper, München

Nuechterlein KH (1987) Vulnerability Models for Schizophrenia: State of the Art. In: Häfner H, Gattaz WF, Janzarik W (Eds): Search for the Causes of Schizophrenia. Springer, Berlin

Nuechterlein KH, Dawson ME (1984) A Heuristic Vulnerability-Stress Model of Schizophrenic Episodes. Schizophrenia Bulletin 10: 300-312

Nuechterlein KH, Zaucha KM (1990) Similarities Between Information-Processing Abnormalities of Achively Symptomatic Schizophrenic Patients and High-Risk Children. In: Straube ER, Hahlweg K (Eds): Schizophrenia: Concepts, Vulnerability, and Intervention. Springer, Berlin, pp 77-96

Odum EP (1983) Grundlagen der Ökologie. Bd.1. Thieme, Stuttgart

Pflug B (1987) Circadian Rhytmus and Depression. In: Rensing L, an der Heiden U, Mackey MC (Eds) Temporal Disorder in Human Oscillatory Systems. Springer Series in Synergetics, Vol.36. Springer, Berlin, pp 194-201

Rensing L, an der Heiden U, Mackey MC (Eds) Temporal Disorders in Human Oscillatory Systems. Springer Series in Synergetics. Vol.36. Springer, Berlin

Sarbin TR, Manusco JC (1982) Schizophrenie. Urban & Schwarzenberg, München

Schiepek G (1986) Systemische Diagnostik in der Klinischen Psychologie. PVU, München

Schiepek G (1991) Systemtheorie der Klinischen Psychologie. Vieweg, Braunschweig

Schmid GB (1991) Chaos Theory and Schizophrenia: Elementary Aspects. Psychopathology 24: 185-198

Schuster HG (1988) Deterministic Chaos. An Introduction. VCH, Weinheim

Seifritz W (1987) Wachstum, Rückkopplung und Chaos. Hanser, München

Strauss JS (1989) Intermediäre Prozesse in der Schizophrenie: zu einer neuen dynamisch orientierten Psychiatrie. In: Böker W, Brenner HD (Hrsg) Schizophrenie als systemische Störung. Huber, Bern, S 35-50

Strauss JS, Carpenter WT (1981) Schizophrenia. Plenum Press, New York

Süllwold L (1986) Basisstörungen: Instabilität von Hirnfunktionen. In: Böker W, Brenner HD (Hrsg) Bewältigung der Schizophrenie. Huber, Bern, S 42-46

Tienari P, Lahti I, Sorri A, Naarala M, Moring J, Wahlberg KE (1989) Die finnische Adoptionsfamilienstudie über Schizophrenie: Mögliche Wechselwirkungen von genetischer Vulnerabilität und Familien-Milieu. In: Böker W, Brenner HD (Hrsg) Schizophrnie als systemische Störung. Huber, Bern, S 52-64

Tretter F (1989) Humanökologische Medizin. In: Glaeser B (Hrsg) Humanökologie. Westdeutscher Verlag, Opladen, S 209-224

Trulson ME, Preussler DW (1984) Dopamine-containing Ventral Tegmental Area Neurons in freely Moving Cats: Activity during the Sleep-waking Cycle and Effects of Stress. Experimenta Neurology 83: 367-377

Tyson JJ (1976) The Belousov-Zhabotinsky-Reaction. Springer, Berlin

Wehr TA, Goodwin FK (1979) Rapid Cycling in Manic-Depressives Induced by Tricyclic Antidepressants. Arch Gen Psychiatry 36: 555-559

Zubin J (1989) Die Anpassung therapeutischer Interventionen an die wissenschaftlichen Modelle der Ätiologie. In: Böker W, Brenner HD (Hrsg) Schizophrenie als systemische Störung. Huber, Bern, S 14-26

Zubin J, Spring B (1977) Vulnerability: A New View of Schizophrenia. J Abnorm Psychol 86: 103-123

Dynamical Systems and the Development of Schizophrenic Symptoms – An Approach to a Formalization

Brigitte Ambühl, Rudolf Dünki, and *Luc Ciompi*

With 6 Figures

Abstract. Based on the hypothesis of irregular dynamics in the evolution of schizophrenia (Ciompi et al., 1991), we explored the possibility of describing these processes in terms of dynamical systems (chaos) theory. By analyzing time series of a single schizophrenic patient we found support for the existence of a strange attractor. A short discussion of methods (Grassberger-Procaccia algorithm) that can be applied to dynamical systems completes the theoretical part. This formalization is a basis for further experimental studies as well as for testing and developing models and simulations.

1. Introduction

Clinical data on the development of schizophrenic symptoms show irregular patterns. Neither can a simple classification be made nor is it possible to predict exactly the development of the illness. This is obvious in two aspects: a) in observations of the long term behavior of schizophrenics over thirty years (Ciompi et al., 1976; Hubschmid et al., 1986), and b) in observations of the daily fluctuations of psychotic symptoms (Fig. 1; Aebi et al., 1989).

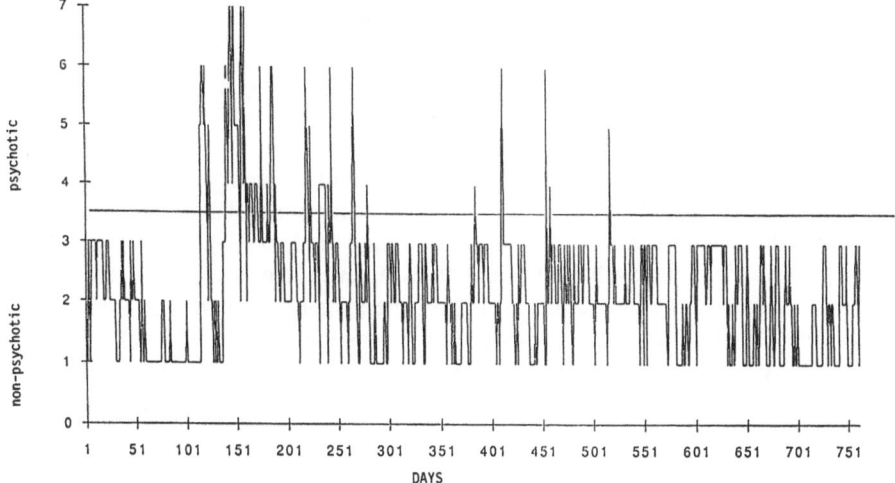

Fig. 1. Daily fluctuations of psychotic symptoms. The medication during the first seven days was 80 mg Clopenthixol in total, from day 194 to day 751 in total of 790 mg Flupenthixol in equal dosages per day. In between there was no medication

195

Phenomenologically similar time series are known in physics (Bénard instability), chemistry (Belousov-Zabotinsky reaction), mathematics (Cantor set, Julia set, Koch's curve), biology (Volterra, Lotka), and meteorology. This similarity raises the question of how such dynamical processes and methodologies could be useful in schizophrenia research, too (Ciompi, 1991; Ciompi et al., 1991). All these processes have the following features in common: they relate to simple dynamical laws, they show irregular patterns, and their time evolution cannot be fully predicted. The dynamics of these processes can be formalized with a system of coupled differential equations, where the variables define the system's phase space. Often there are sets in the phase space of these systems, into which the solutions of the system converge; they remain there in the absence of further disturbance. These sets are known as attractors (Eckmann, 1981). Important quantities (topological invariants) for the characterization of an attractor are the fractal dimension (a geometric property), Liapunov exponents and the Kolmogorov entropy (dynamical properties) (Schuster, 1989).

2. Definition of the Problem

Starting from these considerations we can formulate the questions mentioned above more precisely:

1) Can we formalize the development of schizophrenia and describe it with a mathematical model?

2) If there is such a model, are there ranges of the parameters in which the system has attractors? If the answer is yes, how are these attractors characterized? (see Grebogi et al., 1984).

3) Can we test the predicted properties experimentally?

3. Methods

We start with our observations of the development of schizophrenia of young adult schizophrenics hospitalized for the first time in a special environment (the therapeutic community of Soteria, Berne; Ciompi et al., 1990). We registered the daily fluctuations of psychotic symptoms by a seven item scale that included the various stages of the illness. One rating of the symptomatology was made every day (Fig. 2), and the worst state of each day was recorded (Aebi et al., 1989).

The resulting time series of the development of psychotic symptoms were analyzed with the Grassberger-Proccacia algorithm (Grassberger & Proccacia, 1983). Grassberger & Proccacia introduced a so-called correlation integral, C(r):

$$C(r) = \frac{1}{N^2} \sum_i \sum_j \theta(r - |\vec{x}_j - \vec{x}_i|) \ . \tag{1}$$

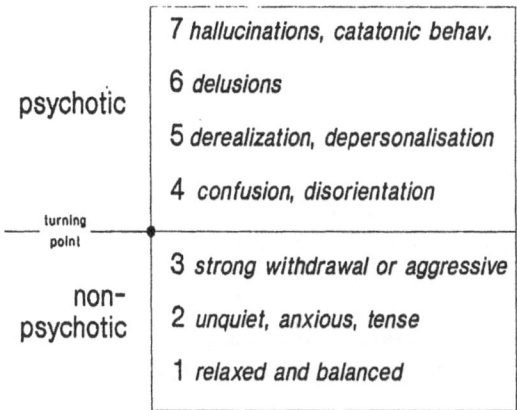

psychotic	7 *hallucinations, catatonic behav.*
	6 *delusions*
	5 *derealization, depersonalisation*
	4 *confusion, disorientation*

turning
point

non-psychotic	3 *strong withdrawal or aggressive*
	2 *unquiet, anxious, tense*
	1 *relaxed and balanced*

Fig. 2. Seven-item scale for daily fluctuation of psychotic symptoms. Testing the reliability of this scale between different raters yielded a correlation of .67 (p<.01)

Vector x is a point on the attractor and θ(s) the Heavyside distribution:

$$\theta(s) = \begin{cases} 0 & \text{for} & s \leq 0 \\ 1 & \text{for} & s > 0 \end{cases}$$

In (1) the term within brackets denotes a norm, i.e. C(r) is the normalized sum of all points separated from one another by less than r. The points on the attractor are generally not known. However, if we know the time evolution of a single coordinate we can reconstruct the attractor in an m-dimensional euclidian space (Takens, 1981) with coordinates $x(t_i)$, $x(t_i+\tau)$, ... $x(t_i+(m-1)\tau)$, where τ denotes an appropriately chosen delay time. With this construction we have for the correlation integral C(r) (Grassberger & Procaccia, 1983; Dünki, 1991)

$$\lim_{r \to 0} \lim_{m \to \infty} C(r) \propto r^D c^{-m\tau k} , \tag{2}$$

i.e. the correlation integral contains information on the (fractal) dimension and the Kolmogorov entropy K.

4. Practical Aspects

a) **Analysis of a time series.** Over a period of 751 days we observed the evolution of psychotic symptoms of one schizophrenic patient (Figs. 1, 2). The first 200 days were discarded from the analysis because of possible non-stationarity of the time series. This corresponds to the onset of constantly dosed medication which started on day 194 (Fig. 1).

b) **Choice of an appropriate delay time.** To choose the delay time one of the following criteria is often used: first minimum of the autocorrelation function, first zero of the autocorrelation function, or $\tau = t_c / m$ (t_c is a typical cycling time of the system; Atten et al., 1984).

c) Calculation of C(r) for a series of embedding dimensions m. In a $\log C(r)$ vs. $\log r$ plot for m sufficiently large and r small enough, the dimension D follows from the slope of the curves, and the Kolmogorov entropy K from their separation.

d) Important remarks about the methods. In our study we could not use thousands of exact data points as one can with physical or chemical systems. However, it has been shown that low dimensions can be estimated even if the attractors are reconstructed only with some hundreds of delay time vectors (Abraham et al., 1986). One must avoid confusion between points in time and points in a reconstructed phase space. The latter requires an appropriate choice of the delay time.

Our own tests with such systems with low resolution (3 bit) still returned reasonable estimates for the dimension and K-entropies. The slopes of the $\log C(r)$ vs. $\log r$ plots have been estimated by fitting quasi Hermite polynomials. These polynomials approximate the smooth curve that one might draw by hand through

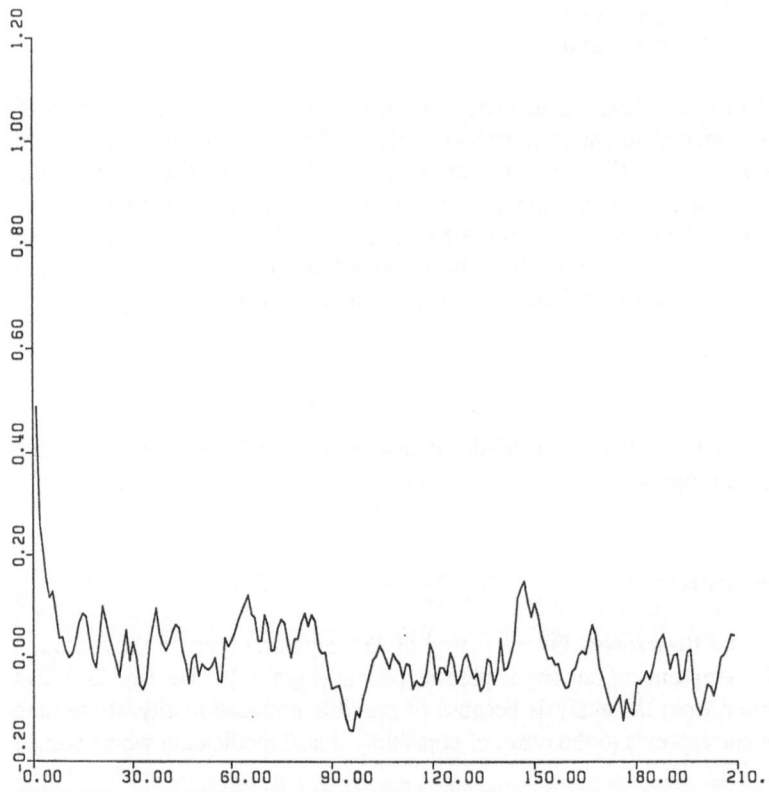

Fig. 3. Autocorrelation function. The first 200 points were discarded from the analysis because of a possible nonstationarity of the time series

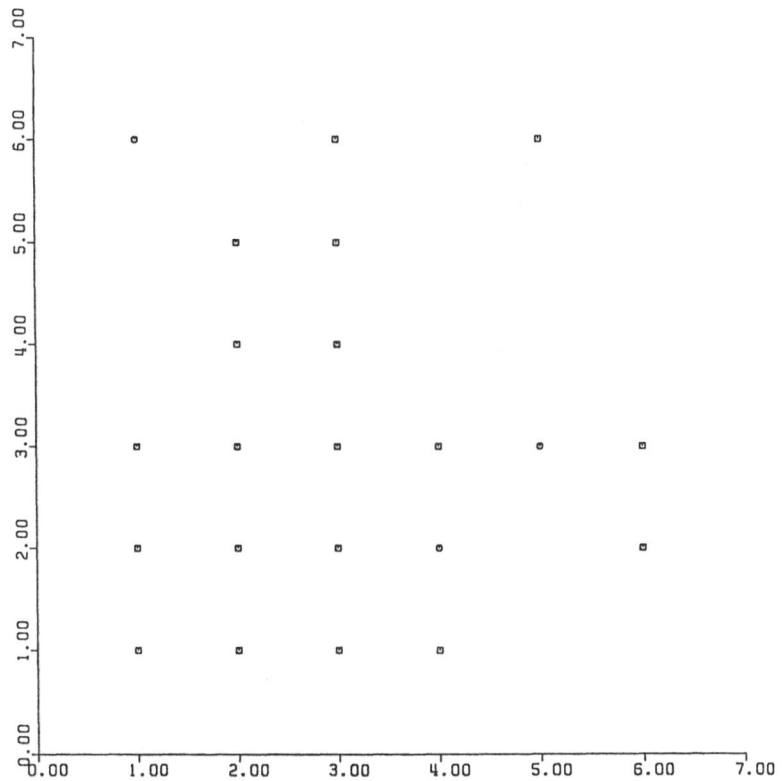

Fig. 4. Phase diagram. 400 plotted points

a set of points (Figs. 5, 6). Estimating the slope of such a discretized system is always a problem and the influence of the method is being considered in a study currently in progress. The quasi Hermite polynomials, however, have shown good results in the test systems. Furthermore we took care – as in the case of schizophrenia data – to avoid counting of autocorrelated vectors (Theiler, 1986) by the algorithm we used (Theiler, 1987).

The influence of stochastic processes must also be considered. It has been shown that the superposition of such processes onto a deterministic dynamics influences the dimension and the K-entropy estimates. In general the dimension estimate will overshoot the true dimension (Atten, 1984; Abraham, 1986) and the K-entropy estimate will be inaccurate, too (Dünki, 1991). The construction of the vectors $x^m(\tau)$ provides equidistant intervals between data points. In our construction this is not the case. However, taking only the maxima of the daily fluctuation of psychotic symptoms defines a cross-section through the attractor (Shaw, 1981). Hence our reconstructed vectors give the reconstruction of the dynamics of the cross-section where each maximum can be interpreted as an iterative step. The dimension of the real dynamics will be approximately one higher than the dimension we find by this procedure.

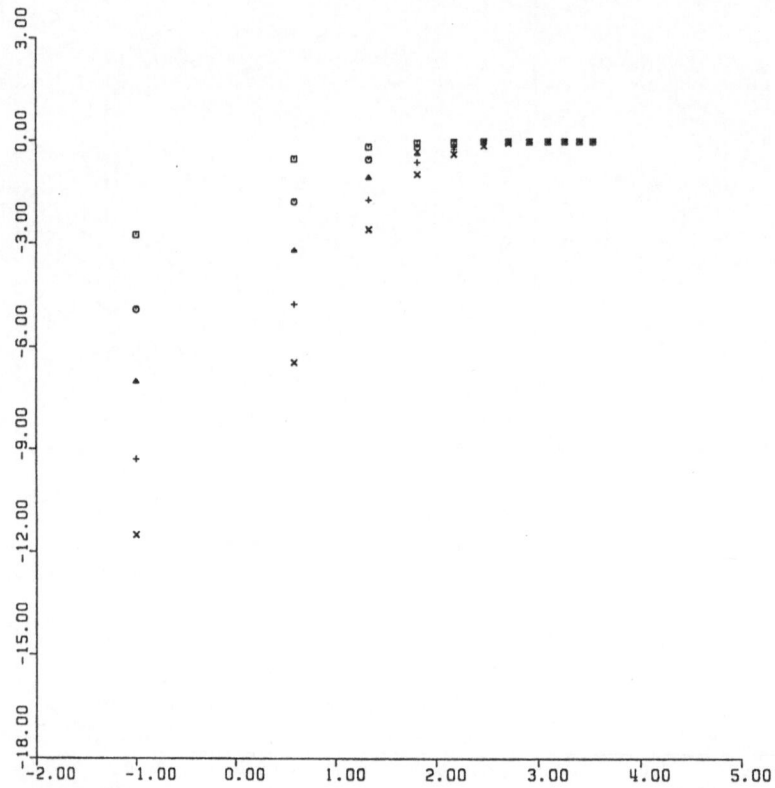

Fig. 5. Dimension analysis. 400 vectors, embedding dimensions 2, 4, 6, 8, 10 (□,○,△,+,×)

5. Results

In characterizing the time series of one schizophrenic patient (Fig. 1) we can search for the existence of an attractor. A first clue is found in the phase diagram (Fig. 4). The set of points does not fill the whole phase space and seems to have some structure. The phase diagram consists of only 7 x 7 points. Within the framework of our hypothesis, we regard this as a discretized form of a continuous phase space. This plot does not show detailed fine structure (a lot of which is lost during the discretization). However, this figure is still clearly distinct from the same kind of plot for a purely stochastic process. Our analysis was made according to the description of Sects. 3 and 4. The autocorrelation function (Fig. 3) decays quickly and shows some oscillations with a mean cycle time of 15 days. Therefore we used the delay time $\tau=1$. For small values of the parameter r and an embedding dimension m=4, the first derivatives (Fig. 6) converge towards a common value (≈ 2.2), while the separation of the curves corresponds to a K-entropy of about one

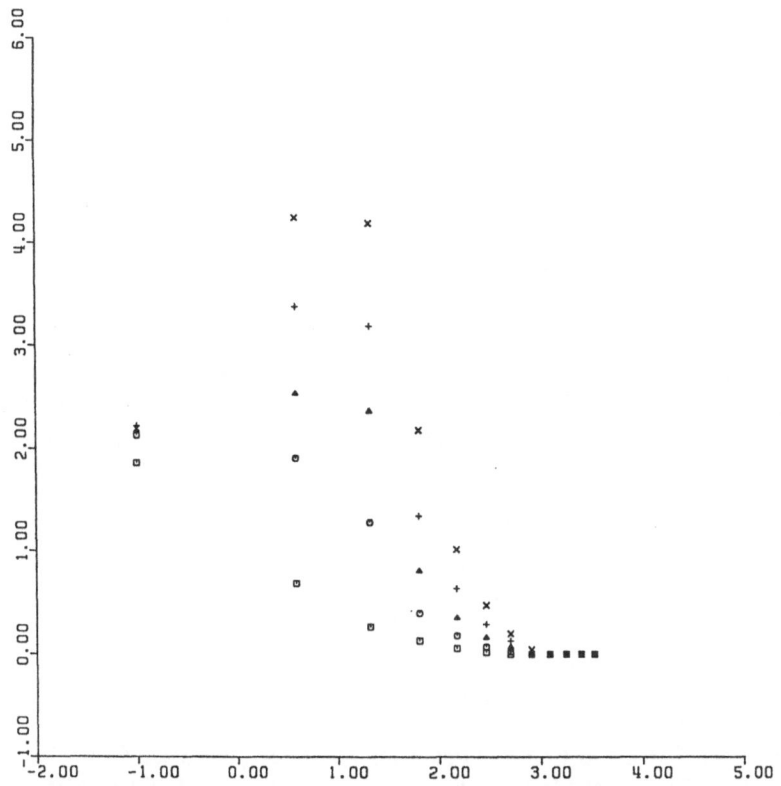

Fig. 6. First derivatives of Fig. 5. 400 vectors, embedding dimensions 2, 4, 6, 8, 10 (□,○,∆,+,x)

bit per iteration (Fig. 5). These findings are consistent with our hypothesis of an underlying chaotic attractor in the development of schizophrenia.

6. Discussion

This result is remarkable even though we are aware that this is only pilot study and must keep in mind the remarks made in Sect. 4. The indication of the existence of a deterministic dynamics for the time evolution of schizophrenic symptoms opens new perspectives for the modelling. We should be able to describe the dynamics of the development of schizophrenia with a mathematical model containing only a few degrees of freedom. Although we do not provide an explicit model, the methods described should allow an experimental verification (Ciompi et al., 1991). Basically such a model would contain free parameters like coupling strengths or stochastic elements, for example noise.

The classification of the dynamics of a nonlinear system depends on the range of the parameters. Thus, the same model shows different types of dynamics for different values of the parameters. Possible types of dynamics are fluctuations, regular or damped oscillations and chaotic dynamics. Based on these properties we could probably define classes of prognoses for the development of schizophrenia. This could be verified with clinical data and long term records of the illness (Ciompi et al., 1976; Hubschmid et al., 1986). Further studies to support this hypothesis are planned.

References

ABRAHAM, N.B., ALBANO, A.M., DAS, B., GUZMAN de, G., YONG, S., GIOGGIA, R.S., PUCCIONI, G.P. & TREDICCE, J.R. 1986. Calculating the dimensions of attractors from small data sets. Phys. Lett. A, **114**, p. 127.

AEBI, E. & CIOMPI, L. 1989. Everyday Events and the Temporal Course of Daily Fluctuations of Psychotic Symptoms. Schizophrenia Research, Vol. 2, Nos. 1-2.

AEBI, E., REVENSTORF, D. & ACKERMANN, K. 1989. A Concept of Optimal Social Stimulation in Acute Schizophrenia – Time Series Analysis of Psychotic Behavior. The Third European Conference on Psychotherapy Research, Berne, September 5-9.

ATTEN, P., CAPUTO, J.G., MALRAISON, B. & GAGNE, Y. 1984. Détermination de dimension d'attracteurs pour différent écoulements. J. Méc. Théor. Appl. Numéro spécial, p. 133.

CIOMPI, L. 1991. Affects as Central Organising and Integrating Factors. A New Psychosocial/Biological Model of the Psyche. British Journal of Psychiatry, **159**, 57-105.

CIOMPI, L. & MÜLLER, C. 1976. Lebensweg und Alter der Schizophrenen. Berlin: Springer.

CIOMPI, L., DAUWALDER, H.P., MAIER, Ch., AEBI, E., TRÜTSCH, K., KUPPER, Z. & RUTHISHAUSER, Ch. 1990. The pilot project "Soteria Berne". Clinical experiences and preliminary results. Lecture presented at the Third International Schizophrenia Symposium. Berne.

CIOMPI, L. AMBÜHL, B., DÜNKI, R.M., THOMAS, R. 1991 (in press). Schizophrenia and Chaos Theory. Exploratory Investigations in the Dynamics of Complex Psycho-Socio-Biological Systems.

DÜNKI, R.M. 1991 (in press). The estimation of the Kolmogorov entropy from time series and its limitations when performed on EEG. Bulletin of mathematical biology.

ECKMANN, J.P. 1981. Roads to turbulence in dissipative dynamical systems. Rev. Mod. Phys., **53**, p. 643.

GRASSBERGER, P. & PROCACCIA, I. 1983. Characterization of strange attractors. Phys Rev. Lett., **50**, p. 346.

GREBOGI, C., OTT, E., PELIKAN, S. & YORKE, J.A. 1984. Strange attractors that are not chaotic. Physica 13D, p. 261.

HUBSCHMID, T. & AEBI, E. 1986. Berufliche Wiedereingliederung von psychiatrischen Langzeitpatienten. Soc. Psychiatry, **21**, 152-157.

SCHUSTER, H.G. 1989. Deterministic Chaos. An Introduction. Weinheim: Verlagsgesellschaft mbH.

SHAW, R.S. 1981. Strange Attractors, Chaotic Behaviour and Information Flow. Z. Naturforsch., **36a**, p. 80.

TAKENS, F. 1981. Detecting strange attractors in turbulence. Lecture Notes in Mathematics., **898**, Berlin: Springer, p. 361.

THEILER, J. 1986. Spurious algorithms applied to limited time-series data. Phys. Rev. A, **34**, p. 2427.

THEILER, J. 1987. Efficient algorithm for estimating the correlation dimension from a set of discrete points. Phys. Rev. A, **36**, p. 4456.

Psychiatric Disorders: Are They "Dynamical Diseases" ?

Hinderk Emrich and Charlotte Hohenschutz

1. Introduction

The introduction of the concept of "dynamical diseases" into theoretical medicine by M.C. Mackey and U. an der Heiden (1982), addressing especially the possible role of dynamical systems and the interpretation of "transitions" as bifurcations within them, raises the question of whether psychiatric disorders may be interpreted in this way. A speciality of psychiatric disorders may be that these disorders differ from other nosologies insofar as there appears to exist an unescapable dichotomy between physiological variables and the "mentalese" (cf. Table 1). The type of "language" in which one may try to clarify the difficulties of the mind-body problem is based here on the terminology of Donald MacKay, using the distinction between "mind talk" and "brain talk" (1965). A reconstruction of the "missing link" between these two phenomenological types of "languages" may be interpreted as a "hypersystemic" code which contains the ability to generate "intentionality". Such a code may be interpreted as a "hyperlanguage" (Emrich, 1991) which cannot be described in conventional terms, since it is "a priori" in regard to the constitution of human mind, and may be defined – in the terminology of Hesch (1989) – as a multiperspective integration of all "dynamic codes" – in a similar fashion as Thomas Nagel (1976) construed the "self" as the "view from nowhere", which is

Table 1. "Languages" and the mind-body problem

Heuristic level	Code	Observation
Mental System (thoughts, feelings)	Intentionality code ("mind talk")	Self-observation
Neuronal System	Intensity code ("brain talk")	Physiological and neurochemical measurements
Hypersystem	Intentionality generating code ("bráïn talk")	Approximation only by re-construction

Springer Series in Synergetics, Vol. 58
Editors: W. Tschacher, G. Schiepek, and E.J. Brunner

Self-Organization and Clinical Psychology
© Springer-Verlag Berlin Heidelberg 1992

perspectiveless in its extrapolation. An application of the concept of "dynamical diseases" to psychiatry would require that these intentionality-generating hypersystemic codes are already uncovered, i.e. it would mean that the mode in which the internal "dialogues" within the brain generate mental states is elucidated. The disturbances of psychic functions as "pathological changes in the behavior", resulting from the "underlying physiological control system" which "may still be intact and unaltered, with the exception of a single parameter being out of its physiological range" (Mackey & an der Heiden, 1982) can be analyzed in regard of the pathogenetic mechanisms only, if the underlying physiological processes have been described or have at least a chance of being describable. Under these perspectives the question of the possible relevance of the concepts of dynamical diseases – and thus of "chaos theory" – in theoretical psychiatry has an obviously aporetic character.

On the other hand, several arguments point to the view that – besides the above-mentioned difficulties raised by the mind-body dichotomy – there may be some parallels between functional psychoses and dynamical diseases, which will be discussed in the following. A comparison between the time-course of the oscillatory pattern of breathing frequency in a patient with Cheyne-Stokes breathing and, e.g. the phase chart of a patient with an affective psychosis, yields the observation that in those cases states of abnormal function alternate with complete normality. Thus, the question is raised of whether this phenomenon represents only a superficial homology or whether there exists a substantial parallelism between hte two phenomena, making such a comparison sensible.

Recent chronobiology-oriented psychiatric basic research which tries to elucidate the regularities of time-courses of affective psychoses is mainly influenced by the sinus-oscillator model of Halberg (1968) in which the superposition of such rhythms leads to "beat" phenomena which are used to interpret the functional breakdowns during affective psychoses. Following this concept, referring to the "desynchronization hypothesis", the phase-advance hypothesis, and related hypotheses (for a review see Wehr & Goodwin, 1983), several functional systems which exhibit circadian rhythms (e.g. temperature, sleep-wake rhythm, motor activity, cortisol secretion, etc.) were measured and the oscillatory patterns were analyzed in terms of their interaction and their relation to the time-course of psychiatric symptoms. The results obtained have to be characterized as more or less discouraging; however, one must ask whether this result is due to the application of a too restricted form of a general theory of dynamical systems.

Addressing again the characteristics of the Cheyne-Stokes paradigm, it becomes obvious that a functional variable (breathing pattern) is related to a regulated biochemical variable, e.g. the CO_2 concentration in plasma. In the regulation of breathing – complex though it is (besides CO_2 and O_2 partial pressure, local pH and other variables are important) – at least several of the relevant variables are known; thus the application of a theory of dynamical systems can at least in principle be realized. But in affective disorders these interactions are mainly unresolved; on the one hand, since the relevant variables are not measurable (e.g. changes of local monoamine concentrations in CNS in their temporal pattern), on the other, since the relevant variables are as yet undiscovered. In other words, as stated above, the establishment of a "dynamical theory of psychiatric disorders" would require a knowledge of the CNS "patterns" on the basis of which e.g. a depressive state is expressed (for a review see Emrich, 1990).

2. The Effects of Sleep Deprivation, the Diurnal Variation of Mood, and Their Possible Functional Relationship

Wu and Bunney (1990) have recently summarized the biological findings regarding the antidepressant effects of sleep deprivation: In 61 publications dealing with a total of 1700 patients, it was documented that more than 50% of the depressives showed an antidepressant reaction after sleep deprivation. In those patients who were unmedicated with an antidepressant, in 83% of the cases a depressive relapse occurred after nocturnal sleep. It has also been documented that a nap during the day can induce a severe relapse. From the point of view of systems-theory, these findings may be discussed under two perspectives: Firstly, the process-"S" hypothesis of Borbély may be applied according to which during wakefulness a sleepiness-induced process "S" is initiated which – after reaching a critical value – induces sleep; on the other hand, the hypothesis of Wu and Bunney has to be considered, according to which sleep induces a depressiogenic process. According to Borbély's model (1982), the process "S" is slowed down in depressive patients, leading to later onset of normal values and thus an improved mood, whereas, according to Wu and Bunney, in depressive patients sleep induces an increased release of a depressiogenic substance with the consequence that these patients show an improved mood not before the afternoon. From a systems-theoretical point of view, however, an alternative model may also be discussed, an interesting variant which may be a compromise between Borbély's model and that of Wu and Bunney: During the day a mood improving substance – e.g. an endorphin – may be synthesized which is released and metabolized during sleep. In such a model the sleep-induced reduction of opioid tone would have a depressiogenic effect with the consequence that during the day – if the synthesis rate is reduced during depression – the time-lag until mood is improved again is relatively high (mood improvement during the evening). In agreement with these hypotheses is the finding that opioids represent efficient antidepressants in therapy-resistant depression (Emrich et al., 1982) and that electroconvulsive treatment induces an activation of opioids (Emrich et al., 1979). These ideas also fit into the concept of Fink (1990) about the existence of an "antidepressin" called mood-modifying peptide, derived from the hypothalamus. Thus, from the point of view of synergetics, the question has to be raised as to whether depressives show a disturbed temporal pattern of synthesis and release of opioids ("dynamic code") and whether the pathophysiology of this disorder may thereby be explained.

3. Systems-Theoretical Aspects of Schizophrenic Psychoses From the Viewpoint of "Chaos Theory"

The pathophysiological concepts regarding affective psychoses discussed above give a hint that these types of psychiatric disorder may be interpreted as "dynamical diseases". Let us consider in a similar fashion the pathogenesis of schizophrenic psychoses. Schizophrenia appears not to represent a homogeneous disorder with a monocausal origin. There are hints that the clinical term "schizophrenia" describes a "final pathway syndrome". The type of the disorder may be described as representing a characteristic abnormal behavior of the human CNS which can also be precipitated in normal subjects under special conditions, e.g. sensory deprivation, protracted sleep deprivation, special ("psychedelic") drugs, etc. On the other hand, there are a lot of hints that schizophrenia-like syndromes do not fit into the same "type of disorder" which has been discussed above for the pathogenesis of affective psychoses under the perspective of synergetics, and that schizophrenia has to be attributed to a constitutional disturbance in which "increased vulnerability" (Nuechterlein & Dawson, 1984) plays a major role. The multitude of possible and already partially documented (Cazzullo et al., 1990) biological factors predisposing a subject for schizophrenia may thus not primarily represent "precipitators" of the disorder but may increase the probability of the manifestation of an "altered state" of the CNS ("vulnerability" concept).

How can this "abnormal state of the CNS" be characterized? The three-component model of psychosis (Emrich, 1989) assumes that perception is principally made up by three components: Firstly, sensory input; secondly, internal conceptualization, and thirdly, censorship. It also assumes that the special interaction between these three components is responsible for the biologically fruitful and efficacious conscious internal representation of the external world during perception and that the equilibrium between these three components is disturbed during psychosis. The question arises as to which equilibrium between these three components is "labilized" during psychotic episodes. Data from experiments using the "binocular depth inversion" paradigm (cf. Emrich, 1990) point to the view that the censorship component in relation to conceptualization shows a malfunction in psychotic states, leading to illusional experiences and hallucinations. Data pointing to a similar view have also been documented by Malenka et al. (1982). Interestingly, in normal Ss the application of THC (tetrahydrocannabinol), the active agent of cannabis, induces a similar disturbance of three-dimensional depth inversion as observed in schizophrenic patients (Emrich et al., 1991). On the other hand, recent cannabinoid research has demonstrated that especially hippocampal structures contain cannabinoid receptors (Devane et al., 1988). This finding is in line with the idea that hippocampal structures play a role in screening "plausibility" of sensory data and that this function may be impaired in schizophrenic patients (Beckmann et al., 1991; cf. also Roberts et al., 1990). In relation to these findings and concepts it is also of interest to mention the so-called "comparator systems" (Gray & Rawlins, 1986), localized in hippocampal structures, which play a major role in the recently elaborated models of anxiety and affective dyscontrol.

Summarizing, one may conclude that schizophrenia represents a diagnostic classification of a syndrome in which apparently not primarily the "dynamic code" as such is disturbed, but – due to structural deficits – the "internal dialogue" between different subsystems is impaired, thus leading to the guess that schizophrenia – in comparison to affective disorders – is less likely to fit into the concept of "dynamical diseases".

4. Aspects of Gene Dynamics

Family, twin and adoption studies clearly indicate that genetic factors make an important contribution to the etiology of affective disorders and schizophrenia (Nurnberger & Gershon, 1984; Baron, 1986). Therefore, genetic approaches might contribute to understanding pathogenetic factors and mechanisms leading to these diseases.

One main strategy of genetic research is to study genetic influences on biochemical alterations thought to be of etiological relevance. Mainly neurotransmitter systems, their receptors and metabolites, responses to pharmacological agents as well as polysomnographic alterations have been studied in families, in twins and in the general population. However, up to now, this strategy has failed to define any definite vulnerability marker, reflecting the problem that genetic research presently cannot refer to defined pathophysiological alterations with etiological relevance for psychiatric diseases.

Therefore another main strategy primarily directed to identify genes involved in mental illnesses has attracted much interest. The most frequently used method is to study linkage or association of psychiatric disorders in multiply affected families with protein markers and more recently with DNA markers: cloned genes or DNA fragments by the use of restriction length polymorphism (RFLP). The aim is to link a phenotypic expression to a certain gene locus and then, in a second step, to characterize structure and function of the gene product using methods of molecular biology. This strategy does not depend on etiological hypotheses but can also study "candidate genes", i.e. genes hypothesized to be of etiological relevance (for a review see Kidd, 1985). However, until now all linkage results in affective disorders and in schizophrenia have, unfortunately, been negative or controversial, and except for a subgroup of patients with manic depression probably linked to chromosome Xq28 (Mendlewicz et al., 1979; Risch et al., 1986), no definite marker gene has been evaluated (for reviews see Byerley, 1989; Mellon, 1989).

One of the basic problems of the linkage approach in such complex diseases as psychiatric disorders is that the strategy can only detect major gene effects. Therefore the linkage approach has examined only families compatible with Mendelian inheritance assuming the possibility that these families may have a major gene responsible for the disorder. However twin, relative and adoption studies in humans as well as selection and inbreeding experiments in animals have shown that genetic factors rarely account for as much as half of the variance of behavioral traits. Studies have addressed IQ, learning disabilities, and other aspects of cognition, features of personality such as neuroticism, extroversion and schizophrenia, affective disorders and other psychiatric diseases. Multiple genes with small effects appear to be involved rather than one or two major genes; nongenetic sources of variance are important. Beside linkage studies looking for major genes of etiological relevance for psychiatric diseases, one promising direction of genetic research on behavior and psychiatric disorders might be to identify genes that account for small amounts of behavioral variation which might help to understand "dysfunctional ones" (for a review see Plomin, 1990).

Study of gene expression in the brain, though today still in its infancy, should help to build valid hypotheses. Recent studies of gene expression of receptors in the brain have found a highly selective and specific expression and a heterogeneity of receptors never expected (for a review see Schofield et al., 1990). Molecular brain research has shown that a very large number of different genes are active in the mammalian brain and perhaps 30% of brain mRNA is brain specific. However, the function, loci of expression and regulation for the vast majority of these genes is still completely unknown (for a review see Gurling, 1985).

Another promising area of brain research, which is also relevant to the topic "dynamical diseases", concerns interactions on a genomic and non-genomic level between neurotransmitter systems, endocrine parameters and peptides in brain. Studies e.g. of the glucocorticoid receptor found many neurochemical interactions between this receptor and noradrenergic and serotonergic systems as well as many genomic effects of this receptor on expression of the gene coding for the corticotropin releasing hormone (CRH), the β_2-adrenergic receptor, the growth hormone, the nerve growth factor, interleukin-1β, prolactin and others (for a review see Holsboer, 1990).

In addition, the rapidly progressing general knowledge of the structure and organization of the human genome might help to understand what kind of genetic defect or alteration leads to the mostly episodically appearing psychiatric diseases. Today it is known that the DNA of a gene has distinct regions that code for parts of the protein ("exons"), intervening sequences ("introns"), the precise function of which are unknown, and flanking sequences. Mutation in the flanking regions can alter the time at which during development and/or the type of tissue in which the gene is turned on. Intervening sequences as well as flanking sequences are all transcribed into RNA. During the processing steps the intervening sequences are spliced out at precisely determined positions. Then the mature mRNA is translated into the protein. The gene expression, e.g. how much of the protein product is produced, is regulated by other DNA sequences (promotor regions) not copied themselves into RNA (for a review see Kidd et al., 1985). The glucocorticoid receptor, for example, regulates, through binding at the promotor region of the CRH gene, the transcription and building of CRH (for a review see Evans, 1988). Therefore, hypothetically, a mutation leading to "vulnerability" for psychiatric diseases might affect the regulatory sequences of genes or may occur in the exon sequences affecting the conformation of the protein or in the intervening sequences or might lead to altered mRNA splicing (for a review see Cloninger et al., 1983). From this point of view, a pathology in genomic-non-genomic interaction, e.g. at the level of promotor regions, may speculatively give rise to an understanding of some of the psychiatric disorders as "dynamical diseases".

In conclusion, future genetic research in psychiatry faces many problems, since psychopathological symptoms are only the "tip of an iceberg" of many influences from a genetic point of view. Therefore, genetic research is expected to benefit both from studies and hypotheses concerning processes that might lead the episodic appearance of psychopathological symptoms in order to find relevant biological markers, and from new molecular methods characterizing genes and gene expression in the brain.

References

BARON, M. 1986. Genetics of schizophrenia: I. Familial patterns and mode of inheritance. Biol. Psychiatry, 21, 1051-1066.

BECKMANN, H., HEINSEN, H. & JAKOB, H. 1991. Cytoarchitectonic abnormalities in rostral entorhinal fields of schizophrenics. Proc. 5th World Congress of Psychiatry, Florence. Amsterdam: Elsevier.

BORBÉLY, A.A. & WIRZ-JUSTICE, A. 1982. Sleep, sleep deprivation and depression: a hypothesis derived from a model of sleep regulation. Hum. Neurobiol., 1, 205-210.

BYERLEY, W.F. 1989. Genetic linkage revisited. Nature, 340, 340-341.

CAZZULLO, C.L., SACCHETTI, E., CONTE, G., INVERNIZZI, G. & VITA, A. (Eds.) 1990. Plasticity and Morphology of the Central Nervous System. Dordrecht: Kluwer Academic Publishers.

CLONINGER, C.R., REICH, T. & YOKOYAMA, S. 1983. Genetic diversity, genome organisation, and investigation of the etiology of psychiatric diseases. Psychiatric Developments, 3, 225-246.

DEVANE, W.A., DYSARZ, F.A., JOHNSON, M.R., MELVIN, L.S. & HOWLETT, A.C. 1988. Determination and characterization of a cannabinoid receptor in rat brain. Molec. Pharmacol., 34, 605-613.

EMRICH, H.M. 1989. A three-component-system hypothesis of psychosis. Impairment of binocular depth inversion as an indicator of a functional dysequilibrium. Brit. J. Psychiatry, 155 (Suppl. 5), 37-39.

EMRICH, H.M. 1990. Psychiatrische Anthropologie. München: Pfeiffer.

EMRICH, H.M. 1991. Systemtheoretische Anthropologie: Auf dem Wege zu einer Strukturtheorie des Bewußtseins auf der Basis eines nichtreduktiven Monismus. In: Fischer, H.R. (Ed.): Autopoiesis. Heidelberg: Carl-Auer Verlag.

EMRICH, H.M., HÖLLT, V., KISSLING, W., FISCHLER, M., LASPE, H., HEINEMANN, H. & von ZERSSEN, D. 1979. β-Endorphin-like immunoreactivity in cerebrospinal fluid and plasma of patients with schizophrenia and other neuropsychiatric disorders. Pharmakopsychiat., 12, 269-276.

EMRICH, H.M., VOGT, P., HERZ, A. & KISSLING, W. 1982. Antidepressant effects of buprenorphine. Lancet, II, 709.

EMRICH, H.M., WEBER, M.M., WENDL, A., ZIHL, J., von MEYER, L. & HANISCH, W. 1991. Reduced binocular depth inversion as an indicator of cannabis-induced censorship impairment. Neurosci. Biobehav. Rev. (in press).

EVANS, R.M. 1988. The steroid and thyroid hormone receptor superfamily. Science, 240, 239-241.

FINK, M. 1990. How does convulsive therapy work? Neuropsychopharmacology, 3, 73-82.

GURLING, H. 1985. Application of molecular biology to mental illness. Analysis of genomic DNA and brain mRNA. Psychiatric Developments, 3, 257-273.

GRAY, J.A. & RAWLINS, J.N.P. 1986. Comparator and buffer memory: an attempt to integrate two models of hippocampal functions. In: Isaacson, R.L. & Pribram, K.H. (Eds.): The Hippocampus, Vol. 4. New York: Plenum, 159-201.

HALBERG, F. 1968. Physiologic considerations underlying rhythmometry with special reference to emotional illness. In: De Ajuriaguerra, J. (Ed.): Cycles Biologiques et Psychiatrie. Paris: Masson et Cie, 73-126.

HESCH, R.D. 1989. Einführung in Hormone und Rezeptoren. In: Hesch, R.D. (Ed.): Endokrinologie. Teil A Grundlagen. München: Urban & Schwarzenberg, 3-49.

HOLSBOER, F. 1990. Neurobiologische Forschungskonzepte für die Pharmakotherapie affektiver Störungen. In: Herz, A., Hippius, H. & Spann, W. (Eds.): Psychopharmaka heute. Berlin: Springer, 13-33.

KIDD, K.K. 1985. New genetic strategies for studying psychiatric disorders. In: Sakai, T. & Tsuboi, T. (Eds.): Genetic Aspects of Human Behavior. Tokyo: Igaku-Shoin, 235-246.

MacKAY, D. 1965. A mind's eye view of the brain. In: Wiener, N. & Schadé (Eds.): Cybernetics of the Nervous System (Progress in Brain Research, Vol. 17). Amsterdam: Elsevier, 321-332.

MACKEY, M.C. & AN DER HEIDEN, U. 1982. Dynamical diseases and bifurcations: Understanding functional disorders in physiological systems. Funkt. Biol. Med., 1, 156-164.

MALENKA, R.C., ANGEL, R.W., HAMPTON, B. & BERGER, P.A. 1982. Impaired central error-correcting behavior in schizophrenia. Arch. Gen. Psychiatry, 39, 101-107.

MELLON, C.H. 1989. Genetic linkage studies in bipolar disorder: a review. Psychiatric Developments, 2, 143-158.

MENDLEWICZ, J., LINOWSKI, P., GUROFF, J.J. & VAN PRAAG, H.M. 1979. Color blindness linkage to bipolar manic-depressive illness: new evidence. Arch. Gen. Psychiatry, 36, 1442-1447.

NAGEL, T. 1986. The View from Nowhere. Oxford: Oxford University Press.

NUECHTERLEIN, K.H. & DAWSON, M.E. 1984. A heuristic vulnerability/stress model of schizophrenic episodes. Schizophren. Bull., 10, 300-312.

NURNBERGER, J.I. & GERSHON, E.S. 1984. Neurobiology of mood disorders. In: Post, R.M. & Ballenger, J.C. (Eds.): Neurobiology and Mood Disorders. Baltimore: Williams and Wilkins, 76-101.

PLOMIN, R. 1990. The role of inheritance in behavior. Science, **248**, 183-188.

RISCH, N., BARON, M. & MENDLEWICZ, J. 1986. Assessing X-linked inheritance in bipolar related major affective disorder. J. Psych. Res., **20**, 275-288.

ROBERTS, G.W., DONE, D.J., BRUTON, C. & CROW, T.J. 1990. A "mock up" of schizophrenia: temporal lobe epilepsy and schizophrenia-like psychosis. Biol. Psychiatry, **28**, 127-143.

SCHOFIELD, P.R., SHIVERS, B.D. & SEEBURG, P.H. 1990. The role of receptor subtype diversity in the CNS. TINS, **13**, 8-11.

WEHR, T.A. & GOODWIN, F.K. (Eds.). 1983. Circadian Rhythms in Psychiatry. Pacific Grove, Ca.: Boxwood Press.

WU, J.C. & BUNNEY, W.E. 1990. The biological basis of an antidepressant response to sleep deprivation and relapse: Review and hypothesis. Am. J. Psychiatry, **147**, 14-21.

Using Multivariate Time Series Models in Systemic Analysis

Klaus Ackermann, Uta Streit, Hansjörg Ebell, Arno Steitz,
Ilse M. Zalaman, and Dirk Revenstorf

With 5 Figures

Abstract. Multivariate autoregressive moving–average modeling of relation-ships between two or more stochastic time series as well as the underlying principles are discussed (Tiao & Box, 1981; Jenkins & Alavi, 1981). The rele-vance of this particular type of time series analysis with regard to identifying systemic relationships among observed process variables is demonstrated by using examples from pain and marital counseling research. This kind of analysis is especially suited to describe time series for which the relationships between variables are thought to be stable or stationary throughout the time of observation.

1. Introduction

Observations based on single cases or individual units play an important role in clinical psychology and psychotherapy. In research work, these ob-servations frequently facilitate the refinement of theory as well as the de-velopment of hypotheses, whereas in therapeutic practise the individual sys-tem tends to be the basic unit of reference. When studying an individual case or a single system autoregressive (integrated) moving average model-ing (ARMA/ARIMA modeling, respectively) can be used in various ways to identify dependencies among variables in series of repeated observations at equally spaced intervals over time. The general objective is to reveal the underlying processes and relationships within and among variables.

The ARIMA intervention models have enjoyed widespread use in the social sciences and clinical research. With these models the effects of ex-ogenous influences or interventions on a single time series can be tested, for example, the impact of a specific treatment on clinically relevant target be-havior or symptoms in the single case (Box & Tiao, 1975; Glass et al., 1975; Revenstorf, 1979), whereas ARMA/ARIMA modeling of mutual relationships between two or more stochastic time series has only received little attention (Gottmann, 1981; Tiao & Box, 1981; Schmitz, 1989). When analyzing multi-ple time series most behavioral scientists preferred to use the simpler cross-correlational method because the more advanced multivariate ARMA models had not yet been properly inaugurated. Since multivariate ARMA analysis is a variant of multiple regression analysis, ARMA models can be consid-ered a more efficient and informative alternative to cross–correlations. The

213

Springer Series in Synergetics, Vol. 58   Self-Organization and Clinical Psychology
Editors: W. Tschacher, G. Schiepek, and E.J. Brunner   © Springer-Verlag Berlin Heidelberg 1992

objective of this paper is to demonstrate some of the possibilities of multivariate ARMA modeling for identifying and describing the mutual relationships among stochastic process variables in clinical psychology.

2. Multivariate ARMA Models

When several variables of a single unit are measured repeatedly at equally spaced time intervals, multivariate ARMA models can be applied to the joint statistical analysis of these multiple time series. Furthermore, no a priori assumptions about correlations among variables have to be made. As a result, multidirectional time–lagged relationships can be identified. Essentially these models reflect time–lagged regressions and thus the mutual dependence between the variables involved. In the following we will briefly describe some of the basic statistical principles in a user–oriented fashion; for a more comprehensive review see Liu (1986) or Schmitz (1989).

With SCA programs for modeling and forecasting (Liu & Hudak, 1989) we employed multivariate ARMA models as outlined by Tiao and Box (1981). Hence, each k series contains n observations taken at equally spaced time intervals and the k observations at time t make up the k–dimensional vector z_t. The vector time series z_t is assumed to be stationary. If not, stationarity can be obtained by appropriate transformations, so–called "differencing" procedures which eliminate linear and polynomial data trends and create the integrated component in an ARIMA model. Nonstationarity due to exogenous influence can be modeled too, and can consequently be eliminated from the individual series before applying multivariate modeling.

The general ARMA(p,q) model representation of the mean centered vector time series z_t takes the form

$$\mathbf{\Phi}(B)\mathbf{z}_t = \mathbf{\Theta}(B)\mathbf{a}_t \tag{1}$$

where $\mathbf{a}_t$ is a series of k–dimensional vectors of random shocks (white noise) independently and identically distributed as multivariate normal $\mathbf{N}(\mathbf{0}, \mathbf{\Sigma})$. The matrix polynomials in the backshift operator B (where $BY_t = Y_{t-1}$) can be expanded to

$$\mathbf{\Phi}(B) = \mathbf{I} - \mathbf{\Phi}_1 B - \ldots - \mathbf{\Phi}_p B^p$$

$$\mathbf{\Theta}(B) = \mathbf{I} - \mathbf{\Theta}_1 B - \ldots - \mathbf{\Theta}_q B^q$$

where $\mathbf{\Phi}_m (m = 1 \ldots p)$ and $\mathbf{\Theta}_n (n = 1 \ldots q)$ are $k \times k$ matrices containing the AR and MA regression coefficients of z_t on z_{t-m} and $\mathbf{a}_{t-n}$, respectively, for significant lags.

For example, the general ARMA(1,1) model can be represented by

$$(\mathbf{I} - \mathbf{\Phi}B)\,\mathbf{z}_t = (\mathbf{I} - \mathbf{\Theta}B)\,\mathbf{a}_t \tag{2}$$

(cf. Liu, 1986). In the simplest case of multiple time series, where $k = 2$, this compressed matrix notation can be expanded to

$$z_{1t} - \Phi_{11}z_{1(t-1)} = a_{1t} - \Theta_{11}a_{1(t-1)} + \Phi_{12}z_{2(t-1)} - \Theta_{12}a_{2(t-1)}$$
$$z_{2t} - \Phi_{22}z_{2(t-1)} = a_{2t} - \Theta_{22}a_{2(t-1)} + \Phi_{21}z_{1(t-1)} - \Theta_{21}a_{1(t-1)}$$

where the autoregressive coefficients Φ_{11} and Φ_{22} and the corresponding coefficients of the moving average portion Θ_{11} and Θ_{22} represent the effects of immediate past values and random shocks (lag 1) on the time series *itself*. The off-diagonal elements $(\Phi_{12}, \Phi_{21}, \Theta_{12}, \Theta_{21})$ describe the effects *between* variables. With all variables standardized to a mean of 0 and a standard deviation of 1, the coefficients represent the change in the current value of the predicted variable.

The objective of modeling is to identify the Φ and Θ coefficients in order to adequately describe the underlying stochastic process. In the identification and diagnostic evaluation of potentially relevant models, the user usually employs auto– and cross–correlation, that is, the time lagged correlations of the time series with itself and the other series. Partial autoregression coefficients, explained variance, the Akaike information criterion (AIC), and a chi–square statistic indicating deviation from independence can also be utilized. As a rule, model building is an iterative procedure, where the three steps — tentative identification, estimation, and diagnostic evaluation — may have to be applied repeatedly until model requirements are met. This means first and foremost that significant entries into the residual cross–correlation matrices can only be few and that the residual series must be random.

3. Study 1: Systemic Variables in Pain Management

3.1 Data

This single case study involved the analysis of four time series consisting of 70 observations each. The time series represented daily ratings of four variables by a female patient who was suffering from cancer pain. The patient rated the variables on visual analog rating scales ranging from 0 to 100. The variable "pain intensity" (PAI) reflected the rating of both episodic and continuous pain. These ratings were combined by taking the higher value because episodic pain generally exluded the presence of continuous pain and vice versa ($M = 53.47, SD = 10.69$). The other variables were "well being" (WEL; $M = 47.04, SD = 14.06$), "suffering" (SUF; $M = 55.44, SD = 11.04$), and "coping expectations for tomorrow" (CEX; $M = 46.79, SD = 8.75$). The two missing values were replaced by approximation based on neighboring observations. Figure 1 shows the four input series.

Because the patient had received two forms of hypnotic treatment, intervention models were fit to the series to eliminate slight and rarely significant level changes. This was done to prevent possible spurious correlations based on level changes caused by the treatments.

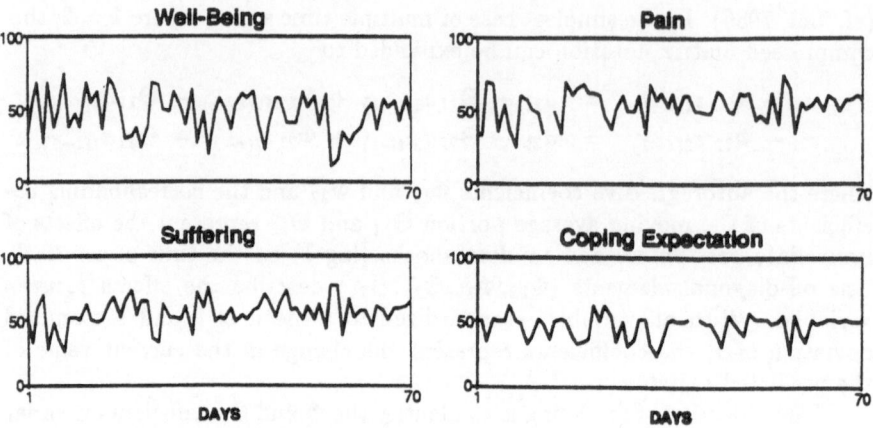

Fig. 1. Input series for pain patient

3.2 Model Identification and Results

Tentative model identification. The objective of tentative model identification is to use statistics which can be readily calculated from the data in order to facilitate the selection of an appropriate model. Since chi-square statistic testing for serial independency and the Akaike information criterion suggested the inclusion of lag 2, AR(1), AR(1,2),MA(1), MA(1,2), MA(2) models were computed for lag 1 and lag 2. No constraints were put on non-significant elements of the parameter matrices. All lag 1 models revealed that the variable "coping expectations for tomorrow" had a significant effect on how the variables "well being", "suffering" and "pain" were rated the next day. Moreover, all lag 2 models demonstrated that pain intensity experienced two days ago had a significant influence on the variable "coping expectations for tomorrow".

The moving–average model reflecting auto– and cross–regressive influences of random fluctuations from lag 1 and 2 (MA(1,2)) produced the best model in terms of residual cross–correlations and explained variances. Table 1 also lists the explained variances of the other models. On average, the input variance was reduced by about 20 percent in the MA(1,2) model, other models generally being less effective in terms of explained variance. However, for the variables WEL and SUF an AR(1) or MA(1) model would have been sufficient in terms of explained variance.

Process equations and "system rules". Even though the overall significance of a specific model may be informative, regression coefficients should receive the primary attention. Figure 2 depicts the significant relationships between the process variables and their past random shocks in the MA(1,2) model ($p < .05$).

Table 1. Prediction in tentative models

	Model														
	AR(1)			AR(1,2)			MA(1)			MA(1,2)			MA(2)		
Variable	R^2	F	df	R^2	F	df	R^2	F	df	R^2	F	df	R^2	F	df
WEL	0.189	3.73**	4,64	0.179	1.61	8,59	0.162	3.14*	4,65	0.161	1.46	8,61	0.003	0.05	4,65
SUF	0.081	1.41	4,64	0.179	1.61	8,59	0.152	2.91*	4,65	0.188	1.77	8,61	0.040	0.68	4,65
PAI	0.108	1.94	4,64	0.148	1.28	8,59	0.123	2.28	4,65	0.234	2.33*	8,61	0.102	1.85	4,65
CEX	0.032	0.53	4,64	0.159	1.39	8,59	0.071	1.24	4,65	0.279	2.95**	8,61	0.221	4.61***	4,65

Note. *p<.05, **p<.01, ***p<.005.

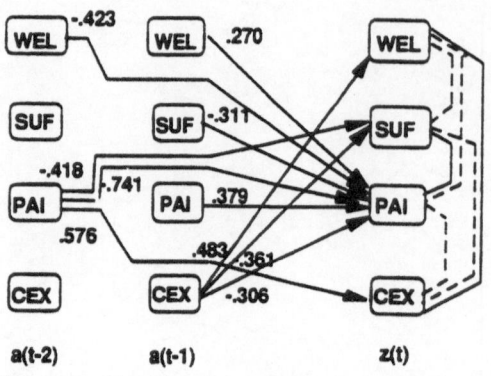

a(t-2)   a(t-1)   z(t)

Fig. 2. Significant coefficients in MA(1,2) model

The following process equations and concomitant rules for significant findings may be stated as

$$WEL(t) = .483 \times a(CEX; t-1) + a(WEL; t)$$

— which shows that the patient would feel better today if yesterdays coping expectations had been high.

$$SUF(t) = -.361 \times a(CEX; t-1) - .418 \times a(PAI; t-2)$$
$$+a(SUF; t)$$

— indicates that the patient would suffer less today if she experienced a lot of pain two days ago.
— Also the patient's suffering would be less today if she thought yesterday that she would be able to cope with her pain.

$$PAI(t) = .270 \times a(WEL; t-1) - .311 \times a(SUF; t-1)$$
$$+.379 \times a(PAI; t-1) - .306 \times a(CEX; t-1)$$
$$-.423 \times a(WEL; t-2) - .741 \times a(PAI; t-2)$$
$$+a(PAI; t)$$

Relationships among variables for observations made at time t-1 demonstrate that:

— Pain intensity would be rated higher today if it had received a high rating yesterday.
— If the patient suffered a lot of pain yesterday pain intensity would be rated lower on the following day.
— If coping expectations were met yesterday then pain intensity received a lower rating on the following day.
— The patient would feel more pain today if she felt better yesterday.

218

Observations at time t-2 show that:
- — The higher pain intensity was rated two days ago the lower it would be rated today.
- — The pain would be less intense if the patient felt better the day before yesterday.

$$CEX(t) = .576 \times a(PAI; t-2) + a(CEX; t)$$

- — Coping expectations for tomorrow are expected to get a high rating if the patient gave pain intensity a high rating two days ago.

Contemporaneous relationships. Contemporaneous correlations of the residuals were computed after multivariate modeling. The residuals were also analyzed exploratively by standard multiple regression analysis. Table 2 shows the results. Compared to time lagged relationships contemporaneous covariations explained a greater proportion of the common variance for the variables "well being", "suffering", and "pain intensity". On the other hand, the variable "coping expectations for tomorrow" was more strongly related to the past, especially to "pain intensity" experienced two days before.

3.3 Conclusions

The contemporaneous correlations and the regression analysis performed on the ARMA residuals demonstrated that appraisals of "suffering" and "pain intensity" correlated highly with each other, and only moderately with the variable "well being", which in turn correlated with the variable "coping expectations". In order to gain more insight into underlying processes and systemic aspects of pain management we examined time–lagged relationships between variables and their stochastic fluctuations (random shocks).

There is strong evidence that the variable "expectation of coping with pain tomorrow" related negatively to tomorrow's rating of the variables "pain intensity" and "suffering" and positively to the variable "well being". Moreover, coping expectations were also related to pain intensity felt two days ago. Thus one can say that pain intensity and coping expectations displayed a negative feedback pattern, that is, pain facilitated coping expectations, which reduced pain.

Table 2. Contemporaneous correlation and prediction for residuals MA(1,2) model

Variable	1	2	3	R^2	F	df
1. WEL				.354	12.03**	3,66
2. SUF	− .55**			.610	34.39**	3,66
3. PAI	− .53**	.73**		.568	28.93**	3,66
4. CEX	.30**	− .38**	− .23*	.169	4.47**	3,66

Note. $^*p < .05$, $^{**}p < .01$.

Furthermore, all variables were related to the appraisal of pain intensity. For example pain intensity related positively to its own random shocks. A more complicated relationship was revealed between the subjective appraisals of suffering or well being and pain intensity, that is, the patient would give pain intensity a higher rating on the next day if on the previous day she had felt well and suffered less.

The data also revealed relations to time t-2 (two days ago): If the patient experienced more pain two days ago she would suffer less and feel less pain two days later. These findings point towards an antagonistic relation between pain intensity at lag 2 and current state in this individual. However, a negative contribution of well being at lag 2 to current pain intensity suggests a concordant relationship between the feeling of well being two days ago and the pain experienced now supplementing the previously mentioned feedback pattern.

4. Study 2: Description of Marital Relations

4.1 Data

The second time series study analyzed marital relations. The reported time series comprise daily ratings from a couple who participated in a multicomponent marital counseling training program (Revenstorf et al., 1991). The couple had been married for 22 years. They have a son (19) who was then doing his military service and consequently he did not spend much time at home, except when on leave. The couple participated in this program because of their violent quarrels. They were asked to separately rate seven aspects of their marital relationship over a period of 126 days. Due to the missing data the ARMA model was based on 14 time series each having 70 observations (Fig 3). The ratings ranged from 1 to 6 ("not at all" to "very much"). The variables were "time spent together", "longing for the spouse", "attention paid to the spouse", "tolerance for the spouse", "tenderness experienced together", "feeling accepted by the spouse", and "being content with the relationship". The mean values and standard deviations of the series showed that the husband in general tended to demonstrate a more positive attitude towards his wife than vice versa (Table 3).

4.2 Model Identification and Results

Tentative model identification. A multivariate AR(1) model was applied in order to identify the time–lagged relationships between the 14 variables (7 for each spouse), since the data exhibited a typical pattern of decaying significant values for the cross–correlation matrices. The diagnostic checking of the residual cross–correlations confirmed the model chosen. Aside from estimating the saturated model, additional analyses were done by setting unequivocal nonsignificant regression coefficients iteratively at zero. This

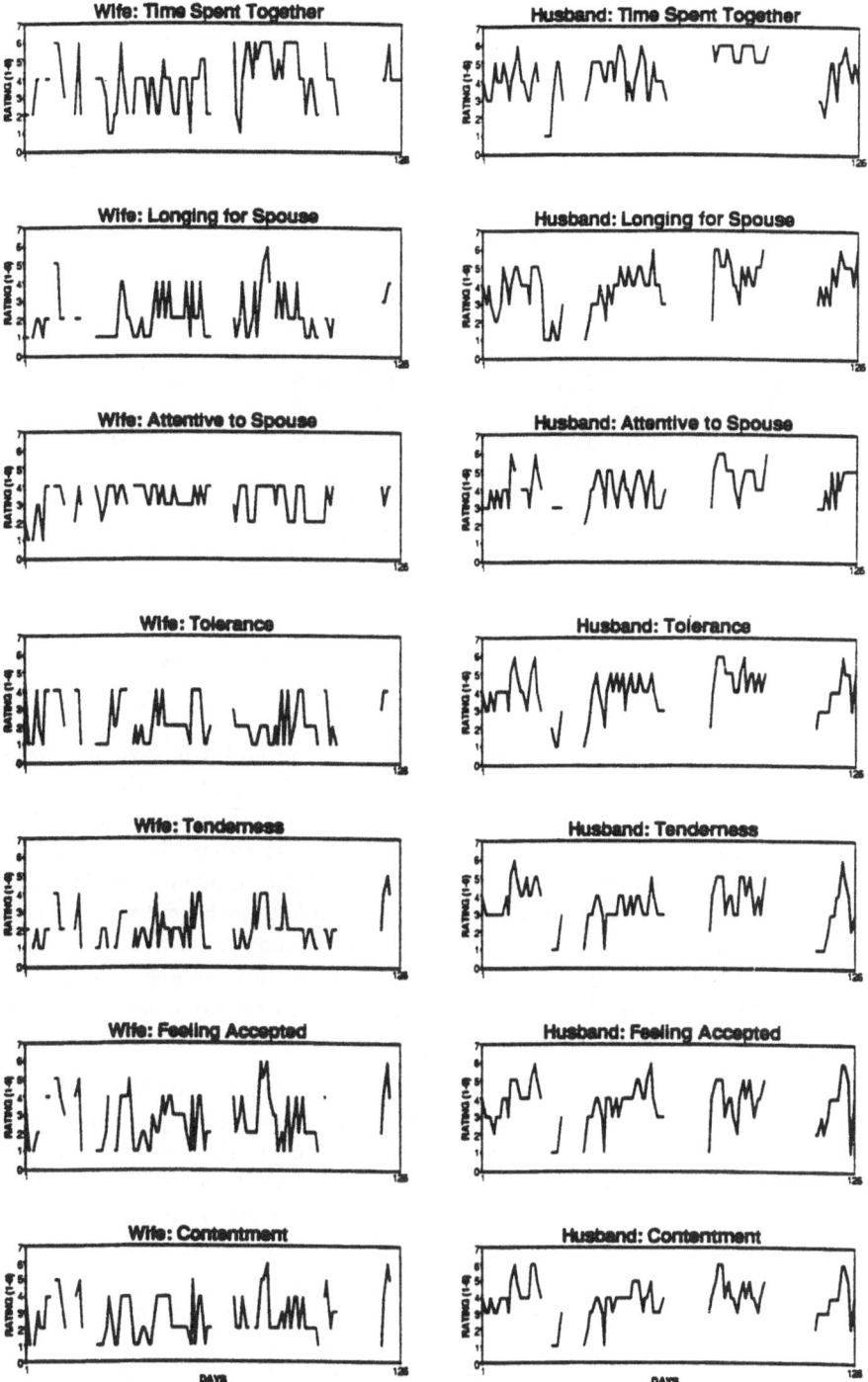

Fig. 3. Time series for variables of marital relationship

Table 3. Mean scores on scales of marital relationship

Variable	Day 1 – 126			Day 1 – 63			Day 64 – 126		
	N	M	SD	N	M	SD	N	M	SD
Wife									
TIME	94	3.83	1.51	53	3.34	1.39	41	4.46	1.39
LONG	91	2.20	1.25	55	2.00	1.15	36	2.50	1.16
ATTE	92	3.23	0.88	54	3.28	0.83	38	3.16	0.70
TOLE	91	2.33	1.22	54	2.41	1.30	37	2.22	1.25
TEND	90	2.08	1.05	53	1.96	1.02	37	2.24	1.04
ACCE	88	2.80	1.41	53	2.66	1.39	35	3.00	1.37
CONT	92	2.77	1.38	55	2.62	1.38	37	3.00	1.37
Husband									
TIME	88	4.38	1.23	55	4.02	1.16	33	4.97	1.18
LONG	88	3.88	1.28	56	3.48	1.25	32	4.56	1.29
ATTE	85	4.11	0.96	52	3.87	0.89	33	4.48	0.90
TOLE	86	3.98	1.16	53	3.74	1.13	33	4.36	1.16
TEND	86	3.43	1.19	53	3.38	1.08	33	3.52	1.11
ACCE	86	3.66	1.21	53	3.60	1.18	33	3.76	1.22
CONT	86	3.83	1.17	53	3.66 ·	1.21	33	4.09	1.25

Observation period

yielded a so-called "restricted" model. The husband and wife variables were also analyzed separately which allowed an educated guess of the spouses' "cross determination" in terms of explained variance. Figure 4 and Table 4 show the amount of explained variance for the different models applied to each variable. When comparing the separate and the common models it becomes obvious that the spouses exert a considerable time–lagged influence on each other. The findings also suggest that all the models used account to a greater extent for the husband's appraisal of the seven variables than for the wife's. This may indicate that the husband's behavior was more consistent in terms of the variables under consideration.

Lagged regressions. Figure 5 contains all the process equations and the probabilistic "system rules". Arrows indicate the 17 out of 196 possible significant regressions ($p < .05$) between the 14 variables in the saturated AR(1) model. The "restricted" model has 35 significant coefficients and this rise may be attributed to the moderating or suppressing effects in either one of the two models. However, nearly all of the significant coefficients of the saturated model were also significant in the restricted model.

In the following some basic relational aspects as revealed by the saturated model will be discussed. The husband's "longing for his spouse" seemed to have had a positive effect on tomorrow's rating of "tolerance" and "feeling accepted" by both spouses and on the tenderness and longing felt by the

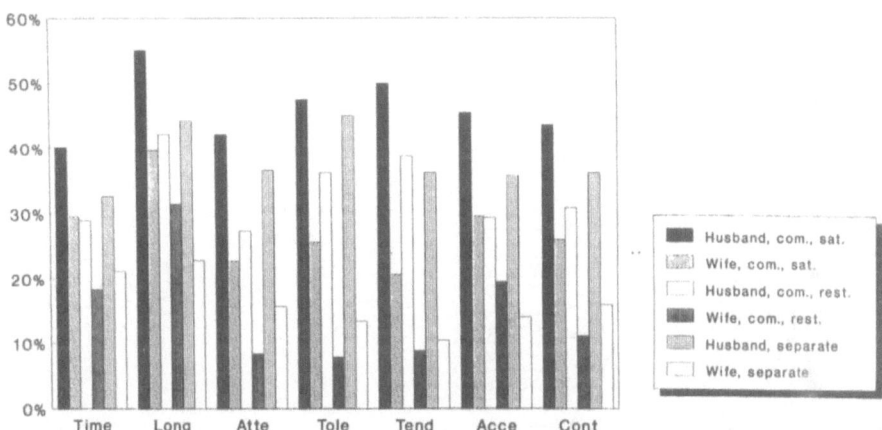

Fig. 4. R^2 in different models for variables of marital relationship

Table 4. Prediction in various models of marital variables

| | Common Models | | | | | | Separate Models | | |
| | AR(1), saturated | | | AR(1), restricted | | | AR(1) | | |
Variable	R^2	F	df	R^2	F^a	df	R^2	F	df
Wife									
TIME	0.295	1.61	14,54	0.183	4.85	3,65	0.211	2.33*	7,61
LONG	0.396	2.53**	14,54	0.314	9.92	3,65	0.228	2.57*	7,61
ATTE	0.227	1.13	14,54	0.084	6.14	1,67	0.157	1.62	7,61
TOLE	0.256	1.32	14,54	0.079	2.83	2,66	0.134	1.35	7,61
TEND	0.206	1.00	14,54	0.089	2.12	3,65	0.105	1.02	7,61
ACCE	0.296	1.62	14,54	0.195	5.25	3,65	0.141	1.43	7,61
CONT	0.260	1.35	14,54	0.112	8.45	1,67	0.160	1.66	7,61
Husband									
TIME	0.401	2.58**	14,54	0.289	8.81	3,65	0.326	4.21**	7,61
LONG	0.551	4.73**	14,54	0.421	15.75	3,65	0.442	6.90**	7,61
ATTE	0.421	2.80**	14,54	0.273	12.39	2,66	0.366	5.03**	7,61
TOLE	0.475	3.99**	14,54	0.362	12.29	3,65	0.450	7.13**	7,61
TEND	0.500	3.86**	14,54	0.388	13.74	3,65	0.362	4.94**	7,61
ACCE	0.455	3.22**	14,54	0.294	9.02	3,65	0.358	4.82**	7,61
CONT	0.436	2.98**	14,54	0.309	14.76	2,66	0.362	4.94**	7,61

Note. ^{a}Because model restrictions were made a posteriori, these F – values were not taken for formal tests of significance.
Note. $^*p < .05$, $^{**}p < .01$.

husband. Consequently the husband would pay more attention to his wife when he felt content. On the other hand, the spouses would spend less time together on the next day if the husband felt more accepted by the wife.

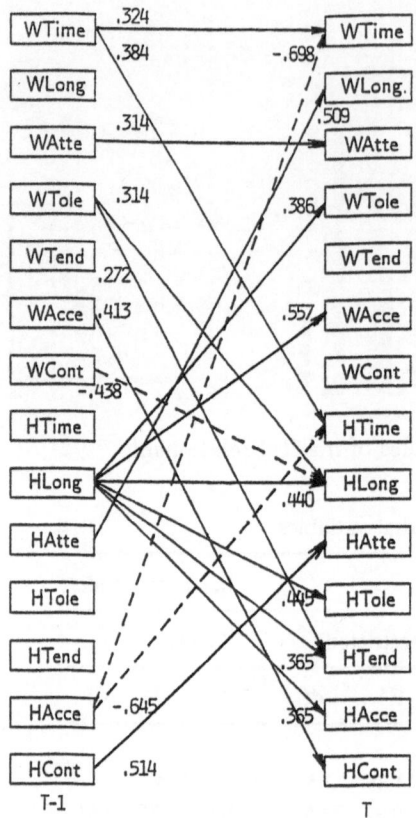

Fig. 5. Significant coefficients in saturated AR(1) model

A similar relationship was displayed by the wife regarding being content. When the wife felt content her husband would experience less yearning for her. In general, however, the wife's variables by comparison tended to exert less influence in terms of explained variance. Most of her lagged regressions referred to her husband's variables except for the autoregressive relations of "time spent together" and "being attentive to the spouse". Her acceptance of him made him content, her tolerance toward him increased his tenderness towards, and longing for, her. However, while the wife's assessment of "time spent together" was postively related to itself and to the husband's appraisal of "time spent together" for the next day, the husband's appraisal of "time spent together" did not relate to any of the other variables. We might discern a positive feedback loop from these different relations: Thus the husband's appraisal of his yearning in turn enhanced her tolerance for his short–comings which in turn increased his yearnings for her. Since her intolerance for his short–comings was the central issue, therapy should explore those events which facilitate tolerance. Another positive chain of events, which could be explored during therapy, is the fact that when she felt accepted by her husband he would feel more content which made him more attentive and in turn increased her longing for him.

Contemporaneous relationships. To reduce the complex information provided by the contemporaneous correlation matrix a principal–component analysis was applied to the data (P technique, see Cattel 1957). Three factors with an eigenvalue greater than 1.0 were extracted which accounted for about 75 percent of the total variance. After a varimax rotation the factor explaining 35 percent of the variance could be imputed unequivocally to the husband. The factor which accounted for 23% of the total variance was attributable to the wife. The third factor, which explained 19 percent of the total variance, comprised both spouses' appraisal of "time spent together", the wife's rating of "longing for her spouse" and of "being attentive to the spouse" were considered indicator variables.

4.3 Conclusions

In this study we used both the principal–component analysis and ARMA modeling in order to gain more insight into the relevant aspects of a marital relationship. The principal component analysis clearly identified the two main independent sources of variance, namely husband and wife, although the wife accounted for less variance. Time–lagged relationships between variables characterized the systemic aspects involved. For example the ARMA analysis showed that beneath the surface some relational resources still remained although overtly this couple was enmeshed in a vicious circle of marital discord.

Corroborating clinical judgement the husband's variables "longing for his spouse" and "feeling accepted" influenced the other variables substantially while past appraisals of the wife only had a moderate effect. However, her appraisal of tolerance for her partner's short–comings was related to how he would rate "tenderness" and "longing for his spouse" tomorrow.

5. Discussion

Hypotheses about systemic relationships between specific behaviors, feelings, and cognitions in clinical psychology and psychotherapy frequently originate from casuistic observations or reasoning. However, observations from single cases do not defy statistical analysis as long as they are sufficient in number.

In our studies we have been able to demonstrate that multivariate ARMA or ARIMA models can identify time–lagged relationships between specific variables and future short-term behavior can be predicted relatively well with this type of analysis. Beyond their predictive use multivariate ARMA models also point out directions of temporal sequences. Feedback relationships and in some cases "key variables" can also be identified statistically. The results of such a procedure for the individual case could be of major interest to clinical psychology and psychotherapy research, not only in terms of specific interventions but also in terms of providing empirical support to scientific reasoning.

ARMA models — as with all mathematical modeling — are subject to idealized assumptions about the real world. In addition to assuming linearity one primary assumption is that the coefficients of ARMA models inherently refer to stable, that is, stationary processes. Therefore, observational data undergoing ARMA modeling should be based essentially on stable relationships. With this in mind, what significance does multivariate ARMA modeling have for clinical psychology, and, in particular, for psychotherapy? For one, clients' problems could be viewed as stable, repetitive patterns of dysfunctional belief systems, affects, and behaviors. Such variables can be subjected to multivariate ARMA modeling. However, dysfunctional patterns may undergo a rapid reorganization or even be replaced by others as soon as the client engages in effective psychotherapy. In order to monitor nonstationary behavior, Keeser and Bullinger (1984) suggest that data should be sampled before, during, and after the actual intervention phase and that separate models should be computed for each phase. This kind of a procedure would reveal changes in the structural patterns and relationships of the observed variables. In addition, such a design could be of heuristic value in that it would allow the identification of favorable target patterns in the client.

On the other hand, constituent effects between variables remain, even though the various levels may change from dysfunctional to functional depending on conditions. In this case nonstationarity would be restricted to the level aspect (not to covariation) and can consequently be accounted for with intervention models.

In cases where model stability is indeed doubtful certain precautionary measures can be applied. A frequently recommended procedure for testing model stability is to exclude the last few observations from model estimation and to analyze the model's predictive power for these last observations. If this "cross validation" results in a good fit then the tentative model can be accepted.

Another procedure is to compare cross–correlation coefficients from stepwise displaced time windows for relevant pairs of time series. Unfortunately, from an inferential point of view, some aspects of this method are rather poorly defined. The advantage of this method, however, is the fact that such an analysis may reveal more information concerning the evolving dynamics of ongoing systemic processes.

The stability of the regressive relationship between two variables in a longitudinal single case design can shed some light on corresponding aspects of cross-sectional studies using classical group statistics. Aside from the fact that they tend to neglect individual differences, cross–sectional correlational data implicitly suggest that the underlying relationships are quite stable across time. Although there is good reason to believe that individual fluctuations are compensated by the law of large numbers, we are of the opinion that longitudinal analyses used as an adjunct strategy to interpret cross–sectional parameters would be of great use, especially when conflicting results have been found. On the other hand, findings from single unit time series analyses may be combined and analyzed with group statistical procedures. The results would then have nomothetical status (Keeser & Bullinger, 1984).

Another problem of modeling is the selection of the appropriate variables. Domain oriented knowledge is needed to select those variables which reflect the hypothesized system in the best and most useful way. Also a model comprising only a few variables is easier to handle and its results are more interpretable than those of a model with many variables. Another important step concerns the selection of the time interval for successive observations. According to Schmitz (1989) special care should be exercised, since the interval itself will exert considerable influence on time–lagged dependency detection.

For therapeutic purposes further assessment of systemic modeling based on ARMA models should concentrate on at least two objectives. First a more systematic examination regarding the convergence of model and clinical judgement should be done in terms of validation. Although relatively few studies have been done on this subject matter, the data available indicate a good correspondence between model and clinical expert judgement. Secondly multivariate ARMA modeling to facilitate concrete intervention planning in therapeutic settings should be studied. Since clients are frequently put on waiting lists when seeking therapy, the waiting period could be used for data collection; that is, the patient keeps a standardized diary. If multivariate time series models should indeed prove to facilitate the identification of target variables, then therapeutic interventions could be aimed directly at the variable under consideration, thereby forcing the system to destabilize and to reorganize at a faster rate, which may expedite treatment.

Acknowledgment

Preparation of this article was supported in part by a grant from the Deutsche Forschungsgemeinschaft (RE 402/9-3). We gratefully acknowledge the assistance of Marika Moeck in data collection.

References

BOX, G.E.P. & TIAO, G.C. 1975. Intervention analysis with applications to economic and environmental problems. Journal of the American Statistical Association, 70, 70-79.

CATTELL, R.B. 1951. On the use and misuse of P, Q and O-techniques in clinical psychology. Journal of Clinical Psychology, 7, 203-214.

GLASS, G.V., WILLSON, V.L. & GOTTMAN, J.M. 1975. Design and analysis of time series experiments. Boulder: University Press.

GOTTMAN, J.M. 1981. Time-series analysis. A comprehensive introduction for social scientists. New York: Cambridge University Press.

GREGSON, R.A.M. 1983. Time series in psychology. Hillsdale, N.J.: Lawrence Erlbaum Associates.

JENKINS, G.M. 1979. Practical experiences with modeling and forecasting time series. St. Helier: Gwilym Jenkins & Partners.

JENKINS, G.M. & ALAVI, A.S. 1981. Modeling and forecasting multivariate time series. Journal of Time Series Analysis, 2, 1-47.

KEESER, W. & BULLINGER, M. 1984. Process-oriented Evaluation of a cognitive-behavioral treatment for clinical pain: A time-series approach. In: B. Bromm (Ed.): Pain measurement in man. Amsterdam: Elsevier, 417-428.

LIU, L.-M., & HUDAK, G.B. 1989. The SCA statistical system reference manual for forecasting and time series analysis (Version IV.1). DeKalb: SCA.

LIU L.-M. 1986. Multivariate time series analysis using vector ARMA models. DeKalb: SCA.

REVENSTORF, D. 1979. Zeitreihenanalyse für klinische Daten: Methodik und Anwendung. Weinheim: Beltz.

REVENSTORF, D., Knapp, K., & Moeck, G. 1991. Multimodales Paartherapieprogramm. In Vorbereitung.

SCHMITZ, B. 1989. Einführung in die Zeitreihenanalyse. Modelle, Softwarebeschreibung, Anwendungen. Bern: Verlag Hans Huber.

TIAO, G.C. & BOX, G.E.P. 1981. Modeling multiple time series with applications. Journal of the American Statistical Association, 76, 802-816.

Interactional Bifurcations in Human Interaction – A Formal Approach

Otto Rössler

Abstract. Dynamical systems can undergo a change of qualitative behavior as an external parameter is varied. This is called a bifurcation or "function change." A striking example is the explanation of morphogenesis proposed by Turing: Two symmetrically coupled systems ("cells") undergo a bifurcation under a symmetric coupling even though – due to the symmetry of the coupling – the coupling has no effect whatsoever as long as both cells are in the same state. This counterintuitive result deserves the name "interactional bifurcation." Two new proposals are made. First, autonomous optimiziers (formal brains) can undergo interactional bifurcations, too. Second, the resulting function change can be radical: A previous "autistic" type of functioning may give way to a "thoughtful" one which includes rational and moral behavior. A working computer model has yet to be given. Potential applications to the real world are based on the finding of van Hooff that only the human primate exhibits a symmetric type of interactional cross coupling (a happy laugh mimicking a friendly smile). The proposed importance of this fact is consistent with an empirical finding of Fraiberg. Therapeutic consequences in the spirit of Bateson suggest themselves.

1. A Phenomenological Problem

One of the sciences which are basic to human ecology is anthropology in a functional sense (Knötig, 1972). Two major biological views can be distinguished. One stresses more or less "quantitative" differences between human and animal performance (as in exploration, manipulation, planning, imitation, cooperation, communication, tradition formation), cf. Lorenz (1973); Kawai (1975). The formation of the human culture is then considered as an overcritical collective phenomenon, occurring beyond a certain level of some or all of these properties, reached in a sufficient number of individuals. The second view, first clearly formulated by G.H. Mead (1934), is monistic in comparison, stressing another set of performances as reflecting a single "qualitative" difference of man (showing courtesy, devising a joke, trading, making a commitment, asking an opinion, conceiving of benevolence). These traits are thought to reflect a new biological property, "ultraperspective" (as Mead's "capability to evoke in himself the very attitudes one evokes through his actions in the other" may be called for short (Rössler, 1975)). The two views certainly are not incompatible. Man's leap forward in the first set may be a partial consequence of his acquisition of the second and vice versa. Phenomenologically, a decision may be very hard to reach.

Springer Series in Synergetics, Vol. 58   **Self-Organization and Clinical Psychology**
Editors: W. Tschacher, G. Schiepek, and E.J. Brunner   © Springer-Verlag Berlin Heidelberg 1992

In the following, an abstract roundabout way is proposed. It leads to the conclusion that ultraperspective may be an enigmatically acquired trait. Specifically it can arise through an interactional bifurcation.

2. A Deductive Approach to Brain Function

A comprehensive answer to the question "What is a brain for?" is still impossible, even with respect to that of a small bony fish, say. Yet in the analysis of complicated systems, such global questions are sometimes the most important ones (Rosen, 1972). Some time ago, a biological answer became possible within an adaptation-theoretical context (Rössler, 1974).

While adaptation theory can usually "explain" a biological design a posteriori only, by relating it to some survival-important environmental factor (which, however, might conceivably have been coped with in some other manner too), there exists one class of environmental factors (so-called "second-level factors") which uniquely determine the minimum performance of *any* design coping with them. One such factor is "modulation-in-space" (of some primary factor or factors). The only adaptive solution possible is translation in space. The rules for optimum motion (for example, locomotion) then are abstractly fixed, waiting to be implemented in a (loco-)motion-control device. If "the" task of a brain is precisely the actualization of these rules, a comprehensive deductive theory of "brains" (in this sense) is possible, up to functional equivalence. Mathematically, all these "brains" are optimizing systems (Rössler, 1974), cf. Rosen (1967). For illustration (and later use), the properties of an advanced "formal brain" may be summarized briefly, cf. Rössler (1978). It is an optimizing system observing the following five specifications:

1) **Continuity.** (The space of control parameters is continuous; the objective functional is a continuous function of position in control space; the optimizing motion in control space is continuous and follows the local gradient of the objective functional.)

2) **Autonomy.** (The control parameters are parametrized by *physical space*; the optimizing motion in control space is accomplished through locomotion in physical space; the objective functional is, rather than being exogeneously imposed, endogeneously computed on the basis of information about the control parameters gathered while traveling in the actual environment.)

3) **Time dependency.** (The objective functional is time dependent in at least some of its components.)

4) **Adaptability.** (The mapping between environmental inputs and control parameters, as endogeneously computed, is modifiable in response to contingencies; Tsypkin (1973).)

5) **Built-in option for path-optimization.** (This is possible through "internal" local optimization on the basis of, (i) a short-circuit between input and output, (ii) a short, first-in, first-out memory used in a reafference-type way, and (iii) subsequent modification of the local objective functional.)

For more details see Rössler (1974, 1976a, 1978, 1979, 1981a,b, 1983), Rössler et al. (1990).

Phenomenologically, the preceding five points mean that the system moves around spontaneously, approaches (and avoids) varying targets, makes internal compromises, learns, and shows spatial insight. If these properties are analogues to certain invariant performances of vertebrate brains, there is obviously *no* analogue implied to the functional property of ultraperspective so characteristic of the human brain.

3. Ultraperspective Cannot Be Taught to a Formal Brain

In principle, the presence of "ultraperspective" (Rössler, 1975, 1976b) in a formal brain is not inconceivable. One could speak of the presence of this (analogous) property if a formal brain were to take into account in its actions the fact that there exists an objective functional behind the partner system's moves. The assumed capabilities of "learning" (point 4) and "internal simulation" (point 5) appear at first sight sufficient for the formation of such an internal representation. However, the acquisition of this particular response through learning is actually impossible on logical grounds.

A formal brain, being a dynamical system that responds differently to different environments, is a classifying system in the sense of Rosen (1972). If it could be taught ultraperspective, it would be a classifying system capable of classifying an activity isomorphic to its own (classifying) activity. Such a system is, however, analogous to a set containing itself as an element beside other, non-empty, elements. Such a set is paradoxical, see Löfgren (1968), because any one of those "other" elements can be an element of the set only if it is not an element of the set.

This result is unexpected. It seems to leave no room for ultraperspective acquisition in any case. This conclusion is not inescapable, however, because it overlooks the possibility that a formal brain may undergo a transition toward a dynamical system of a different type with respect to which no logical paradox is involved any more.

4. Hysteresis, Bifurcation, and Interactional Bifurcation

Formal brains are nonlinear dynamical systems (cf. the textbooks by Rosen (1970) and Hirsch and Smale (1974)). If, for example, specification (5) had been omitted, they would even belong to the simplest class of gradient-type dynamical systems. Nonlinear dynamical systems are in many cases capable of several different types of qualitative behavior, one in each "basin" of state space (Rosen, 1970). In each qualitative mode, the system is in fact a *different* system (because of its belonging into a different equivalence class of systems). In order to push a system from one qualitative mode into another, a parametric change is needed during which a bifurcation threshold in parameter space is surpassed. If the system stays in the

231

new qualitative mode even after the parameters have all been returned to their original values, one speaks of hysteresis (after-duration). From a practical point of view, one type of bifurcation induction is especially simple to realize in many cases, justifying the special name "interactional bifurcation": One takes two specimens of a system in which one is interested, and couples them together in a symmetrical fashion. On increasing the coupling, bifurcations can be elicited in many cases. The question of whether their effects remain present after decoupling (hysteresis) or not is then a secondary question again.

Interactional bifurcation was first employed by Turing (1952) who showed that in this way two initially identical abstract cells can be induced to spontaneously differentiate in different directions (principle of interactional morphogenesis). Analogous models can be set up for carcinogenesis, cf. Pitot and Heidelberger (1963) and perhaps also for ultraperspective acquisition ("interactional anthropogenesis"). Just as two extremely simplified "formal cells" sufficed for the demonstration of the morphogenetic principle, two very simple "formal brains" appear to be sufficient in the last-mentioned case.

5. A Possible Model

The basic idea, originally derived from biological phenomenology, is as follows. The two systems are supposed to be coupled in such a manner that the value of the objective functional of each system acts (via an appropriate signal channel) as a specific, objective-functional-increasing, environmental input to the other system (Rössler, 1968, 1975, 1976b). (This value may by thought of as represented by the intensity of the human smile; see Musterle and Rössler (1986) for a successful computer simulation and van Hooff (1972) for biological background information.) Then, a bifurcation occurs beyond a "supercritical" value of the product of the two coupling parameters. Unlike other bifurcations leading to a positive feedback (as occur quite frequently in the "ordinary" functioning of a formal brain; cf. Zeeman's (1971) "behavioral catastrophes") in the present case the functioning of both systems is transformed in a strange, "mixing", manner:

Basically, a "mirror" is created through the coupling. Specifically, any environmental situation in which the objective functional of one of the two systems is increased will trigger the positive feedback (supposing that the involved sub-objective functionals had enough time to regenerate). In potentially competitive environmental situations, each system will in the long run just as frequently be the "giver" (because the other system's objective functional responded first to the environmental stimulus) as the "taker". The more symmetrical the situation, the more contradictory will be the two equiprobable outcomes. Due to the presence of learning (point 4) on each side, the result will be "erratic" behavior of mounting complexity. This is an example of chaos (cf. Rössler (1987) for a simulation of an analogous, much simpler case). Nonetheless, the complexity can be cut once and for all if both systems start exploiting the "exchange" symmetry of the situation in a new way. If the time constants are such that several analogous events – albeit with an interchange of the active and the passive partner – are possible within the simulation span assumed (point 5), a "blurring" of simulated actions becomes possible within each system. It has the effect that the system confounds its own next (planned) action with the partner system's presently occurring analogous act – just as if it had planned this latter act itself. This false interpretation (of controlling two spatial executives rather than one) gives rise to a much more economical interpretation of the overall situation on both sides. Among the set of objectively false, but mutually reconfirming, internal representations formed on each side, there is one in which a "benevolent" other "self" is posited on the other side. Ultraperspective is thereby implied.

The preceding description is incomplete. The most important point to be added is, perhaps, that the two falsehoods (one on either side) cancel each other, generating a truth.

6. Concluding Remarks

Suppose the preceding model had been corroborated by quantitative evidence (numerical simulation results, for example; so far not even a single formal brain has been simulated): Would not the observation that so "dry" a mathematical structure as an optimizing system (or dynamical system) produced an interesting result cause an uneasy feeling in the observer, much more so than in the case of Turing's analogous "reduction" of morphogenesis?

A first point to make is that the present approach is still quite insufficiently formalized. It is improbable that a straightforward numerical elaboration will yield the expected, and only the expected, results (to judge from simpler analogous design problems). Thus, the approach is still in the stage of a glass-bead game.

A second answer is that dynamical systems are in no way trivial. Perhaps the most striking evidence is provided by modern physics which showed that the dynamical process of measurement is something that needs to be reevaluated, a development which is only now beginning to extend itself into the realm of dynamical systems theory (Rosen, 1968, 1978). The very notion of potential ("force"), which was a philosophical problem at Hegel's time still, has become a mystery again: Who (or what) is at the "center" of the optimizing process?

Even more interesting, however, is the above-indicated possibility of "self" (and "truth") emergence. It would be misleading to assume that the "content" of such entities, when they arise in an artificial system, would need to be empty. Thus, the preceding approach is a challenge to human tolerance.

Four possible practical consequences may be mentioned finally.

a) Human ecology, the science of man-environment interaction, is possibly the basic science for a functional understanding of human nature (Bateson, 1972).
b) In psychiatry, there is a rare disease (early infantile autism) in which ultraperspective acquisition appears to be blocked in one of its steps ("bribability" by a smile). Compare Rössler (1975) and, independently, Fraiberg (1977).

c) Sociology is interested in systems-theoretic models of "socialization" (Rössler, 1968, Habermas, 1970, Prewo et al., 1973).

d) "Motivation interaction" models in the spirit of Rashevsky (1947) may become relevant in psychotherapeutical thinking.

7. Summary

In analogy to the well-known dynamical hypothesis of interactional morphogenesis, a hypothesis of "interactional anthropogenesis" has been presented. It focuses on explaining a single functional property of human beings: ultraperspective. The mathematical framework employed is the theory of formal brains, a special class of optimizing systems. Under certain conditions, two symmetrically coupled formal brains are predicted to show a bifurcation leading to a new mode of symmetric functioning which implies ultraperspective. The model is possible because the property of ultraperspective is formalizable. At first sight, all it takes is to exploit the "objective reality of perspectives" (Mead, 1926). However, this exchange-symmetry exists only when the different optimizers are identical; only then can the vehicle of "vicarious" optimization (in the sense of Tolman, 1948) be used across individuals. If, however, the optimizers are each endowed with a different optimality functional (self), then the latter needs to be given up totally before the symmetry can be exploited. Only a potentially infinitely strong force can fuel the involved ultimate sacrifice. A possible mechanism ("positive cross-optimization") has been suggested. It is amenable to empirical and eventually numerical scrutiny. In spite of its still "soft" state of development, the present theory opens up new possibilities of thought in philosophy, biology, and psychology.

Acknowledgments

I thank Wolfgang Tschacher for his kind help in updating this paper. I also thank Helmut Knötig, Jan Kryspin and the late Gregory Bateson for discussions.

References

BATESON, G. 1972. Steps to an Ecology of Mind. New York: Ballantine.

FRAIBERG, S. 1977. Insights from the Blind. Comparative Studies of Blind and Sighted Infants. New York: Basic Books, Chap. 11.

HABERMAS, J. 1970. Towards a theory of communicative competence. Inquiry **13**, 360-376.

HIRSCH, M.W. and S. SMALE 1974. Differential Equations, Dynamical Systems, and Linear Algebra. New York: Academic Press.

KAWAI, M. 1975. Precultural behavior of the Japanese monkey. In Kurth, G. and I. Eibl-Eibesfeldt (Eds.): Hominisation und Verhalten, 32-55. Stuttgart: G. Fischer.

KNÖTIG, H. 1972. Bemerkungen zum Begriff "Humanökologie". Humanökol. Bl. 1 (H. 2/3), 3-140.

LÖFGREN, L. 1968. An axiomatic explanation of complete self-reproduction. Bull. Math. Biophysics 30, 415-426.

LORENZ, K. 1973. Die Rückseite des Spiegels, Versuch einer Naturgeschichte des menschlichen Erkennens (The Flip-Side of the Mirror, Essay on a Natural History of Human Understanding). München: Piper.

MEAD, G.H. 1926. The objective reality of perspectives. Proc. 6th Int. Congr. Philos., 75-85.

MEAD, G.H. 1934. Mind, Self, and Society. Chicago: University of Chicago Press.

MUSTERLE, W. and O.E. RÖSSLER 1986. Computer faces: The human Lorenz matrix. BioSystems 19, 61-80.

PITOT, H.D. and C. HEIDELBERGER 1963. Metabolic regulatory circuits and carcinogenesis. Cancer Res. 23, 1693-1700.

PREWO, R., J. RITSERT und E. STRACKE 1973. Systemtheoretische Ansätze in der Soziologie (System-theoretic Approaches in Sociology). Hamburg-Reinbek: Rowohlt.

RASHEVSKY, N. 1947. A problem in mathematical biophysics of interaction of two or more individuals which may be of interest in mathematical sociology. Bull. Math. Biophysics 9, 9-15.

ROSEN, R. 1967. Optimality Principles in Biology. London: Butterworth.

ROSEN, R. 1968. On analogous systems. Bull. Math. Biophysics 30, 481-492.

ROSEN, R. 1970. Dynamical Systems Theory in Biology, Vol. 1. New York: Wiley.

ROSEN, R. 1972. Relationale Biologie (Relational Biology). Fortschritte der experimentellen und theoretischen Biophysik, Vol. 15. Leipzig: Thieme.

ROSEN, R. 1978. Fundamentals of Measurement and Representation of Natural Systems. New York: North-Holland.

RÖSSLER, O.E. 1968. Zum Tier-Mensch-Problem aus der Sicht der theoretischen Verhaltensbiologie (On the animal-man problem from the vantage point of the theoretical biology of behavior). Schweizer Rundschau 67, 529-532.

RÖSSLER, O.E. 1974. Adequate locomotion strategies for an abstract organism in an abstract environment – A relational approach to brain function. Lecture Notes in Biomathematics 4, 343-369. Berlin: Springer-Verlag.

RÖSSLER, O.E. 1975. Mathematical model of a proposed treatment of early infantile autism – Facilitation of the "dialogical catastrophe" in motivation interaction. In Martin, J.I. (Ed.): Proceedings of the San Diego Biomedical Symposium 1975, Vol. 14, 105-110. San Diego: San Diego Biomedical Symposium Press.

RÖSSLER, O.E. 1976a. Prescriptive relational biology and bacterial chemotaxis. J. Theor. Biol. 62, 141-157.

RÖSSLER, O.E. 1976b. Cross-stimulation: Theoretical implications. In Wauquier, A. and E. Rolls (Eds.), Brain Stimulation Reward, 600-605. Amsterdam: North-Holland.

RÖSSLER, O.E. 1976c. A relational approach to the problem of man-animal behavior comparison. In Cranach, M. von (Ed.): Models of Inference from Animal to Human Behavior, 103-117. Aldine: Chicago.

RÖSSLER, O.E. 1978. Deductive biology – some cautious steps. Bull. Math. Biol. **40**, 45-58.

RÖSSLER, O.E. 1979. Path optimization on the basis of direction optimization. In Trappl, R., F. de P. Hanika and F.R. Pichler (Eds.): Progress in Cybernetics and Systems Research, Vol. 5, 546-549. New York: Hemisphere-Wiley.

RÖSSLER, O.E. 1981a. An artificial cognitive map system. BioSystems **13**, 203-209.

RÖSSLER, O.E. 1981b. An artificial cognitive-plus-motivational system. Progr. Theor. Biol. **6**, 147-160.

RÖSSLER, O.E. 1983. Artificial cognition-plus-motivation and hippocampus. In Seifert, W. (Ed.) Neurobiology of the Hippocampus, 573-588. New York: Academic.

RÖSSLER, O.E. 1987. Chaos in coupled optimizers. Ann. N.Y. Acad. Sci. **504**, 229-240.

RÖSSLER, O.E., G. KAMPIS, W. NADLER, W. MUSTERLE, B. SCHAPIRO and P. URBAN 1990. Highly parallel implemenation of autonomous direction optimizer with cognition. Biophys. J. **57**, 194a. (Abstract.)

TOLMAN, E.C. 1948. Cognitive maps in rats and men. Psychol. Rev. **55**, 189-208.

TSYPKIN, Ya.Z. 1973. Foundations of the Theory of Learning Systems. New York: Academic.

TURING, A.M. 1952. The chemical basis of morphogenesis. Phil. Trans. Roy. Soc. London, Series B, **237**, 37-72.

VAN HOOFF, J.A.R.A.M. 1972. A comparative approach to the phylogeny of laughter and smiling. In Hinde, R.A. (Ed.): Non-Verbal Communication, 209-241. Cambridge: Cambridge University Press.

ZEEMAN, E.C. 1971. The geometry of catastrophe. Times Literary Supplement, 10th December 1971, 1556-1558.

Part III

Self-Organizing Processes
in Psychotherapy

Synergetics of Psychotherapy

Günter Schiepek, Bernd Fricke, and Peter Kaimer

With 10 Figures

This article outlines a synergetic model of self-organizational processes in psychotherapy. Following a short introduction, the fundamental conditions required for self-organization to occur are indicated. The therapeutic process is described in terms of the interaction between the client's lifestyle scenario, the therapeutic system and the context of therapy. The aim of therapy is to deform the potential landscape created by the client's lifestyle and to amplify the fluctuations in the behavior of the client system. The third section demonstrates how a concrete therapeutic approach, that of the Brief Family Therapy Center (Milwaukee, USA), can be interpreted using the concept of synergetic self-organization. The fourth section presents empirical methods which allow the description of change in the lifestyle scenario and the therapeutic system.

1. Introduction

In the last decade, it has been possible to identify significant trends in the field of psychotherapy research and theory development:

a) A shift of emphasis has taken place from outcome research, focussing primarily on the differential effectiveness of psychotherapies, towards process research (Greenberg & Pinsof 1986; Orlinsky & Howard 1986; Rice & Greenberg 1984), or combined outcome process research and its concentration on differential change patterns (Arnold & Grawe 1989; Grawe et al. 1990 a-d).

b) A rather technological conceptualization of psychotherapy, i.e. psychotherapy as the application of specific methods, has given way to a more heuristic conceptualization (Caspar & Grawe 1989) which emphasizes the importance of adaptability within the client-therapist relationship, on the one hand, and of the client's eigendynamics and autonomy, on the other. From this point of view, the function of psychotherapy is to establish boundary conditions that are tailored as optimally as possible to the client and his development, in order to allow for processes of psychological and social self-organization. (Such boundary conditions include the timing of specific interventions, the manner in which the therapeutic alliance is handled, the process whereby the therapist joins with the client's idiosyncratic expressions and his or her emotions, and getting the client to make use of his own resources; see for example Grawe 1987; Schiepek & Kaimer 1988; Goudsmit 1989).

This progress is the result of intensive efforts aimed at introducing new, innovative concepts and theoretical perspectives, and corresponds to Kiesler's challenge that "psychotherapy research needs new life and directions" (Kiesler 1985, p.528) so that it may attain a more adequate level of description along with a more detailed understanding "of what is actually occurring in therapy" (Rice & Greenberg 1984, p.8; see also Greenberg 1983). The quest for better understanding is accompanied by the re-orientation of research practice towards the improvement of qualitative methods and the closer examination of single-case studies of therapy (Grawe 1988).

2. Principles of Self-Organization in Psychotherapy

2.1 Preliminary Remarks

There are a number of reasons why the concept of self-organization suggests itself for an understanding of psychotherapeutic processes:

1) Clients seem to be in a relatively stable, yet problematic state of order, which is to be overcome. This state of order may be characterized by the activation of certain negative emotional schemata, certain constructions of reality, patterns of behavior as well as experience, and certain patterns of interaction and ecological situations, all of which are actively generated either by one person individually or in interaction with significant others, as is usually the case. With respect to these phenomena, psychotherapy aims to achieve a phase transition to a different state of order of the system (Stadler & Kruse 1990; Kriz 1990; Mahoney 1980).

2) There is no superordinate "control-center" regulating this phase transition. Nor are there any agencies within the client — conscious or unconscious — nor the therapist himself to determine this complex biopsychosocial process of psychotherapy. Rather, this process is governed by the synergetic effects of conscious and unconscious, psychological and biological, communicative and ecological feedback processes (Revenstorf 1990; Varela 1979; Horowitz 1987).

3) Practically every therapeutic school assumes that psychological and social systems function in an autonomous, non-trivial manner (von Foerster 1985; Luhmann 1985). It is a striking paradox that it is precisely within the context of attempts to explain trance-induction and hypnosis, phenomena requiring the client to allow the therapist to assume control over him, that constructivist models of explanation are suggested which grant the client maximum autonomy regarding the manner in which he constructs reality and relates himself to this reality (Revenstorf 1990).

4) Both practical experience and detailed analysis of therapeutic processes demonstrate that it is possible to view patient change as a nonlinear phase transition (e.g. Strauss 1989). Phases of stagnation or even deterioration are followed by sudden unforeseen improvements, despite there having been

no change in the therapeutic efforts. Sometimes minimum fluctuations, often picked up only after a substantial delay, are sufficient to trigger significant changes in a client's cognitive, emotional and behavioral organization; at other times in the therapeutic process, on the other hand, such fluctuations exert no influence whatsoever. Moreover, a therapist's increased efforts to effect change often coincide with the client placing stronger emphasis on his problems and their invariability. Finally, nonintended oscillations can be observed, for example in connection with the client's well-being, the markedness of his/her symptoms, his/her choice of main topics, or the regulation of closeness and distance in a relationship, such as that between husband and wife.

In summary, we can say that *psychotherapy* can be conceptualized as a *process of autonomous order transition*. This conclusion justifies the efforts being made to establish an empirically based synergetics of psychotherapy in the coming years.

2.2 General Conditions for Self-Organization

Before turning to the field of psychotherapy, it will be shown in general terms, what prerequisites must be fulfilled within the theoretical frame of synergetics for self-organization to take place:

1) A *thermodynamically open system* is essential. This precondition may be taken for granted in the case of psychotherapy, since we are concerned here with human beings - that is, with living systems.

2) The occurrence of processes *far from equilibrium* is to be assumed. Within therapy, the client's motivational energy to achieve change generates nonequilibrium conditions.

3) A *microscopic level* is to be assumed, comprising a large number of elements with a corresponding number of degrees of freedom. The elements referred to should not be conceived of as being material, such as atoms or molecules. Rather, within the framework of psychological synergetics, the microscopic level comprises all those biological, psychological and social processes which constitute a system. Furthermore, these processes interact in a recursive manner. This microscopic level is regarded as the *hypothetical limit of the maximum degree of resolution*. In practice, however, the description of these processes cannot reach this maximum degree of resolution, which is not necessary either.

4) *Nonlinearity* (a) within the system and (b) between the system and its environment.

(a) Nonlinearity is based on cross- and autocatalytic cycles, which means that recursive interaction of the relevant biopsychosocial processes takes place within the system (circular causality). Positive, negative or, most importantly, mixed feedback may occur within this process of interaction.

(b) Nonlinearity in the relationship between system and environment implies that a linear change of parameters in the environment of a system (control parameters) may cause non-linear changes within the system, showing spontaneous phase transitions (behavior change) under certain instability-conditions on the basis of the processes mentioned under (a). Crucial sensitivity to minimum fluctuations in the external milieu of the system is to be expected when the phenomenon of critical slowing down occurs (indicating that a bifurcation is being approached).

2.3 Systems Constituting the Process of Psychotherapy

The general characteristics outlined above need to be further specified, however, to allow the conceptualization of a synergetic model of psychotherapy. The following specifications represent an attempt in this direction.

2.3.1 The Biopsychosocial Lifestyle of the Client

"Lifestyle" refers to the state of order governing the individual's present way of life. The state of order may be characterized by specific enduring negative emotional schemata, together with their complementary positive self-schemata. The activation of certain schemata may cause specific patterns of perception to be activated, especially in the field of interpersonal and emotional perception. These patterns in turn suggest corresponding patterns of behavior which may serve to confirm the corresponding patterns of perception. Thus we find typical emotion-behavior-perception cycles, into which significant others may be intricately integrated. These cycles may also be repeated with people not yet entangled in close relationships (compare Freud's principle of repetition compulsion). It is not usually by chance that others appear in an individual's lifespace. Rather, they are selected consciously or unconsciously so that they fit in as partner, friend, authority, colleague, idol, etc.

However, it is not only intrapsychic schemata which determine the process of experience and the patterns of interaction. Various other psychological, social or even biological processes and their recursive interaction have an impact on the reproduction of life-themes, success and failure, the course taken by one's life with respect to career or health, the selection of partners, conflicts, living conditions etc., all of which may be typical for a certain lifestyle scenario. However, the individual should not be considered solely responsible for the creation of his or her lifestyle. Significant others and their characteristics, as well as social conditions or material circumstances (e.g. air pollution, food pollution, the shortage of housing, structural unemployment), also play an important part. Nonetheless, the daily reproduction of a person's lifestyle can be considered an active achievement, partly conscious but for the greater part unconscious. Alternatives to the actual lifestyle do exist, at least in principle (compare the literature concerning strategic therapy), yet they are neither perceived nor actively taken into consideration. The individual therefore perceives his or her lifestyle to be normal or necessary, or even characterizes it by saying "there is no way out".

The concept of the lifestyle scenario introduced here is similar to the appreciation of psychological functioning as presented in the schema theoretical framework (see Ciompi 1986; Grawe 1987; Zingg 1988). It goes beyond that appreciation, however, as we include a complex system of different — not only psychological — interrelated processes constituting relatively stable dynamics. In particular, immunological, endocrinological and physiological processes must also be taken into account, as they are fundamental to physical and mental health and illness (compare Hesch 1990). The concept of the lifestyle scenario is almost identical to the concept of "state of mind" proposed by M.J. Horowitz (1987). In synergetic terms, it describes the fact that a large number of processes on a microscopic level combine to form a coherent, dynamic pattern (mode) representing a fundamental attractor of the dynamics predominant in a person's life. Attractors can be identified not only in the lifestyle of one individual, but may also depict the pattern of cooperation in systems made up of several individuals (couples, families, teams). (For the most simple case compare the limit cycle attractor in predator-prey systems).

When potential landscapes are utilized as a means of representation, an attractor can be recognized by a potential valley. It can be assumed that the biopsychosocial *potential landscape* is the result of the cooperation of several order parameters. The potential landscape — in mathematical terms a potential function — describes the amount of energy contained in a system which may potentially carry out work. All forces acting in the system are mathematical derivatives of the potential function (see Haken 1983a, pp.105; Haken 1983b; Tschacher 1990). The depth and diameter of a potential valley may vary in accordance with the phases in actual life. Two or more potential valleys may exist at the same time, transitions between them being possible. The ability to make such transitions may be regarded as a criterion of mental health (Fig. 1).

It is to be expected that scale-invariant phenomena (Großmann 1989) can be observed in the lifestyle of an individual: similar behavioral patterns on different levels of behavioral organization as well as on different time-scales (self-similarity). Examples of this are nonverbal signals conveying vagueness and detachment; vagueness of content; discussion of difficulties in decision-making; long-term avoidance of decision-making and acknowledgement of responsibility (Ciompi, private communication).

2.3.2 The Therapeutic System

The therapeutic system comprises all events within the scope of the relationship between client and therapist which are defined as therapy. It involves one or more clients (or one client and one or more related persons not defined as clients) and one or more therapists. The existence of the therapeutic system is temporally and spatially confined. Empirical analysis is possible at different levels. An example is the analysis of immunological and endocrinological parameters before, during and after therapy. However, the

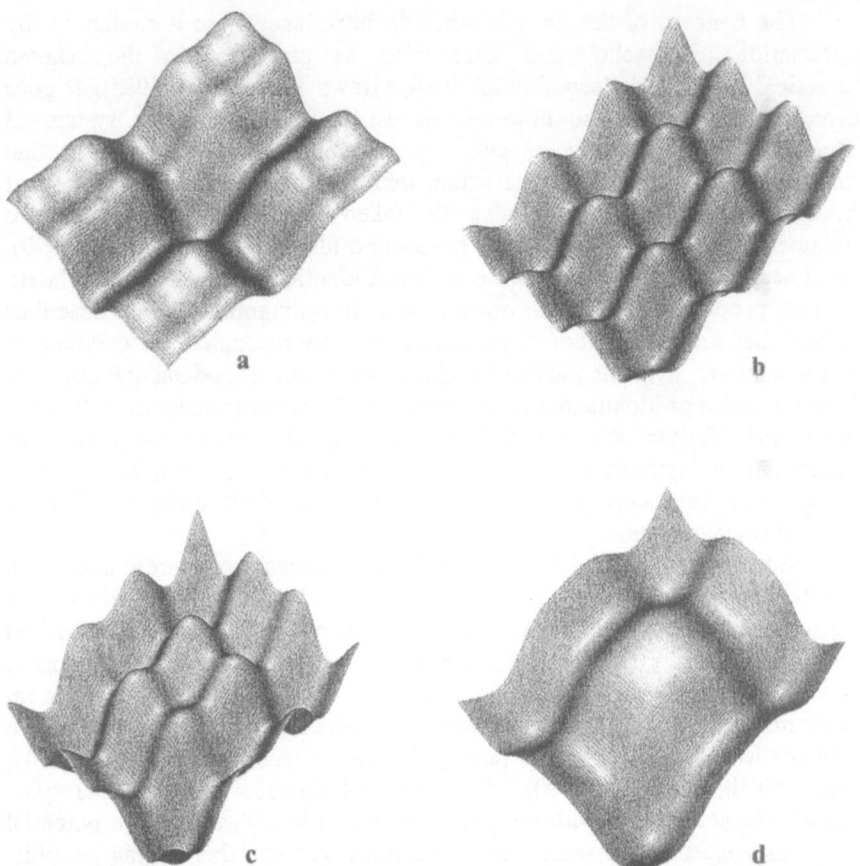

Fig. 1 (a-d): The temporal evolution of a potential landscape (representing a client's lifestyle scenario): A central potential valley is changing its shape in favor of several other potential valleys.

Technical remarks for the production of these figures: Each potential valley shown in figures (a)-(d) is a function of two variables $z(x,y)$ over the x-y plane:

$$z(x,y) = a \sin(x/4) + b(\cos x + \cos y) + c (\cos 2x + \cos 2y)$$

The plane is constituted by $-1.5\ \pi \leq x \leq 1.5\ \pi$ and $-1.5\ \pi \leq y \leq 1.5\ \pi$. The parameter values a, b, c of the individual figures are: (a) $a = 0$, $b = -1$, $c = -0.5$; (b) $a = 0$, $b = 0$, $c = -1$; (c) $a = -0.5$, $b = 1$, $c = -1$; (d) $a = -1$, $b = 1$, $c = -0.25$. (We are indebted to W. Lorenz of the University of Stuttgart, Institut für Theoretische Physik und Synergetik, for providing the figures presented here.)

methods proposed here aim at an analysis of data on two different levels. These are (a) the level of the psychological systems of client and therapist, which are autonomous, operationally closed, but structurally coupled (for details, see Schiepek 1991, Chap. IV); and (b) the level of the social or — following Luhmann (1984) — communication system constituted through therapy. The communication system evolves on the basis resulting from the functioning of the participating psychological systems; however, it evolves over and above these to constitute an emergent reality.

a) Concerning the level of psychological — i.e. cognitive-emotional — processes, a method will be proposed below which allows the modification of topics focussed on during therapy and their semantic association to be traced over the course of therapy. Therapy transcripts are used to demonstrate how the structures of the topics introduced change during a continuous process of co-creation (Reconstruction of Cognitive Representation, RCR). This can be shown for the therapist as well as for the client. Another method aims at analyzing, at a level of high temporal resolution, the sequence of activation of interactional plans (Caspar 1989) of each individual system in the course of their structural coupling (Sequential Plan Analysis, SPA).

Although both methods rest on the analysis of communication (RCR: transcripts of verbal expression; SPA: videotapes of verbal and nonverbal expression), the focus of attention is on analyzing how therapeutic interaction is produced by an internal processing of the participating psychological systems and their mutual perturbation of one another's internal processing.

b) The method to be applied in the process analysis of the communicative therapeutic system abstracts from the psychological systems of the participants and is directed at analyzing how the self-referential sequence of communicative proceedings constitutes and modifies the social system "therapy" with its corresponding structures, limits, identities etc. (that is to say, how this social system self-organizes and creates disorder-to-order transitions; for similar considerations concerning self-organization of the legal system, compare Teubner 1987; 1989). The coding scheme for relevant communicative proceedings is called "Process Analysis of Communicative Systems" (PACS).

Self-similarity can also be expected when we compare an individual's lifestyle scenario and the therapeutic system in which he or she is involved. As far as clients are concerned, this phenomenon has been termed "transference" in psychoanalysis (for a recent conceptualization, see the plan-concept (in German: Test-Konzept) proposed by the Mount Zion Psychotherapy Research Group; cf. Silberschatz et al. 1989). Such phenomena also occur with respect to therapists (counter-transference) and are made the subject of encounter groups and supervision.

2.3.3 The Social Context of Psychotherapy

Finally, all attempts at describing the field of psychotherapy remain incomplete if they do not take into account the context in which therapy takes place. The context can result for example from the fact that therapists are supervised, that therapy is carried out within an institutional frame of varying complexity or in cooperation with a team of therapists, as envisaged by many so-called systemic approaches. The systems involved are thus interrelated so as to create more or less useful or problematic ways of cooperation. Interrelational patterns can be produced within each of the systems, or between the systems involved, similar to those produced by the clients and significant others, or to those observable in the therapeutic system (self-similarity). The recursive interactions between the various institutions, teams and therapists involved on a particular case may close to form a hypercycle, for which the client — theoretically the central figure — constitutes merely the trigger enabling significant eigendynamics to occur (for details see Schweitzer 1989; Imber-Black 1990; Schiepek & Reicherts, this volume).

The therapeutic system, together with the interacting systems forming the context of a therapy, may take on more or less significance for the lifestyle of a client (a couple, a family). For clients who define themselves as part of a therapeutic system over a long period of time, the therapeutic system may become a constitutive part of their lifestyle. For people living in a mental health care unit (a children's home, a psychiatric clinic), the institution becomes nearly identical with the "lifespace" and will determine their lifestyle in a fundamental way.

Figure 2 shows the systems described above which are basic to the understanding of self-organization in psychotherapy.

2.4 Self-Organization: The Deformation of Potential Landscapes

How is the process of self-organization in psychotherapy to be conceived?

Usually the reason for seeking therapy lies in the continuing existence of a specific state of order experienced by the client or his/her significant others as unpleasant, distressing or intolerable. Furthermore, the state of order tends to be relatively stable; attempts to overcome it on one's own have failed

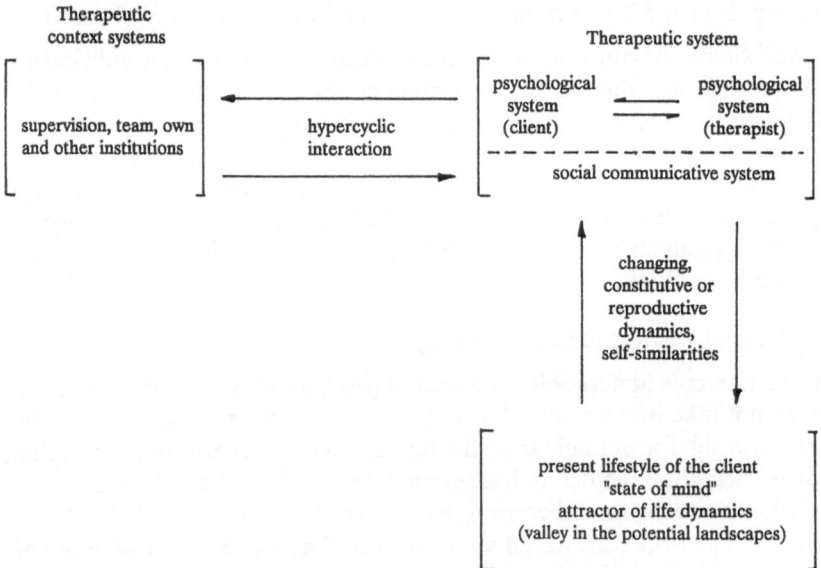

Fig. 2: Systems constituting the process of psychotherapy

repeatedly. The client finally seeks professional assistance, which he/she hopes will help towards the achievement of a different state of order.

Thus, successful therapy can be conceived as an order-to-order transition (phase transition). In the terminology introduced above, a state of order is represented by a certain present lifestyle scenario. The present lifestyle constitutes a powerful attractor of the biopsychosocial behavior of an individual and despite every effort, it is not possible to escape from the basin of its attraction. If we describe this phenomenon in terms of potential landscapes, a dominant lifestyle is represented by a deep potential valley, while the system can be viewed as a particle resting at the bottom of it. The fluctuations of the system are not sufficient to allow it to overcome the potential barriers and leave the potential valley.

Psychotherapy can therefore follow two different courses, which will usually be combined:

a) It may contribute to the deformation of the potential landscape by flattening the dominant potential valley (i.e. by decreasing the dominance of the present state of mind), by lowering the potential barriers, and by forming alternative potential valleys (see Fig. 1).

b) It can encourage more intensive fluctuations of the system, i.e. of the particle resting in the potential valley. From the viewpoint of the client, this means trying to act differently from before, experimenting with new ways of behavior and new ideas in certain situations, or confronting oneself with situations or emotions so far avoided.

When the previously deep potential valley becomes flatter, phenomena characteristic of instability will occur, such as critical slowing down: feedback processes do not take the system back to the stable state as quickly as before. If certain parameters defining the potential function become less than zero, then a symmetry-breaking instability occurs: minimal fluctuations are sufficient to allow the particle to roll into one of two new equiprobable potential valleys (Fig. 3). It is at this point that chance and nonpredictability come into play. With regard to psychotherapeutic practice, analysis of the stability of the system dynamics would be desirable, allowing an assessment of the degree to which the system dynamics have approached the boundaries of instability, whether creative or dangerous.

Following this model, two different approaches are possible in psychotherapy. Either an amplification of the fluctuations of a system is attempted, so that it moves on from one potential valley into another, existing within the lifespace but currently not in use (such "jumps", which are often not precisely aimed, are turbulent in nature); or efforts are made to deform the potential landscape by using minimum irregularities of the surface to create new potential valleys via processes of deviation-amplifying feedback. As time passes, continued use of the new potential valleys may further shape them. These two approaches are complementary, rather than mutually exclusive.

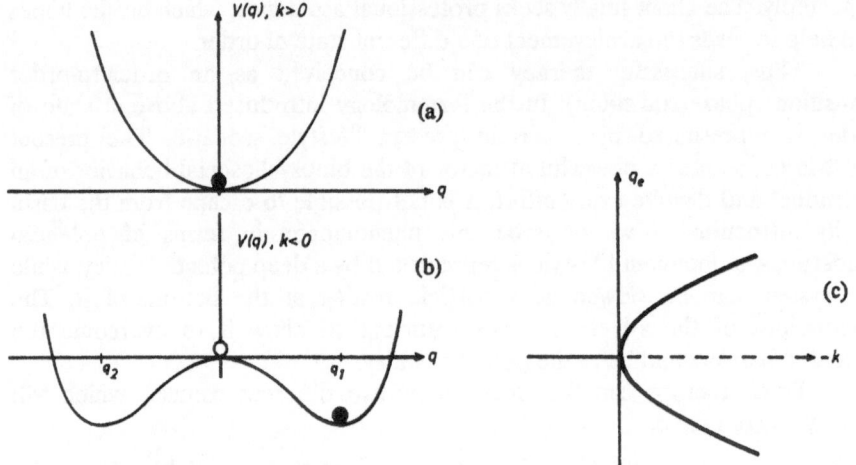

Fig. 3: The transition from a symmetrically stable state (a) to two equiprobable potential valleys (b). In this process, the particle rolls into one of two potential valleys following a symmetry-breaking instability. The potential function is: $F(q) = -kq - kq^3$, where q is the elongation from the equilibrium position and k is the bifurcation parameter (control para-meter); with $k > 0$ (a) and $k < 0$ (b). This is the so-called anharmonic oscillator which contains a cubic term besides a linear term in its force F.

(c) The bifurcation diagram of the system (anharmonic oscillator), with the equilibrium co-ordinate q_e as function of k. For $k > 0$, $q_e = 0$, but for $k < 0$, $q_e = 0$ becomes unstable (dashed line) and is replaced by two stable positions (solid fork).

However, the client's cooperation and active participation are fundamental prerequisites.

It is the function of therapy not only to transfer the system dynamics into the regime of a different attractor but also generally to increase accessibility to alternative attractors (states of mind, lifestyles). Thus, the focus of attention is not merely reliability within another fixed lifestyle, leading once again to a low degree of fluctuations and deep potential valleys, but adaptability and ease of switching, requiring large fluctuations and flat potential curves (compare Haken 1983a).

An interesting question, and one which is still to be answered, concerns which of the *order parameters* are in fact those governing the system's states of order. In view of the enormous complexity of biopsychosocial systems, the determination of order parameters proves to be much more difficult here than in physical systems (compare Stadler & Kruse 1990). For lack of a mathematical model for the synergetics of psychotherapy, the states of order of a system (here: lifestyles) will be characterized not by a quantifiable value of one or more order parameters but by the description of the respective lifestyle patterns. The method to be applied here is the construction of an idiographic model of the system, one which recursively interrelates the relevant processes and variables constituting a specific, dynamic lifestyle pattern (Schiepek 1986; 1991; Schiepek & Tschacher, this volume).

The *control parameters* of therapeutic order-to-order transitions are assumed to be inherent in the processing of the therapeutic system, which means that a change in the value of the relevant control parameter is a result of the interaction between therapist and client. Empirical results concerning clients' receptiveness (Ambühl & Grawe 1988; 1989; Orlinsky & Howard 1986) show that the techniques employed by the therapist, and even the therapist's behavior generally, exert no significant influence on therapeutic change unless precisely adjusted to the client and his specific receptiveness. One possible control parameter might be the amount of motivational energy invested by the client, influencing the degree of fluctuations and the deviation-amplifying feedback processes effective in the deformation of the potential landscape. Everything that happens during therapist-client interaction which effectively raises the level of motivational energy also raises the value of this control parameter. Another possibility for the identification of relevant control parameters could be the search for either one or a small number of variables typical for the client's lifestyle which would respond to treatment. These characteristics are symbolized by the core variables in the idiographic systems model of the client's lifestyle. It is also possible to rely on the client's mentioning his/her central complaints and goals, as contained in the approach put forward by the Brief Family Therapy Center (BFTC); this will be presented in Sect. 3.

Finally, it is possible to regard as a control parameter that part of the client's communicative acts (out of the total of communicative acts within the therapeutic system) considered as indicative of therapeutic change either by a specific therapeutic school or by the therapist involved. With regard to the BFTC approach, this corresponds to the ratio of "change talk" to "problem talk".

A word of caution should be voiced, however. The term "control parameter" should not be misunderstood to mean that there is a parameter independent of the client which can be manipulated by technical means, that a therapist need only make use of his/her professional techniques to modify this parameter and effect a phase transition in the client's lifestyle. Such a conception would allow a technical-algorithmic model, which therapy research has only overcome through great effort, to re-enter by the back door. To repeat: The concept of control parameters refers to psychological or social processes taking place within the frame of the therapeutic system or outside — at any rate, however, produced by the processing of the therapeutic system and thus stimulated by the cooperation of therapist and client.

2.5 Self-Organization of the Therapeutic System

Apart from order-to-order transitions in the lifestyle scenario, there is another self-organized process which is of importance with regard to psychotherapy, namely the self-organization of the therapeutic system as a communicative system. When therapist and client meet for the first time, the situation is characterized by double contingency (cf. Luhmann 1984). The symmetry of this situation has to be broken, and emergent entities will evolve from the interaction of the participant psychological systems which may serve as elements of a different self-reproductive system: communicative proceedings as elements of the social system "therapy" (Markowitz 1991; Teubner 1990).

The self-reference of communication creates the social system of therapy along with all the components required: boundaries, identity, structure, environment, expectations and rules, definition of membership (Teubner 1987, 1990). It thus represents a self-constituting and self-organizing social system which describes itself. Unlike the order-to-order transitions in the client's lifestyle scenario, the transition here is from disorder to order: the self-reference of communicative processing allows an order and structure to evolve which did not exist before.

3. Illustration of Principles of Self-organization: The Concept of the Brief Family Therapy Center (BFTC - Milwaukee, Wisconsin, USA)

The principles of self-organization in psychotherapy outlined in Sects. 2.2-4 can be seen as inherent in quite different ways of therapeutic practice. In the following the approach devised by the BFTC (de Shazer 1985, 1989; de Shazer et al. 1986) will be used to illustrate some of these principles. Although the development of the BFTC concept has been based on practical therapeutic work rather than on a specific theoretical framework, the principles underlying this approach may be expressed in synergetic terms as follows. Stable problem patterns are to be stimulated through deviation-amplifying feedback so that they may be transferred into other states of order (order through fluctuation). Then there seems to be a "natural fit" between the practical approach of the BFTC and the "perspective theory" (in German: Perspektiventheorie, Foppa 1984) made available by synergetics as an interdisciplinary concept. This will be explained in greater detail.

Therapeutic work at BFTC can best be described as radically solution oriented, which means that the problems that lead clients to seek therapy play a relatively minor part. As a consequence, no detailed problem analysis is drawn up, nor is a diagnosis made in the traditional sense. Attention is focussed as early as possible on exceptions to the problem and the possible goals to be reached. The therapist works with the client(s) for approximately 40 minutes, then takes a brief time out (approximately 10 minutes) to discuss further proceedings with the team behind the one-way mirror. Finally, the

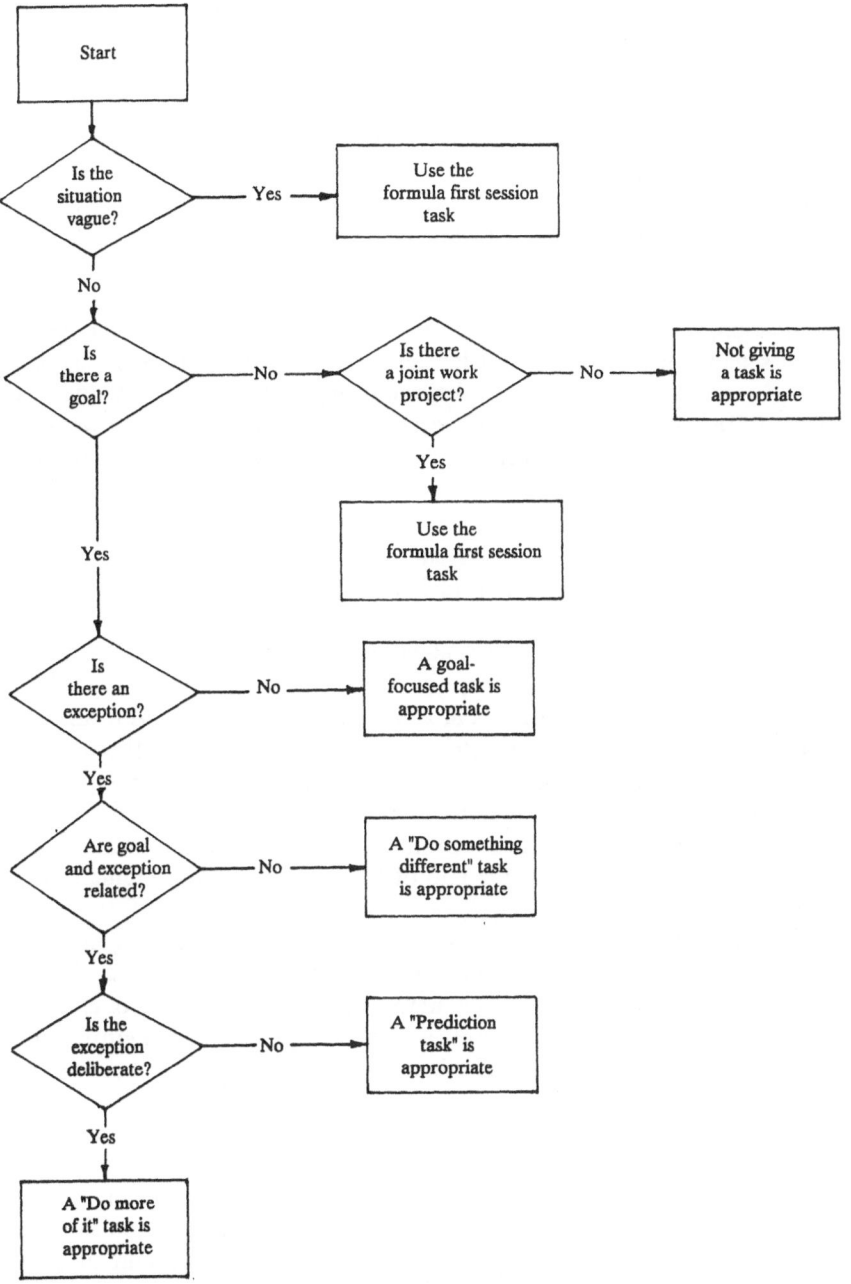

Fig. 4: Flow chart for intervention design (searching for "clues") within the conceptual framework of the BFTC.

client is given feed-back consisting of "compliment" and "clue" (Fig. 4). The team at BFTC works with individuals, couples, family units etc., depending on

251

whom the people concerned believe to be able to contribute to the solution of the problem.

1) No stabilization of the problematic potential valleys.
Therapists at BFTC try to avoid everything that could possibly result in the potential valley which the clients want to leave growing even deeper. In particular, there is no assessment or diagnostic process in the traditional sense. Thus, labelling of the clients, blaming someone for the problem or for establishing the problem are all prevented. Equally importantly, however, client and therapist progress very quickly to a solution-oriented form of communication, leading away from the potential valley. On the other hand, limits are set to this line of action when clients first need to have somebody accept and respect their complaints, their suffering, their difficult situation. If the clients were refused this feeling of acceptance, they would have to work to achieve it themselves — and experience shows that this results in a stabilization of their lifestyle scenarios.

2) Energizing processes to keep the system far from equilibrium.
The following starting-points and principles of treatment correspond to a theoretical energizing process:
a) Current status of motivation. Every client is seen as motivated and ready to cooperate; there is no such concept as resistance. The important question is to find out what exactly the client is motivated to achieve. Motivation is regarded as a combination of intrinsic processes, aspects of the current lifestyle and — most important — the therapeutic relationship. The therapist's job is to attune himself to the motivation of the client. The distinction between a visitor-, complainant- and customer-relationship leads to appropriate clues at the end of the session, corresponding to the client's motivation.
b) Talking about goals is seen as a concretizing, expectation-raising process. Thus, goals are helpful in finding orientation and limiting the aim of therapy. It is most important to join in with the preexisting goals of clients.
c) Searching for possible resources gives rise to the belief that help can be found within the resources of the client as well as within his or her context (other helping persons).
d) The therapist pays attention to confirming the client's central interactional plans. This prerequisite must be met if one is to allow for the development of a stable client-therapist relationship, which in itself bears energizing qualities. Two different therapeutic techniques serve to accomplish this: First, every client is given a series of compliments at the end of each session. These are statements from the therapist and/or team about what the client has said that is useful, effective, good, or fun. This helps to promote client-therapist fit and thus cooperation on the task at hand. Secondly, very often clients are given a standard homework task at the end of the first session. They are asked to observe what in their life they

consider useful, worthwhile or good that they want to remain unchanged in the course of therapy.

e) Therapists adapt themselves to a very high degree to clients' receptiveness and characteristics. This refers to the specific ways of communication (speech characteristics, dialect, intellectual level, idiosyncrasies), and emotional states (ranging from sincere sympathy to provocative humor) of clients. They are also alert to signs of increased receptiveness which require immediate therapeutic action (Ambühl & Grawe 1988; 1989). Therapeutic responses take into account the resources and dynamics of the specific client (Jiu-Jitsu principle).

3) Deviation-amplifying feedback

Therapy can be viewed as a process of (a) deforming the potential landscape and (b) consistently amplifying spontaneous fluctuations crossing the limitations of the actual potential barriers.

The following procedures constitute the feed-back mechanisms necessary in this process:

a) Search for exceptions. Together with the client, the focus of attention is shifted to exceptions to the rules of the problem. If such exceptions can be found, the distinction is made as to whether they are spontaneous or deliberate. Consequently, tasks are recommended that will increase the frequency of exceptions to the problem, so that finally the exception will turn into the rule.

b) "Do Something Different" task. At times, clients report exceptions to the problem and goals reached, yet it is not possible to describe what made the difference leading to either the exception or the problematic behavior. If this is the case, they are recommended to "do something different" in a consistent and systematic way in specific situations. However, what exactly is done differently is basically irrelevant. This is utilized to create fluctuations and exceptions similar to the actual goals desired.

c) Construct hypothetical, fantastic realities. There are two decisive reasons as to why the therapists at BFTC promote alternative views of reality. The first - energizing - has already been mentioned. Another reason lies in the fact that fantasizing a changed reality gets clients to look past the edge of their potential valley. It is precisely in this way that concrete, changed realities (potential valleys) are produced.

d) Search for coping strategies. A "list of what works" is drawn up on the basis of the exceptions that have been found. The client is given the task "simply to do more of what works".

The amplifying mechanisms in effect are, for example, feed-forward processes in the context of induced expectations and focussed attention, positive reinforcement during and outside of therapy, actual experience outside the therapy setting, and even emotional processes.

4) Control parameters

Therapists at BFTC try in various ways to bring about "change talk" together with their clients, the purpose of which is to create communication that will help them to escape from their problematic lifestyle (potential valley). In particular, this means exploring different states (real or hypothetical), prompting them to risk new experiences or design different constructions of reality (e.g. reframing). This process in which therapist and client are involved is, of course, a co-creative one. The ratio of "change talk" to "problem talk" could be regarded as an important control parameter.

5) Stability vs. Instability

In most cases, the situation in which a client seeks therapy can be described as relatively stable (with the exception of crisis intervention, for example). Destabilizing processes are at the core of therapeutic efforts. When bifurcations are approached, symmetry-breaking fluctuations necessarily produce unpredictable developments towards new states of order. Thus therapy inevitably involves taking risks. However, the client may first merely fantasize about different ways of acting in given situations (inner rehearsal) and so experiment with various alternative behaviors within the protective therapeutic setting. This provides at least a certain a degree of security. In addition, the therapist assesses stability intuitively, on the basis of his/her therapeutic experience. The development of stricter criteria to be applied to such stability analysis is a desirable goal for future work.

The approach put forward by BFTC is suitable for scientific research for the following reasons: The strict limitation to 10 sessions presents a clear time frame, one within which therapy happens in a very condensed fashion. The problems and solutions dealt with are more accurately seen as a "talk of" problems or solutions, rendering the therapy observable, something which is an advantage in terms of methodological features of psychotherapy research. Furthermore, the BFTC concept presents a relatively clear-cut frame as to the actions taken by the therapists, making treatments readily comparable. Finally, it is easy to conceive "how it works".

4. Methods for Empirical Research on the Synergetics of Psychotherapy

The following puts forward possible methods for the investigation of systems and processes as conceptualized in Sect. 2. The focus is on (a) the lifestyle scenario as a coherent pattern emerging out of the complexity of biopsychosocial life processes, and (b) the therapeutic system.

The method of idiographic systems modeling, which allows the depiction of dynamic biopsychosocial patterns, is used to describe the lifestyle scenario (see Schiepek 1986; 1991; Schiepek & Kaimer 1988). An observation-coding scheme (Process Analysis of Communicative Systems, or PACS) is

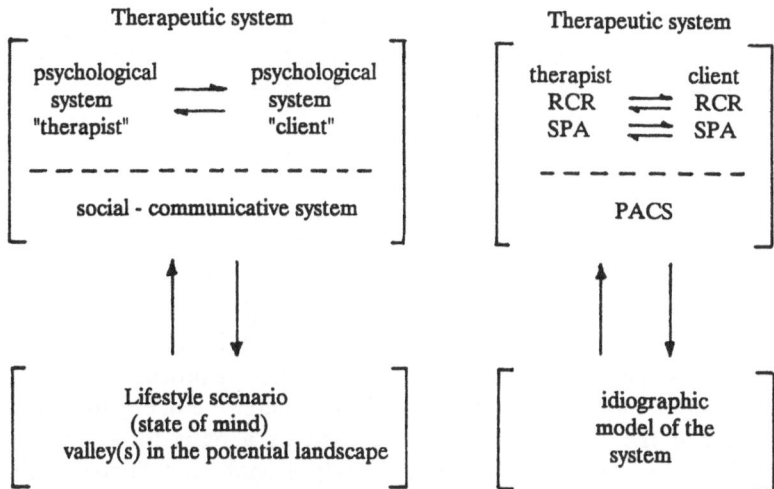

Fig. 5: Synopsis of the proposed empirical research methods for systems-process analysis of psychotherapy, related to the theoretical model in sect. 2.3.

developed which refers to the basic operations of self-referential communication processes as described by Luhmann (1984), in order to analyze the communicative social system of therapy. The interaction of the psychological systems involved in the process of therapy is analyzed by means of two different methods: first, Sequential Plan Analysis (SPA), based on the method of plan analysis (Caspar 1989) developed by Grawe and Caspar, and further elaborated to allow for analysis of a sequence of plans; secondly, an analysis of the client's and therapist's cognitive representations based on transcripts (Reconstruction of Cognitive Representation, RCR), which is basically an analysis of the topics introduced in therapy sessions and the resulting semantic networks. These methods can be related to the schema of Fig.2 as shown in Fig. 5.

Both PACS and SPA are at present undergoing development and refinement. Accordingly, only the basic concepts will be outlined below. The method of idiographic systems modeling, for its part, has been described in detail and illustrated through examples elsewhere (see Schiepek 1986; 1991; Schaub & Schiepek, this volume). We will therefore not discuss it at length here. The method of RCR will be explained in somewhat greater detail below (compare Fricke 1990).

4.1 Idiographic Systems Modeling

This method is a metastrategic approach to model building. The models worked out are recursively interrelated, symbolic systems, the components of which, i.e. theory-based constructs, are determined anew for each individual case. The choice of components depends on the desired degree of resolution and the reference levels of the model (e.g. does it refer to biological, psychological and/or social processes?). Theories referred to in the formulation of subhypotheses also determine the selection of components. These subhypotheses relate constructs via "if-then" rules, or even via mathematically precise functions. The theories and findings of various disciplines can be regarded as the stock that can be drawn upon for the generation of subhypotheses in the construction of the concrete model. Thus, different theoretical perspectives are introduced into the model by lineal subhypotheses which combine to form a complex, recursively interrelated network. In this way, processes at different levels of the system and their relations can be integrated into one model. It should be possible to operationalize the components (constructs) of the model via process indicators (variables). Multiple time series result from these recursive models, the same result occurring if the system dynamics are used as a basis for computer simulations (compare Schiepek & Schaub 1991; Schaub & Schiepek, this volume; Schiepek & Schoppek, 1991).

The metastrategic approach contains the following prescriptions for the construction of the model:

a) Cybernetic principles for structure formation (e.g. positive/negative/ mixed feedback; identification of damping or accelerating components; determination of limit and threshold values etc.) which serve as construction rules for the model,
b) a questionnaire to survey the patterns and recursive interrelations of the "area" (e.g. a lifestyle scenario), and
c) criteria to guide the selection of subhypotheses and the construction of system boundaries, together with those governing the breaking off of model construction (for details see Schiepek 1986).

System models are able to depict the dynamic pattern of a lifestyle scenario because (a) they can deal with the recursive interrelation of the biological, psychological and social processes contributing to a concrete lifestyle scenario through their employment of a number of different theoretical perspectives, (b) they are able to generate dynamic patterns, and (c) they can be computer-simulated, which is the precondition for the study of their attractor and stability characteristics, at least on the level of simulations.

With regard to the description of changes in a lifestyle scenario, there are at least two possibilities: The first is to modify the system model in a qualitative way. This is done by omitting or adding subhypotheses, or by modifying their contents (compare Schiepek & Schaub 1991), corresponding to the fact that a changed lifestyle scenario is at least partially determined by different biological, psychological and/or social processes. The second is to transform the system model into a system of nonlinear equations, and to modify the relevant parameters. Nonlinear change in the dynamics of the system should result, if these are control parameters. This and the subsequent change of the order parameters of the dynamic pattern would then represent the change of a lifestyle scenario (compare Kriz 1990; Schiepek, Schoppek & Tretter, this volume; the concept of dynamical diseases, Mackey & an der Heiden 1982; see also the deformation of the potential landscape effected by the modification of the parameters of the potential function, Fig. 1).

4.2 Process Analysis of Communicative Systems (PACS)

The aim of the Process Analysis of Communicative Systems is to study the self-organization of the therapeutic communicative system. Empirical data is to be gathered on the processes of the system's internal differentiation and structural changes, its construction of identity and the system/environment relations, in short, all the characteristics of self-organization in social systems (Teubner 1990; Luhmann 1984). The observation-coding scheme consists of a number of categories, the operationalization of which is being developed at present. These categories refer to those communicative acts which ensure the continuation of communication. The following is a selection of categories (for a detailed presentation see Schiepek 1991, Chap. II):

a) Acceptance vs. rejection of a communication (information) by the recipient. The third possibility taken into consideration is that of keeping a communication offered pending.
b) Continuation of a current subject vs. changing the subject (in Luhmannn's terminology: reciprocal selection).
c) Confirmation vs. non-confirmation of understanding of a communication.
d) Reference to information vs. reference to communication. Does the recipient respond to the content aspect or to the psychological quality (meaning,impact), the relationship aspect of a communication?
e) Identity vs. differentiation. Is there a differentiation introduced into communication?
f) Necessity vs. contingency. Are things described as necessarily and practically given the way they are or are they conceived of as possessing alternatives, as being changeable?
g) Introduction of expectations. Are expectations, rules or guidelines conveyed, whether implicitly or explicitly?
h) Basic self-reference vs. self-reflection (in German: Basale Selbstreferenz vs. Reflexivität). Are we concerned with communication which simply joins communication or communication about previous communication?
i) Reflexivity (in German: Reflexion): Does the social system reflect its own identity by way of describing itself in relation to its environment?
j) Reference to types of systems. Are events related to biological, psychological, social-communicative or other levels of reference and thereby explained?
k) Actual genesis vs. historiogenesis. Are events explained by simultaneous or past events?
l) Questions vs. statements.

The list of categories may be enlarged by formal aspects such as the duration of each communication, or by the characteristics of therapeutic communication such as solution-oriented vs. problem-oriented communication. Finally, the coding scheme may provide for a differentiation as to whether the content of communication refers to previous content, actions, thoughts or emotions. As can be seen, most of the categories provide for a binary code, which means that it has to be decided (a) which of the categories and (b) which of the two alternatives is applicable at each moment in the course of communication.

As the Process Analysis of Communicative Systems outlined here is designed to register features of communication according to their occurrence, its practical use does not require the identification of communicative units. This is of great pragmatic value, since one of the most difficult problems with which empirical communications research is faced is that of reliably and validly determining the delimitations of communicative acts. Moreover, it is especially with regard to the communication theory proposed by Luhmann, which is based on the assumption that communication, information, and understanding (in German: Mitteilung, Information und Verstehen) form an integrated unit, that there are serious doubts as to whether it is at all possible for an external observer to identify such communicative units.

It is possible to describe the process of communication within therapy by means of a sequence of categories or their combinations. It is assumed that the state of each category at time t depends on the states of different categories at times t-1,...,t-n, and also on previous states of the same category as well (conjoint probabilities and conditional transitional probabilities). Thus, we are concerned here with Markov chains possessing a "memory".

These provide the basis for the next possible step, the simulation of therapeutic communication processes (via production systems, for example). This type of computer simulation adopts the basic principle of self-referential systems, namely that the production of components is a function of the interaction of the system's components (represented by means of conjoint and transitional probabilities).

4.3 Sequential Plan Analysis (SPA)

The method of plan analysis (Caspar 1989), which has so far been applied mainly for purposes of status assessment, will be modified so as to allow for the registration of processes of mutual coordination with respect to the self-presentation of, and the relationship between, the participants of an interaction (client and therapist, for example). In its traditional form, plan analysis is a method of generating a model of an individual's functioning. Hierarchially organized structures of hypothetical plans, consisting of a goal and a means component, are inferred from the interactional behavior of a person. At first, the plan structure is usually inferred inductively, which means that the existence of one or a number of superordinate or guiding plans is inferred from the observation of several ways of behavior considered to represent means of achieving a certain goal. Later on in the process, it is possible to examine in a deductive manner whether different ways of behavior or subordinate plans are in accordance with a hypothesized superordinate plan. Of particular importance here is the principle of "multiple determination". This corresponds to the fact that on the one hand several different behaviors or sub-plans serve one guiding plan, whereas on the other hand a specific behavior is usually determined by several guiding plans at the same time.

For the purposes of Sequential Plan Analysis, the procedure will be modified as follows. The video recordings of therapy sessions are subdivided into units of approximately 10-15 seconds. Within each time unit, the coding of up to five interactional plans for each person involved is envisaged. All persons participating in the interaction are monitored simultaneously, to allow for an analysis of mutual coordinating processes governing interpersonal relations and the interplay of emotions. This method offers the advantage of allowing for a more detailed analysis of nonverbal behavior, as opposed to the PACS. The sequential patterns of the interactional plans identified, together with their duration, indicate the evolution of communicative states of order.

4.4 Reconstruction of Cognitive Representation (RCR)

This method presents an attempt at investigating the internal orientation and coordinating processes of psychological systems in the context of psychotherapy. As the focus of attention is on psychotherapeutic processes, there are a number of more general questions, such as:

— What are the notions and subjective theories used by clients to explain the problems giving rise to their seeking therapy? What are the models constructed by therapists to describe these problems?
— What are the co-evolutionary processes, similarities and common features of the cognitive representations of the problem(s) as presented by therapist and client respectively? In what respect do the conceptions of the problem(s) by therapist and client differ (with regard to various aspects such as content, degree of differentiation, degree of complexity and interrelatedness, or stability) at the beginning of therapy and during each succesive session?
— What is the relation between the changes in these parameters and therapeutic success? Are there wide-range modifications of previously utilized frames of reference, or even "personal revolutions" (Mahoney 1980)? Does the representation of problems show a high degree of stability within unsuccessful therapy?

In order to obtain answers to the above questions we have developed Reconstruction of Cognitive Representation (RCR). This method is directed at an analysis of the contents of communication in the therapeutic process by

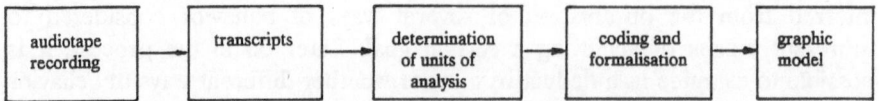

Fig. 6: Assessment steps taken in the implementation of the method of RCR.

means of a specific coding scheme. As a result, it is possible to create a graphic model depicting the cognitive organization of the problem presented in therapy, which may serve as a basis for a detailed single-case study. Fig. 6 shows a survey of the steps taken in the application of this method.

As a first step in the course of the evaluation of therapy transcripts, every verbal utterance (Schindler 1989) made by therapist and client, which is connected with the problems presented in therapy, is coded. Certain of the categories envisaged by the coding scheme for interactions in psychotherapy designed by Schindler et al. (1986) are used by the rater as guidelines in determining distinct units for analysis.

For the second step, leading to the graphic model of cognitive representations of problems a procedure has been developed similar to that of the formalization of structures of knowledge by means of so-called propositions (Huber & Mandl 1982; Schnotz 1982; Kluwe 1987). The basic notions of this concept focus upon a breaking down of facts and their logical interrelations as presented in written material into propositions, which are then graphically illustrated by means of lists or semantic networks (compare Sect. 1.4 of Schaub & Schiepek, this volume).

Up to now, this concept of research has been applied to analyze relatively simple written texts. Due to its high level of resolution, this method is hardly suitable for the investigation of highly complex and extensive "subjective structures of knowledge" (Schnotz 1982). As shown by Huber and Mandl (1982), it requires the use of as many as four propositions to represent a relatively simple statement such as: "Psychoanalysts concentrate on unconscious intrapsychic complexes to heal neuroses". This example makes it possible to conceive the extent of the lists of propositions required for the coding of verbal utterances of longer duration by therapist or client. This is the reason for the selection of a procedure characterized by a lesser degree of formalization than the method by which subjective structures of knowledge are reconstructed; it therefore has to rely explicitly on the interpretations of the rater. The codings of the verbal utterances derived from the transcripts are then integrated to form statements, themes and descriptions of characteristics according to specific rules (for a list of coding rules see Fricke 1990).

After this step has been taken it is possible to represent graphically the characteristics indicative of a problem and their interrelations. One graphic representation which contains all the statements referring to the problems presented within therapy is drawn up for both therapist and client for each therapy session. Excerpts from the cognitive representations of the problems presented by a client during the second and the eighth session out of a total of 12 sessions are shown in Figs. 7 and 8.

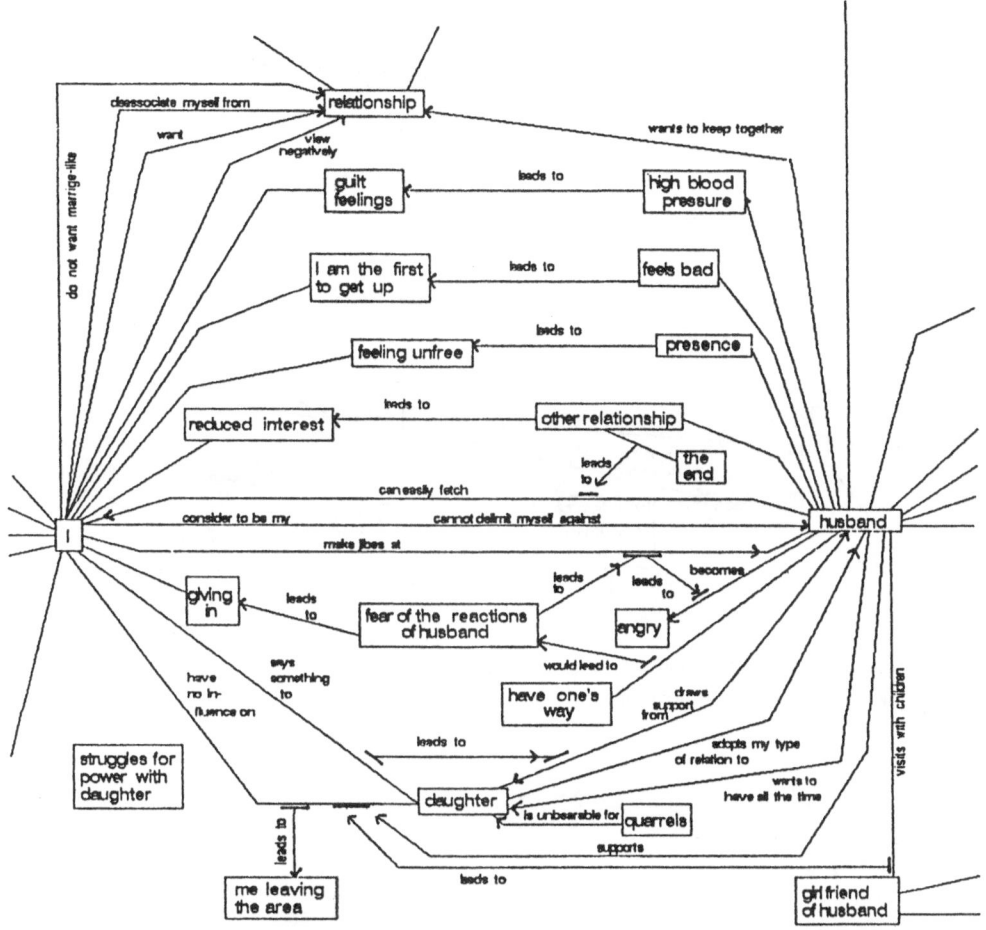

Fig. 7: Reconstruction of cognitive representations presented by a female client, second therapy session.

A comparison of these two structures of the cognitive representations of problems indicates differing degrees of the variables "number of different topics mentioned" (diversity) and "degree of cross-linking". These data allow an assessment of the complexity and stability of the pertinent problem representations. It is also possible to identify potential fluctuations and processes of change. However, evaluation is carried out not only by means of the quantitative comparison of certain parameters (such as degree of cross-linking, diversity, complexity) but also through a detailed qualitative analysis (including an assessment of the contents of cognitive representations for each therapy session; a comparison of the contents and their rank order as obtained by therapist and client; an evaluation of significant changes in the core themes over the course of therapy). Two possible modes of assessment of the graphic models (degree of cross-linking and congruence of therapist's and client's problem representations) are shown in Figs. 9 and 10.

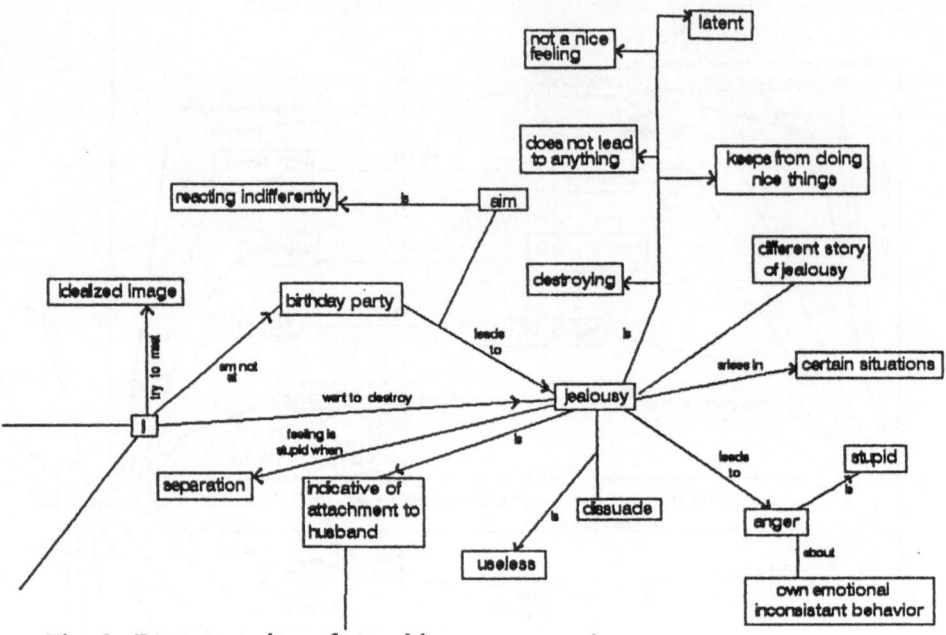

Fig. 8: Reconstruction of cognitive representations presented by a female client, eight therapy session.

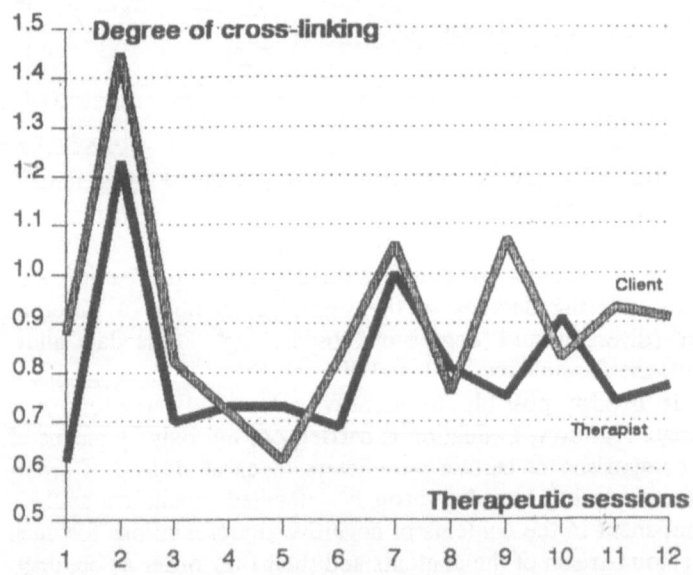

Fig. 9: Degree of cross linking of cognitive problem representations of therapist and client over the course of 12 therapy sessions. This variable is obtained by computing the quotient number of interrelations (N_i) divided by number of variables (N_v) (according to Vester & von Hesler 1980).

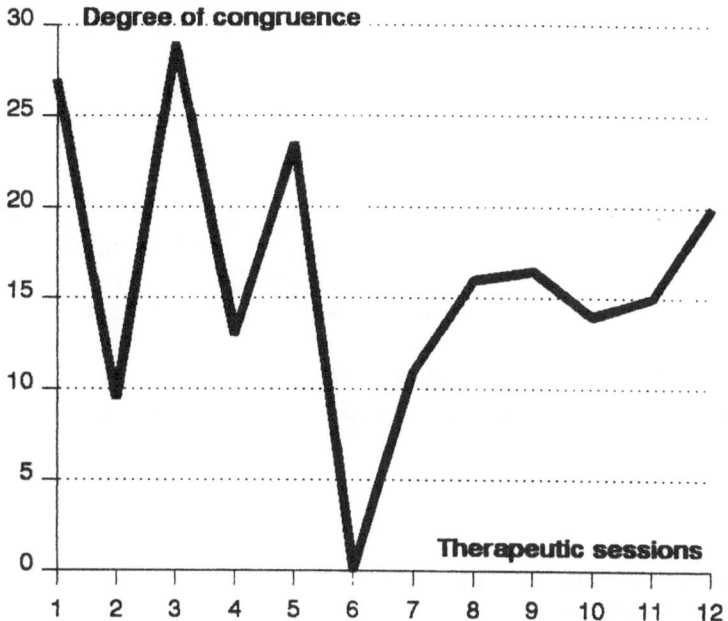

Fig. 10: Degree of congruence of therapist's and client's cognitive problem representations over the course of 12 therapy sessions. The ratings of the degree of congruence between therapist and client regarding topics of central, intermediate and marginal importance are weighted and added up for each therapy session.

On the basis of the assessment procedures described, as well as further and more detailed analyses, it is possible to describe a process of "co-ontogenesis" of the psychological systems involved in therapy, together with its relevance regarding the outcome of therapy.

The implementation of the method of RCR has been attempted within the scope of one single-case study (Fricke 1990). Its application is accompanied by a great number of methodological and theoretical problems with which psychologists have been faced ever since the beginning of research into the cognitive foundations of human behavior. In addition, there is disagreement about the transferability of the classical test criteria to qualitative assessment procedures (Tergan 1987). Special attention was paid to the design of the assessment steps and the coding rules so as to prevent any bias to the greatest possible degree, whether caused by the rater or by the method itself.

The implementation of the method of RCR in the field of psychotherapy process research is of assistance in attaining a number of methodological and theoretical goals. On the one hand constructs such as "subjective lifespace", "structural coupling", "area of consensus", "assimilation and accomodation", etc. are rendered empirically accessible; on the other hand these constructs also undergo further refinement and integration into a metatheory of psychotherapeutic processes.

It was shown that synergetics of psychotherapy as outlined here is to be viewed as an empirical approach, which does not mean quantification at any price, however. Rather, synergetics as a metatheory was adapted to the field of psychotherapy in a way that may lead to a more profound understanding of psychotherapeutic processes in general as well as for the single case. This is why "soft" methods are also accepted.

The application of an exact mathematical model to psychosocial interaction — some of the fundamental apects of which can hardly be quantified — and the attempt to make this a heuristically fruitful endeavour creates the exciting field of tension inherent to synergetics of psychotherapy.

Acknowledgement. We would like to thank Dipl.-Psych. Martina Weisensel (Staatl. gepr. Übersetzerin) for the translation of the present contribution.

References

Ambühl H, Grawe K (1988) Die Wirkungen von Psychotherapien als Ergebnis der Wechselwirkungen zwischen therapeutischem Angebot und Aufnahmebereitschaft der Klient/inn/en. Zeitschrift für Klinische Psychologie, Psychopathologie und Psychotherapie 36: 308-327

Ambühl H, Grawe K (1989) Psychotherapeutisches Handeln als Verwirklichung therapeutischer Heuristiken. Psychotherapie und medizinische Psychologie 39: 1-10

Arnold E, Grawe K (1989) Eine Strategie zur Untersuchung von Wirkungszusammenhängen in der Psychotherapie. Zeitschrift für Klinische Psychologie, Psychopathologie und Psychotherapie 37: 262-276

Caspar F (1989) Beziehungen und Probleme verstehen. Eine Einführung in die psychotherapeutische Plananalyse. Huber, Bern

Caspar F, Grawe K (1989) Weg vom Methoden-Monismus in der Psychotherapie. Bulletin der Schweizer Psychologen Nr 3: 6-19

Ciompi L (1986) Zur Integration von Fühlen und Denken im Licht der "Affektlogik". Die Psyche als Teil eines autopoietischen Systems. In: Kisker KP et al (Hrsg) Psychiatrie der Gegenwart, Bd 2. Springer, Berlin, S 373-410

Foerster H von (1985) Sicht und Einsicht. Vieweg, Braunschweig

Foppa K (1984) Operationalisierung und der empirische Gehalt psychologischer Theorien. Psychologische Beiträge 26: 539-551

Fricke B (1990) Untersuchung von Selbstorganisationsprozessen psychischer Systeme im Kontext von Psychotherapie mittels einer Methode zur Rekonstruktion kognitiver Problemrepräsentationen (RKR). Universität Bamberg, Diplomarbeit

Goudsmit AL (Ed) (1989) Self-Organization in Psychotherapy. Springer, Berlin

Grawe K (1987) Psychotherapie als Entwicklungsstimulation von Schemata. Ein Prozeß mit nicht vorhersehbarem Ausgang. In: Caspar F (Hrsg) Problemanalyse in der Psychotherapie. Forum 13, dgvt, Tübingen, S 72-87

Grawe K (1988) Zurück zur psychotherapeutischen Einzelfallforschung. Zeitschrift für Klinische Psychologie XVII: 1-7

Grawe K, Caspar F, Ambühl H (1990a, b, c, d) Die Berner Therapievergleichsstudie. Zeitschrift für Klinische Psychologie XIX (4), 292-376

Greenberg LS (1983) Psychotherapy Process Research. In: Walker CE (Eds) The Handbook of Clinical Psychology I. Theory, Research and Practice. Homewood, Dow Jones-Irwin, pp 169-204

Greenberg LS, Pinsof WM (Eds) (1986) The Psychotherapeutic Process: A Research Handbook. Guilford Press, New York

Großmann S (1989) Selbstähnlichkeit: Das Strukturgesetz im und vor dem Chaos. In: Gerok W (Hrsg) Ordnung und Chaos in der unbelebten und belebten Natur. S Hirzel, Wissenschaftliche Verlagsgesellschaft, Stuttgart, S 101-122

Haken H (1983a) Synergetics. An Introduction. Springer, Berlin (3. Aufl.)

Haken H (1983b) Advanced Synergetics. Instability Hierarchies of Self-organizing Systems and Devices. Springer, Berlin

Haken H (1990) Synergetik. Eine Einführung Springer, Berlin (2. Aufl)

Hesch RD (1990) Sein im deterministischen Chaos: Ansätze einer neuen Systemtheorie der Medizin. Vortragsmanuskript, Hannover

Horowitz MJ (1987) States of Mind. Plenum Press, New York

Huber GL, Mandl H (1982) Verbalisationsmethoden zur Erfassung von Kognitionen im Handlungszusammenhang. In: Huber GL, Mandl H (Eds) Verbale Daten. Beltz, Weinheim, S 11-42

Imber-Black E (1990) Familien und größere Systeme. Im Gestrüpp der Institutionen. Auer, Heidelberg

Kiesler DJ (1985) The Missing Link in Psychotherapy Research. Review of Rice & Greenberg (Eds) (1984) Patterns of Change. Contemporary Psychology 30: 527-529

Kluwe RH (1987) Methoden der Psychologie zur Gewinnung von Daten über menschliches Wissen. In: Mandl H, Spada H (Hrsg) Wissenspsychologie. Urban & Schwarzenberg, München, S 359-385

Kriz J (1990) Synergetics in Clinical Psychology. In: Haken H, Stadler M (Eds) Synergetics of Cognition. Springer Series in Synergetics, Vol 45. Springer, Berlin, pp 393-404

Luhmann N (1984) Soziale Systeme. Suhrkamp, Frankfurt am Main

Luhmann N (1985) Die Autopoiese des Bewußtseins. Soziale Welt 36: 402-446

Mackey MC, an der Heiden U (1982) Dynamical Diseases and Bifurcations: Understanding Functional Disorders in Physiological Systems. Funkt Biol Med 1: 156-164

Mahoney MJ (Ed) (1980) Psychotherapy Process. Current Issues and Future Directions. Plenum Press, New York

Markowitz J (1991) Referenz und Emergenz: Zum Verhältnis von psychischen und sozialen Systemen. Systeme (Interdisziplinäre Zeitschrift für systemisch orientierte Forschung und Praxis in den Humanwissenschaften) 5: 22-46

Orlinsky DE, Howard KI (1986) The Relation of Process to Outcome in Psychotherapy. In: Garfield SL, Bergin AE (Eds) Handbook of Psychotherapy and Behavior Change. Wiley, New York, pp 311-381

Revenstorf D (1990) Zur Theorie der Hypnose. In: Revenstorf D (Hrsg) Klinische Hypnose. Springer, Berlin, S 79-99

Rice L, Greenberg LS (Eds) (1984) Patters of Change. Intensive Analysis of Psychotherapy Process. Guilford, New York

Silberschatz G, Curtis JT, Nathans S (1989) Using the Patient's Plan to Assess Progress in Psychotherapy. Psychotherapy 26: 40-46

Stadler M, Kruse P (1990) The Self-Organization Perspective in Cognition Research: Historical Remarks and New Experimental Approaches. In: Haken H, Stadler M (Eds) Synergetics of Cognition. Springer Series in Synergetics, Vol. 45. Springer, Berlin, pp 32-52

Schiepek G (1986) Systemische Diagnostik in der Klinischen Psychologie. Psychologie Verlags Union, München

Schiepek G (1991) Systemtheorie der Klinischen Psychologie. Vieweg, Braunschweig

Schiepek G, Kaimer P (1988) Von der Verhaltensanalyse zur selbstreferentiellen Systembeschreibung. Familiendynamik 13: 240-269

Schiepek G, Schaub H (1991) Als die Theorien laufen lernten... Ein Simulationsmodell zur Depressionsentwicklung. In: Schiepek G, Systemtheorie der Klinischen Psychologie. Vieweg, Braunschweig, Kap. VIII

Schiepek G, Schoppek W (1991) Synergetik in der Psychiatrie: Simulation schizophrener Verläufe auf der Grundlage nicht-linearer Differenzengleichungen. In: Niedersen U, Pohlmann L (Hrsg) Selbstorganisation. Jahrbuch für Komplexität in den Natur-, Sozial- und Geisteswissenschaften, Bd. 2. Duncker & Humblot, Berlin, S 69-102

Schindler L (1989) Das Codiersystem zur Interaktion in der Psychotherapie (CIP): Ein Instrument zur systemischen Beobachtung des Verhaltens von Therapeut und Klient im Therapieverlauf. Zeitschrift für Klinische Psychologie XVIII: 68-79

Schindler L, Müller U, Sieber E, Hahlweg K (1986) Codiersystem zur Interaktion in der Psychotherapie (CIP). Manual für den Beobachter. Manuskript. München, Max-Planck-Institut für Psychiatrie

Schnotz W (1982) Rekonstruktion von individuellen Wissensstrukturen. In: Huber GL, Mandl H (Hrsg) Verbale Daten. Beltz, Weinheim, S 220-239

Schweitzer J (1989) Professionelle (Nicht-)Kooperation: Ihr Beitrag zur Eskalation dissozialer Karrieren. Z system Ther (7)4: 247-254

Shazer S de (1985) Keys to Solution in Brief Therapy. Norton, New York

Shazer S de (1989) Der Dreh. Auer, Heidelberg

Shazer S de, Berg IK, Lipchik E, Munally E, Molnar A, Gingerich W, Weiner-Davis M (1986) Kurztherapie - Zielgerichtete Entwicklung von Lösungen. Familiendynamik 11: 182-205

Strauss JS (1989) Intermediäre Prozesse in der Schizophrenie: zu einer neuen dynamisch orientierten Psychiatrie. In: Böker W, Brenner HD (Hrsg) Schizophrenie als systemische Störung. Huber, Bern, S 35-50

Tergan SO (1987) Qualitative Wissensdiagnose - Methodologische Grundlagen. In: Mandl H, Spada H (Hrsg) Wissenspsychologie. Urban & Schwarzenberg, München, S 400-422

Teubner G (1987) Episodenverknüpfung: Zur Steigerung von Selbstreferenz im Recht. In: Baecker D, Markowitz J, Stichweh R, Tyrell H, Willke H (Hrsg) Theorie als Passion. Suhrkamp, Frankfurt am Main, S 423-446

Teubner G (1990) Hyperzyklus in Recht und Organisation. Zum Verhältnis von Selbstbeobachtung, Selbstkonstitution und Autopoiese. In: Krohn W, Küppers G (Hrsg) Selbstorganisation: Aspekte einer wissenschaftlichen Revolution. Vieweg, Braunschweig, S 231-264

Tschacher W (1990) Interaktion in selbstorganisierenden Systemen. Asanger, Heidelberg

Varela FJ (1979) Principles of Biological Autonomy. North Holland, New York

Vester F, Hesler A von (1980) Sensitivitätsmodell. Regionale Planungsgemeinschaft Untermain, Frankfurt am Main

Zingg MB (1988) Schematheoretische Sichtweise psychologischer Störungen. Inauguraldissertation, Universität Bern

Theories of Self-Organizing Processes and the Contribution of Immediate Interaction to Change in Psychotherapy

Henri Schneider

Abstract. A model of change in a relationship schema is specified in the framework of Piagetian theory and theories of self-organizing processes. This model is illustrated using three passages taken from three consecutive sessions from the middle phase of a short-term psychotherapy. An analogy is drawn between the process of change in a relationship schema and the process of nest-building in termites in order to highlight the aspects of immediate therapeutic interaction which are pertinent to understanding change in intrapsychic structures as a self-organizing process.

1. Introduction

Theories of self-organizing processes may prove to be particularly suited to the investigation of psychotherapeutic phenomena which, so far, it has not been possible to conceptualize. As no scientific models were available, these phenomena could only be described in an "intuitive" manner by psychotherapists. Immediate interaction in psychotherapy may well turn out to be such a phenomenon. Bacal, a psychoanalyst, articulates the difficulty of expressing this aspect of the psychotherapeutic process in words:

> "... there is considerable noninterpretive, optimal responsiveness in the course of everyday analytic work – sometimes transitory but sometimes quite prevalent – that is of significant therapeutic benefit. Most analysts know in their heart that this is a crucially therapeutic aspect of all analyses much of the time. *But they seldom talk about it, and it is almost never written about* [italics added], ..." (Bacal, 1988, p. 130)

The basic idea underlying the research approach to be documented in this article is that below the "surface" of the psychotherapeutic process there are patterns, or pathways, of change to be detected which offer a more thorough understanding of change in psychotherapy. To this end, a theoretical model is set up which specifies a "mechanism" by which change is thought to come about. This model then allows passages to be selected from therapeutic material (i.e., video recordings, transcripts, etc., of psychotherapy sessions) and to be analysed in a detailed manner. Careful analysis of these passages, in turn, helps to refine the theoretical model. Using this kind of methodological procedure, a particular aspect of therapeutic knowledge, which has remained largely implicit for lack of theoretical models, is rendered explicit (see Schneider, 1991b).

Springer Series in Synergetics, Vol. 58   Self-Organization and Clinical Psychology
Editors: W. Tschacher, G. Schiepek, and E.J. Brunner   © Springer-Verlag Berlin Heidelberg 1992

The model of change in a relationship schema presented in this article has been developed by Schneider & Wüthrich (1991; see also Schneider, 1983; Wüthrich, 1987) on the basis of Piaget's theory (Piaget, 1974, 1975) and Prigogine's theory of *Order Through Fluctuations* (Prigogine & Stengers, 1984). This model will be outlined first. It will then be illustrated by three sequences taken from a short-term psychotherapy (29 sessions) of a male patient in his late twenties, referred to as the "Berne" therapy[1].

In the main part of this article, the question as to how immediate interaction in psychotherapy may contribute to change in the client's intrapsychic structures will be addressed. An analogy will be drawn between the process of change in a relationship schema and the construction of a termites' nest conceptualized as a self-organizing process in the model of Kugler, Shaw and Turvey (see, e.g., Kugler & Turvey, 1987; Kugler & Shaw, 1990). This model has been chosen as an "interface" between an intrapsychic process (to be understood as a process of self-organization) and a biological process (for which a model in terms of self-organizing processes already exists) for two reasons. On the one hand, earlier versions of the model of nest-building in termites (Prigogine, 1976) played an important role in the development of the model of change in a relationship schema (see, e.g., Schneider, 1986). On the other hand, Kugler & Shaw (1990, p. 313) themselves seem to support applications of their model such as the one attempted in this paper:

> "If we are correct in our hypothesis, then the paradigmatic properties of this socially complex system exploit externally, and thus make visible, the same principles that govern the less visible, inner workings of individual systems with complex interiors (i.e., with central nervous systems)."

The approach presented in this paper will then be discussed with respect to what further questions may be asked in order to obtain a more comprehensive picture of the client's self-organizing process of change.

---

[1] This is one of the two short-term therapies being investigated by more than twenty groups of psychotherapy researchers from Austria, Germany and Switzerland in the framework of the PEP project (*Psychotherapy Process Research on Single Cases*) directed by Klaus Grawe and Horst Kächele. An analysis analogous to the one presented in this paper has been conducted on the "Ulm" therapy (see Schneider & Wüthrich, 1991). The clinical material was provided by the Ulmer Textbank. The protection of personal data makes it impossible to reproduce the source in its entirety. As far as scientific interest demands, it is possible to view the material at the Department of Psychotherapy of the University of Ulm, Am Hochsträss 8, 7900 Ulm, Germany.

2. The Model of Change in a Relationship Schema

The model of change in a relationship schema (Schneider & Wüthrich, 1991) describes the phase of the change process of a relationship schema in which a relationship schema of the client is enacted in the interactions with the therapist (for an attempt to understand how this "middle" phase of change in a relationship schema is prepared, and how the newly developed relationship structure becomes stabilized later on, see Schneider, 1989). The assumption is made that certain affective states, such as (in the Berne client) feeling that it is possible to express discontent and anger without being abandoned or rejected, correspond to far-from-equilibrium states. Some plausibility is lent to this assumption by a model developed by Skarda & Freeman (1987) on the recognition of odours in rabbits. In this model, the information corresponding to a particular odour is encoded in the form of a pattern of neural activity, or global activity state of the neurons, which is specific for each odour. In mathematical terms, these patterns of neural activity are called "attractors" (Skarda & Freeman, 1987, p. 168).

These affective states are thought of as "records of the far-from-equilibrium dynamics" (cf. Briggs & Peat, 1989, p. 150) that originally brought about these states in dyadic interaction. This means that in a person's actual experience these affective states are generated by an intrapsychic structure which has developed from the relationship structure that once induced the significant other to respond in the way that led the person to experience the desired state. This intrapsychic structure is called a "relationship schema" (cf. Grawe, 1988; Safran, 1990ab). In order for such an affective state to come into existence, a person has to experience the feeling to be generated by the relationship schema in an actual interaction with a significant other at least a few times. What is needed is a kind of confirmation or validation of the affective state – very much in the sense of Stern's (1985) *attunement*. Later on, the same state may be experienced by the underlying relationship schema being activated "autonomously", i.e., without a significant other having to respond in a way that produces this state in the person.

In psychotherapy, these affective states are thought to evolve from the amplification of "fluctuations" in the client's experience. A fluctuation in the client's experience may be characterized as a deviation from what the client is "used to" experiencing, pointing in the "right" direction, i.e., in the direction of a fuller experience of the developing affective state. This "new" feeling may only be noticed fleetingly by the client at specific moments in therapy. Such a fluctuation in a client's experience is thought to result from incremental changes in the relationship schema underlying the client's exchanges with the therapist, i.e., from new variants (or "mutants" in the sense of evolution theory; cf. Eigen, 1987, p. 73) of the relationship schema generated by the client.

By the amplification of a fluctuation in the client's experience, the underlying relationship schema is assumed to be induced to take another step in changing in the desired way.[2] The next move generated by the relationship schema may be more "direct" with respect to the affective state to be acquired, and the feeling that the client experiences fleetingly may be more articulate. There is an autocatalytic loop[3] in which new variants of the relationship schema are generated and fluctuations arising in the client's experience are amplified in therapeutic interaction in the sense of Kugler & Turvey's (1987) *perception-action cycle* (see below). This loop is carried on until the new macroscopic structure, i.e., the new structure of the relationship schema and the new affective state generated by it, have come into existence.

This model will be illustrated by examining three passages (called "situations") taken from three subsequent sessions of the Berne therapy. However, the way in which these passages have been worked out from the material will first be briefly discussed.

3. Three Situations from the Berne Case as an Illustration

In order to confront the above mentioned theoretical model with therapeutic material, passages in which fluctuations referring to the developing relationship schema had been identified were selected from the material. Three of these passages were then analysed by means of a checklist, referred to as a "frame" (for further information on the "frame procedure", see Schneider, 1990).

A frame is "a sort of skeleton, somewhat like an application form with many blanks or slots to be filled" (Minsky, 1985, p. 245). In artificial intelligence and cognitive science, frames are used to represent persons, objects and situations, etc., in a systematic manner. The slots of the frame designate different aspects of the persons, objects or situations to be represented.[4] In the frame presented in Table 1, the slots designate the core aspects of the theoretical model of change in a relationship schema.

In filling in the frame for a particular passage, the episode is first "divided up" into the aspects represented by the different slots of the frame. Then care is taken to render the passage taken from the therapy transcript in such a way as to enable the reader to reconstruct the episode without having to infer too much. This is to say that although the description of the particular situation reflects the respective

---

[2] The feelings experienced by the client are thought of as a kind of inner environment for the evolutionary process of the relationship schema, in the sense of Piaget's (1974) phenocopy model (Schneider, 1989).

[3] "Autocatalytic or self-amplifying processes may be understood as simple feedback loops in which the same operation is carried out repeatedly, the output for one operation being the input for the next one." (Stadler & Kruse, 1990, p. 43)

[4] In the case of a chair (Stillings et al., 1987, p. 151), for example, slots would be defined for the number of legs, the number of arms, the kind of seat, the style, etc. The information specific to a particular chair is entered into these slots, such as that a particular chair has four legs and two arms, that the seat is cushioned and that the style is high-tech.

Table 1. Frame representing the model of change in a relationship schema

---

A fluctuation arises: the client attempts a new move, and/or experiences a "new" feeling.

How does the situation evolve? What does the therapist do? Is the fluctuation amplified?

Does the client experience the evolving situation as an amplification of the fluctuation?

---

aspect of the theoretical model represented by the slot, the description itself ought to be as free from theory as possible. Thus, using a frame to represent passages thought to be relevant for change allows the description to be separated from the theoretical interpretation, the relation to the theoretical model being constantly maintained by the slots. From the point of view of scientific methodology, using this kind of a frame means that the way in which the theoretical model is linked to the material is made transparent (cf. VanLehn et al., 1984). Anybody may comment, for example, on whether the slot has been formulated in such a way as to express the corresponding aspect of the model, and whether the description of a situation adequately reflects the meaning of the passage taken from the material.

Using the frame presented in Table 1, three passages taken from the 15th, 16th and 17th sessions, respectively, have been analysed in which a new feeling pertaining to the developing relationship schema arises in the client's experience. These passages have been described according to the three slots of the frame. The resulting analysis is presented in Table 2. In characterizing the client's relationship schema enacted with the therapist, the old pattern, i.e., the pattern "naively" re-enacted with the therapist, is distinguished from the slowly emerging new pattern. The old pattern may be formulated in the following way: The client refrains from initiating any action to let others know how he feels or what he might like to do, for fear of being rejected. The newly emerging pattern may be described as follows: The client is able to stand up for his needs and wishes, or to get angry, without being too afraid of being rejected. The fluctuations pertaining to the developing relationship schema arising in the client's experience may be characterized as a "compound" feeling that it is possible to express discontent and anger without being abandoned or rejected (being looked down upon, etc.) by a significant other, or as a feeling resulting from being taken seriously when acting according to his own needs and wishes.

The course of change in the client's relationship schema enacted with the therapist may be described briefly as follows. In the 15th session, the client may feel safe enough for the first time in expressing his impatience about not getting anywhere by ruminating over the same problem over and over again. Thus, when the client complains about discussing the same problem over and over again, the relationship schema may be said to be enacted in the relationship with the

Table 2. Three situations from the Berne therapy as extracted from the therapy transcripts by means of the frame procedure (B... standing for the number of the therapy session, P... or T... indicating the number of client's or therapist's turn, respectively). The slots are printed in bold characters, while the description of the situation is in normal typeface

---

B15 P 148

A fluctuation arises: The client attempts a new move, and/or experiences a "new" feeling.
The 15th session starts with the client telling the therapist what has been going on since the last session. In T 147, the therapist states that it might be useful to look more carefully at one of the situations the client has just reported on.
The client answers that there is nothing that would suggest itself. "... the situation with my girlfriend, I've been ruminating over it again and again."

How does the situation evolve? What does the therapist do? Is the fluctuation amplified?
The therapist (T 151) responds by saying: "'Ruminated over' – this sounds like repeating it over and over again – but nothing really comes out of it?" He encourages the client to keep to his reproach, asking him whether the result has not been what he had expected from therapy (T 157).
In T 171, the therapist is quite explicit: "It might well be possible that you could say to me: 'I'm not satisfied about what has happened here.' How would this be?"

Does the client experience the evolving situation as an amplification of the fluctuation?
In P 180, it is possible for the client to formulate his objection openly: "I've been thinking about it – we've been going back and forth for such a long time now, racking our brains about these problems..."

B16 P 152

A fluctuation arises: The client attempts a new move, and/or experiences a "new" feeling.
The client is reporting on an incident that he had experienced as a child with his grandfather, which often comes to mind. Grandfather and grandmother had a quarrel, and the grandfather chased the grandmother around a table. He even grabbed a chair. The client "was afraid of moving at all" (P 124).
The therapist asks the client how he feels when he thinks back on this situation now.
"It's strange," the client answers (P 152), "... one would have liked to offer resistance, and still, one couldn't – just the fear. I'm scared even now when I think of it."

How does the situation evolve? What does the therapist do? Is the fluctuation amplified?
In T 159, the therapist is taking up the client's feeling of helplessness: "You couldn't stop your grandfather in his anger...," adding, "as a child, I mean... In the

meantime, you have become an adult, and this is still in you, as an impression, as a fear?"

The client answers (P 162) that even now he would scarcely dare to oppose his grandfather.

In T 165, the therapist offers to help the client face his grandfather. The therapist's question in T 171 initiates an "active" enactment: "Could you imagine letting yourself into this situation and finding out in what way you might stop your grandfather?"

Later on, the therapist says: "Could you imagine planting yourself in front of him, here now like this, and telling him, telling him something? It may only be one word, 'Stop', or 'No'?" (T 249).

In T 253, the therapist holds out a cushion to the client. "Just imagine that this is the force of the grandfather that you would have to tame" (T 263).

The client starts to beat the cushion and to tell his "grandfather": "Stop it!", "Leave grandmother alone!", etc.

Does the client experience the evolving situation as an amplification of the fluctuation?

In P 292, the client reports on the feeling he has experienced during the "fight": "I just felt strong." The client seems to experience this feeling as a new inner state which it is possible to experience in such a situation.

B17 P 144

The client, an electrician by profession, is telling the therapist about his experiences in 2127 (i.e., a foreign country about one hour's flight away), where he had been working for four weeks, installing some kind of industrial robots. He had been staying there with two colleagues, one of whom was an electronics engineer.

By what the client is telling him, the therapist feels urged to ask (T 143): "I could imagine that at times you had the feeling 'Fuck you!'"

A fluctuation arises: The client attempts a new move, and/or experiences a "new" feeling.

The client laughs and starts to report on an episode in which he had become angry. After he had been staying at 2127 for about two weeks, one of the machines broke down completely. He nearly cracked up. He yelled at the electronics engineer. He would have been capable of returning home, if the electronics engineer had not told him what to do next.

How does the situation evolve? What does the "therapist" do? Is the fluctuation amplified?

In the evening, the electronics engineer told him that he had indeed noticed that the client's anger was not directed at him.

(Here the assumption is made that the electronics engineer had already pointed this out to the client during his "fit" of anger in some way or other, e.g., by signalling to him that it is all right for him to be angry under these circumstances.)

Does the client experience the evolving situation as an amplification of the fluctuation?

The client experiences that he is not being disapproved of, or rejected by, his colleagues when he expresses his anger.

In the therapeutic session, the client is retrospectively conceptualizing this experience: that it had been possible for him to express his anger because he and his colleagues were on such friendly terms with each other (P 152).

---

therapist[5] and to be starting to change. The "active" enactment in the 16th session (lasting about 30 minutes) may be thought of as a method to artificially compile the elements which the schema needs to develop further: the reconstitution of the old situation with the grandfather, the safety provided by the therapist's presence, the therapist's support, etc. The aspect of the client's newly developing affective state which is validated in this enactment is his feeling of being strong enough to stand up against a powerful opponent. The "fit" of anger that the client reports on in the 17th session, and the subsequent reaction of his workmate, may be looked at as a confirmation of other aspects of the client's developing affective state, namely that he is not being disapproved of for having been angry, and that relationships don't break up for such reasons.

4. Immediate Interaction and the Formation of Intrapsychic Structure

How can the theoretical model outlined in this article contribute to understanding the meaning of immediate interaction in psychotherapy? In order to be able to work out as clearly as possible the connection of the model of change in a relationship schema to theories of self-organizing processes as well as its contribution to the understanding of the meaning of immediate interaction in psychotherapy, a model about the development of a new structure "imported" from biology will be introduced first. As already indicated, the model developed by Kugler & Turvey (1987; Kugler & Shaw, 1990) to describe the process of nest-building in termites (see, e.g., Prigogine, 1976) has played an important role in elaborating the basic model of change in a relationship structure (see Schneider, 1986, for an earlier version of this model).

Termites transporting material to a building site are attracted to the uppermost region of a forming pillar, called a "singularity", from which a pheromone is diffused. However, it is the termites themselves that transport the material to the building site, impregnating it with the pheromone during their flight, and thus at the same time augment the singularity by their building activity. This autocatalytic loop is called an "action → perception → action ... cycle" (i.e., force field (muscular activity) → flow field (pheromone control constraints) → force field (muscular activity)... and so on; Kugler & Shaw, 1990, p. 319), meaning that "the

---

[5] This expression of the client's anger may have been prepared in the 13th session by the therapist saying (B13 T 343), "... I mean, it is possible that you might explode in here as well sometimes; for me this would pose no problem..."

insect behavior both contributes to the structure of the pheromone field and is oriented by the structure of the pheromone field" (Kugler & Shaw, 1990, p. 318).

Later on in this process, when two pillars start bending towards each other, the singularity to which the termites are attracted may be only "virtual" (i.e., have no physical existence) in that termites flying towards the building site from certain angles may be attracted by a point between two forming pillars at which pheromone density is highest for them. This new singularity is called a "saddle-point".

> "The saddle-point is a flow-field property that specifies a virtual kinetic property, namely, a nonexistent (but incipient) concentration of mass lying between two pillars at a distance above the ground surface that is roughly the average height of two pillars... In strictly nest-building terms, the significance of the saddle-point is that it introduces a depositing bias toward a region between two pillars; loosely speaking, the saddle-point encourages arch construction." (Kugler & Turvey, 1987, p. 77)

Thus, new properties orienting the termites' flight (such as the saddle-point mentioned) may come into existence on the basis of "established" properties of the pheromone field (i.e., the uppermost regions of the pillars attracting the termites) on the one hand and the termites' continued transporting activity on the other hand.

If we now apply this model to the development of a new relationship structure in the client, the pheromone field is likened to the "affective" field resulting from client-therapist interaction, while the acts of transporting the building material are taken to be analogous to the client's new moves bringing about slightly different constellations of the affective field in each cycle of "significant" interaction with the therapist (i.e., in each perception-action cycle). What the termite analogy may help us understand is that a client's attempts at trying out new interactive patterns are oriented by feelings arising from his (former) exchanges with the therapist (that have been evaluated positively), while, at the same time, these interactions contribute to the structure of the affective field (to paraphrase the above quotation from Kugler & Shaw), each cycle confirming or strengthening the "singularity" which "represents", in the client's experience, the new affective state to be acquired.

The mechanism by which immediate interaction drives the process of change in a relationship structure proposed in this model may be described in the following way. In client-therapist interaction, a specific affective field is generated by a singularity coming into existence in the client's experience which represents the new affective state that the client would like to acquire. As this singularity is augmented by the therapist validating the new feelings arising in the client's experience, it attracts further exchanges with the therapist by which the desired feeling may – however fleetingly – be experienced by the client (i.e., exchanges which are based on the relationship schema that has undergone a slight change in the "right" direction). Thus, it is the client's experience of a new singularity (i.e., an anticipation of the new affective state that the client would like to acquire) that is driving the process of generating new variants of the underlying relationship schema.

The most important point to be made on the basis of this mapping of therapeutic interaction and structure formation onto the model of nest-building in termites is hidden in the distinction that Kugler & Turvey draw in order to characterize the two "times" of the perception-action cycle, namely the distinction between the "low-energy informational constraints" (i.e., the pheromone field orienting the termites' flight) and the "high-energy force outputs" (i.e., the muscular activity of the termites transporting building material):

> "The intrinsic, dissipative geometry of the chemical flow field (that is, patterned layouts of singular and gradient states) provides a set of low-energy informational constraints that classify or restrict the high-energy force ouputs generated by the insects's movement system." (Kugler & Turvey, 1987, p. 82)

This means that an insect is not driven by external forces, but is oriented by internally generated forces that move the insect relative to the topological properties (e.g., the singularities) of the pheromone field, i.e., the configuration of the pheromone field "'induces' a dynamics in the insects (modulating their trajectories in the direction of preferred locations)" (Kugler & Turvey, 1987, p. 73).

Thus, the model may help us understand how a new structure can arise in the client without it being present in the affective field from the outset. A new affective state first comes to be experienced by the client as a "singularity" which then orients his or her further "structure-building" activity. From this vantage point, the therapist's activity consists in providing the client with an adequate affective field, or therapeutic environment, in order to facilitate his or her "self-organizing" process. Thus, the model may shed new light on the question of "internalization" of intrapsychic structures and the much debated issue of "corrective emotional experience" (e.g. Jacobs, 1990).

5. Discussion

The process described by the theoretical model presented in this article, i.e., the amplification of fluctuations related to the new interactional pattern towards which the relationship schema enacted with the therapist is evolving, is thought to be an important aspect of change in all kinds of therapy in which interpersonal aspects are dealt with in immediate interaction. However, the model concentrates on what happens from the moment at which a fluctuation arises in the client's experience onwards. It does not spell out the conditions that may have to be fulfilled in order for fluctuations to arise in therapeutic interaction. In the 16th session of the Berne therapy, for example, it may well have been important that the client felt understood by the therapist, as opposed to what may be described as a lack of care on his parents' part about what the client experienced. In B 16 P 68, for example, after the therapist had just summarized some of the client's childhood experiences that he thought relevant for his present difficulties, the client said, "I'm astonished that you know all this, that you remember it after such a long time...", and a little later in this session (B 16 P 108), the client pointed out to the therapist, "When I told my parents about this (i.e., his grandparents' fight), my parents never believed me."

In the research project on which this article is based (Schneider & Wüthrich, 1991), a specific aspect of change in a relationship schema, namely the validation of new feelings arising in the client's experience in therapeutic interaction, has been investigated. In order to obtain a more. comprehensive understanding of change processes in psychotherapy, further theoretical models will be developed and subsequently refined by analysing passages selected from therapeutic material. Drawing on theories of self-organizing processes (see, e.g., Yates, 1987; Goodwin & Saunders, 1989) or possibly also on "connectionist" theories from cognitive science (see, e.g., Rumelhart et al., 1986; Rumelhart, 1989; Varela, 1988) in order to develop a theory of change in psychotherapy allows pertinent problems to be discussed by psychotherapists as well as researchers from the domain of self-organizing processes. On the one hand, these researchers may take up theoretical ideas emerging from research on change in psychotherapy and develop them further (e.g., by formalizing them) and thus point out to the therapists patterns that have not yet been noticed in the material. Whether, on the other hand, the investigation of change in psychotherapy may help to better understand, at a general level, the emergence of new structures out of interactions, remains an open question. However, let me briefly illustrate the idea that mechanisms of evolution and change might be very general by quoting from Kugler & Turvey (1984, p. 74):

> "Pillars are included among the time-independent structures that constrain the dynamics of nest-building. These structures, however, are erected on the basis of *time-dependent* geometries carried in diffusive flow fields. *Is this a crude blueprint, a strategy, for the evolution of biological constraints in general?*" [italics added]

In the domain of psychotherapy, the problem would have to be stated in terms of "experiences" turning into "intrapsychic structure". This formulation, of course, immediately calls to mind Piaget's approach (e.g., Piaget, 1967), on which much of the work described in this article is ultimately based.

What are the next steps with respect to refining theoretical models by investigating change processes in therapeutic material? An imminent task is to trace change in a relationship schema over an extended period of time, and in a larger number of relationship schemas which may be enacted in the therapeutic relationship one after the other, or even partly in parallel. In addition, models in which the process of change in a client's relationship schemas is specified for a particular phase of this process (see Schneider, 1991a, for a preliminary model of the initial phase) may be used, or in which the model of change in a relationship schema is enlarged to encompass another kind of fluctuation, namely "perturbations", i.e., aspects of the client's experience that have been discarded or neglected and that make themselves felt fleetingly to the client. The basic idea of the model of the negative class (Schneider, 1985, 1989) is that it takes the amplification of fluctuations of both kinds (i.e., of "new" feelings that the client is not used to experiencing, on which the above analysis has concentrated, as well as of perturbations) to bring about change in a relationship schema. In theoretical terms, the interplay of the amplification of new feelings and perturbations is thought to enlarge the basin of the attractor corresponding to the developing relationship schema (Abraham & Shaw, 1983, p. 45; L. Ciompi, private communication, October 2, 1990). In experience-near terms, this means that the amplification of both new feelings and perturbations leads to situations in which the client comes to experience – however fleetingly – the relationship to the therapist in the manner of the newly developing relationship schema and at the same time becomes aware of an aspect of the old pattern (e.g., by being reminded of an episode in which he or she had a "negative" experience with a significant other). Thus, two "threads" of change may be related to each other, corresponding to the validation of the client's newly emerging feelings and the taking up of hitherto neglected aspects of the client's experience, respectively (cf. Schneider, 1991c).

This sketch of the model of the negative class remains rather schematic (see Schneider, 1989, for a fuller elaboration). However, the intention has been to point out that inspiration may be drawn from theories of self-organizing processes in order to obtain a pattern to be used as a heuristic for identifying passages in therapeutic material thought to be relevant for change (cf. Rice & Greenberg, 1984; Safran et al., 1988). The theoretical model may then be refined by analysing these passages in a detailed manner, using the frame procedure (Schneider & Wüthrich, 1991).

Theories of self-organizing processes may turn out to be pertinent to a new understanding of change in psychotherapy for two major reasons. First, they represent a theoretical framework which allows the emergence of something new to be conceptualized in scientific terms. Secondly, theories of self-organizing processes keep at our disposal a large range of concepts which make amenable to research the question as to how change in psychotherapy takes place in individual steps.

References

ABRAHAM, R.H. & SHAW, C.D. 1983. Dynamics – the geometry of behavior. Part 1: Periodic behavior. Santa Cruz, Ca.: Aerial Press.

BACAL, H.A. (1988). Reflections on "optimum frustration". In: Goldberg, A. (Ed.): Learning from Kohut. Progress in self psychology, Vol. 4. Hillsdale, N.J.: The Analytic Press, 127-131.

BRIGGS, J. & PEAT, F.D. 1989. Turbulent mirror. An illustrated guide to chaos theory and the science of wholeness. New York: Harper & Row.

EIGEN, M. 1987. Stufen zum Leben. Die frühe Evolution im Visier der Molekularbiologie. München: Piper.

GOODWIN, B. & SAUNDERS, P. (Eds.) 1989. Theoretical biology. Epigenetic and evolutionary order from complex systems. Edinburgh: Edinburgh University Press.

GRAWE, K. 1988. Heuristische Psychotherapie. Eine schematheoretisch fundierte Konzeption des Psychotherapieprozesses. Integrative Therapie, **4**, 309-324.

JACOBS, T.J. 1990. The corrective emotional experience – Its place in current technique. Psychoanalytic Inquiry, **10**, 433-454.

KUGLER, P.N. & SHAW, R.E. 1990. Symmetry and symmetry-breaking in thermodynamic and epistemic engines: A coupling of first and second laws. In: Haken, H. & Stadler, M. (Eds.): Synergetics of cognition. Berlin: Springer, 296-331.

KUGLER, P.N. & TURVEY, M.T. 1987. Information, natural law, and the self-assembly of rhythmic movement. Hillsdale, N.J.: Erlbaum.

MINSKY, M. 1985. The society of mind. New York: Simon & Schuster.

PIAGET, J. 1967. Biologie et connaissance. Essai sur les relations entre les régulations organiques et les processus cognitifs. Paris: Gallimard.

PIAGET, J. 1974. Adaptation vitale et psychologie de l'intelligence. Sélection organique et phénocopie. Paris: Hermann.

PIAGET, J. 1975. Phenocopy in biology and the psychological development of knowledge. In: Gruber, H.E. & Vonèche, J.J. (Eds.): The essential Piaget. London: Routledge and Kegan Paul, 1977, 803-813.

PRIGOGINE, I. 1976. Order through fluctuation: Self-organization and social system. In: Jantsch, E. & Waddington, C.H. (Eds.): Evolution · and consciousness. Human systems in transition. Reading, Ma.: Addison-Wesley, 93-133.

PRIGOGINE, I. & STENGERS, I. 1984. Order out of chaos. Man's new dialogue with nature. New York: Bantam.

RICE, L.N. & GREENBERG, L.S. (Eds.). 1984. Patterns of change. Intensive analysis of psychotherapy process. New York: Guilford.

RUMELHART, D.E. 1989. The architecture of mind: A connectionist approach. In: Posner, M.I. (Ed.): Foundations of cognitive science. Cambridge, Ma.: MIT Press, 133-159.

RUMELHART, D.E., SMOLENSKY, P., McCLELLAND, J.L. & HINTON, G.E. 1986. Schemata and sequential thought processes in PDP models. In: McClelland, J.L., Rumelhart, D.E. and the PDP Research Group: Parallel distributed processing. Explorations in the microstructure of cognition. Vol. 2: Psychological and biological models. Cambridge, Ma.: MIT Press, 7-57.

SAFRAN, D. 1990a. Towards a refinement of cognitive therapy in light of interpersonal theory: I. Theory. Clinical Psychology Review, **10**, 87-105.

SAFRAN, D. 1990b. Towards a refinement of cognitive therapy in light of interpersonal theory: II. Practice. Clinical Psychology Review, **10**, 107-121.

SAFRAN, J.D., GREENBERG, L.S. & RICE, L.N. 1988. Integrating psychotherapy research and practice: Modeling the change process. Psychotherapy, **25**, 1-17.

SCHNEIDER, H. 1983. Auf dem Weg zu einem neuen Verständnis des psychotherapeutischen Prozesses. Bern: Huber.

SCHNEIDER, H. 1985. Modèles piagétiens comme point de départ d'une théorie du processus psychothérapeutique. Schweizerische Zeitschrift für Psychologie und ihre Anwendungen, **44**, 161-171.

SCHNEIDER, H. 1986. Piagets Begriff der majorisierenden Äquilibration: Kognitive Entwicklung als selbstorganisierender Prozess. In: Stolz, F. (Ed.): Gleichgewichts- und Ungleichgewichtskonzepte in der Wissenschaft. Zürcher Hochschulforum, Band 7. Zürich: Verlag der Fachvereine, 57-67.

SCHNEIDER, H. 1989. Toward a more detailed understanding of self-organizing processes in psychotherapy. In: Goudsmit, A.L. (Ed.): Self-organization in psychotherapy. Berlin: Springer, 72-99.

SCHNEIDER, H. 1990. A tool for the tracing of change sequences in psychotherapy (TSP): Guidelines for investigation into sequences relevant to change in psychotherapy (version of the 28th May 1990). Unpublished manuscript.

SCHNEIDER, H. 1991a. How does change start in the client? A preliminary model based on theories of self-organizing processes. Paper presented at the 21st Annual Meeting of the Society for Psychotherapy Research. Wintergreen, Virginia, USA, June 26-30, 1990,

SCHNEIDER, H. 1991b. Psychoanalyse und Cognitive Science. In: Mertens, W. (Ed.): Schlüsselbegriffe der Psychoanalyse. Weinheim: Verlag Internationale Psychoanalyse.

SCHNEIDER, H. 1991c. 'Order Through Fluctuations' as a heuristic for identifying "threads" of change in psychotherapy. Paper presented at the the 22nd Annual Meeting of the Society for Psychotherapy Research. Lyon, France, July 2-6, 1991.

SCHNEIDER, H. & WÜTHRICH, U. 1991. A model based on theories of self-organizing processes as a tool for the investigation of change in psychotherapy. In: Leuzinger-Bohleber, M., Schneider, H. & Pfeifer, R. (Eds.): "Two butterflies on my head". Psychoanalysis in the interdisciplinary scientific dialogue. Berlin: Springer.

SKARDA, C.A. & FREEMAN, W.J. 1987. How brains make chaos in order to make sense of the world. Behavioral and Brain Sciences, **10**, 161-195.

STADLER, M. & KRUSE, P. 1990. The self-organization perspective in cognition research: Historical remarks and new experimental approaches. In: Haken, H. & Stadler, M. (Eds.): Synergetics of cognition. Berlin: Springer, 32-52.

STERN, D.N. 1985. The interpersonal world of the infant. New York: Basic Books.

STILLINGS, N.A., FEINSTEIN, M.H., GARFIELD, J.L., RISSLAND, E.L., ROSENBAUM, D.A., WIESLER, S.E. & BAKER-WARD, L. 1987. Cognitive science. An introduction. Cambridge, Ma.: MIT Press.

VANLEHN, K., BROWN, J.S. & GREENO, J. 1984. Competitive argumentation in computational theories of cognition. In: Kintsch, W., Miller, J.R. & Polson,

P.G. (Eds.): Method and tactics in cognitive science. Hillsdale, N.J.: Erlbaum, 235-262.

VARELA, F.J. 1988. Cognitive science. A cartography of current ideas. (Kognitionswissenschaft – Kognitionstechnik). Frankfurt am Main: Suhrkamp.

WÜTHRICH, U. 1987. Über Veränderungsprozesse in der Psychotherapie. Eine Konkretisierung des Schema-Ansatzes. Unpublished doctoral dissertation, University of Berne, Switzerland.

YATES, F.E. (Ed.). 1987. Self-organizing systems. The emergence of order. New York: Plenum.

System-Theoretical Prerequisites Concerning Paradoxical Intervention

Arno Schöppe and Ewald Johannes Brunner

Abstract. In order to understand paradoxical interventions in clinical practise the therapeutical communication is analyzed by means of the systems theory of N. Luhmann. It is assumed that systems of communication and systems of consciousness have to be strictly distinguished from one another. The client's system is described as a self-constituting system; it acts, operationally speaking, as a closed unit. This amounts to postulating that therapy, which must be thought of implicitly as a controlling intervention in another system, is not possible. The only possibility of contact made with the "environment" of a system can best be indirectly described with the concept of "structural coupling". According to these prerequisites a (paradoxical) intervention only appears to be possible if the autopoiesis of the client's consciousness provides a high level of "reception-readiness" in the sense of achieving a structural coupling with the therapeutical communication system.

1. Introduction

In psychotherapeutic counselling, at least in systems-oriented work, paradoxical intervention has long been standard practice. This technique of intervention has a long tradition and has left a correspondingly broad trail of discussion[1] which has mainly been restricted to publications in the field of psychology. Such an apparently successful technique of behavior change naturally arouses interest and raises hopes in the neighboring disciplines of the social sciences. After all a successful therapy can be seen as evidence that the manipulation[2] of a person (client) by another (therapist) has borne fruit. The means used in this manipulation, paradoxical intervention, are moreover obvious and open to research. A modal point seems to have been discovered here which elucidates an access to a person's psyche capable of causing change by showing how changes can so to speak be forced.

It is all the more surprising to note how heterogeneous the academic assessment of this effect has been. It seems as if research is still confronted with the proverbial black box of which one admittedly knows what is put in and what

[1] Representative works are Weeks & L'Abate (1982) or Seltzer (1986).

[2] Intended in the sense of "appropriate, skilled handling" without any derogatory overtones.

Springer Series in Synergetics, Vol. 58   Self-Organization and Clinical Psychology
Editors: W. Tschacher, G. Schiepek, and E.J. Brunner   © Springer-Verlag Berlin Heidelberg 1992

comes out transformed on the other side but which does not reveal the secret of the transformation.

The following work inspired by systems theory is a venture to illustrate why it is that paradoxical intervention cannot be explained. In a countermove, however, the explanation deemed impossible is to be attempted after all. This paradoxical manner of working seems in its way appropriate to the subject matter. The resolution of this contradiction lies in the method of the approach. The same will of necessity have to be conducted metatheoretically in so far as the bird's eye perspective comprises not only the object but beyond that reflects its own efforts to observe itself. In this respect a higher level of abstraction can be expected in order to do justice to the method. The terrain can in this way be viewed and the intractability of the object to traditional attempts at explanation can be located. On the other hand the researcher is enabled in this way to spy out positive chances for explanation in the neighborhood of the resistance.

2. The Object

Traditional attempts at explaining paradoxical intervention start by regarding people as *individuals* producing acts of communication with which they can make themselves understood by others and with the help of which, however, they also intervene in the therapeutic sense. Attempts to assess the person as a *communicating* system can also often be found in the work of interpreters influenced by systems theory. This way of seeing persons leads however to a causal dilemma. So long as the exercise of influence via communication appears successful, the justification of theoretical interpretation of this act of persuasion is corroborated. The theoretically predicted normal case thus corroborates itself. However, as soon as success is denied (the popularity of supervision and in-service counselling sessions confirms the everyday character of this snag) considerable effort has to be undertaken to explain the failure while remaining within the causal process. In most cases it is assumed that *perturbations* must have crept into the process unnoticed. As a result the chosen strategy for theoretical explanation does not have to be considered since the villain is hiding in the noise interference of reception or in the faulty coding or de-coding of the signal[3] . The fault is not ascribed to the explanation strategy but to irregular and therefore improper boundary conditions or unprofessional observation or even to the dilletante way in which the causal strategy was carried out. The disadvantage of this theoretical procedure lies in the fact that failure cannot be explained with the means available to this theory and that its occurrence is a source of surprise.

For this reason in the following a different weight is to be given to the factor of *perturbation* and perturbation is to be seen as part of the programme, as the

[3] Cf. Shannon & Weaver (1949).

normal case[4]. Not only this, but the intention of this article is, as mentioned, to search first of all for the reasons why influencing an individual by communication appears impossible even if only for the purpose of forming a provocative hypothesis which draws attention directly to the boundary area of the "normal" therapeutic effort.

This intention can most easily be achieved when it is postulated that communication is an area totally distinct from the "individual"[5]. The "object" under discussion may well be sufficiently metatheoretically defined with the concept "system" and suggest an equality between "individual" as a system and "communication" as a system. An equality of this kind can, it must be said, only apply formally and never materially, otherwise the two areas would only be arbitrarily distinguished one from the other and the possibility of contact between the elements of the systems which had been labelled as distinct would suggest itself. This, however, is to be avoided. This statement offers the operational advantage that attention can be concentrated on the place at which communication and individual potentially meet. This of course applies not only to the traditional linking point between the communication emitted by a transmitter and the receiver, but already to the transmitter himself and his message. In what follows it will be useful to leave behind the traditional transmitter–receiver nomenclature which is too clearly linked to a specific theory in order to take the new facts into account. With this we venture the leap into the sociologically influenced design of systems theory according to Luhmann.

[4] Not only in the psychotherapeutic tradition but above all in social psychology we realize that the influence of persons on other persons represents an unquestioned postulate: either specific settings are studied in which the influence can be optimized (for example group pressure), or methods of direct influencing are discussed (for instance, the concept of persuasion). The picture of human beings which is handed on here maintains the fiction of the mastering of human individuals by other individuals. Where (social-) psychological theory cannot support the postulate of reciprocal exercise of influence (e.g. through lack of empirical verification), interfering variables are made responsible, a manner of explanation which goes well with the method just described.

[5] The postulate of regarding communication and individuals as separate entities may at first seem strange since in colloquial language communication is never thought of as being independent of persons The simplest reflection, however, will show that communication can occur independently of persons (for instance in the form of written messages): here the communication is emitted by an individual but acquires its own character and, as the case may be, its own dynamics and is to be treated as a self-organized phenomenon. Compare with this the independence of the science of "communications", which itself deals with nothing but the "media" via which communications are transported: here, too, pieces of communication are regarded as independent units.

3. Operational Closure

If one assumes that systems, in this case communication and consciousness[6], have to be strictly distinguished from one another, one is forced to name a criterion for the distinction between the two. This criterion is determined by virtue of the way the elements of a system are determined from outside, that is to say for example by another system, but by the very system within which they occur. With this the manner of operating which can achieve such a result becomes the focus of our interest, i.e. elements only become elements of a system by virtue of the fact that they are called on by that same system, thereby being accepted as belonging. This happens in its simplest form when the elements constituting a system are produced by that same system. By producing, however, we are not referring to any material production but rather to a qualitative one. Here producing (cf. Luhmann, 1984, p. 40) in contrast to the usual sense of the word merely means *choosing*. This process is comparable to the process of choosing players in children's games with the help of counting-out rhymes. The rhyme itself represents the coding instrument which is in a position to select which elements are to take part in the game and in which manner. The game itself provides for there to be elements that belong and those that do not. Thus the players are a precondition for the game but only because the game or rather the rules of the game dictate this. If play-elements threaten to drop out, the game caters for the regeneration of these elements or for the recasting of the rules. Thus, this self-maintaining effect ensures that the game shall continue; in the opposite case the game does not exist. This continuous reproduction of the components of the system by the system itself is called, as is well known, *autopoiesis* (cf. Luhmann, 1984, p. 60 ff).

The criterion for distinguishing whether or not an element belongs to a system is therefore of a formal nature. It is represented by the operation which determines the elements. In this respect there can be no firm borders to systems since these might lose their relevance in the next act of determining. Borders are defined only on an occasional, partial and temporary basis. One can only speak of a border from the point of view of this operation and then only in a formal sense in that defined and undefined (but potentially definable) are distinguished from one another in the sense of difference of complexity at an arbitrarily defined position placed between previously organized (system-) complexity and unorganized (environment-) complexity (cf. Luhmann, 1984, p. 249 ff)[7].

[6] The concept "individual" is to be replaced by the word "consciousness" from this point onwards to stress that it refers to an independent system type.

[7] In the following we shall be returning to the opportunities which a concept "border" opens up.

Let us take a closer look at the aforementioned operation. What strikes us is that it acts according to an inbuilt *code* which undertakes *inclusions* and thereby automatically *exclusions*. This effect can most easily be understood in the example of the selective(!) attention of the perceptive apparatus[8]. In order to recognize patterns the consciousness distinguishes between foreground and background[9]. In a pre-stage of recognition they are not perceived as different. Only when what is defined (that is to say given a border; finis = border) as foreground is named can the act of recognition be completed. In other words: a difference is produced in which belonging is distinguished from not belonging and is defined as such (cf. Spencer Brown, 1979).

This description, however, is hardly sufficient to do justice to consciousness. The naming on the basis of a distinction is not achieved without certain prerequisites. It is based, rather, on a *time-dependent* status quo from which the operation starts. By this is meant the site which in the case of the consciousness is marked materially by the concept "experience". In formal terms one could speak of *"structure"*[10]. A structure, it must be pointed out, has to be conceived of as something that can be changed, otherwise there would be no sense in speaking of a *time-dependent* (historical) status quo. A structure carries out selection but it must also be elastic (Maturana & Varela, 1987, p. 182) enough in face of possible failures in the selections made. The game has an in-built drive, to return to the previous example, to keep itself going even if circumstances have changed in the meantime. It must be able to react with a self-restructuring if it does not want to break down. A self-organized (autopoetic) system must be able to find the "courage" to do without components (of structure) which had up till then successfully handled the ascription (selectivity) of what was neccessary and useful for the continuance of the system. It has to be flexible enough to reorganize structure. It follows that structure is subject to a (historical) *process of change* which has the purpose of testing codes for their chances of success in continuing the operation and of applying them and of revising them. In this sense one can

[8] Processes of perception are – from a psychological viewpoint – always processes of selection. Since potentially myriads of units of information from his surroundings are constantly impinging on the human individual, and he can never hope to take in, let alone digest this quantity, the individual *must* select at all moments and choose from the superabundance.

[9] Cf. von Weizsäcker (1973).

[10] In place of the concept "structure" the word "schema" could also be used, as introduced into psychology above all by Jean Piaget.

justifiably speak of an evolutionary *adaptation* to the environment[11]. It must be pointed out, however, that the concept of adaptation should not be overinterpreted as if there were an impulse emanating from the environment to controll the system's selections. The case is rather that the environment can only in a diffuse way provide everything to which the system can have recourse in its selecting. This reaching out to select is based however solely on the active scope of the system's code. This in turn is based solely on the system. In this way the system works in *contact with itself*. It acts, operationally speaking, as a closed unit. It reacts only to its own inner states.

With this we have described a system as a self-constituting order which has no rule-bound contact to the outer world. Everything that a system does is done for its own sake. Strictly speaking this amounts to postulating that therapy, which must be thought of implicitly as a controlling intervention into another system, is not possible. Worse still, even observation from outside, the prerequisite for an evaluation of symptoms (on the basis of a case history), seems pointless because the operations which the observing system (therapist) carries out take place entirely according to the structures of the self same observer[12]. With this approach it becomes possible to describe the phenomenon of misunderstanding and the frequently quoted resistance to therapy, but at the same time (positive) therapeutic contact is unlikely. For this reason the remainder of this work will be devoted to the search for supplementary conditions which may yet resolve what is apparently impossible: the elucidation of intersystemic contact.

4. The Environment of the System

It seems natural to assume that in the surroundings of our fictitious system there should be further systems. Such an assumption forbids the entry of solipsistic

[11] Codes are tested for their chances of success in continuing their own operation and they are processed correspondingly. This corresponds to the idea of the equilibration process in Piaget's cognitive psychology in which the cognitive schemata are examined as regards the adaption to the environment and in the process are retained or changed. According to this view the organism does not behave passively in the acquisition of experience but changes either the contents of the environment by imprinting its own structure on them (Piaget calls this incorporation of objects "assimilation") or changes its schemata by restructuring them according to the "objective" requirements (this aspect Piaget calls "accommodation"). As Piaget notes the individual is capable of various cognitive accommodations, but only within certain limits which are set by the necessity of conserving the corresponding assimilation structures (Piaget, 1970, p. 706 ff.). Assimilation guarantees the continuity of the structures and the integration of new elements.

[12] Think of the satires which mock the differentiated clinical diagnoses that grew out of a single gesture on the part of the client.

contraction. But how can a system know of the presence of another if it acts on the basis of self-constitution? To answer this question a small mental experiment using previously described means may be a help. The coding of the "attention" of the system operates as shown with the differentiation between what belongs and what does not. Instead of this being a distinctive event produced by the system, the aforementioned difference could register itself, that is to say as a whole. In doing that, an attempt would be made to mark off the whole of the self as against something other. As against what? The concept of self-reproduction comprises the notion that only what the operation itself produces will be of any relevance; but this is precisely what is to be registered by this differentiating operation. The result is a paradoxical exclusion by the self of its own operational basic principle. This operation is blocked de facto; as a by-product the consciousness of self-blocking emerges. We can no longer treat this phenomenon as an example of exclusion, it is at best an "excluded inclusion" or an "including exclusion". The dilemma can only be resolved if one pretends that the self could be seen as a distinct entity in the traditional manner from outside, virtually from the view-point of a supposed alien system. An alien system of this kind is admittedly not proven by this play of thought, but an "idea"[13] of it, in the Kantian manner, is born to which one can link one's further codings and to which one can *sensitize* them. A contact with the outside world is so to speak simulated and, moreover, with the system's own means.

This argumentation "explains" a further "impossibility" which was seen only as a disturbing factor or not noticed at all by users of the traditional approach. If contacts with the outside world are self-reproductive occurrences then a whole range of apparently non-empirical contacts, whether of transcendental or (in some other way) fictitious kind can be explained. The world is accordingly above all a *constructed* world in which the criterion for closeness to reality serves purely as a code with which autopoietic events are sounded out.

The line of argument taken up to this point may represent a consistent way of specifying the problem of self-reproduction and its consequences, but it does not exclude in any way a really existing world, it merely formulates a certain scepticism about the possibilities of recognising a real world of this kind. In the following an environment formed in this manner is to be more closely considered. For this we do not need to give up our point of observation, that is to say the present systems-oriented view. Let us instead imagine that a supposed (see above) contact to an outer world could take place. In what way is this possible to imagine?

Let us posit that the system of consciousness in our example uses a qualitatively new coding to ensure its own structure formation. This new code is called, as already hinted, "belonging to its own system" versus "not belonging to its own system". All occurrences are translated by this means into structure-changing impulses, which can be registered by this code. In this way the code has

[13] What is meant is a transcendental idea which stems not from the empirical but from the intelligible sphere.

to select the occurrences according to the binary scheme which is predetermined by the sensors, which means that the ascription of the label "non-systemic" will be inevitable. This will happen with the same inevitability with which increasingly nowadays any occurrences are subjected to the "man/woman" code. As can be seen, even subject matter which seemed far removed is made to "fit" and, by the way, with not insignificant consequences.

It will certainly be useful at this point to introduce a number of new concepts, on the one hand to characterize the system of consciousness more closely and in a psycho-logical manner, and, on the other hand, to be better able to estimate the point of contact between the system and the environment.

The first reconceptualization concerns the subject of "attention" of a system of consciousness as regards its environment. This is to be symbolized in the pair of concepts *"expectation"* and *"imagination"*. In general terms, expectation would be a form of orientation which scans the contingencies of the environment in reference to itself (cf. Luhmann, 1984, p. 362). Contingency in this context means "the possibility of being different" of a phenomenon. The fact that something can be different will depend upon the preconception which the system has **pre**viously formed of the object. Accordingly, an event is screened by the use of an observation raster or, in other words, a code, and subsequently entered in a manner yet to be more closely determined. It is here that the structure which is being processed, that is to say being perceived as different from the expected event, becomes apparent. In the process we meet the effect that the structural expectation either coincides with the event or does not. That is to say the expectation is fulfilled or disappointed. In the case of disappointment the notion of "the possibility of difference" receives confirmation and the orientation adopted has to be either interrupted or changed. This is the crucial starting-point for a learning, socialization or therapy theory.

A further reconceptualization must also be taken into account. The form in which the contingental character of the environment presents itself can only be an undifferentiated *"noise"* never a *perturbation*. The concept of a perturbation is seen here consistently as dependent on the constant reproduction of the system, that is to say on the operations of the system. A perturbation must consequently be interpreted as a self-perturbation. The way this occurs is that the result of differentiation by means of the use of a code, that is an "insight", disappoints the expectation represented by the system's own structure. As a result the structure is forced to adjust[14]. This can happen in one of two directions. Either the consciousness revises the content of its expectations or the content of its perception. In both cases a false ascription is given to an interpretation of the world created by the system itself with the aim of creating a new relation between the two spheres according to the measurement system of the counterpart. In this

[14] The process of differentiation and the ensuing ascription "belonging to the system" or "not belonging to the system" with its consequences for the possible revision of the content of expectations can also be described with the help of the TOTE-scheme of Miller, Galanter and Pribram.

way the consciousness generates a continuous, unbreaking series of "preconceptions" about the "real world", preconceptions of structure-building potential. A set of such preconceptions can be seen as "the way a person sees the world". Out of these ideas new expectations arise which pre-form openings on the principle of similarity. These in turn confirm or stabilize in feedback already existing notions, thus keeping them invariant over a certain length of time. This constant comparison of what already exists as a structure and what is expected as a confirmation of structure ensures that the closed mode of operation is always being opened a little again and again. In accordance with the basic evolutionary theme of the restabilization of selections out of a pool of variations of states of consciousness, what emerges is a reasonable flexibility towards the unexpected, i.e. means towards the disappointment of expectations. The consciousness adjusts itself by a process of learning to the demands made on it.

After this re-formimg of concepts, the explanation of contacts with the environment can be approached. The possibility of conscious contact made with the environment can best be described indirectly with the concepts of *"resonance"* and *"structural coupling"*[15].

Resonance in the technical sense means the "sounding together of several bodies". Contained in this description is the key to understanding intersystemic contact. The concept of "resonance" is drawn from physics and comprises on the one hand a source of sound in the environment of a body and in addition the physical qualities of that body, which is to be compared for the sake of elucidation with a system of consciousness. The qualities of a body which enable resonance to take place could, in the case of a psychological system be interpreted as the specific readiness to navigate expectations in regard to the interference of the environment in such a way that the selections of the system produce regularities. The concept of initiating describes two things: firstly, the system's own state of expectation is conditioned, secondly, this conditioning produces a regularity which can lead to stimulation on the same pattern for another resonant body. A tuning in to form a chord of this kind is the model for the emergence of language. Language can be interpreted as an acoustic signal into the blue, but also as heightened attention, that is to say the readiness to select, directed towards the interpretation of potential signals (by means of expectation). This means that via just such heightened attention an increase in the readiness to be disturbed, i.e. perturbability, is possible by reduction in the range of a system's own selection. The basis for perturbability of this kind is the system's own structure, which is to say a re-weighing of the system's own structural preconditions with the expectations which are either fulfilled or disappointed in the course of a coded scanning of environmental "noise".

If the expectations are to be able to stabilize themselves unchanged over a period of time something is needed to keep the range of selections deliberately narrow, which implies undertaking exclusions. *Communication* does just this. It acquires the character of a system by regulating these selections autopoietically and

[15] The concept comes from Maturana (1982).

by filtering out the environmental "noise" of psychological systems of its own accord, thereby attuning itself to resonance. Just how complicated this process has to be will be apparent in the example of two fictitious radio transmitters who know nothing of one another and yet expect to make contact with one another in the medium of electromagnetic waves by means of alternating transmission and reception. We imagine them doing this by repeatedly interpreting the noise (of the ether) experimentally as "sense"-bearing signals and in a countermove sending out this interpreted sense with the help of their own range of instruments. As can easily be imagined this contact could only succeed by virtue of an (evolutionary) chance, in no way could it be forced. If this contact could be forced it would lie within the reach of the system's own operations. Since this is not the case the "structural coupling" must be assumed to be an extra-systemic effect which will always occur when selections are tuned into one another. Structural coupling is not produced, it is either simultaneously existent or not existent.

In the same way communication must be taken as assumed from the viewpoint of a system of consciousness. In the course of its biographical development a system of consciousness learns to adjust itself by structural coupling to the interference of the communication system. It learns as if it were to extracting patterns from the interference, just like a voice medium from the "white noise" of a spatial interval between two radio stations.

5. Paradoxical Intervention

From the above it will be clear that a therapeutic contact which is aimed at controlling is based on a cause and effect interpretation of structural coupling. An interpretation of this kind describes what happens in terms of time as a before and after version of a cause-effect relation. It may well be possible, using the model of simultaneous structural coupling, to explain why inter-systemic contact is not out of the question; but it must be pointed out that this contact, if it is to succeed, has to overcome a double hurdle of autopoetic interpretation. On the one hand it has to have created resonance in the communication system, i.e. it must have become a *topic*, and on the other hand it must be structurally linkable according to the autopoiesis of another participating system of consciousness.

According to these prerequisites an intervention only appears to be possible if the autopoiesis of the client consciousness provides a high level of "reception-readiness" in the sense of achieving a structural coupling with the communication system[16]. Furthermore, the theme which the communication selects must be capable of being linked to the reproductive requirements of the client consciousness. This means that the transition from system of consciousness to social system must show capacity for resonance on both sides. On the part of the psychological system this means that the operations (indications on the basis of distinctions made from the pool of the system's noise) of the social system ensure a ready capacity for linking

---

[16] Cf. Ambühl & Grawe (1988, 1989).

in the sense of the reproduction (of further operations) of the system of consciousness. Distinguishing indication operations of this kind can also be defined in Luhmann's sense as *"observations"* (cf. Luhmann, 1984, p. 596 ff). The capacity for linking then means that one observation must be able to induce another. The differentiation undertaken must therefore contain a sufficient reference-"surplus" to be able to feed back structurally guided preconceptions.

When we apply these observations to paradoxical statements we are faced with an effect which is positively surprising, given the linking requirements of autopoietic reproduction. The reference-tendency to continue operations of the consciousness is seen to be turned in on itself (self-reference). This takes place moreover in the shape of a disappointment of expectation, for the reference back to itself negates what it is referring to although what is referred to is a prerequisite of the operation in process. Once the operation has begun there is no way of extinguishing it and in this way it oscillates between its own possibility, which represents the condition for its impossibility, and the impossibility, which in its turn is seen as the condition for its own possibility. In other words the operation of distinguishing, that is to say observation, is blocked although it is in fact continued and has to look for new horizons for the continued flow of the autopoietic stream before boredom causes the system to collapse. In this way a paradox forces the system into a shortage of time and induces rapid decisions.

Applied to the therapy situation, what the therapist intended almost inevitably takes place. The blocked system will attempt to rid itself of the burden of paradox weighing on its reproduction complex, i.e. to embed it in an interpretation complex which prophylactically avoids the apparently mortal trap, by means of restructuring already structured approaches which had led the system into oscillation. This means, in general terms, in every case a change as compared with the position at the outset and in a narrower sense a change of the system context (in the case that the symptom was paradoxed). From now on the world will have to be seen with fresh eyes: the sounding out of the interference having hit on new wave-length areas, these in turn open new perspectives. What these perspectives will be, however, is not determined.

This change in its turn takes place by a process of contact with self. Therapy in this sense takes on the meaning of *self-therapy*. What, then, is the purpose of therapeutic communication? From what has been said it will be clear that the way the client system allowed itself to be inveigled into self-disturbance was made possible by a co-evolutive effect. Both the system of consciousness and the social system have drastically reduced the contingencies of their own operations in order to decode the interference correspondingly. This was a process of rule-bound self-inflicted limitations. Those who wish to take part in the play of resonance must bow to these constraints. Such a system will find itself enslaved by these constraints (rules). No wonder, therefore, if these rules lead the system's operations very rapidly to the limits of what is permitted by the state of being bound by rules. This can easily be comprehended in the example of the paradoxes. The intention of taking part in communication compels the system to keep to the rules given by this system; these are syntactical and semantic limitations which move along paths laid down by logic. Paths of this nature can cross immediately when the modes of self-reference and of negation come into play simultaneously. In this case a topic can contradict itself for the very reason that it refers to itself. There can be no question of such a self-reference with simultaneous negation not being allowable, as the strategies of de-paradoxifying, see the case of the theory of types (Whitehead & Russell, 1913). A paradox remains a paradox. De-paradoxifying is an operation which applies to a paradox and for just this reason represents a totally different operation qualitatively. The smuggling of a paradox into a social system of communication will, therefore, only succeed when one fulfills the condition neccessary for a structural coupling. There can admittedly be no guarantee of success here any more than with the retranslation of the social interference back into the operating of the system of consciousness. To this extent therapy itself is a paradoxical affair since it demands that its own operations be closed as a social system[17] for processing communication in order to be able to open itself for inter-systemic flow.

References

AMBÜHL, H. & GRAWE, K. 1988. Die Wirkungen von Psychotherapien als Ergebnis der Wechselwirkungen zwischen therapeutischem Angebot und Aufnahmebereitschaft der Klient/inn/en. Zeitschrift für Klinische Psychologie, Psychopathologie & Psychotherapie, 36, 308-327.

AMBÜHL, H. & GRAWE, K. 1989. Psychotherapeutisches Handeln als Verwirk-lichung therapeutischer Heuristiken. Psychotherapie & medizinische Psychologie, 39, 1-10.

LUDEWIG, K. 1988. Problem – "Bindeglied" klinischer Systeme. Grundzüge eines systemischen Verständnisses psychosozialer und klinischer Probleme. In: Reiter, L., Brunner, E.J. & Reiter-Theil, S. (Eds.) Von der Familientherapie zur systemischen Perspektive. Berlin: Springer, 231-249.

LUHMANN, N. 1984. Soziale Systeme. Grundriß einer allgemeinen Theorie. Frankfurt am Main: Suhrkamp.

MATURANA, H.R. 1982. Erkennen: Die Organisation und Verkörperung von Wirklichkeit. Braunschweig: Vieweg.

MATURANA, H.R. & VARELA, F.J. 1987. Der Baum der Erkenntnis. Die biologischen Wurzeln des menschlichen Erkennens. Bern: Scherz.

PIAGET, J. 1970. Piaget's theory. In: Mussen, P.H. (Ed.) Manual of child psychology. Vol. I. New York: Wiley, 703-732.

SELTZER, L.F. 1986. Paradoxical strategies. A comprehensive overview and guidebook. New York: Wiley.

SHANNON, C.E. & WEAVER, W. 1949. Mathematical theory of communication. Urbana, Ill.: University of Illinois.

SPENCER-BROWN, G. 1979. Laws of form. New York: Dutton.

[17] In this context Ludewig (1988) speaks of problem-processing systems.

WEEKS, G.R. & L'ABATE, L. 1982. Paradoxical psychotherapy. Theory and practice with individuals, couples and families. New York: Brunner/Mazel.

WEIZSÄCKER, V. v. 1973. Der Gestaltkreis. Theorie der Einheit von Wahrnehmen und Bewegen. Frankfurt am Main: Thieme.

WHITEHEAD, A.N. & RUSSELL, B. 1913. Principia mathematica. Cambridge: University Press.

GOLEM[1] – Two Adaptive Systems Communicate

Hansjörg Znoj

With 5 Figures

The main topic of this contribution is the interaction of theoretical concepts of psychotherapy change processes, basic properties of the GOLEM - the name stands for goal-oriented learning model - and their empirical investigation. Starting with psychotherapeutic schema theory a connectionist model is proposed to meet the most important requirements for a successful psychotherapeutic process. It is able to represent environment, to act and to communicate. The second part of this article deals mainly with the empirical analysis of the concepts obtained from the model and with the verification of conclusions drawn from the model. In this model a systematic variation of input signals suffices to activate those signal representations trapped in local minima (clustered-out areas). This corresponds to the assumption in real therapy that variations in how themes as global representations of interactional situations are addressed suffice to bring about change.

1. The Bernese Schema-Analytical Therapy Approach

Presented here is the preliminary result of an intensive study of therapies carried out within the framework of the "schema theory" (Grawe, 1991, 1986).

The above mentioned project dealt in the main with how and to what extent the understanding and outcome of therapy processes can be improved by the therapist's optimally intensive care. Care in this sense means the following: supervision by a group of observers, both plan and schema analyses, and initial as well as accompanying interviews (after every 10 sessions) with the clients with the therapist being informed of the results. The following qualitative instruments are used:

- Extensions: Extensions are an observer's interpretative notes on the verbal and non-verbal therapy process. They consist of periodic video-recorded résumés of therapy sessions.
- Plan structures: Plan structures (Grawe & Caspar, 1984; Caspar, 1989) are hierarchically organised constructs of clients' action plans and behaviour patterns. They are formulated in a strictly functional way, i.e. means-purpose oriented. The plan concept is closely related to the concept formulated by Miller et al. (1960).

---

[1] GOLEM stands for Goal-Oriented Learning Model

Springer Series in Synergetics, Vol. 58   **Self-Organization and Clinical Psychology**
Editors: W. Tschacher, G. Schiepek, and E.J. Brunner   © Springer-Verlag Berlin Heidelberg 1992

- Schema analyses are, by our definition, exact descriptions of clients by means of an extensive questionnaire devised on the lines of the theoretical assumptions made in the schema theory by Grawe (1986). A biographical interview is conducted to elicit additional information which aims to collect as much information on the origins of psychological developments as possible prior to the therapist's starting his/her work proper. This information is also integrated into the schema analysis.

These instruments are to be understood as representing various steps on the road to a comprehensive understanding of the processes involved in genesis and change in clients' psychological constellations.

Schemata are considered to be both a basis and a product of individual-environment interaction. The part of the schema construct most relevant to psychotherapy is its motivational component which energises and controls the psychological processes. The action-controlling component of schemata are called plans (cf. above). Schemata have a self-perpetuating tendency; perceptions (and actions) are continually concerned with reproducing themselves within the schemata activated at any one time. Even though, according to Prigogine's theory (1977), the longer-term result of a dissipative development process cannot be predicted, possible change-inducing interventions may be derived from it: "*The process is brought into motion when information cannot be assimilated to the existing schema. Usually individuals respond initially to this "disturbance" with denial, other forms of defense, or an increased attempt to deal with the situation in terms of their old schema.*" (Grawe, 1991, p. 9) In a second phase, a limited accommodation process tends to take place: the schema is reorganised in such a way that the previously disturbing factor becomes a variation within the new structure. Basically, we distinguish two types of schemata:

- *Self-schemata*: these are schemata displaying an activity aimed at attracting or maintaining positive goals, they thus have a need-fulfilling effect.
- *Negative emotional schemata*: they are aimed at avoiding certain negative emotional states. They have to be deduced from certain avoidance strategies as it is frequently impossible to reach any understanding of the content of these schemata.

For therapists, the following basic intervention strategy emerges: they have to establish a relationship complementary to the clients' positive goals and to enable them to experience perceptions in the sense of these goals as frequently as possible. This conception of the change process is related to Neisser's (1976) perception cycle: Schemata are anticipations, they are the means by which the past influences the future; information already assimilated determines what is to be taken in next. Clients perceive their environment in preventing them from negative emotional states. To break this continuity, therapists have to disturbe this cycle, they have to disturb the patients' negative emotional schemata[2].

An intensive preoccupation with therapies — as a therapist as well as an "observer" — has stimulated me into pursuing some questions of my own, looking for ways in which changes can be induced and determined in concrete therapies.

2. A Goal Oriented Learning Model

In the following, a model is introduced which attempts to redefine Grawe's schema theory in a connectionist way. As a purely functional model, it is meant to uncover the minimal conditions required in order to effect changes by means of psychotherapy. In addition, it also aims at determining process variables.

The preliminary result of my research consists of a model of the psychotherapeutic process called *GOLEM* (Goal Oriented Learning Model)[3]. I would like to briefly present this GOLEM. It is envisaged to devise a progam of GOLEM to enable it, as a dynamic model, to supply data. Independently of programming, however, it can already serve to formulate new questions relating to psychotherapeutic practice. The model is based on a neural network architecture and has the following properties: It is a network for dealing with a finite number of "signals" possessing, in addition, certain "action possibilities". These actions are divided into the basic classes approaching and avoiding or enhancing and suppressing. Signals are combinations of properties. Properties can be recognised by the initial layer. The signals are detected according to "properties"; these properties which are directly linked with the pertinent "action structures" are filed in a first layer. The nodes within the network are mutually linked in an associative fashion; the degree of association

---

[2] It goes without saying that a number of different intervention strategies can be derived from the schema theory; to enumerate all the implications would take up too much space here. Grawe (1986) lists the following points which ought to be taken into heuristic consideration by therapists:
- the enhancement of the "reflecting abstraction", which means becoming aware of important parts of one's own functioning as a favourable condition for accommodative processes;
- the working through of emotions; and
- the increase of competence to create a life which is satisfying in terms of the most important needs of an individual.

[3] At this point, I want to mention that GOLEM wouldn't exist (even as an idea) without my friend Werner Zaugg. He forced me to make my intuition clear. Moreover his ingenious technical understanding helped a lot to create the neural network, which is the prerequisite for the interactional model.

is governed by a forgetting function (parameter). Each connection is reduced by a factor of one with each step in the calculation. This serves mainly to facilitate the learning of new signals and to create new patterns, i.e. it weakens in time. However, there is an increase when the link is confirmed with the help of the simultaneously active nodes involved (hebbian learning rule). It is a multilayer network with a theoretically unlimited ceiling. The increasing "height" in the network is interpreted as "degree of abstraction". Ideally, a "generic term", or rather a signal integration, is formed when two signal properties are activated simultaneously with their interrelationship exceeding a threshold value. The generic term is then constructed (or rather activated) in the next layer above the relation. The network constructs itself by means of this generating of terms. The model has a learning history[4].

The differentiation of the internal signal representations is a result of adapting the original "external signals" simulating the environment in the model.

The passage through the signals takes place in line with Hebb's rule (Hebb, 1949), which stipulates that closely connected signal properties stimulate an accelerated passage through the signals; loosely connected signal properties inhibit it. The weighting of relations (associative power) is mathematically described as the so-called "mexican hat or dog function" (double of gaussian). This permits the function of "clustering out", or simulates actively inhibiting connections[5]. In Fig. 1. a simplified graph of the internal representation shows the weights and relations after several passages through signals. The GOLEM architecture consists of both a "bottom up" and a "top down" network which are, to a large extent, computed independently of each other. The model is also supposed to be capable of modifying the environment by itself! The top down computation models the system's "neediness". We distinguish between afference (activity field) and efference (activation field). An activation field can be seen as a chreode within an "epigenetic landscape" (Waddington, 1942) which is satisfied by incoming signals. An activation field is triggered by a

[4] Here, learning means "unsupervised learning" (as opposed to most neural networks); no learning goals are fixed, success is the criterion for learning: the signal can, after succeeding in an adequate representation of any signal - when the corresponding need situation arises - be reconstructed or the environment be dealt with actively. Needs or rather their fulfilment play an evaluatory role: if the representation of the environment is satisfactory, imbalances inherent in the system can be balanced by consuming the corresponding "signals" from the environment.

[5] An important effect of this function in our model is the simulation of repression.

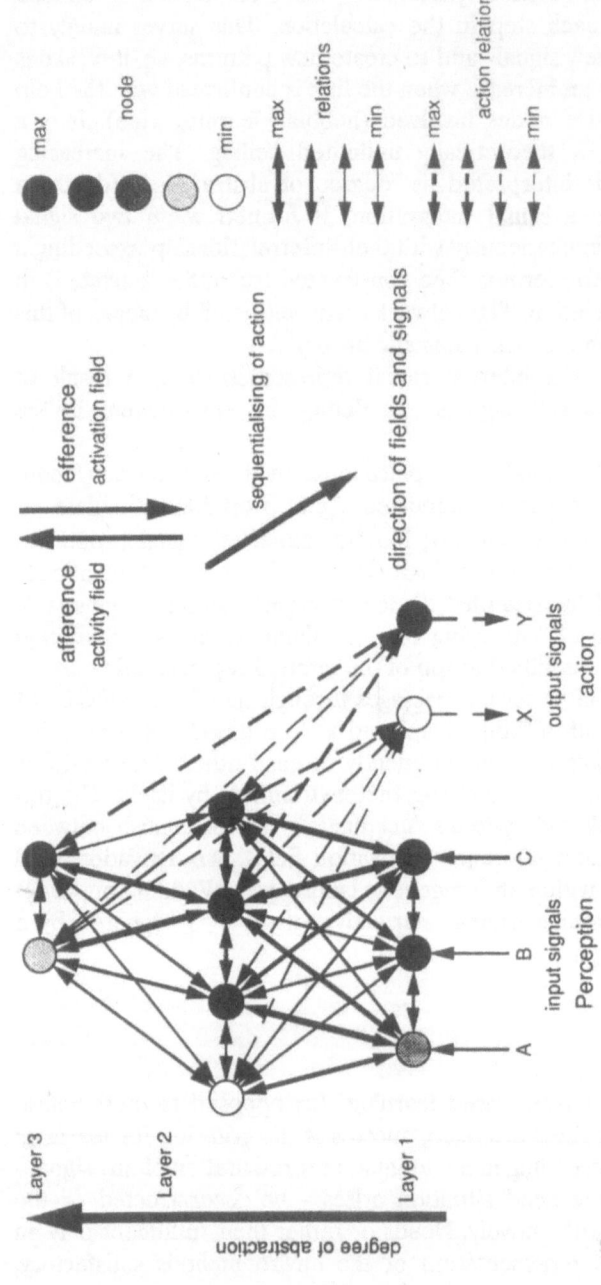

Fig. 1. GOLEM – Layer Model (Zaugg, 1991). Simplified graph of the internal representation after several passages through signals. We have two signal-processings through the same nodes: the activity field (afference) and the activation field (efference). Each signal has its own way to activate the different nodes and their relations. Action relations are the relations triggered by the "needs" of the system, signal–relations are dependent of the system's environment. Together, they form what we call a schema.

"need"[6] spreading over nodes and connections. Depending on the differentiation in this activation field, corresponding actions associated with the nodes are triggered; however, ideally this happens only when the model perceives such signals as consider these actions to be sensible[7]. The perceptions belonging to the actions form a field of activity which covers, or rather fills in, the activating field. Thus, signals are perceived or consumed in a particular (selective) way. This chreode, transposed into the model, corresponds to what we call schema. The model does not act unless the action appears as functioning in accordance with its goals.

Individual objects form property clusters, i.e. the model learns (is supposed to learn) that in the "world" there are objects possessing certain typical properties. Note: objects do not mean physical entities but psychological constructs. It is, however, also possible to enter objects possessing changing or inconsistent properties. These are responsible for the dynamics within the picture as well as the action structures of the model. Such inconsistent or violently contradictory signals (traumata), give rise to so-called "inhibited areas". They are areas within the environment representation which can no longer be accessed due to the links with other nodes being inverted as a result of the mexican hat-function.

The characteristics of GOLEM can be summarized as follows:

In this model, unlike other neural networks, action goals are not fixed right from the start. The goals arise from the degree of differentiation of the representations, on the one hand, and the fortuitously arisen "need", on the other hand. These properties permit the GOLEM to be extended to a model of the process of psychotherapy.

3. Information and Communication

If two models are connected with each other, i.e. a communicative link is established, one model is then defined as therapist (GOpist) and the other as

---

[6] Needs: needs arise by chance (i.e. in a circadian/fluctuative fashion) and/or the influence of leftover chreodes (these are former "needs" not fully dealt with, they can, however, also arise in a circadian way via property clusters). If a junction is activated by a "need", those junctions linked with it by association (representations) will also be activated. In the model a need is artificially produced by activating a cluster of junctions.

[7] Let us remember, in this context, the motivation model by Bischof (1985), i.e. the search for security and the search for variety (curiosity behaviour) and their derivatives.

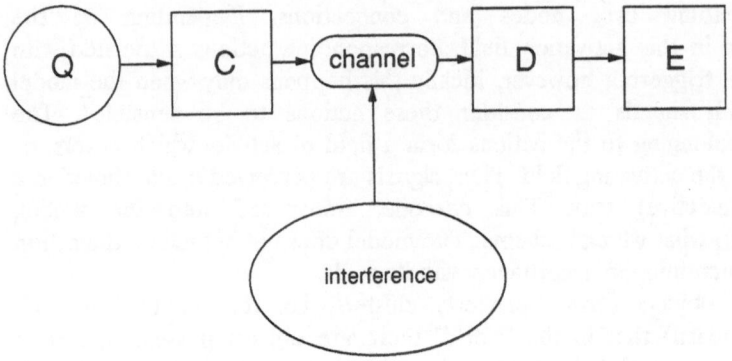

Q = source of information

C = coding agent

D = decoding agent

E = information recipient

Fig. 2. Model of transmission of information. This is a theoretical minimal model of information (here, only a unidirectional information channel is shown), containing the necessary functional groups to which Shannon's theory of information applies – the recipient is passive, and it is generally assumed that the recipient is ideal, i.e. able to optimally evaluate the information received; the recipient is considered infinitely intelligent and possessing an infinite storage capacity. In this case, depending on the source of interference, the quantity of information detectable by the recipient can be derived (for more information, including a bi-directional example, see Marko, 1966).

client (GOlient), this fusion allows for a model representation of the therapeutic process[8].

This model is functional in an abstract sense; it is not a matter of simulating real therapy processes. The value of a model of this kind lies in the heuristic gain it supplies. Normally, the exchange of information between two systems is viewed as shown in Fig. 2. Thanks to the "individual" signal processing of GOLEM no source of disturbance is necessary simulating difficulties of communication; it is assumed that a sufficient number of sources of error arise from the differing representation of the environment; a reasonably adequate model of therapeutic reality is thus created. Fig. 3. shows an illustrated metaphor of this process.

Communication — in the model this includes the entire range of contentual exchange — is understood to correspond to acting (Winograd, 1983). Acting

---

[8] The therapist system may need to be given at least partial access to greater differentiation so that the client system may be subjected to a systematic variation.

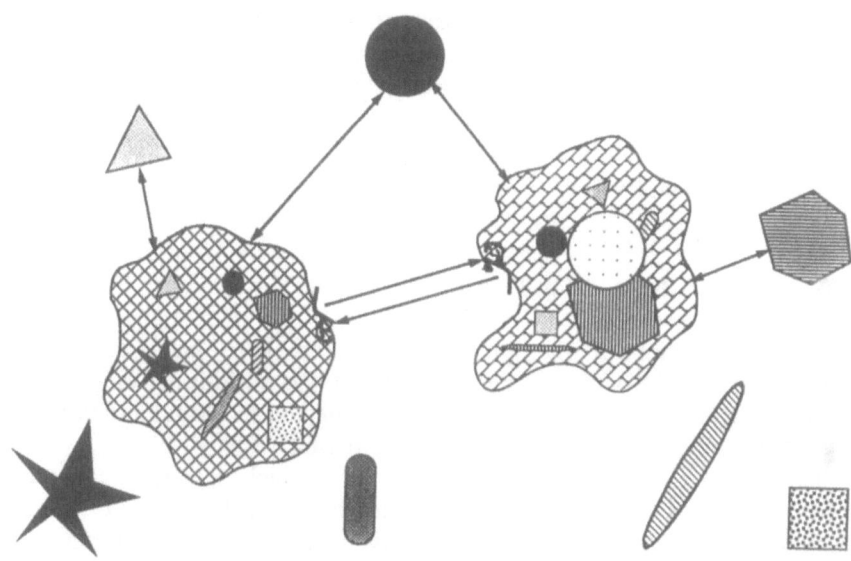

Fig. 3. Two GOLEMS communicate.... Information can also be viewed from a semantic aspect (i.e. the question of how accurately the signs transmitted convey the desired meaning) or from a pragmatic aspect (the question of effectiveness, i.e. how effectively the meaning received influences behavior in a desired way). These latter ways of viewing are psychologically highly relevant and apply especially to the therapeutic situation. The therapeutic setting can be seen as a special case of communication, insofar as the two communicating systems are not really concerned with exchanging direct instructions or information. Rather, the situation is such that one part of the system of interaction suffers from being either not at all or else not adequately able to receive information and/ or being unable to translate the information received from the "world" into such behavior as would be of personal benefit. In this context, benefit means that the individual acts in accordance with his/her needs, i.e. the needs can be satisfied.

by communicating does not, in principle, distinguish itself from other types of acting in the world, with the only difference that the "world" consists of the remarks of the partner. The talking partner thus supplies the signals which are then filed in a network "spread over the top" as speech signal properties. Representations can have the property of being transmittable by talking.

What kind of data are supplied by GOLEM, or rather, how can communication between two systems be made transparent? In the model, actions or objects of actions are to be, in part, represented by "language". Language is considered to be an action (it corresponds to a temporally structured working off of a chreode); it follows that "language production" is a chain of language-labelled nodes which, depending on their force and temporal flow, is expressed as a sequence. An additional threshold ensures that only sufficiently activated nodes can be expressed. We should, therefore, obtain as data a sequential flow of a spatial process. This sequence can thus be seen as a Markov chain with the force of representation of a node being interpreted as corresponding to the probability of an expression. To clarify this a little: In essence, a GOLEM interaction is an exchange of "thought" in which the activity of the system itself causes extreme interference. The sequence obtained must then be interpreted by the partner in the interaction, i.e. by either an artificial system or a human interface[9].

4. Application of the Model in Therapy

The starting point of the premise presented here is a constructive understanding of the world (cf. e.g. Segal, 1986), the world is not objective but is experienced subjectively and the world is reflected in the subject in such a way as the latter's survival-ensuring functions permit.

The model expresses a way of observation which sees the therapeutic process as an attempt to alter the client's "viewpoint" by means of specific interventions. In this context, specific interventions means that utterances by the therapist must concentrate on those areas of representation of clients which either cannot be accessed at all or only with difficulty, because they contain stressful material. In the language of the model, these are termed clustered-out areas. Themes or rather represented segments of environment are seen as global signals. In the last analysis, changes in the structure of representations are effected by constantly varying thematic expressions; this causes "local minima" to be cancelled thus permitting access to clustered-out areas.

5. Themes in Real Therapies: An Empirical Investigation

This section deals with a *possible operationalisation* of the therapeutic model outlined above. Global themes (e.g. partnership) are extracted from the therapeutic process in their function as represented environment segments and investigated from different points of view. The therapeutic situation is seen as a special case within a comprehensive communicative behaviour. The special case results from the theme or the content (semantic aspect) of communication. The content is not a knowledge of the environment to be integrated, but rather the "being within a certain type of environment" of the person in

---

[9] To make this clearer: we have to imagine the "speech act" as a string of activated nodes and their connections. To visualize the content of a single utterance one imagines the represented numerical data (degree of activation, weight of connections) converted into pictures. However, the interpretation of these data is the task of the receiving system and underlies the constraints of an hermeneutic process.

question. This includes just about everything concerning this particular person most, i.e. not only his present but also his past environment insofar as the latter continues to exert a significant influence on his present experience.

The purpose of this therapy model is to alter non-functioning environment representations by combining the poorly functioning apparatus with a functioning (or better functioning) one. The individual environment representations are changed by the "Gopist" emitting signals which enable the "Golient's" system to reorganise itself. A possible starting point for operationalising the concept described here may consist of, first of all, seeking out such themes or environment representations, of then labelling them and waiting to see whether and in what way changes arise in these areas. Themes are considered to be environment segments of the client within which (complex) problems[10] have to be solved. Talking about themes in real therapies is interpreted as an exchange of signals.

The above can be illustrated by the following results of a small-scale investigation (Bischof & Lovey, 1990), concerned with two cases taken from the above-mentioned psychotherapy project.

The empirical material this investigation is based on are the extensions (i.e. the interpreted résumés) of therapy sessions. These extensions were checked over according to themes (professional activity, leisure time, family, partnership, relationships and self-concept) and, in a further step, divided into three main categories, namely "emotional involvement", "focus" and "differentiating behaviour".

These form sheets were the basis for a further step: In order to answer the question as to how a client's means of coping (with themes or "world segments") change in the course of therapy an instrument was developed[11] containing the following six features helping to investigate the coping with themes:

- *Self-reflection*: this is the ability to distance oneself from a set of circumstances and to report on it critically. The degree of self-reflection is dependent on the number of variation possibilities adopted by the clients when talking about problems.
- *Perception of one's own needs*: Here it is a matter of establishing to what degree the clients are aware of their own needs and desires.

[10] In this context problems are defined as tasks to be solved in areas of action or possibilities, e.g. the satisfaction of the need for human closeness or sexual fulfilment. The term "task analysis" used by Greenberg and Rice (1984) describes the idea expressed above pretty accurately.

[11] It is not entirely by chance that the reader will be reminded of the FINBE system by Sachse (1988); the LCCP approach by Toukmanian (1986) also shows similarities to the above mentioned categories. Here, too, it is a question, after all, of investigating the clients' cognitive processes during therapy, i.e. their interaction with the therapist.

- *Emotional involvement*: The emotional involvement is above all expressed in non-verbal and para-verbal behaviour and was already characterised above in the creation of the thematic structure. It is significant to know to what degree the emotional situation of a client really corresponds to that which is claimed by him or her.
- *Interactional conflict ability*: We are interested to know whether clients are able to experience and describe their relationships, a partnership, their family or professional situation in terms of a conflict. Conflicts with other people should be recognised as such and come up for discussion in therapy (during the theme of self-concept the ability to express conflicts corresponds to the "feeling of self-esteem").
- *Self-determination*: This is on the lines of the concepts "locus of control" by Rotter (1966) and the "self-efficacy expectation" by Bandura (1977). Roughly speaking, during therapy, the clients' feelings of influencing their behaviour must be increased in order to motivate them into changing themselves.
- *Resistance*: Resistance means all those forces a client consciously or unconsciously directs against advances in therapy. Caspar and Grawe (1981) maintain that resistance has three sources (contentual, methodical and interactional). Here, only resistance against therapeutic goals, i.e. contentual resistance, is taken into consideration. A rating questionnaire was developed in a next step, measuring these ways of coping.

There is no space here to discuss in detail the contentual (qualitative) changes in the mechanisms of dealing with the different themes of the two therapies; these have been discussed elsewhere (Bischof & Lovey, 1990). It can, however, easily be demonstrated that whenever a certain theme is activated in a session, the different ways of dealing with the particular theme are involved to a differing extent.

6. Some Empirical Data Gained of two Therapies of the Bernese Project

The plots presented in Fig. 4. show how some of the themes that were treated in two different therapies of the Bernese project (Grawe et al., 1991). The modalities under which the themes were rated are the categories mentioned above. Because the implementation of GOLEM into an input/output device would take more resources (in time and power) than the author has even dreamed of, these empirical data have no equivalent to any simulated data.

Furthermore, and this seems significant as well, themes are incorporated into the therapy "in waves". Depending on the importance the client attaches to a particular theme, this is visible to a greater or lesser extent. A superficial explanation would ascribe this simply to either interest or topicality. However, a oscillation might be expressed in such instances which corresponds to a basic pattern of change. It is even possible, that such a rythmic change of topics is a prerequisite for change in therapies. Palm (1988) points out, that neural associative nets have global threshold functions in order to improve learning.

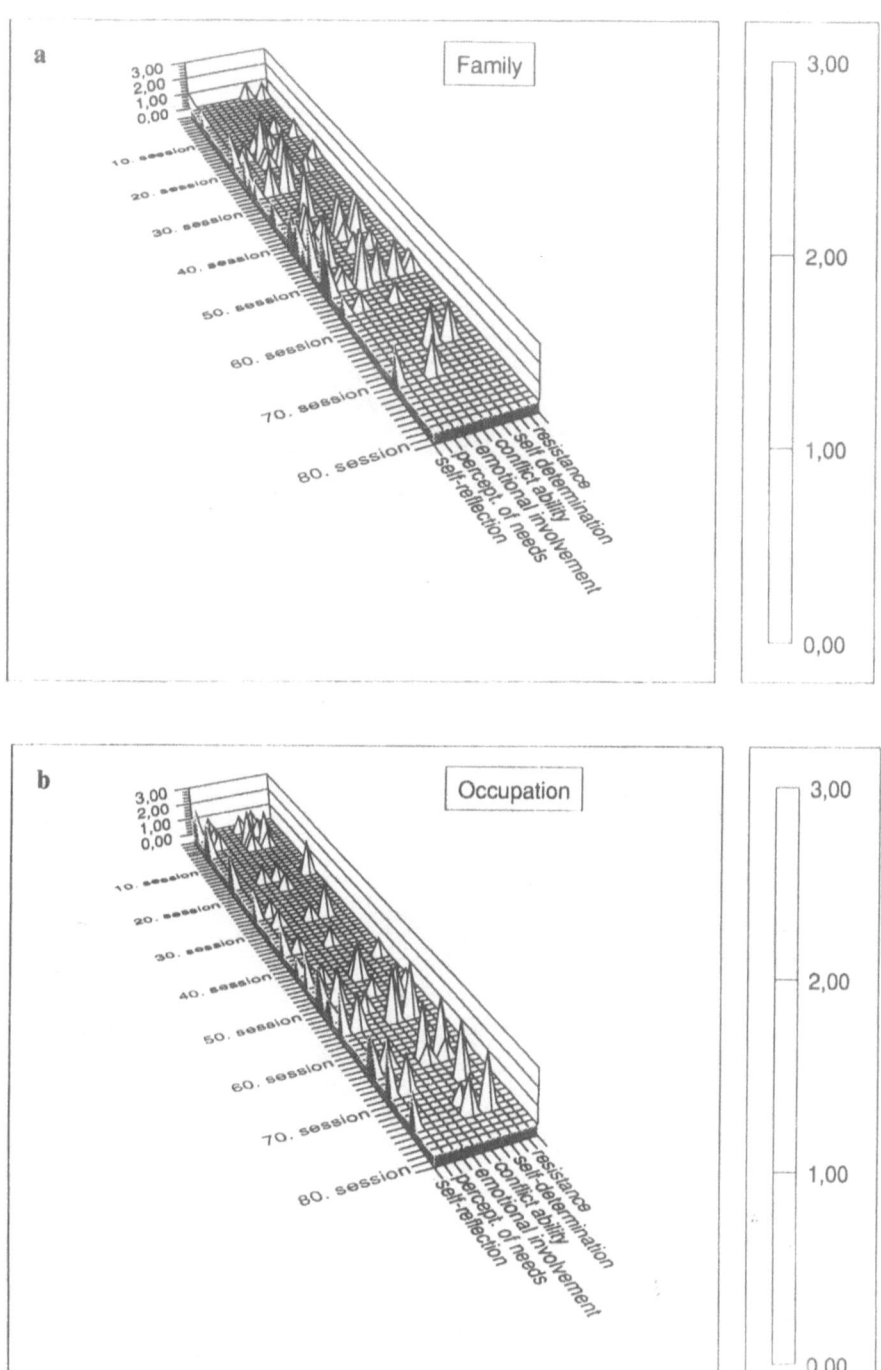

Fig. 4a,b. For caption see p. 308

307

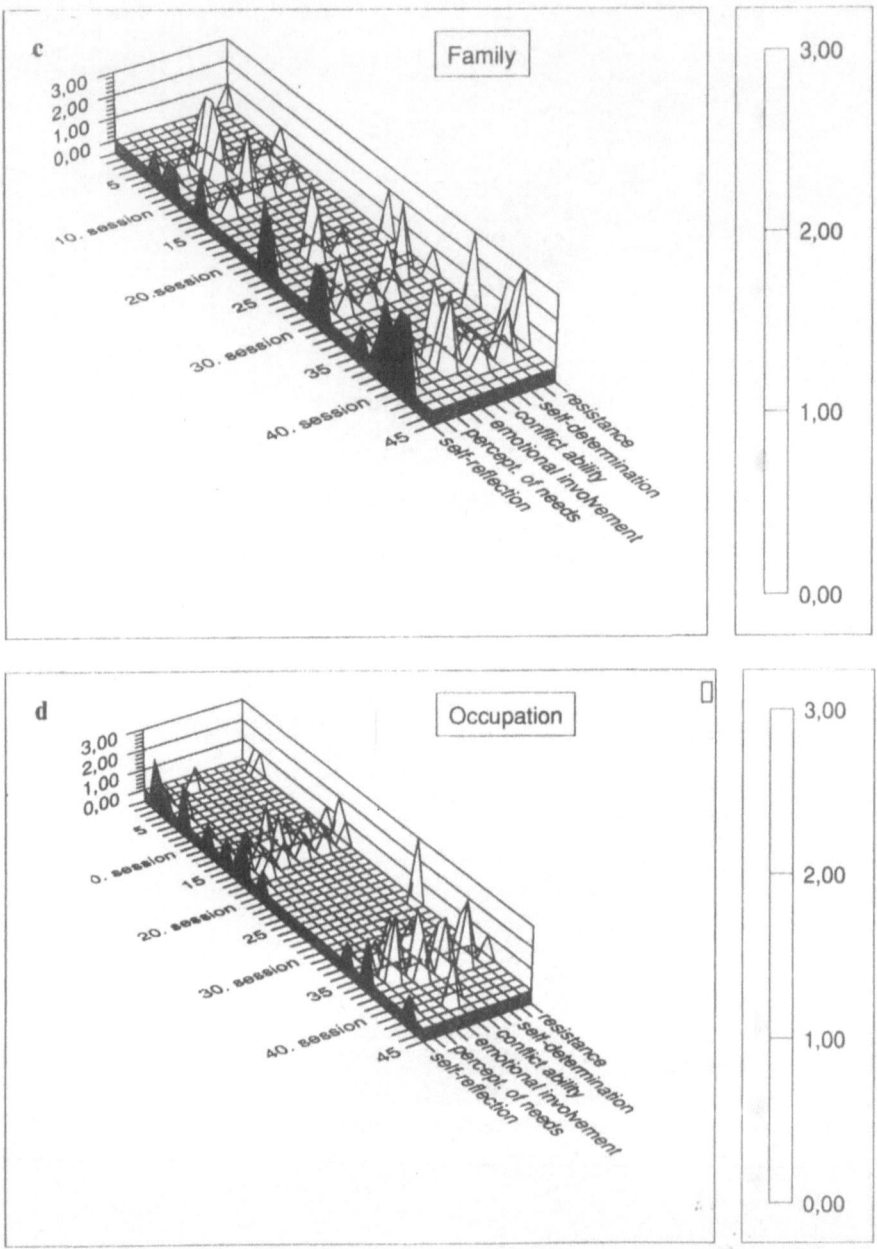

Fig. 4 a-d. Examples of the treatment of various themes: These plots show a selection of two themes (occupation and family) of two clients (male, 27 years old and female, 40 years old). The same topics come again and again in therapies and they differ in various ways. Here it is only shown that themes correspond to various ways of coping and vary in the degree to which certain aspects are discussed.

The above-mentioned theme investigation is based on a contentual macro-analysis. As it becomes evident time and time again that macro- and micro-structures belonging to the same system are subject to similar rules and that things emerge in macro-structures which then also become visible at a micro-level (e.g. material wave of dualism, fractals, chirality and spiral forms in organisms), I would like to suggest the following points for discussion: in psychotherapy, changes are initiated by the exchange of signals (or information). One of the main difficulties encountered in psychotherapy research is identifying those critical situations which are considered immediately relevant to change (cf. Tschuschke and Czogalik, 1990; Fiedler and Rogge, 1989; Elliot, 1984). The computer simulation of Karplus and McCammon (1986) concerning so-called hinge molecules with different functions in different configuations has led me to stipulate the following analogy: In the same way as in the molecular world the surrounding cells rarely become active in "turning" such a molecule - the main effort being left to Brown's movement - therapists rarely intervene intensively and frequently permit the client to talk freely (relaxation). The energy expense is minimised in both instances (path of least resistance). This is illustrated by an (translated) example taken from part of a therapy:

Part One

Cl: Yes, I can imagine the situation exactly, I am looking over my own shoulder... so I am revealing something disadvantageous to me, why disadvantages I don't know, but...

Th.: Is my hacking around on it unpleasant?

Cl.: Yeees, but I really can't say exactly (cl. hits her knees with both fists), yes, it is unpleasant... I am revealing something of myself which... something bad, yes, I feel it's bad. And I know for certain that the other colleague says, too: I won't stay forever, but she only says that to me, therefore I see how she does it and I have the feeling the others don't want to know it from me, either.

Th.: Mmh but it's also unpleasant where I am concerned when I keep hacking around on this point.

Cl.: Yees.

Th.: And what happens?

Cl.: No, you must ask me, I don't know what, how, I am glad when you ask me.

Th.: I don't know whether there are moments when you tell yourself, leave me alone with all this.

Cl.: Mm, no, I have written here (gets out her pocket diary) I would tell, and then I wanted to know what you would say, I am applying for a job which I don't want at all.

Th.: Mmh.

Cl.: But it would also be a matter (looks very competent) to calm down and to wait and see what I want.

Th.: Mmh.

Here, the client touches a "hot" spot. The therapist asks whether the situation is unpleasant for the client and eventually grants a change of subject after having thematised (interpreted) the difficulty.

It becomes evident here, as in the next example as well - it is from the same therapy session - that, apart from the side-stepping, themes (subthemes) are touched on time and time again. A theme variation could be interpreted as learning by redundancy. Invariances (stable properties) crystallise themselves by constant repetition and variation of experiences; these invariances form the basis for a new approach to "reality".

Furthermore, there is a connection with the schema theory: Reflecting abstraction or the derivation of a rule which can lift this mutual experience onto a new level: Repetition and variation result in a fluctuation which in turn gives rise to a crystallisation of new or more complex organisational forms of signal processing.

In this context, further unanswered questions remain regarding the consequences of interaction which might be explained by a simple motivation model (e.g. Bischof, 1983). The approach and avoidance which is centred around the assumed central conflict (the inhibited area) might be simulated by means of a motivational (security and curiosity) "control".

Part Two

Cl.: Hm, I knew that even if I change my job all this is over, and then I go to new... I already go to the trade union but don't feel committed.

Th.: Hm.

Cl.: Well, looking at it that way... yes... well, if I was really interested (th. makes an undefined noise) I would see how it goes.

Th.: Mmh. There is now another problem like that, you say you can't... tell this to anybody or now, with the new job... or the thing with your mother...

Cl.: Well, I feel (touches her head) if I were to tell this to anybody... then that person would no longer... well, if I were to say that I am a woman of trust, you can bring me your things, I will send the things on, but I am no longer interested, really...

Th.: Mmh.

Cl.: Well, who am I then?

Th.: Mm, who are you then?

Cl.: Yes, I don't know (finger next to her mouth)... well, yes, then I would be someone who has the courage to admit it, but pfhh, I find this very difficult.

Th.: Mmh.

Cl.: I don't know, I have, after all, told you that I work in the trade union and so (lively gesturing) and yes... (shrugging of shoulders) maybe you have found it good that I do something and don't say... yes...

Th.: Mm, now the question, who else am I...

This is illustrated metaphorically by the epicyclic model shown in Fig. 5.

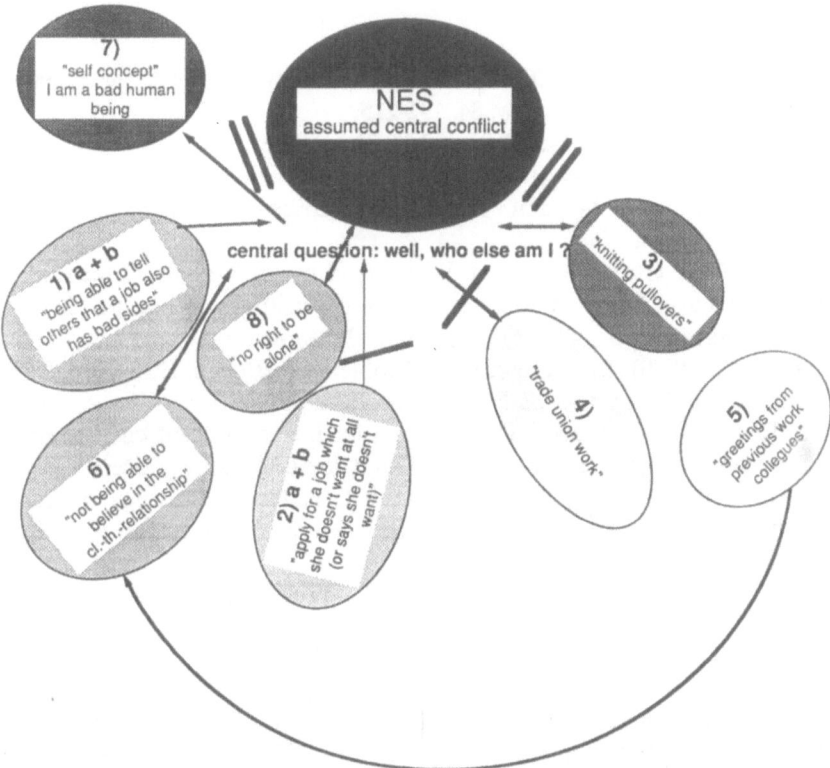

Fig. 5. Epicyclic model: The epicyclic model attempts to give a two-dimensional representation of the temporally graduated sequence of sub-themes. A circling around the stipulated central conflicts (negative emotional schema) becomes evident. Thematic progression time and time again causes the central conflict to be touched on. As a rule, the subject is changed when this occurs, sometimes (rarely) the central conflict is formulated, as in the above example, where the client asks: Well, who am I then?

7. Conclusion

It remains the concern of a process-oriented research to reveal processes that can be explained by means of a variety of theoretical approaches, and to discover the hidden laws governing them. GOLEM does not yet exist as a program. It might never exist in the complexity described above: the next steps will consist of programming a simple system (without direct communication). At the same time, various projects exist whose aim is to clarify the above statements on theme development. The transposing into a model of therapeutic situations as done with GOLEM provides a possible explanation and simultaneously a starting point for further questions.

References

Bandura, A. (1977). Self-efficacy: Toward a unifying theory of behavior change. Psychological Review, 84, 191-215.

Bischof, N. (1985). Das Rätsel Ödipus. München: Piper.

Bischof, S., & Lovey, M. (1990) Erfassung der Veränderungen bei der Bearbeitung von Themen in der Psychotherapie. Vordiplomarbeit, Universität Bern.

Caspar, F. M., & Grawe, K. (1981). Widerstand in der Verhaltenstherapie. In H. Petzold (Eds.), Widerstand - ein strittiges Konzept in der Psychotherapie (pp. 349-384). Paderborn: Junfermann.

Caspar, F. M. (1989). Beziehungen und Probleme verstehen. Eine Einführung in die psychotherapeutische Plananalyse. Bern: Huber.

Elliott, R. (1984). A discovery-oriented approach to significant change events in psychotherapy: Interpersonal recall and comprehensive process analysis. In L. N. Rice & L. S. Greenberg (Eds.), Patterns of change. Intensive analysis of psychotherapy process. (pp. 249-286). New York: Guilford Press.

Fiedler, P., & Rogge, K.-E. (1989). Zur Prozessuntersuchung psycho-therapeutischer Episoden. Ausgewählte Beispiele und Perspektiven. Zeitschrift für Klinische Psychologie, 18(1), 45-54.

Grawe, K., & Caspar, F. (1984). Die Plananalyse als Konzept und Instrument für die Psychotherapieforschung. In U. Baumann (Eds.), Psychotherapie: Makro- und Mikroperspektiven. (pp. 177-197). Göttingen: Hogrefe.

Grawe, K. (1986). Schema-Theorie und interaktionelle Psychotherapie (No. 1986/1). Universität Bern.

Grawe, K., Amstutz, B., Balmer, R., Braun, U., Doblies, G., Heiniger, B., Thierstein, C., & Znoj, H. (1991). The Bernese standardized case history approach: the schemaanalytic therapy of Ms. E. Paper presented at the Society for Psychotherapy Research 22nd Annual Meeting, July 1991, Lyon, France.

Greenberg, L. S. (1984). Task analysis. The general approach. In L. N. Rice & L. S. Greenberg (Eds.), Patterns of change. Intensive analysis of psychotherapy process. (pp. 124-148). New York: Guilford.

Hebb, D. (1949). The organisation of behavior. New York: Wiley.

Karplus, M., & McCammon, A. J. (1986). Das dynamische Verhalten von Proteinen. In P. Sitte (Ed.), Die Moleküle des Lebens (pp. 74-84). Heidelberg: Spektrum der Wissenschaft.

Marko, H. (1966). Die Theorie der bidirektionalen Kommunikation und ihre Anwendung auf die Nachrichtenübermittlung zwischen Menschen (subjektive Information). Kybernetik, 3(3), 128-136.

Neisser, U. (1976). Cognition and reality. Principles and implications of cognitive psychology. San Francisco: Freeman.

Palm, G. (1988). Modellvorstellungen auf der Basis neuronaler Netzwerke. In H. Mandl & H. Spada (Eds.), Wissenspsychologie. München: Psychologie Verlags Union.

Prigogine, I. (1977). Self organisation in nonequilibrium systems: from dissipative structures to order through fluctuations. New York: John Wiley.

Rotter, J. B. (1966). General expectancies for internal vs. external control of reinforcement. Psychological Monographs, 80.

Sachse, R. (1988). Finbe-System: Manual für formale, inhaltliche und Bearbeitungs-Analyse von Klienten- und Therapeuten-Äusserungen (Berichte aus der Arbeitseinheit Klinische Psychologie No. 65). Ruhr-Universität Bochum.

Segal, L. (1986). Das 18. Kamel oder die Welt als Erfindung. Zum Konstruktivismus Heinz von Foersters. München: Piper.

Toukmanian, S. G. (1986). A measure of client perceptual processing. In L. S. Greenberg & W. M. Pinsof (Eds.), The psychotherapeutic process: a research handbook (pp. 107-130). New York: The Guilford Press.

Tschuschke, V., & Czogalik, D. (1990). Psychotherapie - Welche Effekte verändern? Zur Frage der Wirkmechanismen therapeutischer Prozesse. Berlin: Springer.

Waddington, C. H. (1942). Canalization of development and the inheritance of aquired characters. Nature, 150(3811), 563-565.

Winograd, T. (1983). Language as a cognitive process. Reading, Mass.: Addison-Wesley.

The Relation Between Mental and Social Systems

Bernd Nissen

Abstract. In the following essay I shall try to elucidate the relation between mental and social systems. In my opinion this is one of the most interesting aspects of psycho-social work, because the interactional intersection is important for the understanding of dyadic relationships, for example in therapist-patient relationships, and also in institutional contexts (e.g. social-psychiatric networks (see Nissen, 1991)). I have focussed the following study on the mental system within the interpenetration of mental and social systems and show the necessity of psychodynamic processes for a complete understanding of the interpenetration; i.e. the reduction of the mental system to a cognitive or conscious system is not possible.

Before I start to discuss my system-theoretical point of view (sect. 2), I want to describe a short sequence from a group therapy session, in which the empirical presentation is strongly selected to illustrate the problems and questions concerning the general systems theory touched on in the third part (sect. 3).

1. Empirical Example from a Group Therapy Situation

First some notes about the therapeutical setting, which may be helpful in understanding the very short sequence:

The group was part of a psychiatric rehabilitation programme for patients showing schizophrenic symptoms. All of the 6-8 participants of the group had been in a psychiatric hospital several times. At the beginning of the treatment the number of acute psychotic episodes oscillated between 3-15 (on average 7-9); the first schizophrenic disorder could be back-dated 3-12 years. All patients suffered in the acute episodes from multiple accessory symptoms, especially in paranoid hallucinatory, but also in catatonic forms. Although in the meantime the patients had returned to a state of integration and only a few irreversible defective states were visible, the official diagnosis was chronic psychosis or process schizophrenia in 80% of the participants, so that it is not surprising that 80% of all patients are receiving a disability pension or public assistance.

The group was devised as a slow-open one. The period of participation was between 2 and 4½ years (except for members who only took part for a few months) – an exceptional circumstance for this population. The group met once a week for 1½ hours beginning and ending punctually. Obligatory participation was a basic condition. Furthermore, the willingness (not the obligation) to show

314

Springer Series in Synergetics, Vol. 58   Self-Organization and Clinical Psychology
Editors: W. Tschacher, G. Schiepek, and E.J. Brunner   © Springer-Verlag Berlin Heidelberg 1992

mutual assistance in crisis was expected, e.g. visiting at home, in psychiatric hospitals, etc.

The group session organization was clearly structured: Following an individual statement concerning the important aspects of the last week, the work continued with a supportive therapy and ended with some final feedback.

All relevant individual conflicts, problems and questions were the subject of the group counselling. Besides cognitive-behavioral treatment, the central and basic theme of the group work, which was accepted by every group member and which was the focus of nearly every session, was the acceptance of schizophrenic disorder. This included all aspects of shame, narcissistic mortification, mourning, despair and rage, but also facets of fascination and attraction, interdependent strategies of succumbing to the attraction, i.e. strategies of defence, and especially schizoid and manic mechanisms of defence[1].

Besides the individual therapy in the group, the therapeutic work included the "mediation" and structuring between the individuals and the group (as a social system), that is, between mental and social processes, or dynamics and realities[2].

This short description maybe helpful in order to give an impression of the group work, from which the following example was taken:

Six patients participated, two of them in a prepsychotic state with the subsequent effect of psychomotoric excitement. From the beginning the group atmosphere was characterized by a high level of anxiety. The group work started with a report by a patient, Mrs. M (who had been a member for two years), concerning a conflict with her partner, which had escalated until he insinuated that she would become crazy again – an allusion that had hurt Mrs. M very much and made her angry. Remembering that conflict, the patient began to speak about the deep narcissistic mortification of her "madness", and talked about her most difficult acute psychotic episode. She asked herself agonisingly: "Why me?!" (Why have I become psychotic?), crying violently, and then, ashamed, despairing and imploring, she retold a short event of the last psychotic episode:

Mrs. M: "And then – you have to imagine that! – I threw my cat – I loved it so much – out of the window of the 4th floor..."

[The group atmosphere increased in tension and paralysis.]

Mr. J: (group membership for three months, at that time in a prepsychotic state): "Once I had cats too, yes, I had 14 cats!"

[Group atmosphere of "confusion" could be noticed.] It is evident that a therapeutic intervention followed.

Therapist: "That's really a terrible recollection, which is still very painful: 'That could not be me!? How could I have treated my dear cat in that way!?' ... At the same time it also seems to me that the group has problems enduring that presentation, and thus in seeing 'the aspects beyond'."

---

[1] The named aspects of acceptance need a treatment of at least 2-3 years.

[2] Social and psychic reality in the sense "psychical reality" is used in the psychoanalytic theory.

I will use this empirical example to demonstrate some aspects of the "coupling" between mental and social systems in a theoretical way[3].

2. Some Relevant General System-Theoretical Considerations

The short sketch given above is focussed fundamentally on the interpenetration. Interpenetration means that systems place "inter-systemically", reciprocally complexity for building up their own structure (see Luhmann, 1987). Complexity is defined that, based on immanent limitations, every component can no longer be connected with every other one. Consequently, complexity conditions selection. Complexity is an indicator of a deficiency of information (Luhmann, 1987). In the context of observation/self-observation it is not possible to describe a complex system completely. Furthermore, in my opinion, complexity includes the possibility in a developed system, that there exist basal structures, which are irreversible or barely reversible. This is postulated especially for mental systems, but also for social systems. Mental and social systems are conceived implicitly as self-referential or autopoietic ones. Maturana maintains that "there are systems that are defined as unities as networks of production of components that 1. recursively, through their interactions, generate and realize the network that produces them; and 2. constitute, in the space in which they exist, the boundaries of this network as components that participate in the realization of the network" (Maturana, 1981, p. 217). He calls them autopoietic systems (see also Luhmann, 1987, p. 59).

Self-referential mental and social systems reproduce themselves as dynamic stability. They proceed in a (self-)transforming, recursive, but not replicative process. They operate self-referentially based on their specific actual structure (determined by structure) and process their specific self-organisation in qualitatively different ways: Social systems – as Luhmann postulates – by communication; mental systems – as a proposal, referring to S. Freud – by presentations and affects.

Social and mental systems are meaning (*Sinn*) operating ones[4]. As a result they are recursively closed for each other and remain environment for each other. Even so, the processing of the other distinct systems is necessary for their own reproduction, e.g. for mental systems their own organic autopoiesis, as well as other mental and social reproductions, are environmental action, but are in fact connected by interpenetration. The consequences of this factual situation must be taken into consideration and are still unsolved (e.g. problems of psychosomatics).

[3] In the following I presuppose a knowledge of basic terms of the general systems theory.

[4] Meaning is a basic (philosophical) concept in the general system theory, used as an epistemological category (*Weltform*) (see Luhmann, 1987). In my opinion, Luhmann's concept of meaning is too logical, tending towards a rationalistic, integrative comprehension of systemical structures and complexity.

E.g., in the case of Mrs. M all organic, constitutional factors, including those causing the schizophrenia, together with all group members and the social system "group", are environment for her.

But how should one describe the actual operation of mental and social systems?

For a self-referential closed system there is neither input nor output. The system can only process itself by the basal circular operations based on its own status and develop an environment relationship in that way. The facts and events of the environment merely have the quality of a perturbation for the system. This perturbation will be changed in information nearly simultaneously, based on the developed structures of the system. In the moment of the "assimilation" of the perturbation, namely in an element which could be handled by the system as an information, using the interdependent combined processes of selection and reduction, a difference is established. In this difference the transformed perturbation is an element of the system, but also simultaneously in relation to the system, thereby choosing the special status of the system. This constituting difference must be adapted in the system as changed information, potentially specifying the determined structure. The elements of those systems must be interpreted as temporalized events and are in a permanent decomposition process.

If the moment of self-observation[5] – as a condition of the self-reference of mental and social systems – is added again, the possibility follows that the system will condense the pieces of information (as transformed differences) into structures by processing. Operating with, and reproducing the differences of external and internal references, the system creates its own specific structure as an identity and is able to deal with higher complexity.

Taking into consideration this view of system complexity, we have to deduce three theoretical terms:

a) Perturbation: The communication act as an "objective entity", being a perturbation for the systems involved.

b) Element: The changed perturbation as a temporalized element of the system involved, which constitutes a difference. This means, based on the developed structures, the element is already reduced self-selectively and changed into information. At the same time it chooses the status of the system [see point (c)], but nevertheless is in relation to it. The difference between structure and element helps to process and specify the system.

c) Component: The components within a system as structural, specifically related concentrations, building up an individual identity (the term "component" does not mean any material substratum).

Returning to the empirical example, the whole statement of Mrs. M could be used as a perturbation which (in an extremely cursory reflection) Mr. J reduced – based on his actual mental (prepsychotic) structure – to the term "cat", choosing a specified status of system and resulting in his statement.

[5] Observation is defined as the use of a distinction, with which one or other of the "sides" can be defined (see Luhmann, 1985, p. 407).

The basic idea of the self-referential closure and the specific version of the terms "element" and "component" forces the system to manage the complexity, which is permanently reconstituted and potentially increased by differentiation of the system. The differentiation limits and increases the complexity: Limits, because new structures are established with the result that there is a restriction of new potential relations; increases, because the structural capacity of connecting the elements increases in the whole system.

But how can we understand the coupling between mental and social systems in this context?

Mental and social systems belong to different logical classes, each indicated system as a unit remains environment for each other. The process of interpenetration, in which, as mentioned before, systems place complexity for building up structures for each other's disposal (Luhmann, 1987, p. 290), means a convergence into the same temporalized events. These events, in which mental and social systems converge, are communicative acts which are an "objective entity" in the moment of production. Systems that are determined by structure and are able to make observations and self-observations continually give these events their own connectability in the difficult process of self-selective reduction, which transforms a perturbation into a difference. In that process, the individual structure develops.

With this version of temporalized events every communicative action could be divided into a communicative act and a subsequent connecting act. The processing of communication, which automatically establishes social systems, allows an observer to define for each communicative act a relation to at least one social system and to at least two individuals. This defined relationship need not correspond to the subjective represented intention of the participating systems (see Selvini-Palazzolli et al., 1985 and the term "territoriality"). Indeed, there will often be a better understanding of the interaction if systems, not present and maybe not even existing, are taken into consideration. Returning to the empirical example: An observer can define a relation between Mrs. M's communicative act and the following participating systems: the social system group, the seven group members (therapist and six patients) and maybe Mrs. M's partner. That is to say, social systems require individuals who are able to interact effectively. Every substantial social system[6] can be set in relationship to a certain number (n) of individuals (but these persons are neither elements nor components of the social system, because that is an emergent class based on another qualitative level!). The postulate of mental and social systems as self-referential, and the convergence of mental and social systems into the same events, has the following consequence for logical research and scientific theory – we have to distinguish in my opinion between three different types of description of the interpenetrative connection between mental and social systems:

a) the internal dynamics, that is the observation of the system as a unit in the context of internal operations.

b) external dynamics, that is the observation of the interaction of one system with systems in its environment (see Maturana & Varela, 1987)[7].

(a) and (b) are logical consequences of the postulate of mental and social systems as self-referential and determined by structure. They have to be sharply distinguished from each other in the process of research.

c) interpenetrational dynamics (see Nissen, 1991; coupling dynamics). Based on the fact that every social system supposes (n) interacting individuals, one should try with the help of the construct "interpenetrational dynamics", to fundamentally describe and define social systems as emergent classes seen by the observer.

Interpenetrational dynamics is the analysis of the relations between complex individual, communicative act, and social system(s). This analysis implies the investigation of double contingency between me and other(s). Contingency is defined as alternatives of actions (degrees of freedom) to be employed in a special situation (see Willke, 1987, p. 20). Double contingency is defined as the orientation of the selection of actual operation states by the contingency of other systems (see Luhmann, 1975, p. 171). Only double contingency allows the understanding of social systems as emergent classes. That means the interpenetrational dynamics allows an observer the comprehension of the entity "communica-

[6] Substantial social systems are defined as systems which have built up identity in the process of continuous communication; the self which is referring to itself is the system. This means (see Luhmann, 1987, p. 617):
a) the system has to act as a unit in its environment, it has to self-selectively reduce the complexity of the environment based on its actual developed structure. Consequently, the system has to be able to recognize systems in its environment and must be identified by those systems;
b) it has also to keep its identity for a longer latent period. That means it has to operate with the self-observation;
c) a certain number of individuals can be related as identifiable actors to that system;
d) distinct social systems are able to use special codes for constituting themselves as autopoietic subunits (a problem which is not solved e.g. for a therapeutic group).

[7] The extreme complication of the specific operation with differences in the context of the external analysis, as well as the complication of the apprehension of the dynamic stable patterns, which will be raised selectively from the intersystemic interaction, cannot be discussed in detail.
But it is necessary, because the systems (e.g. S(A) and S(B)) indicated each other as distinct, to understand the re-entry of the difference of internal and external reference into the internal working of the system, so that the external reference by itself (the other social system in the environment has to be identified as a distinct one in the case of the context of external dynamics) has to be taken as a unit of a difference, namely e.g. for the social system S(B): The unit of the difference communicative act K(A') - S(A') observed by S(B).

tive act" in its complexity and shows convincing and provable ways to relate mental systems to a social system and for the indexicalisation of social systems.

It is important that for all participating systems the communicative act constitutes an "objective entity" in its nascent moment. (For the indivdidual who produces it, the communicative act as an entity is no longer an element of the mental system.) Based on the individual developed structures of the participating systems, which still provide the environment for one another, this communicative act will be connected in the context of the internal dynamic process, but self-selectively reduced.

Thus, the systems involved remain environment for one another, but because of this fact they are able to reproduce and change themselves structurally by using the same communicative acts, which have to be transformed from complexity into system-specific order. That is, systems are in a context of constitution, in which their own stable dynamical complexity is developing structurally.

This concludes the short sketch of general system-theoretical considerations.

3. Problems and Questions

Refering to the above I want to discuss some problems and questions using the empirical example for illustration. This example could be roughly characterized as follows.

One person, Mrs. M, psychiatric hospitalized several times, remembering a conflict with her partner, made the above mentioned statement during group therapy. It was intended for the therapist, but, once produced, de facto addressed to all seven participants (also to herself, because once the communicative act was out of the mental system, it became an objectified perturbation) and also to the substantial social system (slow open group). That is to say, based on double contingency, the synthetized communicative act was addressed to several persons and one social system. This communicative act was already by itself a high-grade complex one, but only to be comprehended completely by taking into consideration the individual development of the "mental system" Mrs. M. Among other things, it seemed to include a destructive and anxiety-causing quality, which also activated or reactivated specific patterns and states in the group.

Mr. J related to the produced communicative act – as a perturbation – only one word, just an atom in the complexity of this communicative act, namely "cat". Thus Mr. J produced a subsequent act, transforming the complexity of the perturbation in a complicated interlinked process of reduction and selection, based on his actual developed mental structure. That act was also addressed to all seven persons and the social system group. That act contained – noting the group atmosphere – the potentiality of a "maddening" confusing dynamics for all systems involved (individuals and group).

This shows that, in metaphorical language, the group dynamics oscillated between the states of an "implosive" paralysis and an "explosive" confusion.

My therapeutic intervention was an attempt – surely not an entirely successful one resulting from the immediate helplessness – to understand empathically Mrs. M, while at the same time paying attention to the group atmosphere (as a social dynamics), and to "translate" the statements from Mr. J and also from Mrs. M.

In this short reformulation of the event some problems of the psychodynamic processes within the interpenetration of mental and social systems are clarified.

The general systems theory, still a metatheory, has not devised empirical models and a (descriptive) language which can coherently illustrate the complex process of interpenetration. The understanding of meaning (*Sinn*) as a rationalistic and integrative parameter of order, and the understanding of mental and social systems as meaning operating, prevents the comprehension of the complexity of the interpenetration, including that of the systems and their development. Put more concretely: How can this momentary process of the transformation of the perturbation into an element be described? How will one special subsystem of the complex structure activate or reactivate? How can multiple highly complex models of the internal (and external) dynamics of the mental system be evolved which go far beyond the actual operating of the system and are not reduced unproductively to the consciousness and the thoughts as elementary operations, but which include unconscious mental processes? How will components be condensed in the internal process?

To make the above difficulties explicit, I return once more to the empirical example: Using a psychoanalytic vocabulary, Mr. J's statement could be understood speculatively as a complex defence, in which Mr. J – being in a prepsychotic state, that means among other things that heavy endopsychic conflicts could imminently break through – tried to defend the destructive and overwhelming tendencies against loved beings in Mrs. M's statement. This happened in a highly complex process, which included diverse agencies and which was largely unconscious. This communicative act synthesized the subjectively perceived group dynamics as well as interpersonal aspects. Thus, the actually developed, but high-grade unstable structure constituted the element, which again chose the system states, in respect of the destructive and overwhelming tendencies, which included the relevant defensive mechanisms. This actual operative structure constituted "its" element with the nearly simultaneous consequence of activating specific system states. That means, speculatively illuminating our example further: The self, which was referring to itself, was an unstable, highly threatened one, which – based on its actual system state (state 1) – perceived the communicative act of Mrs. M as a preverbal, affective loaded thing presentation, thereby being an element activating new system states. With the transforming self into system state 2, which included among other things the mobilization of sadistic impulses and the defence of them, the disavowal of the named thing presentation started. This structure, which had processed itself by actual operating, now reinforced – and that was the pathological paradox – the threat and helplessness of the transformed self. Consequently – as an attempt at rescue – the thing presentation, which was just self-selectively reduced, was violently condensed into the word presentation "cat". To stabilize the pretended rescue the subsequent communicative act of Mr.

J was formulated (system state 3). This subsequent communicative act, which was an objectified perturbation, interpenetratively connectable and, among other things, able to process simultaneously the social system, had a fatal effect even for Mr. J, who also noticed, by my observation, the dramatic alteration of the group dynamics.

In this difficult description of the empirical situation it becomes clear that the dynamics of connection of communicative acts is not to be comprehended as an unequivocal contoured process but as a high-grade "complex condensed oscillating" process. In other words, as well as the change of the perturbation into an element of the system, the activating of the system states is an "indistinct" process and not a determined or an exactly predictable one.

With respect to some considerations about the intra- and intersystemic dynamics of mental and social systems, there are still a lot of questions and the general system theory has not found relevant answers. To be more specific: The general systems theory has still not created (for mental processes) an adequately usable and applicable language, in which we can find concepts and answers. But the potential of the metatheory of self-reference or autopoiesis is to be found, above all, in the resulting cognitive attitude, which includes, as an important heuristic moment, the necessity and possibility of distinguishing fundamentally highly complex processes and thereby making these processes feasible.

References

LUHMANN, N. 1975. Soziologische Aufklärung II. Opladen: Westdeutscher Verlag.

LUHMANN, N. 1984. Die Wirtschaft der Gesellschaft als autopoietisches System. Zeitschrift für Soziologie, **13**, 308-327.

LUHMANN, N. 1985. Die Autopoiesis des Bewußtseins. Soziale Welt, **36**, 402-446.

LUHMANN, N. 1987. Soziale Systeme. Frankfurt: Suhrkamp.

MATURANA, H. 1981. Autopoiesis. In: Zeleny, M. (Ed.): Autopoiesis: A Theory of Living Organization. New York: North Holland, 21-35.

MATURANA, H. 1985. Erkennen: Die Organisation und Verkörperung von Wirklichkeit. Braunschweig: Vieweg.

MATURANA, H., VARELA, F.J. 1987. Der Baum der Erkenntnis. Bern: Scherz.

NISSEN, B. 1991. Entwicklungsdynamik psycho-sozialer Systeme. Aktions- und systemtheoretische Überlegungen. Göttingen: Hogrefe.

SELVINI PALAZZOLI, M.; ANOLLI, L.; DI BLASIO, P.; GIOSSI, L.; PISANO, J.; RICCI, C.; SACCHI, M., & UGAZIO, V. 1985. Hinter den Kulissen der Organisation. Stuttgart: Klett-Cotta.

WILLKE, H. 1987. Systemtheorie. Stuttgart: Fischer.

Clinical Constellations:
A Concept for Therapeutic Practice[1]

Ludwig Reiter

1. Introduction

In reviewing the current state of the art regarding the origin, development, and the therapy of psychological and psychosomatic disorders, one readily notices the diversity and complexity of the collected literature. Different areas of research, from the biological sciences to social and cultural studies, have contributed to the understanding of the causes and etiologies of these disorders. Multiple proposals and ideas have been generated towards the goal of developing more effective treatment strategies. Most of these theories are theories of a "middle range", in Merton's sense of this term, which attempt to describe and explain only partial aspects of clinical phenomena. Far-reaching, all-inclusive or integrative theories which regard both the biological as well as psychosocial aspects of these disorders have not been fully developed. A reference or starting point for this development was Engel's (1977) suggestion that medical science adopt a *bio-psycho-social paradigm of illness and treatment* in place of the dominant biomedical model.

Although there is so much diversity in the theoretical literature, the practical work of therapy often tends to be a very reductionistic endeavor. Frequently, one or just a few factors get taken out of their complex contexts and then become a primary focus in the therapeutic process. (For example, the exclusive employment of antidepressant medication is based on the reductionistic assumption that depression is essentially a disturbance of the neurotransmitter system.)

This paper is the result of my own efforts to develop *practice-related clinical concepts* which both encompass the current theoretical developments as well as allow for the integration of new knowledge into everyday therapeutic practice. The concept presented below is intended to be of didactic use in promoting discussion while teaching and training of psychotherapists regarding new developments in theory and research.

[1] For valuable assistance with this manuscript, I would like to thank the editors of this volume, my co-workers at the Institute for Marriage and Family Therapy and my wife Stella Reiter-Theil. Translation from original German: Margit Kellenbenz Epstein, Dipl.-Psych., Eugene K. Epstein, MSW.

Springer Series in Synergetics, Vol. 58   Self-Organization and Clinical Psychology
Editors: W. Tschacher, G. Schiepek, and E.J. Brunner   © Springer-Verlag Berlin Heidelberg 1992

2. Systemic and Integrative Concepts

Some of the more recent theoretical developments which attempt to overcome reductionistic thinking and practice can be subsumed under the term "Systems Theories" (Kratky & Bonet, 1989). The different systemic therapy models are derived from and very much related to these developments in systems theories (Ludewig, 1988a,b; Reiter et al., 1988; Reiter & Ahlers, 1991; Schiepek, 1988; Steiner & Reiter, 1989). The development of simulation models, as presented by the Bamberg Group (Schiepek & Schaub, 1989, 1990), has introduced interesting insights into the complex conditions of etiology and course of psychological disorders. Since the term "systemic therapy" generally relates to those therapies which, in adopting the assumptions of radical constructivism, are most often concerned with the clients constructions of reality, I would propose for my work, to use the term "systemic-integrative therapy". This term refers to the ways in which therapists apply knowledge from differing therapeutic schools in a pragmatic manner, while using systems theory as a frame of reference. I have described these issues elsewhere using the example of the depressive disorders (Reiter, 1988). This distinction, of course, does not mean that only *one* therapist, by him or herself has to intervene on all three levels (biological, psychological, and social). Rather it relates to a general clinical posture of cooperation (an illustrative case is presented by Reiter, 1990). To describe therapy in systemic terms does not specify what therapists really do (Schiepek, 1988).

Over the last few years an international movement has been established with the intent of promoting integrative thinking in psychotherapy. The "Society for the Exploration of Psychotherapy Integration" (SEPI) has provided a forum for the discussion of these issues (and since the beginning of 1991 has published a journal related to these issues). Most of what is written in this article has been influenced by the thinking developed out of this movement. In my own development I have been led from psychoanalytic to structural family therapy, then to systemic therapy, and finally to systemic-integrative therapy. I have arrived at this last perspective because, in my opinion, the systemic-integrative perspective seems to me to be the most appropriate when treating severe psychological disorders. As will be discussed later, this systemic-integrative perspective does not require the use of a complex treatment arrangement.

3. The Concept of the Clinical Constellation

In order to develop an understanding with clients and their relatives (or perhaps better, with members of the problem system in the sense of Goolishian & Anderson, 1988 and Ludewig, 1988b), about the origins and development of the problem, and to help them understand my proposed treatment approach, I looked for a metaphor which was plausible, could be readily understood and also has a close relationship to current theories and practice. After a long search which was also stimulated by the presentations of my clients, I chose the metaphor of the "constellation". Important was the integration of several developments from the psychiatric and psychotherapeutic fields. I have presented the multiple other sources of professional inspiration in an other paper (Reiter, 1990). Especially noteworthy however was the concept of the idiographic systems model put forth by Schiepek and his colleagues (Schiepek, 1986; Schiepek & Kaimer, 1988; Schiepek & Schaub, 1989)[2].

Some definitions of the term constellation should exemplify important aspects of the term which are relevant to the concept being described here. Constellation is defined, among other things, as:

"the totality and grouping of factors relevant for a condition or process"
"the determination of a current condition through preceding events"
"the position and characteristics of individual parts or factors within a unit or whole"

These three definitions may suffice to demonstrate that the term constellation refers to the following four conceptual issues:

a) **Complexity.** The concept constellation requires us to search for a variety of determining or causal factors. This is a central issue among current professional discussions about the determinants of psychological disturbances. Therapists adopting this stance look, together with the members of the problem system, for different factors which, when taken together, explain the incidence of a clinical phenomenon.

The question of how much complexity is necessary for successful therapy as well as the question of how much complexity is useful (or too much), is of particular relevance. For practitioners, there is surely a maximum level of complexity that can be effectively managed in their work. Therapy always deals with the reduction or the transformation of complexity. Böse and Schiepek (1989) in reference to Luhmann (1984), propose:

to define certain stages of a therapeutic process as the effort of a social system (consisting of therapists and clients) to transform orderless, system-specific complexity as well as incomprehensible environmental complexity. Structural complexity (e.g. accurate problem formulation) develops in the therapeutic system through giving meaning to selected environmental events while other events remain meaningless (p. 81).

As Cramer (1989) pointed out, the term complexity is useful in explaining highly organized systems like human systems. Complexity however, is a *relative* term which has to be seen in the context of the system and its functions, and depends primarily for its meaning, on the concepts or assumptions one holds regarding an object. This implies that ones understanding of complexity is influenced by the agency, the cognitive complexity, and the information-processing capacity of the

---

[2] I thank Günther Schiepek for pointing out the similarity of the terms constellation and system (system derives from the Greek word "synhistamein", which means to stand together; see Böse & Schiepek, 1989, p. 186).

observer. The clinical constellation as an *heuristic concept* should allow the therapist to find a useful reduction in the level of complexity without ignoring or overlooking the current state of theory and clinical knowledge.

b) Causality. The theory of dynamical systems and systemic therapy has said farewell to simplistic conceptions of causality. The view that big causes have big effects and small causes have small effects was given up in favor of the understanding that seemingly unimportant events are able to produce major consequences (e.g. the "butterfly effect"). This makes every effort to find a single cause which explains the existence of a clinical phenomenon impossible. On the other hand, the concept of constellation does imply an assumption of causal relations (which is important to many clinicians in their work), without necessarily implying a reductionistic causal form of thinking (see Riedl, 1981). I consider it important to point out that in current research into the social origins of psychological illness, questions regarding causality predominate (Brown, 1985; Brown & Harris, 1986; Mirowsky & Ross, 1989). Systemic therapies stress that the traditional conceptions of linear causality have been replaced by notions of circularity. Through this shift, new and important insights have been gained, which have been acknowledged and utilized in research. But one cannot throw out the baby with the bathwater and suggest that *everything* is circular. This is not a useful radicalization. Luhmann (1984, p. 608) posits that in fully temporalized (e.g. psychic, social) systems which use events as elements, no causal circularity can exist, since these fleeting and temporary events do not exist long enough for one to provide feedback. In addition, the investigation of causality always relies on a selective view of reality. Practitioners can decrease their own possibilities for intervening when taking such an extreme position, especially when ignoring asymmetrical relationships (Cierpka, 1989).

c) Dynamics. The term constellation should be understood as a dynamical concept. While searching for an integrative concept it was important to find one which could be linked with dynamical theories. The term dynamics in the context of psychotherapy is not new. Freud's psychoanalysis has used the concept of psychodynamics. What is new, however, is the close connection of systemic therapies to other fields which focus on dynamical systems (i.e. chaos theory, synergetics, etc.). The understanding of dynamical complexity is a focus of research for many sciences from physics to cultural studies. For much too long the effort was made to interpret human systems as stable or homeostatic entities (Bühl, 1990, p. 6). There has recently been a shift away from a linear towards a non-linear conceptualization of the term dynamics (Tschacher, 1990).

d) Uniqueness. The current trend of the so-called "outcome studies" in psychotherapy supports the standardization of therapy, thus risking the danger of neglecting the uniqueness of the individual human being. The biologically dominated medical sciences were accused of merely seeing illnesses and not human beings. Systemic therapies should especially promote the idea that human systems are unique. The difficulty of connecting the unique with commonality (e.g. in diagnosis) is not a focus of this article. For a discussion of this issue, see Reiter (1983). It is especially important to note that the concept of constellation stresses the uniqueness within each particular problem. While linear systems theory leaves no room for the history of a system, non-linear systems theory focuses on irreversible processes (Bühl, 1990, p. 216). Complex systems can be understood through their histories. The concept of constellation serves to remind therapists to be aware of the particularities of the problem, and to provide clients with the understanding that their circumstances are both special and unique. It does not seem sufficient to me to just remind therapists of the ethical imperative to see each case as unique. Rather the concepts and heuristics that therapists utilize need to convey this conviction and attitude convincingly.

4. "Constellation" as a Term of Everyday and Scientific Language

The term constellation is often used in everyday language. One talks and writes about political, economic, and cultural constellations. These uses refer to the fact that constellations are complex phenomena whose components and causal conditions are not always observable and fully comprehensible. In one sense, one can talk about a "holistic" concept in everyday language.

In the discourse of the social sciences, the term was often used, for example by Max Weber, Sigmund Freud, and Carl Gustav Jung. Walter Toman is known for his studies of "family constellations" (1965). Helm Stierlin (1988) used the concept of "relationship constellation." Günter Schiepek and Harald Schaub (1991) have discussed the "person-environment constellation" out of which pathological emotions and affects emerge. Katschnig (1986) writes about "stress-vulnerability constellations" in psychiatric life-event research. The concept of constellations has also found its way into statistical analyses of pathological variables (Rubenstein et al., 1989). It was the usage of the term in both science and everyday language that prompted me to adopt it.

5. Conceptual Frame

Table 1 presents the conceptual framework of the clinical constellation. There are three components or levels which have to be distinguished in this framework:

  a) The level of knowledge and ability
  b) The clinical constellation as a bridging concept
  c) The level of the diagnostic-therapeutic process

The level of knowledge and ability refers to the therapist. His or her knowledge and abilities are person-bound (in the sense of Polanyi, 1985). They are acquired through training and practice. One of the unsolved problems in current therapeutic practice is the difficulty of integrating new insights and developments from research with clinical theory. A most important element of the conceptual framework is therefore the connecting of an individual's particular knowledge and abilities with the overall state of professional knowledge.

Table 1. Integrative Framework for the Concept of "Clinical Constellation"

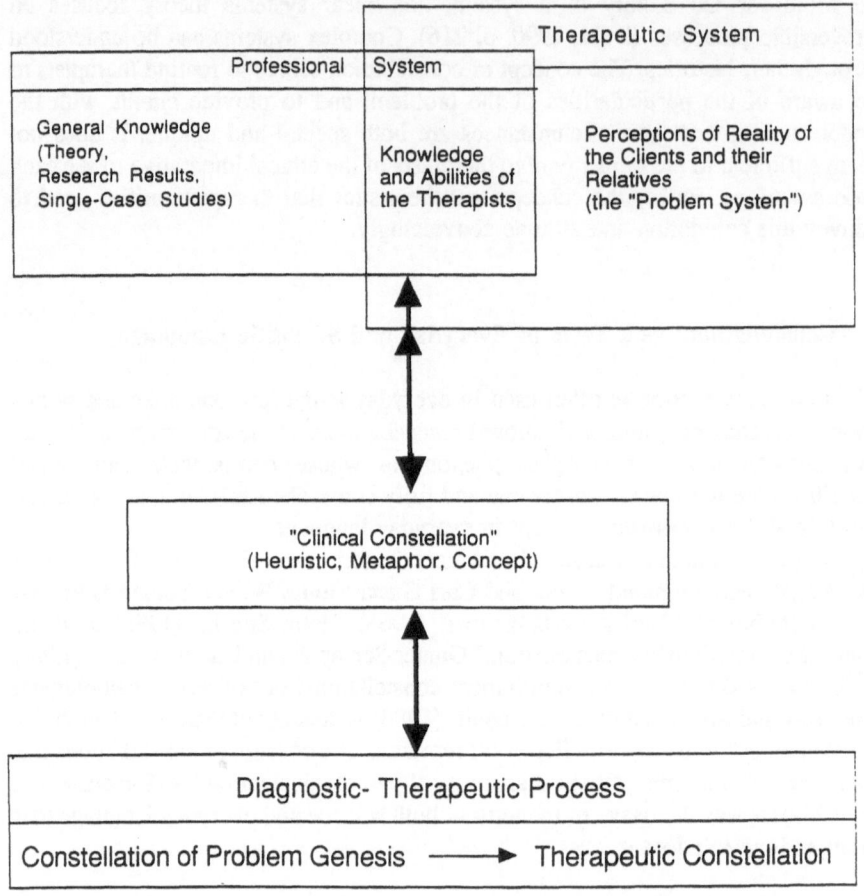

In therapeutic practice the idiographic metaphor is developed by therapists and the problem system (or therapeutic system), based upon common understandings (for example, etiology, etc.), and then used in the diagnostic and therapeutic process. The clinical constellation forms a bridge to the diagnostic-therapeutic process. This means that an etiologically based (pathological) constellation is transformed into a therapeutically based constellation in order to solve the presenting problem.

6. Example: The Depressive-Suicidal Constellation

The sheer volume of compiled research results regarding depression brings about the realization that any effort to achieve a complete overview of the literature would be virtually impossible. Even highly qualified works such as Willners (1985) present but an incomplete range of the research in the field. Theories, too, most often describe only partial aspects of a clinical phenomenon. Because of the vast numbers of studies done on or about the disorder, depression seemed to be a particularly apt field on which to test and apply my concept (Reiter, 1990). The vast quantity of theoretical concepts and research results can be subsumed under the following inclusive factors. A minor modification has been made to my previously developed design (Reiter, 1990):

1) life-events
2) individual factors
   a) individual resources
   b) individual (premorbid) vulnerabilities
3) social factors
   a) protective social factors
   b) social stressors (most often chronic)
4) aspects of diagnosis
5) factors related to chronicity

One readily notes that several factors overlap (e.g. chronic stressors and factors related to chronicity), but more importantly, I believe the overall schema to be pragmatically useful. Since treatment of depression generally acts prophylactically against nonremission, those factors which present a risk for recidivism should be discussed separately. This article cannot adequately address details such as the empirical validation of the proposed schema. For this, the interested reader is referred to the aforementioned article by Reiter (1990).

Point 4, aspects of diagnosis, refers to the different terms used in current psychiatry and psychotherapy. In the context of systemic therapy, these terms may be used as resources as in the sense of a context-related metaphor (see Hinsch & Schörghofer, 1991; Reiter, Ahlers & Hinsch, in preparation). The clinical constellation far exceeds this traditional understanding of diagnosis, since it is a joint construction of therapists, patients and related others who are participating in the process of therapy.

When re-constructing the particular clinical constellation, this listing of relevant factors is meant to be helpful to the therapist, in order to better comprehend those factors contributing to the origins of the disturbance. This design should also be applicable to other psychological and psychiatric disorders, as I will explicate through the example of the problem of suicide. Using parts of a case published previously (Reiter, 1977), the suicidal constellation will be reconstructed.

Ms. S went to an inpatient psychiatric facility in Vienna to make a sleeping cure. Upon seeking admission, she was convinced that outpatient psychotherapeutic treatment would be useful in helping her to cope with her difficulties. She reported in the initial interview that she was originally from Austria, but emigrated to the United States, with her husband, a Swedish businessman, twenty years ago. Her husband became quite wealthy. But over the last few years his business deteriorated due to an economic recession. He slowly lost all of his accumulated wealth. With some rather risky speculation he attempted to recover his lost wealth. This was interrupted by the onset of a rapidly progressive cancer. He died several months after having first been diagnosed. The costs for treatment nearly exhausted their remaining financial assets. For Ms. S, the death of her husband seemed like a stroke of fate which she believed herself to be incapable of overcoming. The demands of caring for her husband exhausted her thoroughly. She had never had financial problems. She felt isolated and was often confronted by creditors and others who sought money from her. She could not see any future for herself. The thought of committing suicide grew stronger for her. The journey to Vienna served two purposes. First, she wanted to visit a sister with whom she had always gotten along well, and second, she expected assistance from a psychiatric treatment to restore her psychological strength. If, after a certain period of time, there were no signs of improvement, she planned to commit suicide. If things improved, she planned to return to the United States to face her problems.

The relevant factors of an idiographic constellation can be shown using this vignette. There is an experience of loss which in turn led to her loss of status. Ms. S not only lost her husband, she also lost her social network. Since she lived in a symbiotic relationship with her husband, she had few remaining outside relationships. In addition, there was the loss of her financial base and the increasing burden of debts. The relatives in Austria with whom she visited, could be seen as a protective social factor. One rather doubtful resource (which later proved true), was her vague hope that the estate of her husband was more substantial than it initially appeared. Her reported lack of experience in practical matters of life can be seen as an individual factor of vulnerability. Her husband had removed all obstacles and difficulties for her in the past, and she now perceived herself to be completely helpless. She described the childless marriage with her husband as completely satisfying. Both partners were closely united against what they perceived to be a potentially hostile environment. In the course of the crisis intervention, Ms. S's ability to ask for and accept the help of others was seen as an individual resource. Of diagnostic relevance was the fact that Ms. S reacted with depression to the situation. This depression was described by the therapist as an abnormal grief reaction or a reactive depression with somatic features, which therefore enabled her to be treated on an outpatient basis. Although she had suicidal ideas, she was not in any immediate danger of hurting herself.

This rather terse clinical sketch (for a more detailed presentation, see Reiter, 1979), outlines clearly the idiographic depressive-suicidal constellation. No one single factor alone led to the patient's depression and suicidal ideology. Rather it was driven by the dynamical synergy of various factors operating on different levels.

Of relevance is that this presentation was not extracted from the patient by my questions. Instead it was a joint reconstruction of her story. What we are dealing with here, is a kind of jointly found narrative[3] which is then used in the process of therapy. This presented therapy is a relatively "simple" case, since only a few people were involved. The process would have been more complex if I had not prescribed the medication myself, but instead asked a colleague to do so. Then a form of cooperation or teamwork would have been necessary. The problems I typically deal with (most often together with co-workers) are much more complex because relatives and frequently institutions are involved. Here it was important to demonstrate the basic concept with the example above.

7. Possible Applications of the Clinical Constellation

The concept presented thus far allows several applications. These applications are not exclusive, but rather complement each other.

a) **Clinical Constellation as a Metaphor.** In my first article (Reiter, 1990) I described how the clinical constellation as a metaphor can be utilized in the therapeutic process. I was most concerned with finding an easily understandable image in order to explain to patients and their relatives what is going on in the therapeutic process, and to stimulate them towards active participation. Since clients have sometimes used the term without being first introduced to it formally by me, it seemed even more applicable in therapeutic communication. The work of my colleagues in our Institute for Marriage and Family Therapy in Vienna on the usage of clinical metaphors, also played a major role in developing the concept (Steiner & Hinsch, 1988). Continuing experimentation with the concept in therapy and training has prompted me to think about expansion of its applications.

b) **Clinical Constellation as a Heuristic.** Like Schiepek and his colleagues, who proposed to use the concept of the idiographic system model as a heuristic concept, it is possible to use the clinical constellation in a similar manner. My primary concern is about how therapists can use the concept in their practical work. To address this point I would first like to give a definition of heuristic. There are several definitions which can be found in the literature:

Heuristic is described as "methodical instruction, for the invention/discovery of new things", and also as "instruction in order to gain new insights." Other

---

[3] Increasingly in the last few years, concepts have been developed which describe therapy primarily as the invention of new realities. These new developments which are connected with the work of H. Anderson, H. Goolishian, B. Hildenbrand, R. Welter-Enderlin, M.B. Buchholz and others, should not lead to a new onesidedness. In my opinion therapy is in essence much more than the invention and telling of stories. It includes *the sum total of all measures taken in order to solve a bio-psycho-social problem*. I do not accept the reduction of the definition of therapy whereby only a small number of interactions remain. This leads again to the splitting of patients into some which are good therapy candidates and the rest who are not. This creates new therapeutic elites. The systemic therapies would then continue with the sorry tradition of having the most difficult cases treated by the least trained professionals. We should remind ourselves that family therapy progressed because the pioneers attended to the most difficult cases.

definitions mention the "theory or science of problem-solving." *Heuristic principles* are "working hypotheses to help in researching preliminary assumptions in order to better understand an issue." Generally heuristic is talked about as the "art of invention."

These multiple, but essentially rather similar definitions clarify what the term means. The concept of the clinical constellation is meant to help the therapist or researcher ask questions and attend to areas which might otherwise have been overlooked. This process then leads to the enrichment of practical therapeutic possibilities because the therapists continue to develop new working hypotheses.

c) **Constellation as a Scientific Concept.** Here we can envision a usage of the term similar to Heinz Katschnig's (1986, p. 254) proposal for life-event research. As the author writes:

> Whereas in terms of practical research and data collection a division between psychosocial and biological variables is unavoidable, it seems important that future controversies in life-event research do not focus on the traditional Cartesian mind-body dichotomy. While life-events are clearly of a psychosocial nature, pathogenic processes can certainly not be attributed to *either* one *or* the other of the two once artificially formed and now reified spheres of reality, the *psychosocial* and the *biological* realm. Future controversies should rather center around the postulation of *specific "life stress-vulnerability" risk constellations* and *specific pathogenic processes*, regardless of whether biological or psychosocial factors are involved. A strict dichotomy between mind and body may, in fact, turn out to be a decisive obstacle to the increase of our knowledge (Ryle, 1949; Engel, 1985; all italics by the author).

This basic approach to research and diagnosis proposed by Katschnig was a central feature in the elaboration of the concept of clinical constellation. As I have previously noted (Reiter, 1988), the systemic perspective has to subsume or include biological as well as psychosocial factors.

A further application is presented in Rubenstein et al. (1989). The authors investigated determinants of juvenile suicidality. They created a table which demonstrated that the statistical risk of juveniles reacting with suicidal behavior was increased by the accumulation of certain factors. This method does not give precise information for the individual case but rather gives hints or suggestions to the practitioner regarding the individual's suicide risk.

8. Connection to Other Concepts

An important and already mentioned reason for introducing clinical constellation is to be found in its connection to other concepts. Some of these opportunities are described below:

a) The Connection of Practice and Research. From within this model, practical knowledge and ability are of equal value to research and general theoretical knowledge. This equality is based upon a mutual respect that is generally lacking presently. David Olson rightly talked about a cold war between family therapists and family therapy researchers. This gap between research and practice is often widened by the ignorance of the other's problems. Since psychotherapy research increasingly acknowledges the need for integrating the problems of the practitioner into the research process, hopefully cooperation will surmount this mutual disregard in the future.

b) The Connection between Nomological and Idiographic Research. Brown and Andrews (1986) pointed to the necessity of connecting epidemiological and idiographic research in their presentation of the current results of depression studies. Among German-speaking researchers, Hildenbrand (1983, 1988) deals with the methodology of biographical studies within the clinical field. A connection of these two traditions of research is increasingly being demanded in order to end this long and fruitless methodological dispute.

c) The Joint Construction of Reality by the Therapist and the Clients. The understanding that therapists need to build collaborative relationships with their clients is becoming more and more commonly accepted. "Paternalism" as a therapeutic principle is increasingly being viewed critically and respect for the client's autonomy is being demanded (Reiter-Theil, 1990). Böker (1990) recently proposed the term "treatment partnership" for the client-therapist relationship. Because the concept of clinical constellation is readily connectable to the clients thinking, due to its universally understood meaning, it allows therapists to explain complex professional analysis in readily comprehensible language. The therapeutic position of "curiosity" (Cecchin, 1988) or "not-knowing" (Anderson & Goolishian, 1990) does not relinquish the therapists obligations to use their professional knowledge and competence and to take responsibility for the therapeutic process and therapeutic interventions. This obligation means among other things, the ability to fruitfully use available general knowledge and research results for the process of therapy. This should not, however, be done in the style of a dominant expert.

d) Team Work Among Therapists and Across Professions. In many therapeutic institutions, and especially with respect to the treatment of severe psychological disorders, there are often several therapists of different professions working together on a single "case". Since the clinical constellation does not relate to a specific therapeutic ideology, it can serve to enhance the professional communication among therapists of differing orientations and educations. Since no superordinate theories currently exist for such cooperative work, pragmatic concepts such as this become meaningful for promoting professional communication.

9. Therapy: From Complexity to Simplicity

Because the concept of clinical constellation refers both to lay knowledge and to professional knowledge, one might tend to assume that the therapeutic processes must therefore be complicated. This, however, is not the case. First, the concept does not imply any preference for any particular school or technique, since each therapist has to develop a focus out of the complexity of client complaints, unless she/he primarily works without a specific goal. Since the majority of therapies done worldwide are short term (Budman & Gurman, 1988), focusing and limiting are the rule. The concept of the clinical constellation presents a plea for not narrowing one's sight prematurely, but rather for allowing as much information as possible in order to build a well-founded focus for the process. Furthermore, the clinical constellation is to be understood as a dynamical concept which enhances the therapeutic system through the development of new insights.

10. Conclusion

This paper presents a generalized description of a practice-related concept, which was developed out of my clinical work with depressive patients and their families. In my therapeutic work and training activities, this concept is a useful metaphor in the sense of de Shazer (1983). Based on the positive reactions of my colleagues to date regarding the depressive constellation, I have been encouraged to put forth this more generalized presentation of the concept. Its applicability should be examined in actual therapeutic practice and in dealing with multiple clinical disorders.

References

ANDERSON, H. & GOOLISHIAN, H.A. 1990. Menschliche Systeme als sprachliche Systeme. Familiendynamik, **15**, 212-243.

BÖKER, W. 1990. Patient, Angehörige und Arzt auf dem Weg zu einer Behandlungspartnerschaft. Kasuistik eines 19jährigen schizoaffektiven Krankheitsverlaufes. Nervenarzt, **61**, 565-568.

BÖSE, R. & SCHIEPEK, G. 1989. Systemische Theorie und Therapie. Ein Handwörterbuch. Heidelberg: Asanger.

BROWN, G.W. 1985. A three-factor causal model of depression. In: Coyne, J.C. (Ed.): Essential papers on depression. New York: University Press, 390-402.

BROWN, G.W. & HARRIS, T. 1986. Establishing causal links: the Bedford College studies of depression. In: Katschnig, H. (Ed.): Life events and psychiatric disorders: controversial issues. Cambridge: Cambridge University, 107-187.

BUDMAN, S.H. & GURMAN, A.S. 1988. Theory and practice of brief therapy. New York: Guilford Press.

BÜHL, W.L. 1990. Sozialer Wandel im Ungleichgewicht. Stuttgart: Enke.

CECCHIN, G. 1987. Hypothesizing, circularity, and neutrality revisited: An invitation to curiosity. Family Process, **26**.

CIERPKA, M. 1989. Das rote Tuch der linearen Kausalität. Antwort auf A. Retzer: Das Problem der Beliebigkeit in der Familientherapie. System Familie, **2**, 223-228.

CRAMER, F. 1989. Chaos und Ordnung. Stuttgart: Deutsche Verlagsanstalt.

ENGEL, G.L. 1977. The need for a new medical model. A challenge for biomedicine. Science, **196**, 129-136.

ENGEL, G.L. 1985. Misapplication of a scientific paradigm. Integrative Psychiatry, **3**, 9-11.

GOOLISHIAN, H.A. & ANDERSON, H. 1988. Menschliche Systeme. Vor welche Probleme sie uns stellen und wie wir mit ihnen arbeiten. In: Reiter, L. Brunner, E.J. & Reiter-Theil, S. (Eds.) Von der Familientherapie zur systemischen Perspektive. Berlin: Springer, 187-216.

HILDENBRAND, B. 1983. Alltag und Krankheit - Ethnographie einer Familie. Stuttgart: Klett-Cotta.

HILDENBRAND, B. 1988. Probleme bei therapeutisch organisierten Ablöseprozessen Schizophrener aus ihren Familien – am Beispiel der therapeutischen Übergangseinrichtung. System Familie, **1**, 160-171.

HINSCH, J. & SCHÖRGHOFER, S. 1991. Krankheit ist auch nur eine Metapher. Zuschreibungen und Versuche zur Auflösung. In: Reiter, L. & Ahlers, C. (Eds.): Systemisches Denken und therapeutischer Prozeß. Berlin: Springer, 154-168.

KATSCHNIG, H. 1986. Prospects for future research. In: Katschnig, H. (Ed.): Life events and psychiatric disorders: Controversial issues. Cambridge: Cambridge University Press, 246-256.

KRATKY, K.W. & BONET, E.M. (Eds.) 1989. Systemtheorie und Reduktionismus. Wiener Studien zur Wissenschaftstheorie 3. Wien: Verlag der Österreichischen Staatsdruckerei.

LUDEWIG, K. 1988a. Nutzen, Schönheit, Respekt – Drei Grundkategorien für die Evaluation von Therapie. System Familie, **1**, 103-114.

LUDEWIG, K. 1988b. Problem – "Bindeglied" klinischer Systeme. Grundzüge eines systemischen Verständnisses psychosozialer und klinischer Probleme. In: Reiter, L., Brunner, E.J. & Reiter-Theil, S. (Eds.): Von der Familientherapie zur systemischen Perspektive. Berlin: Springer, 231-249.

LUHMANN, N. 1984. Soziale Systeme. Frankfurt am Main: Suhrkamp.

POLANYI, M. 1958. Personal Knowledge. Chicago: University of Chicago Press.

REITER, L. 1983. Gestörte Paarbeziehungen. Göttingen: Vandenhock und Ruprecht.

REITER, L. 1979. Psychotherapeutische Krisenintervention: Suizidgefahr – eine Falldarstellung. In: Strotzka, H. (Ed.): Fallstudien zur Psychotherapie. München: Urban und Schwarzenberg, 77-83.

REITER, L. 1988. Auf der Suche nach einer systemischen Sicht depressiver Störungen. In: Reiter, L., Brunner, E.J. & Reiter-Theil, S. (Eds.): Von der Familientherapie zur systemischen Perspektive. Berlin: Springer, 77-96.

REITER, L. 1990. Die depressive Konstellation. Eine integrative therapeutische Metapher. System Familie, 3, 130-147.

REITER-THEIL, S. 1990. "Paternalismus" in der Reproduktionsmedizin. Ein Thema für Familientherapeuten? System Familie, 3, 148-156.

REITER, L. & AHLERS, C. 1991. Systemisches Denken und therapeutischer Prozeß. Berlin: Springer.

REITER, L., AHLERS, C., & HINSCH, J. (in preparation). Der Krankheitsbegriff in der systemischen Therapie. In: Petzold, H. & Pritz, A. (Eds.): Der Krankheitsbegriff in der Psychotherapie.

REITER, L., BRUNNER, E.J. & REITER-THEIL, S. (Eds.) 1988. Von der Familientherapie zur systemischen Perspektive. Berlin: Springer.

REITER, L.& STEINER, E. 1988. Basic modes of interaction and the failure in human communication: Empirical investigation of married couples in therapy. In: Carvallo, M.E. (Ed.): Nature, Cognition and System. I. Dordrecht: Kluwer Academic Publishers, 337-341.

RIEDL, R. 1981. Die Folgen des Ursachendenkens. In: Watzlawick, P. (Ed.): Die erfundene Wirklichkeit. München: Piper, 67-90.

RUBENSTEIN, J.L., HEEREN, T., HOUSMAN, D., RUBIN,C. & STECHLER, G. 1989. Suicidal behavior in "normal" adolescents: Risk and protective factors. American Journal of Orthopsychiatry, 59, 59-71.

RYLE, G. 1949. The concept of mind. London: Hutchinson.

SCHIEPEK, G. 1986. Systemische Diagnostik in der Klinischen Psychologie. Weinheim & München: Psychologie Verlags Union.

SCHIEPEK, G. 1988. Psychosoziale Praxis und Forschung: ein methodologischer Entwurf aus systemischer Sicht. In: Reiter, L., Brunner, E.J. & Reiter-Theil, S. (Eds.) Von der Familientherapie zur systemischen Perspektive. Berlin: Springer, 51-73.

SCHIEPEK, G. & KAIMER, P. 1989. Von der Verhaltensanalyse zur selbst-referentiellen Systembeschreibung. Familiendynamik, 13, 240-269.

SCHIEPEK, G. & SCHAUB, H. 1989. Als die Theorien laufen lernten... Eine Computersimulation zur Depressionsentwicklung. Memorandum der Lehrstühle für Allgemeine Psychologie und für Klinische Psychologie, Universität Bamberg.

SCHIEPEK, G. & SCHAUB, H. 1990. Als die Theorien laufen lernten... Anmerkungen zur Computersimulation einer Depressionsentwicklung. System Familie, 3, 49-50.

de SHAZER, S. 1983. Über nützliche Metaphern. Zeitschrift für systemische Therapie, 1, 21-30.

STEINER, E. & HINSCH, J. 1988. Therapie: Ordungskunst zwischen Finden und Erfinden. Zur Verwendung von Metaphern. Familiendynamik, 13, 204-219.

STEINER, E. & REITER, L. 1989. Family therapy, therapy research, and the theory of self-referential system. In: Dalenoort, G.J. (Ed.): The Paradigm of Self-Organization. Current Trends in Self-Organization. Studies in Cybernetics, 19. New York: Gordon and Breach Science Publishers, 228-239.

STIERLIN, H. 1988. Über die Familie als Ort psychosomatischer Erkrankungen. Familiendynamik, 12, 288-299.

TOMAN, W. 1965. Familienkonstellationen. München: Beck.

TSCHACHER, W. 1990. Interaktion in selbstorganisierten Systemen. Heidelberg: Asanger.

WILLNER, P. 1985. Depression. A psychobiological synthesis. New York: Wiley.

Part IV

Studies of Social and
Mental Self-Organization

Self-Organization in Social Groups

Wolfgang Tschacher, Ewald Johannes Brunner, and Günter Schiepek
With 8 Figures

Abstract. A systemic approach to group dynamics is discussed on the basis of self-organization theory. Groups are conceived of as nonlinear systems characterized by microscopic complexity, circular causality and openness to their psycho-social environments. One possible way of research on group patterns through recursive sculpturing is described; pilot study results with this method are presented. A computational shell for the simulation of social distance regulation is introduced which models attributes of group dynamics by showing different kinds of homeostatic behavior. Finally, consequences of the self-organizational view for the field of management and organizational theory are discussed. Options and restrictions of indirect evolutionary management are inferred from synergetics and recent trends in organizational development.

In the field of psychology a systemic viewpoint may look back upon a long tradition of theories that have been designed in a holistic or Gestalt fashion. Lately, systemic therapy, in particular, has deviated from the conventional personality-oriented thinking and has contributed to call attention to the systemic character of social interaction. In the past years systemic thinking has proved valuable in many clinical areas (family therapy, working with groups in supervision, training, or therapy), whereas the empirical foundations of this field remained insufficient (Wynne, 1988). In this context the concept of *psychological synergetics* turns out to be an innovative way of connecting a systemic approach to clinical practice empirically (Tschacher, 1990; Schiepek, 1991).

Self-organizing systems have been studied extensively in the field of the natural sciences; yet phenomena of self-organization can be observed in the field of psychology and social sciences as well. We expect that a broadened perspective may be gained for these disciplines by new methods and by the interdisciplinary approach of dynamical science.

1. Systems Phenomena in Groups

A systems approach to group dynamics has been applied on various occasions. It is an established point of view among social psychologists that group processes have their own characteristics. A group is not supposed to simply consist of the sum of the characteristics of its members. Group processes are to be regarded as functioning on a higher level of emergence. Lewin (1947) pointed out that there was "no more magic behind the fact that groups have properties of their own,

Springer Series in Synergetics, Vol. 58   Self-Organization and Clinical Psychology
Editors: W. Tschacher, G. Schiepek, and E.J. Brunner   © Springer-Verlag Berlin Heidelberg 1992

which are different from the properties of their subgroups or their individual members, than behind the fact that molecules have properties, which are different from the properties of the atoms or ions of which they are composed." Everyday language calls our attention to this fact; for example a group is said to be in a state of a "tense atmosphere"; or, "there's a good climate in the office".

Social psychologists have described many group phenomena of this kind. Examples range from the formation of group rules and norms to the extreme behavior pattern of "group think" (Janis, 1972). We also consider the development of informal groups or "cliques" to be an emergent feature of human interaction: cliques are found in all organizations – they evolve spontaneously and act seemingly independent of official structures. Informal groups have important functions for the emotional well-being of individuals in organizations (see Sect. 5.1).

A further example can be found in the stages of group decision. According to Bales & Strodtbeck (1951), during interaction, groups tend to shift from a relative emphasis upon problems of orientation, to problems of evaluation, and subsequently to problems of control. Group parameters can thus be characterized as "stationary" over certain periods of time.

Family therapists are concerned with systems processes in "natural" groups. The family system is described by a specific pattern of communication. "It is the rigid, repetitive sequence of a narrow range that defines pathology", as Haley (1976, p. 105) puts it for the case of dysfunctional interaction. Thus, family therapists strive to identify and change these communication patterns rather than individual traits or problems.

2. Towards a Systemic Theory of Group Psychology

As has been indicated, in social psychology a strong emphasis is given to the study of group phenomena. Many theorists in the field of social psychology use a systems perspective in order to describe group processes and group structures. This may be illustrated by a few examples of theoretical approaches.

A group is not merely a collection of individuals. In forming a group two or more individuals – through social interaction – depend on one another to play distinctive roles in the pursuit of common interests or goals (Lambert & Lambert, 1964). The goal-directedness which often is part of the definition of groups is connected with the fact that groups normally develop different positions for their members. Bales & Slater (1955) have shown that there is strong evidence for the natural development of role differentiation in face-to-face groups. The authors postulated for example that "the appearance of a differentiation between a person who symbolizes the demands of task accomplishment and a person who symbolizes the demands of social and emotional needs is implicit in the very existence of a social system responsive to an environment. Any such system has both an 'inside' and an 'outside' aspect and a need to build a common culture which deals with both" (Bales & Slater, 1955, p. 303).

This process of differentiation is – formally speaking – "the division of a unit or structure in a social system into two or more units or structures that differ in their characteristics and functional significance for the system" (Parsons, 1971, p. 26).

There are two ways of designing a theory for this group phenomenon; both can be related to a systems approach:

a) **Groups as interaction systems.** According to Bales & Cohen (1979) the theory of role differentiation can be based upon the interaction theory of small groups. The authors refer to Homans (1961) who emphasizes the exchange character of social interactions by stating that the behavior of persons in a group is related in mutual contingency and that the dynamics of this pattern can be described as circular.

Homans' analysis of interactional behavior fits quite well with systemic assumptions. First of all a social group is considered to consist of elements related in a system. Here we can refer to Hall & Fagen's definition of system ("A system is a set of objects together with the relationships between the objects and between their attributes"; Hall & Fagen, 1956, p. 18). A central point is what the "objects" of a group system might be – are they persons or are they elements of the (inter-) action of persons? Like most theorists, Miller (1978, p. 515) postulates that "the components of groups are animals – human and subhuman". According to Luhmann (1984), however, the components of a social system are the single communicative acts.

Secondly, the character of interaction in the sense of Homans is obviously circular. The basic principle underlying communicative activity is *feedback*, the process by which a system informs its component parts how to relate to one another and to the external environment in order to facilitate the correct or beneficial execution of certain system functions.

b) **Cybernetic approaches to groups.** The concept of feedback is the second possible way of designing a (systemic) theory of group interaction. As mentioned above, groups are considered to be purposive and goal-seeking systems. The recursive character of interaction dynamics is realized by positive and negative feedback loops. Groups are thus conceived of as information-processing systems in which feedback processes maintain the effectiveness of the groups. "When conditions are favorable and the operations are effective, the group not only survives but becomes capable of monitoring itself, altering its direction, determining its own history and learning how to learn to determine its history with the consequence that it accumulates and expands its capabilities or grows" (Mills, 1967, p. 19).

These cybernetic processes can be understood as self-organized. Groups, like all social systems, are not only open systems but also organizationally complex. The postulate of circular feedback processes, and of organizational complexity of group interactions leads us directly to the concept of groups as self-organized systems.

3. Self-Organization Theory as a Framework for Group Psychology

In this section, we will connect group phenomena – like the ones described above – to self-organization theory.

3.1 Prerequisites for Self-Organization

In order to prepare for our enterprise of modeling groups as complex systems, the following general prerequisites for self-organization must be considered.

The concept of self-organization can be applied in a meaningful way only to systems which contain very many (micro-) components; these serve as the potential building-blocks for the assembly of structure and organization. Components are in interaction with one another depending on the system's level of connectivity. From basic thermodynamical considerations we expect that given states of order (assymmetries) will dissolve with time; insofar, all evolution should be *dis*organization and follow the gradient of entropy. The result of such dynamics is a state of maximum symmetry in regard to the localization and behavior of components. This is essentially a statistical argument: all states – ordered or disordered – of a multicomponent system are equally probable in the absence of specific control from outside; so there must be far more disordered states than ordered states (patterns). Thus, macroscopic pattern is extremely unlikely.

The paradigmatic systems used by Haken (laser optics) and by Prigogine (chemical systems) in their first thorough investigations of self-organization were multicomponent systems because of the fine-grained molecular structure of matter; the systems' complexity is provided by the multitude of micro-components (components may additionally be complex in themselves, as for instance the neurons of the brain system or human individuals in societal systems). External influence consists of unspecific flows of energy or matter.

From thermodynamics we would expect that the entropy in open complex systems must increase. So why is there coherent light in the laser and pattern in a chemical solution? It becomes necessary to explain that under certain circumstances a spontaneous formation of order emerges in a system. In the meantime, much research has been conducted on such collective (as seen from the micro level) phenomena of order formation, especially in the disciplines of natural science. But in the field of biology, psychology and social science the evolution and maintenance of ordered states is even more fundamental and ubiqitous (so that this is often no longer experienced as something that needs further explanation). Put formally, in all cases of order formation the emergence of a macroscopic level has taken place (Haken, 1990) or, in terms of Nicolis & Prigogine (1987), correlations of a long range – compared to microscopic correlations – have formed. These correlations can be observed directly; this is why the macroscopic level is sometimes also called "phenomenological".

What are the aforementioned circumstances that facilitate the spontaneous evolution of macroscopic structure? First, self-organized systems are always open systems. Flows of energy, matter and information between the system and its environment allow the system to remain in a state of (thermodynamical) nonequilibrium. Secondly, there is nonlinearity as an essential concept characterizing self-organized systems in several ways. In the case of a system that can be described analytically, the equations that map the system's behavior are nonlinear. Thus, phase transitions and bifurcations can be modeled where the system is sensitive to fluctuations (compare the sensitivity to initial conditions of chaotic processes!). At these critical points little causes may elicit positive feedback loops that rapidly carry the system into new dynamical regimes. The opposite is true for other regions of the system's parameter space; here, self-organized systems are also nonlinear – this time in the stable or homeostatic sense. Again external influences do not simply influence the system additively, but instead are finally erased by the attractor of the system. In both stability and instability the system is phenomenologically "nonlinear" because the relation between a control parameter and an observable of the system is not linear. In essence, all nonlinear phenomena are based upon circular (i.e. again not "linear") causal relations within the system.

3.2 Groups as Self-Organized Systems

Can a group be conceived as a self-organized system along the lines of these statements? To begin with let us consider some usual definitions of group.

In social psychology there is obviously little agreement about how the concept of group is to be defined: there are almost as many different definitions as there are authors. As a rule, though, it is stated that a group consists of a certain number of *persons* being in some kind of *interaction*. Starting from this rudimentary definition we might go into more detail, considering how many persons make up a group. This is a matter of convention – at any rate there will be some upper limit for a group as a social system as soon as direct communication is rendered impossible by the mere number of group members (for the effects of "distance zones" see Hall, 1966; Sommer, 1969). Additional questions arise concerning the term "interaction" in a group definition. By the quality of interaction a group can be distinguished from related concepts in social psychology – like crowd (a gathering or aggregate of individuals without interaction, structure or shared norms), mass (an unstructured large crowd with some common goal), or category (individuals with similar attributes; e.g., see Schneider, 1985; Shalinsky, 1983).

Evidently the rudimentary definition of group is analogous to Hall & Fagen's definition of a "system" as a set of objects or components together with relationships between them. Bunge (1979) defines a system as "a complex object, the components of which are interrelated rather than loose". A social system, accordingly, is "a set of socially linked animals". For many reasons it seems straightforward to choose persons as the components of psychological groups; this is usually done in psychology when a group is conceptualized as a system. Along these lines Brunner (1986) defines a social system generally "as a system, whose members are mutually dependent on one another, so that individual and collective behavior and experience are mutually and simultaneously contingent." This

345

systems concept is an example for an "interactional constellation of individuals" (Schiepek & Tschacher, this volume).

At this point we will try to form a bridge between this conventional conceptualization and synergetic systems. It is manifest at first sight that the former is not consistent with the view of a self-organized system which is necessarily a multicomponent system. One solution seems to be in expecting social systems to be self-organized only from a certain (large) number of members onwards. Haken (1986) points out, that villages or small communities will not produce self-organized patterns because of the small number of individuals they contain; conversely, societies might well be modeled as self-organized systems.

We therefore have to avoid modeling a group as a self-organized system in the same way as we might in the case of a society, even if both may be labelled "social systems". Society is complex on account of the large number of individuals. A model of the self-organization of society may well rest on the notion of individuals being microcomponents like in the approach of population dynamics (Weidlich & Haag, 1983).

On the other hand, we started out with the goal of studying small face-to-face groups as self-organized systems, and have listed some evidence for emergent features of groups above. As a result, we are in need of a more fine-grained resolution regarding the group's micro level. It will not suffice to say that components of a group are the persons that form the group, though this may seem appropriate when viewed from a "anthropomorphic" everyday understanding. We have to shift the focus away from personality-oriented psychological thinking in order to arrive at more fine-grained elements of group dynamics. Some approaches

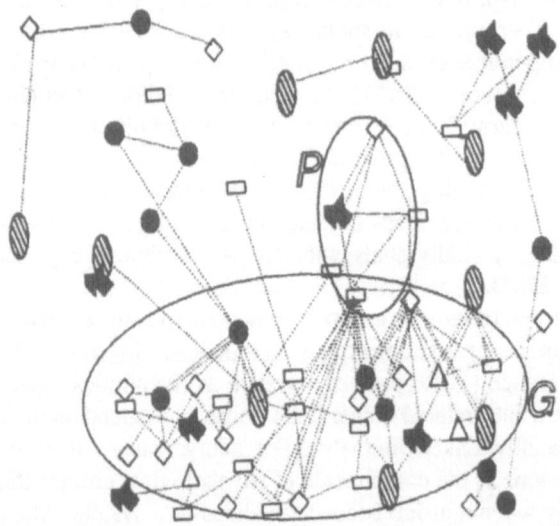

Fig. 1. Chart of a section of the psycho-social world. A group G (or person P) may be defined as an area of increased coupling between psycho-social components

in psychotherapeutic practice have already undertaken this enterprise: in the discussion of family therapy during the last decades there is a tendency to no longer put the person with his or her goals, beliefs, and actions into the center of therapeutic interventions but to reflect the whole system's activity instead. Thus, the persons *within* the system are not viewed a priori as the driving forces of interaction dynamics; in some respect they are not held responsible for group processes. A similar focus on communicative behavior is inherent in the sociological tradition; in this view, social systems build up structures in their own right – like roles and positions. Only then may persons occupy these positions (cf. Barker, 1968).

In our view the following theoretical positions result from the notion that a group is a self-organized entity: the group's micro level consists of emotions and cognitions of group members plus communications (interactions), by which emotions and cognitions are interconnected. Cognitive-emotional and communicative elements of this "stream of group events" can be distinguished from one another only in a superficial way. Therapeutic knowledge tells us that the information "transported" (or rather "transferred") by the various modi of communication between persons is more extensive and more detailed than can ever be consciously experienced. Many of the messages are subsymbolic: their content is not expressed in words (nor in internal language), but is nevertheless effective. The psycho-social micro level of the system "group" is virtual (or hypothetical) in that it cannot be observed directly and in detail (Fig. 1).

Where does self-organization enter into the concept of a psycho-social system with a cognitive-communicative micro level? The formation of standing patterns upon the micro level would be an indication of self-organizing processes. Processes are then not distributed equally and amorphously, but ordered in a way characteristic of a certain group, family or therapeutic system; the interior entropy does not increase. At the same time the pattern that emerges is not determined by the environment (the situational context, the goals, the task) of the system. The pattern is not "organized" or controlled by such parameters (although the degree of environmental control may vary depending on the type of group). Indeed, behavioral patterns of families and groups sometimes seem quite unadapted to the needs and constraints of the environment.

Consequently a group as a self-organized system can be summarized as follows:

i) Group dynamics is always nonlinear, i.e. characterized by recursive causal connections and feedback loops: group process is constituted by positive as well as negative feedback;

ii) a group in development differentiates itself from the environment, i.e. group behavior relates to control parameters in a nonlinear way, too (phase transitions, bifurcations);

iii) groups are open systems (e.g. concerning communication);

iv) group dynamics grows out of a complex micro level (cognitions, emotions and communications);

v) groups are hierarchical in that they form macroscopic (coherent) patterns and structures.

These attributes of groups contain the prerequisites for the emergence of self-organization (points (i), (iii), and (iv), cf. Haken, 1983, 1990; Nicolis & Prigogine, 1987).

For empirical group research a phenomenological approach is recommended because of the inaccessability of most microscopic events. Self-organizational processes can be observed on the macro level, i.e. order parameters are assessed directly through collective variables. In particular, one may study how the homeostatic behavior of a group depends on control parameters. Such macroscopic modeling is proposed by Haken (1988) as the "second foundation of synergetics". Modeling in this vein may also serve as a base for machine-implemented simulations (see below). Further experimental and observational designs that make use of mesoscopic observables have been developed in the context of a dynamical approach to psychology (Tschacher, 1990).

3.3 Group Dynamics and Group Structure

In the previous section we formulated ways in which groups may be conceptualized as self-organized systems: from a micro level perspective, a group encompasses a multitude of psycho-social components; between components structures with long range correlations are built up spontaneously.

Consequently a promising approach for group research can be derived, since the emergence of self-organizing structure presents an interface between group structure and group dynamics. By one such approach we thematized the development of groups (Brunner & Tschacher, 1991). Using the method of "recursive sculpturing" (see Sect. 4.2) we observed the formation of group constellations which arose during the interaction of persons carrying out a task. In our theory there are multiple psycho-social processes running in parallel; these build up self-organized patterns under appropriate conditions. The group's patterns present "free slots" to be filled out with individuals – personalized roles and positions offered and occupied during the formation of a group can therefore be seen as a consequence of self-organizing activity. Thus, roles are emergent features of a group.

Several scenarios of the formation of group structure have been described (Tuckman, 1965; Cissna, 1984; Gersick, 1988). Usually, this was not observed to be a linear monotonic process of increasing group cohesion, performance etc.; rather, sequences of stages with conflict-prone intervals have been reported. Such observations are compatible with models from dynamical systems theory describing "routes to chaos" that take place when control parameters are increased (e.g., the phenomenon of period doubling). "Routes to chaos" are better understood as routes to increasingly differentiated types of order: deterministic chaos is but one special case of self-organization (see Kratky, this volume). At critical points along this route we find bifurcations that function as an (irreversible) shunting during group formation (Fig. 2).

state variables

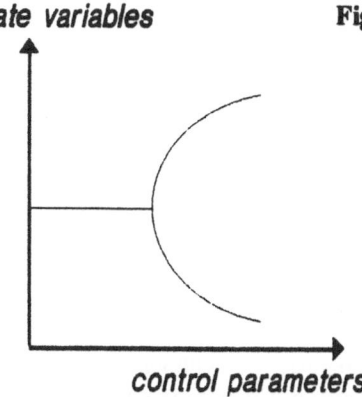

Fig. 2. Bifurcation diagram of group evolution

control parameters

At each stage of group evolution, i.e. along the path between two consecutive bifurcations, a characteristic group structure is realized which is based upon all previous structures. Within this historical dimension of a group the traces of past events are registered in the form of group culture and tradition. The essential aspect of group memory and de facto irreversibility found in the grown structure of psycho-social systems is accounted for by the mathematical model of an irreversible dissipative system.

Thus, the structure and dynamics of a group are intimately connected: i) group structures are emergent features of self-organizing psycho-social processes at a micro level; ii) further, structures may develop irreversibly through many steps in the course of group evolution. Therefore, two corresponding types of processes in groups can be distinguished: first, pattern formation through self-organization and, secondly – on a different time scale – the evolving sequence of group patterns. At critical points in group development, phase transitions have to be expected.

4. Operative Approaches to the Study of Groups

Empirical research on group processes under the auspices of dynamical systems theory has to deal with the restrictions of data acquisition that are characteristic of psychology. Especially the phenomenon of reactivity (what is observed is influenced by the act of observation) has been considered a great problem in psychological measurement. Several theories of social influence apply to the social situation "group experiment" and "group observation": data may be biased because of reactance, dissonance reduction, opportunistic reactions, etc. Competing biases described in social psychology even make prognoses about the direction of effects uncertain. Moreover, in many instances the instruments used for measurement do not satisfy the standards of what is desirable in terms of reliability or data resolution. The acquisition of time series with conventional means like questionnaires, surveys, etc. is further aggravated since repeated applications in short intervals must be ruled out.

Thus, a restricted palette of methods remains for the acquisition of dynamical data on groups; especially observation methods seem appropriate in this respect. We shall not dwell on the broad issue of observation methods here. Rather, a laboratory method will be described that allows studying group development processes (Sect. 4.1). In Sect. 4.2 a simulation approach is presented.

4.1 Circular Causality in Groups

Pattern formation can be observed whenever persons who did not previously know each other come into interaction, either with or without a common goal. Relations previously undefined become defined and structured. Comparing pre-post states gives evidence that order has emerged spontaneously, i.e. without an impact from outside the system; but for research this comparison is insufficient since under the perspective of dynamical systems theory the process of structure formation is essential. Therefore, research methodology has to achieve the following prerequisite standards:

i) Observability and measurability of structure formation; thus,
ii) acquisition of continuous or discrete time series;
iii) minimal reactivity of data acquisition in order to exclude a possible external induction of structure formation.

Substantial reactive effects would have to be expected if an ongoing process of group dynamics were to be interrupted repeatedly, for example by asking members for the completion of sociograms. It would also be difficult to interpret differing assessments of group members. An alternative to this might be in videofilming group processes in order to question members about recorded events later on (video reconstruction). Video reconstruction yields statements on relationship and communication, but also on accompanying cognitions and emotions. Qualitatively distinct phases of relationship formation can be identified quite reliably by this method, but it is difficult to reconstruct time series of quantitative variables out of the interview data.

Our effort to fulfil the above criteria sets out to simultaneously allow the process of group formation and the acquisition of data about this process. Members of a group dynamical experiment (ideally strangers to one another) are asked to position themselves within a confined area, so that distances between subjects (Ss) reflect the experienced closeness (in a third dimension, dominance and submission might be enacted by the use of pedestals, but this was not applied in our experiment). Thus, the distances group members choose lead to a *group sculpture*. Instructions are given to dissolve and reestablish sculptures in short intervals. In this way, distances between members can be measured immediately as indicators of distance regulation or, in other words, as a repeated mapping of relational structure. This method is termed iterative or *recursive sculpturing* (Tschacher, 1990; Schiepek, 1991, Chap. VIII) since each realized sculpture relates to previous sculpturing(s): the structure at time t is a function of the structure at time t–τ. Members get the opportunity to negotiate verbally the formation of the next sculpture; verbal material can be assessed by content analysis and allows for later data gathering by video reconstruction. The group's task is merely to thematize, develop and enact its own group processes.

In this, process and data acquisition coincide insofar as the mapping of the group's dynamics (for the sake of data acquisition) incites and motivates further processes. At first sight, this is the exact opposite of a non-reactive method as stipulated in (iii) – the demonstration of group dynamics catalyzes and alters that which is demonstrated. The recursive-reactive process of sculpturing signifies itself in that both signifier and signified are identical. Methodologically, this is paradoxical: the increased reactivity of data acquisition may even lead to non-reactivity.

The logical form of this paradox is similar to the coincidential figure of so-called performative utterances (Austin, 1961; von Foerster, 1985), i.e. statements that implicitly *do* what they *say*. "I apologize" signifies the act of apologizing and, at the same time, *is* the apology. "I promise" already does the promising. Acting and signifying the action fall into one. As von Foerster puts it, language is usually related only to itself because signs only relate to other signs; the coincidence of sign and signified makes it possible for language to be free from this recursive closure, to reach solid ground.

We propose recursive sculpturing as a method for the gathering of data about self-organizing processes in groups. It can be expected that group dynamical constellations (which may initially be undifferentiated and fluctuating) will converge to a clearly structured constellation with time (i.e., the number of iterations). Groups will produce an "eigensolution" (von Foerster, 1985) or, put differently, group dynamics relaxes to an attractor. Furthermore, if it is correct to conceive of groups as nonlinear dynamical systems then – with changing control parameters – discontinuous transitions between attractors or ordered states (phase transitions) should be observable.

A further expectation – or rather prerequisite for the validity of the method – is that communicative and cognitive-emotional aspects of group processes manifest themselves in spatial distance regulation between persons. There has been much research on this issue in ecological psychology (Aiello, 1987), but also in group dynamics and family therapy practice (Kantor & Lehr, 1975; Schweitzer & Weber, 1983; for a review see Tschacher, 1990).

Figure 3 may serve as an illustration for our method. A sequence of six sculptures of eight Ss (students of pedagogy) is presented. The initial sculpture was formed in a time consuming and hesitative process. The area of the polygon spanned by group members is correspondingly large; it reflects the students' reluctancy. During the next two sculptures distances decreased, until in iteration 4 person C left the field entirely. At the same time the remaining group again increased distances; in iteration 5 three females of the group located themselves back to back in the center of the field. This was declared as the concluding constellation; nevertheless, during the last verbal exchange phase, still another constellation was formed.

It is possible to rate the group members' verbalizations between sculptures.

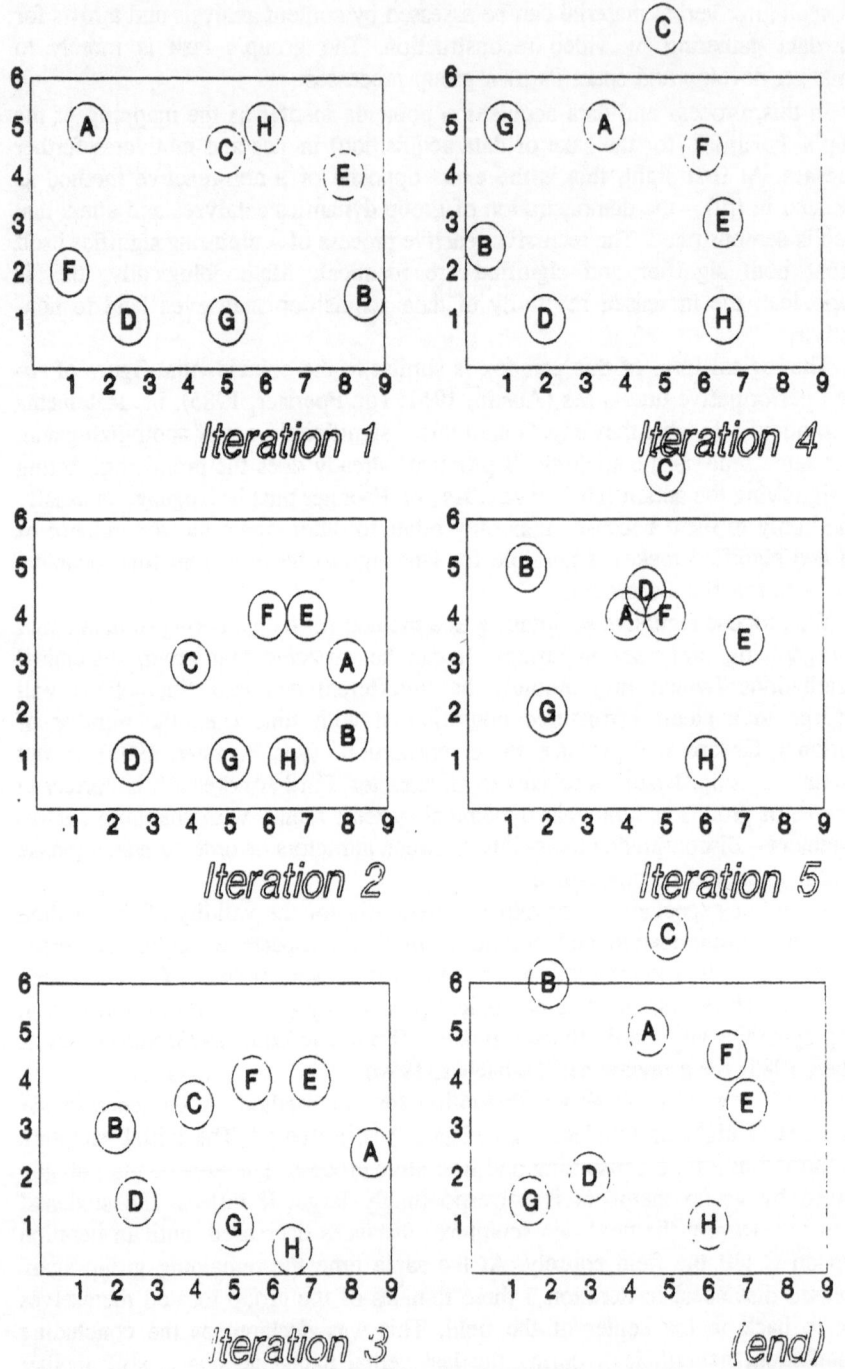

Fig. 3. Group sculptures observed in pilot study (persons A, B, ..., H in a metric field)

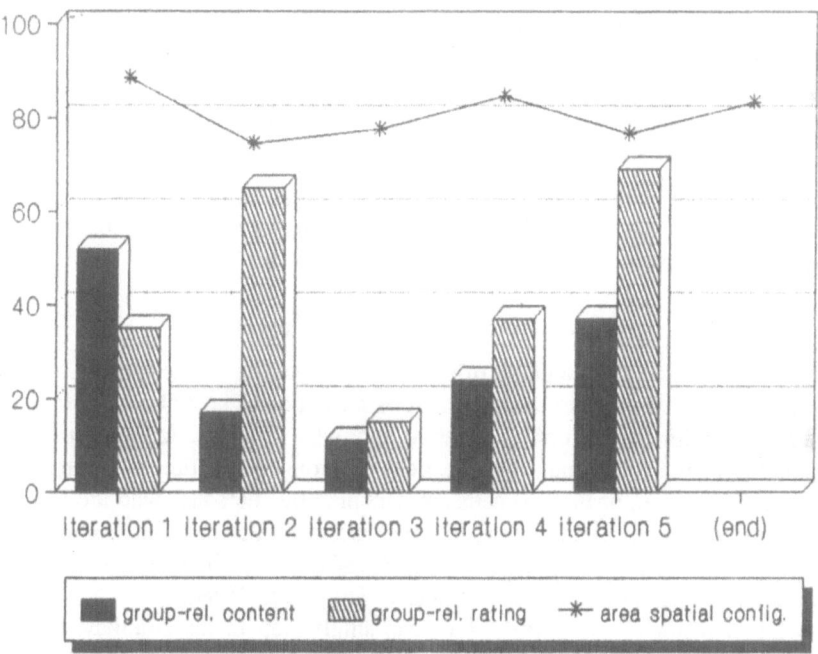

Fig. 4. Diagram of measures of pilot study variables corresponding to group sculptures. From left to right: frequency of group-related content; rating of group-related verbalizations; area of spatial configurations (areas of polygons spanned by persons in sculptures of Fig. 3)

During the five verbal exchange phases in our study, statements were coded as being either self-related or group-related. Two methods were used: i) a content-analytical approach determined the frequency by which certain words (prompts as "group", "we" vs. "I", "my" etc.) appear in the transcripts; ii) two independent raters decided for subsequent parts of the transcript which of the two categories applied (inter-rater reliability was rather low in spite of training: Cohen-Kappa K=.62 and r_{tet}=.78). The results presented in Fig. 4 shall not be interpreted further since there is little convergence between the rating method and content analysis. Moreover, the present data base is insufficient.

Consequentially, the pilot study reported gave rise to several critical questions.

1) Even with short intervals between iterations and highly motivated Ss the time series are not sufficiently long, so that complex methods of analysis (ARIMA, FFT) are not applicable. Above all, the generated series at the present do not permit one to identify group dynamical attractors.

2) Spatial distances are hard to interpret psychologically. Although psycho-emotional dynamics will in principle result in spatial behavior, it remains unknown which agglomeration of motives may lead to which interpersonal distances. For

353

instance, how is a person to behave in the face of both affiliation wishes and social anxiety (approach-avoidance conflict)?

3) Groups in recursive sculpturing reflect upon themselves. There are no goals, tasks or themes like in everyday communication. For this reason, the external validity and generalizability of results gained by this method are unclear.

The method's validity therefore rests more upon "pure" group dynamics realized especially in encounter or therapy groups. This is nonetheless of considerable theoretical interest to social psychology.

4.2 A Simulation Model for Group Development Processes

In order to complement the method of recursive sculpturing we designed a model of self-organizational processes in groups. The formation of spatial configurations was simulated with the help of an iterative computer program. A prototype was implemented in the following way: the field of group interaction is given by a 20x30 array. Eight out of the 600 cells are occupied by "persons" who are free to move according to given optimum distances between persons. The simulation system was inspired essentially by Dewdney (1986). A biocybernetical simulation of spatial and social structure can be found in Hogeweg (1989).

a) **Description of the simulation.** One single step in the simulation is as follows: Person A registers his distance from each of the other persons B through H and in each case computes the difference between the actual distance and the given optimum. The sum of differences serves as an indicator for A's satisfaction with his momentary position. The same computation is carried out for each of A's eight neighboring cells. Then A is moved to the cell with the highest value of satisfaction, i.e. the lowest sum of differences. An iteration is finished after all persons have completed this cycle and have migrated according to their respective values of satisfaction (or have remained in an already optimal position). A "satisfaction value" can be attached to any cell of the field, so that the field as a whole becomes a *potential landscape* for each person (Fig. 5). High satisfaction, i.e. approximately optimal distances to all other persons, corresponds to a valley or sink (an attractor), low satisfaction corresponds to a hill (a repellor).

Interaction within the group of simulated persons is evoked by the competition of eight such potential landscapes. Competition thus avoids the trivial case of all persons directly moving into the potential minima closest to initial positions. Mathematically the simulation is a discretized approximation of a system of 16 ordinary differential equations that define the changes of coordinates of eight persons in the plane. As differential equations, the system is linear since the derivatives of quadratic distance equations contain no nonlinear terms. The nonlinearity of the system stems from discretization. The simulation can be viewed as a variant of a cellular automaton. Cellular automata are discrete dynamical systems that are able to produce complex macroscopic patterns from simple local rules. Von Neumann, pioneer of the serial computer, first used cellular automata with the goal of modeling biological processes such as the self-reproduction of structures (von Neumann & Morgenstern, 1963; Langton, 1989). These automata

Fig. 5. Schematic representation of potential landscape in group simulation

also have been proposed recently as a paradigm of parallel processing (Kemke, 1988). From the part of dynamical systems theory it is noteworthy that cellular automata show phenomena such as breaking of symmetry, order formation, fractality and chaos (Wolfram, 1984; Toffoli & Margolus, 1987; Hayes, 1988).

b) Results of simulations. For each run of the simulation system, a number of values have to be determined: the basic potential functions can be varied widely by setting optimum distances; also the initial state of the system (i.e. initial positions of persons in the field) must be chosen. Thus a wide variety of initial conditions is offered. A phenomenological study of resulting simulations shows that there is also a multitude of discernible categories of behavior. The system depends on initial conditions quite sensibly. To give an example: let the optimum distance matrix (of the form of Table 1) be symmetrical with a common value of $d_{ij}=13$. From certain initial positions for persons a long sequence of changing constellations results, which after some 200 iterations turns into a stable configuration. When in a second run the initial position of person A with coordinates (x/y) is shifted from (5/17) to (6/17), the system enters an equilibrium state after only twelve iterations. Both resulting configurations have a similar shape (an irregular circle) as would be expected from the homogenous matrix; but relative positions of persons A to H are again different (Fig. 6).

Qualitatively different behavior is observed when optimum distances are chosen in a way that they correspond to distances of some actual configuration (for instance, the distances between A, B, ..., H realized in the first constellation of Fig. 3). The matrix is symmetrical then since distance AB=BA, etc.; there are no conflicting values for distance optima between any two persons. In such cases simulations are always *equifinal* – the original configuration (or some rotated variant) will be established from all initial system states. In other words, a point attractor is determined by the matrix that gives the control parameters.

Different behavior results from parameters that contain conflicting distance settings. In the following example, A tries to maintain high distances from all

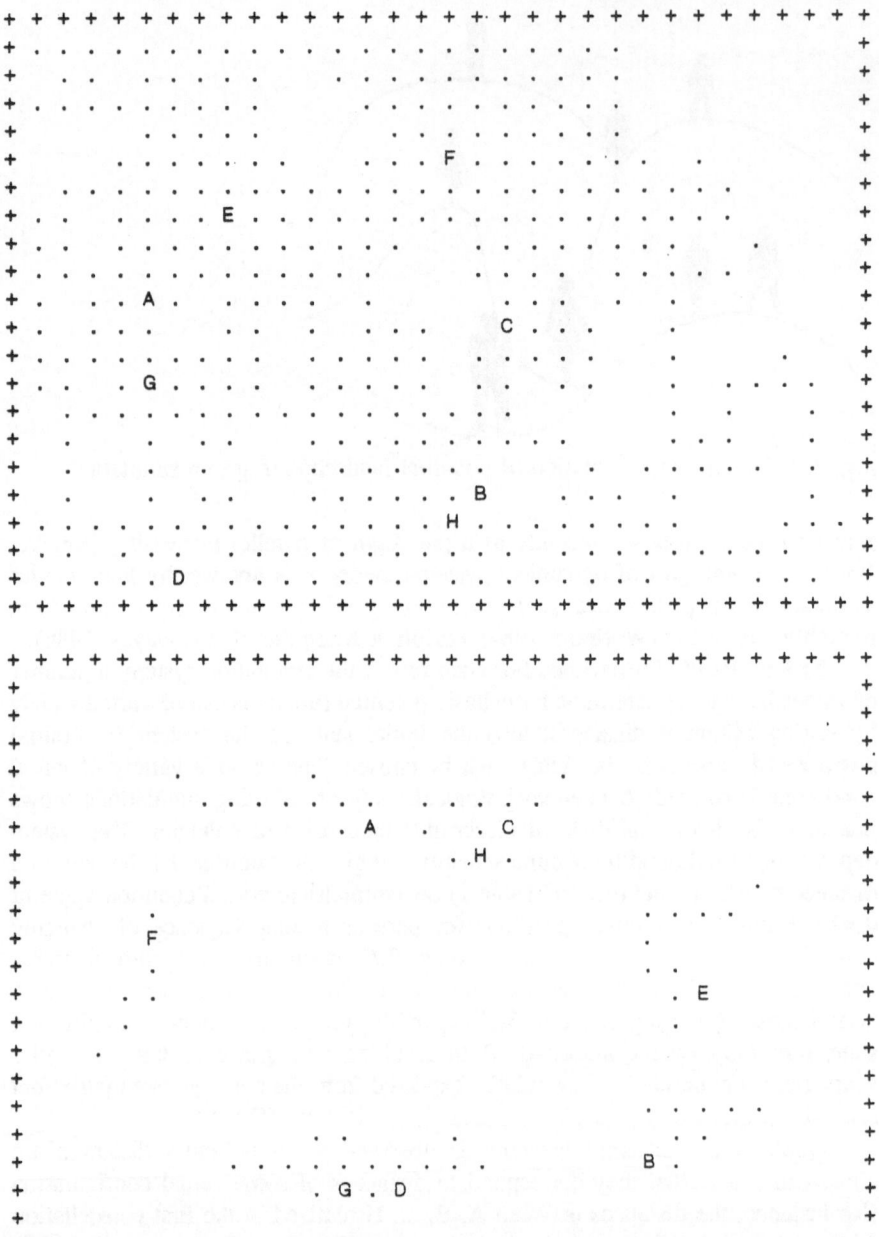

Fig. 6. Configurations of group simulation with two slightly different initial values ("persons" A, B, ..., H); top: after 194 iterations; bottom: after 13 iterations

other persons; B through H as a group aim at small distances from A while having small but conflicting optimum distances among themselves. The resulting asymmetrical matrix is given in Table 1. The simulation takes a peculiar course

Table 1. Matrix of optimum distances of the simulation. Example represents conflict within the group (see text)

	A	B	C	D	E	F	G	H
A	0	11	11	11	11	11	11	11
B	1	0	2	1	3	2	1	2
C	1	2	0	2	2	1	3	2
D	1	2	2	0	3	3	3	3
E	1	1	2	1	0	2	1	1
F	1	2	3	1	2	0	3	2
G	1	2	3	1	1	2	0	1
H	1	2	2	1	2	3	3	0

while not approaching any of the usual stationary states. Persons form a close turbulent unit instead; occasionally the configuration resembles a "glider" (i.e. a stable or oscillating pattern moving through the field in a cellular automaton). The gliders get deformed at the edge of the field, go into a phase of restructuring and start moving in a different direction. A deterministic time series analysis of this example is presented in the chapter by Steitz et al.

A general phenomenological description of the simulation model yields the following categories of equilibrium states: ·

–*fixed points*: stationary constellations where the system remains unchanged;
–*limit cycles*: a terminal constellation oscillates between two or more states;
–*gliders*: constellations wandering through the field, sometimes oscillating;
–*turbulent states* (deterministic chaos not yet proven).

c) Expansion of the simulation model. From a dynamical systems point of view it seems legitimate to study the simulation model as a complex system for its own sake – nevertheless, the system was designed as a model of distance regulation in a developing group. Which is the relation between model and real group events?

The model is obviously very simple because optimum distances will not vary during simulation. In real groups we would expect distance "needs" to change continually depending on interactions during group formation. First explorative studies with groups (see Sect. 4.1 and Tschacher & Brunner, 1990) demonstrate this as a lack of one degree of freedom in the simulation. Yet we think our model shows that even simple model machinery can produce very complex behavioral patterns and equilibrium constellations and may therefore map group processes on a macroscopic level. A deep simulation of group processes is clearly beyond the reach of a computer experiment, but it makes sense to simulate phenomenological patterns with as simple a model as is possible in order to sound out formal analogies of group development. This is especially important because, in the framework of synergetics, we expect that in social systems simple patterns will evolve out of a complex psycho-social micro level.

It is possible to design a more realistic model by adding another recursive loop: distance matrices now constant may depend on the output of the simulation; e.g., closer distances are "learned" after they have repeatedly appeared in previous iterations (in principle, any number of hidden "motivational" variables can be introduced to interact and influence spatial behavior). The potential landscape causing the group's behavior will thus be modified itself by behavior. In this way the simulation model could be expanded as a kind of multidimensional cybernetic model, as in Bischof (1985).

Another extension of the realized version would be allocating different dimensions to the "field" that the group is mapped to. Originally this simulation was designed as a spatial model of laboratory distance behavior, but there is no need to conceive the field of group dynamics as Euclidean. A conceptualization based upon topological representations of psychological and social dimensions relates to Lewin's (1947) "social field". A social field develops from the superposition of the group members' life spaces (Fig. 7).

Lewin sees as a basic property of group life the circular causal processes controlling individual and group action. In the social context perception and action are linked such that perception depends on the way a social situation is changed by action and vice versa. The proximity of this socio-topological approach to system-oriented group psychology is obvious. In the context of our simulation the social field of a group may be modeled as follows: presently potentials are defined as fixed to individuals, moving with their spatial positions; in order to represent a life space or field in the sense of Lewin stationary potentials can easily be defined which influence locomotions additionally. The field then becomes a potential

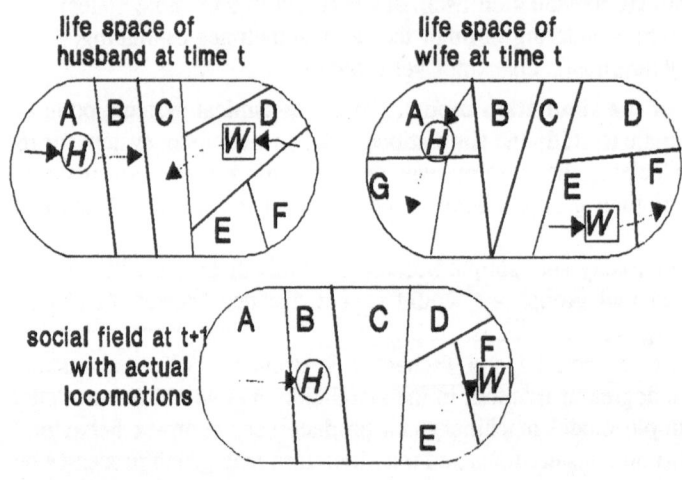

(H) : husband

[W] : wife

Fig. 7. The social field as a superposition of life spaces (after Lewin, 1947). Arrows in life spaces indicate intended (or expected) locomotions

landscape built up from stationary potentials (general laws affecting on the dynamics of a group) and personal potentials (as in the distance matrices above). This will provide a computational shell which can help model therapy situations described with a synergetic potential metaphor (see Schiepek, Fricke & Kaimer, this volume). A further application is planned in connection with the "systems game" approach – here distances can be modeled as social distances (rather than metric ones) by variables of the protocols (e.g. frequencies of contact between persons, announcements of protagonists, etc.; see Schiepek & Reicherts, this volume).

5. Self-Organized Groups in Organizations: Synergy or Paradox?

5.1 Self-Organization Phenomena in the Firm

In this volume it has been stated repeatedly that phenomena of self-organization are relevant to psychotherapy theory and practice. From this point of view we will now try to contribute to the analysis of a quite different issue – the economic organization – where the dynamics of groups is of major importance.

The tension between organization and individual is commonly viewed as central for the understanding of organizations (Wiedl & Greif, 1991). We hold this to be the minor truth since the function of groups in organizations is neglected in this statement. First of all, organizations in general have started as small groups, and are composed of multiple small groups. Big companies can be described as miniature societies consisting of groups that interact in various ways. All groups are interlinked by the members they have in common (Likert, 1961) and by being involved in the same social network, the hierarchy of the organization. Each group tends to develop norms, standards, and traditions essential to the climate for work and achievement. New members that join the group go through the same stages of socialization (McDavid & Harari, 1968). On the organizational level culture and corporate identity may be established, too. Culture encompasses norms and values that characterize an organization; even rituals and behavior patterns are to a degree determined by the style of the company (Schein, 1985; Matenaar, 1989).

Today it is seen as an important aspect of successful organizations that they provide an attractive climate and identity. One of the advantages of Japanese leadership seems to be that the company's philosophy is preserved and continued independently of changes in top management (Bleicher, 1989).

In this context we want to stress the fact that culture must be seen as an emergent phenomenon caused by psycho-social self-organization which cannot be introduced or changed quasi-mechanically, "by decree". Thus, Doctoroff (1977) points out the importance of informal organization in producing *synergy effects*, for example through "unpurposeful meetings". Sprüngli (1981) counts trust, effective communication, rapid feedback and creativity among the prerequisites for positive synergistic effects.

As can be seen, the emergence of norms, standards and culture in groups and in the entire organization is to a large degree based upon self-organized processes of group dynamics. The groups (formal and informal) are fundamental for all emergent attributes of organizations. The emergent aspects (culture, climate, philosophy) strongly influence the creativity and efficiency of the entire social system. It is therefore necessary to regard small group dynamics in questions of management and organizational development. "Not the behavior of single humans, but the behavior of social systems is the subject of management theory" (Ulrich, 1984, p. 87). From a slightly different standpoint, Schein (1985) posits that the "unique and essential function of leadership is the manipulation of culture". This begs the question of what management might look like under this supposition.

5.2 Management of Self-Organized Systems

When we apply self-organization theory to social groups and social organizations we have to keep in mind that these are purposeful and goal-oriented structures. Organizations and enterprises do not rely on spontaneous pattern formation but are founded in order to manufacture services or products in a rational way. Generally, self-organizing dynamics in goal-oriented organizations is regarded as interfering with planned processes and therefore as dysfunctional – self-organization is viewed as misorganization and is not expected to contribute to organizational goals. How can the relations between organization and self-organization be determined in this respect? How do these opposing dynamics connect?

In the social sciences, two quite contrary views are discussed in relation to the question of planning and intervening in social systems.

The extreme position of *non-interventionism* is held by some theoreticians of systemic psychology. Essentially two arguments for non-interventionism are put forward. In non-trivial machines input and output are not directly connected but are regulated by state variables of the machine; from the resulting abundance of combinations of input, state, and output von Foerster (1985) concludes that even simple non-trivial machines must be unpredictable. The second argument refers to self-reference: in social systems interventions are started from inside the system which leads to paradoxes because of a mixing of logical types; rather than observing something else, the system observes its own observation; planning becomes self-planning (Krohn & Küppers, 1990). Finally, there are outcomes from other disciplines, e.g. cognitive psychology, that confirm the difficulty of intervening in multicausal systems. Linear manipulations of complex systems are supposed to lead to negative results; this even seems to follow a "logic of failure" irrespective of the intelligence and motivation of the manipulator (Dörner, 1989).

On the other hand a position of *interventionism* thrives and dominates in everyday practice. Having diagnosed maladaptive behavior (symptoms), an interventionist would try to "cure", transform, or extinguish this behavior directly. Behavior therapy in its early naive phase, for example, posited that the behavior of individuals and groups could be shaped almost unrestrictedly by controlling stimuli and contingencies. The repertoire of methods applied has been greatly expanded in the meantime (e.g. multi-modal behavior therapy and behavior medicine) and the banishment of cognitive constructs has been given up (cognitive learning theory). The linear causal thinking of interventionist theories has nevertheless survived. The same applies to established organizational theories in industrial economics: companies are above all seen as hierarchical systems; here also interventionist thinking is still effective in the shape of action theory and decision theory (Ochsenbauer, 1989).

In our view, both of the above positions become untenable when the results of self-organization research are considered (Tschacher & Brunner, 1992). The non-trivial machine cannot be described analytically as only its unpredictable behavior is given; one is reminded of a deterministic but chaotic dynamics (Bergé et al., 1984). But chaos is not necessarily predominant or typical in complex systems. Therefore von Foerster's implicit argument (if simple variable systems can be chaotic, complex systems should even be more likely to show irregular behavior) is appealing but invalid. Synergetics and self-organization theory have shown that in very complex recursive systems, the opposite phenomenon – an enormous reduction of degrees of freedom – arises. The stunning result of this type of research is rather: the more complex the system, the simpler its behavior. On the other hand interventionist ideas are contradicted by self-organization theory as well: the spontaneous activity of systems is obviously contrary to a primitive can-do! approach as in early behavior therapy.

When tackling the problem of intervention and planning along these lines one is forced to differentiate. The question is not: Can we plan and intervene purposively at all? but rather: What restrictions and options are inherent in interventions into systems that show self-organization?

a) **Restrictions.** We have to consider the irreversibility of self-organized systems. Therapeutic or managerial influences and communications cannot be undone in psycho-social systems. The theory of dissipative structures leads to conclusions that are well-known in systemic therapy: one cannot *not* communicate (Watzlawick et al., 1969), nor can communications be "taken back" by selecting some kind of inverse communication. All the same the system differentiates itself via bifurcations into new behavior patterns, while the same route back is barred. This applies to management decisions too (cf. Ulrich & Probst, 1990). In sum, there are connections between the second law of thermodynamics and axioms of intervention pragmatics.

Secondly, long term planning will be possible only in rare cases. The evolution of a system may undergo surprising branchings since close to instability points chance fluctuations can decide in which way the system evolves further. The problem of prognosis is obviously most acute in the case of chaos because the sensitivity to small perturbations becomes permanent then.

Finally, attractors (equilibria) cannot be influenced directly and at will. Within a certain range of boundary conditions (control parameters) and under the permanent influence of fluctuations the system states fall into the equilibrium again and again (Fig. 8). One of many examples would be the following observation: single employees are often delegated to attend training seminars; back in their groups or departments employees usually "forget" whatever new skills and procedures they had just acquired. Thus, trying to influence behavior directly may

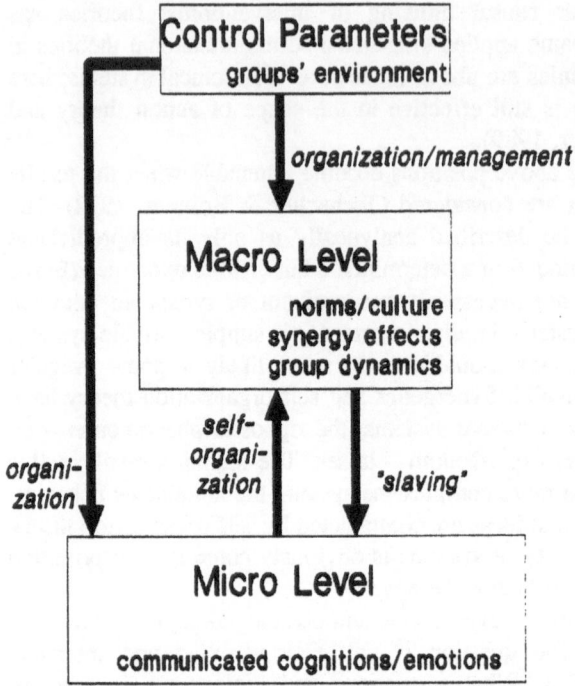

Fig. 8. Scheme of self-organization and organization in a social system

turn out to be a futile fight against the dynamics of an unchanged system; instead of "breaking the resistance" the method of choice will rather consist of indirect action (like intervention on the "meta-level", cf. Malik, 1984). But additionally, there is still self-reference: a manager who sets out to change the culture of an organization is herself or himself culturally biased to an unknown degree (Schein, 1985).

b) **Options.** On the other hand, the systemic view can give some clues to better achievement that rests in self-organized dynamics; these options may be annihilated by linear advances.

One advantage can unfold when the adaptability and creativity of evolving systems is utilized. The trust in Tayloristic specialization and increasing division of labor as a principle of success has been overturned in the last years. In his review of type Z organizations, Ouchi (1981) describes such companies as promoting a coherent culture of egalitarian rather than rigidly hierarchical interaction. Ouchi points to the Japanese ritual of *ringi*, i.e. collective decision making.

If one tolerates solutions that are useless at first sight, unforeseen solutions to problems may open up. From a self-organization perspective it makes sense to follow the present trend towards participative models of leadership and organizational development. The creative potential that is embedded in the dynamics of the constitutive small groups is to be utilized rather than fought as an

362

disturbing "shadow organization". Autonomous work teams and quality circles are well-known examples for promising approaches towards new ways of delegation (Weinert, 1987). In order to utilize synergy effects flat organizational structure and decentralization seem adequate. Less hierarchy has the additional advantage of requiring less interfaces in an organization's information flow.

Especially in the design of groups that are supposed to develop new products or projects, evolutionary aspects should be considered. Here it is important to preserve a broad spectrum of alternatives for some time at the start of the project; immediately aspiring to a fixed (and necessarily fictitious) goal often turns out to be dysfunctional.

Leadership should be understood as an indirect and, at most, strategic intervention. In interventions of a small range we advise use of the sequence of pacing→leading which is a related principle in systemic therapy and hypno-therapy: it is not very effective to try to control the system right from the start – rather one should follow the system's own dynamics first (pacing), and only later get hold of some control out of the impact gathered until then (in the field of management, cf. the "jiu-jitsu principle", Gomez, 1981).

How can we investigate indirect management further? Influencing the attractors of self-organized systems can be achieved primarily by shifting control parameters, i.e. the conditions in the system's environment. Control parameters change the potential landscape, and thereby indirectly change the path that the system will follow. According to our theory, interesting control parameters of groups in organizations will be those variables that establish the distance from thermodynamical equilibrium; this distance is realized by flows of matter, energy and information through the system's permeable boundaries. Correspondingly (as a proposal for further research) pattern formation and pattern change should be studied in parameter spaces spanned by variables that measure the flow of energy (resources, budget, money?) and especially information (which is not easy to operationalize, but say social or political environment, setting of goals, pressure for efficiency and the like). As a result of a research enterprise of this kind we would learn about the bifurcation scenarios of psycho-social systems in organizations and also learn how to manage (and adapt to) the nonlinear life of groups.

Acknowledgments

We are grateful for the contribution of Peter Tritschler who implemented the simulation together with the first author of this article. Thanks go also to participants of the Herbstakademie 1990 for helpful discussions.

References

AIELLO, J. 1987. Human Spatial Behavior. In: Stokols, D. & Altman, I. (Eds.): Handbook of Environmental Psychology, New York: Wiley, 389-504.

AUSTIN, J.L. 1961. Performative Utterances. In: Urmson, J.O. & Warnock, G.J. (Eds.): Philosophical Papers. Oxford: Carendon Press, 220-240.

BALES, R.F. & COHEN, S.P. 1979. SYMLOG. A Manual for the Case Study of Groups. New York: Macmillan.

BALES, R.F. & SLATER, P. 1955. Role Differentiation in Small Decision-Making Groups. In: Parsons, T. & Bales, R.F. (Eds.): Family, Socialization, and Interaction Process. Glencoe: Free Press, 259-306.

BALES, R.F. & STRODTBECK, F.L. 1951. Phases in Group Problem Solving. Journal of Abnormal and Social Psychology, **46**, 485-495.

BARKER, R.G. 1968. Ecological Psychology. Stanford: Stanford University Press.

BERGÉ, P., POMEAU, Y. & VIDAL, C. 1984. Order within Chaos. Towards a Deterministic Approach to Turbulence. New York: Wiley.

BISCHOF, N. 1985. Das Rätsel Ödipus. Die biologischen Wurzeln des Urkonfliktes von Intimität und Autonomie. München: Piper.

BLEICHER, K. 1989. Chancen für Europas Zukunft. Führung als internationaler Wettbewerbsfaktor. Wiesbaden: Gabler.

BRUNNER, E.J. 1986. Grundfragen der Familientherapie. Systemische Theorie und Methodologie. Berlin: Springer.

BRUNNER, E.J. & TSCHACHER, W. 1991. Distanzregulierung und Gruppen struktur beim Prozeß der Gruppenentwicklung. I.: Theoretische Grundlagen und methodische Überlegungen. Zeitschrift für Sozialpsychologie, **22**, 87-101.

BUNGE, M. 1979. Treatise on Basic Philosophy, Vol. 4 (Ontology II: A World of Systems). Dordrecht: Reidel.

CISSNA, K.N. 1984. Phases in Group Development: the Negative Evidence. Small Group Behavior, **15**, 3-32.

DEWDNEY, A.K. 1986. Computer Recreations. Scientific American, **255** (No.2), 10-15.

DOCTOROFF, M. 1977. Synergistic Management: Creating the Climate for Superior Performance. New York: AMACOM.

DÖRNER, D. 1989. Die Logik des Mißlingens. Reinbek: Rowohlt.

FOERSTER, H. v. 1985. Sicht und Einsicht. Braunschweig: Vieweg.

GERSICK, C.J.G. 1988. Time and Transition in Work Teams: Toward a New Model of Group Development. Academy of Management Journal, **31**, 9-41.

GOMEZ, P. 1981. Modelle und Methoden des systemorientierten Managements. Bern: Haupt.

HAKEN, H. 1983. Advanced Synergetics. Instability Hierarchies of Selforganizing Systems and Devices. Berlin: Springer (3rd ed.).

HAKEN, H. 1986. Erfolgsgeheimnisse der Natur. Stuttgart: Deutsche Verlags-Anstalt (4th ed.).

HAKEN, H. 1988. Information and Self-Organization. A Macroscopic Approach to Complex Systems. Berlin: Springer.

HALEY, J. 1976. Problem-Solving Therapy. New York: Harper & Row.

HALL, A.D. & FAGEN, R.E. 1968. Definition of System. In: Buckley, W. (Ed.): Modern Systems Research for the Behavioral Scientist, Chicago: Aldine, 81-92.

HALL, E.T. 1966. The Hidden Dimension. Garden City, N.Y.: Doubleday.

HAYES, B. 1988. Zelluläre Automaten. Spektrum der Wissenschaft, Sonderband Computer II, 60-67.

HOGEWEG, P. 1989. MIRROR beyond MIRROR, Puddles of LIFE. In: Langton, C.G. (Ed.): Artificial Life. Redwood City, Ca.: Addison-Wesley, 297-316.

HOMANS, G.C. 1961. Social Behavior: Its Elementary Forms. New York: Harcourt Brace.

JANIS, J.L. 1972. Victims of Group Think. A Psychological Study of Foreign Policy Decisions and Fiascoes. Boston.

KANTOR, D. & LEHR, W. 1975. Inside the Family. Toward a Theory of Family Process. San Francisco: Jossey-Bass.

KEMKE, C. 1988. Der neuere Konnektionismus. Informatik-Spektrum, 11, 143-162.

KROHN, W. & KÜPPERS, G. 1990. Selbstreferenz und Planung. In: Niedersen, U. & Pohlmann, L. (Eds.): Selbstorganisation und Determination. Berlin: Duncker & Humblot, 109-127.

LAMBERT, W.W. & LAMBERT, W.E. 1964. Social Psychology. Englewood Cliffs: Prentice Hall.

LANGTON, C.G. (Ed.) 1989. Artificial Life. SFI Studies in the Sciences of Complexity. Redwood City, Ca.: Addison-Wesley.

LEWIN, K. 1947. Frontiers in Group Dynamics. Human Relations, 1, 2-38.

LIKERT, R. 1961. New Patterns of Management. New York: McGraw-Hill.

LUHMANN, N. 1984. Soziale Systeme: Grundriß einer allgemeinen Theorie. Frankfurt: Suhrkamp.

MALIK, F. 1984. Strategie des Managements komplexer Systeme. Ein Beitrag zur Management-Kybernetik komplexer Systeme. Bern: Haupt.

MATENAAR, D. 1989. Entwicklungstendenzen der Unternehmenskulturforschung. In: Seidel, E. & Wagner, D. (Eds.): Organisation. Evolutionäre Interdependenzen von Kultur und Struktur der Unternehmung. Wiesbaden: Gabler, 325-340.

MCDAVID, J.W. & HARARY, H. 1968. Social Psychology. Individuals, Groups, Societies. New York: Harper & Row.

MILLER, J.G. 1978. Living Systems. New York: McGraw-Hill.

MILLS, T. M. 1967. The Sociology of Small Groups. Englewood Cliffs: Prentice Hall.

NEUMANN, J. v. & MORGENSTERN, O. 1963. Theory of Games and Economic Behaviour. Princeton: University Press.

NICOLIS, G. & PRIGOGINE, I. 1987. Die Erforschung des Komplexen. Auf dem Weg zu einem neuen Verständnis der Naturwissenschaften. München: Piper.

OCHSENBAUER, C. 1989. Organisatorische Alternativen zur Hierarchie. München: GBI-Verlag.

OUCHI, W.G. 1981. Theory Z. How American Business Can Meet the Japanese Challenge. Reading, Mass.: Addison-Wesley.

PARSONS, T. 1971. The System of the Modern Societies. Englewood Cliffs: Prentice Hall.

SCHEIN, E.H. 1985. Organizational Culture and Leadership. San Francisco: Jossey-Bass.

SCHIEPEK, G. 1991. Systemtheorie der Klinischen Psychologie. Braunschweig: Vieweg.

SCHNEIDER, H.-D. 1985. Kleingruppenforschung. Stuttgart: Teubner.

SCHWEITZER, J. & WEBER, G. 1983. Beziehung als Metapher: Die Familienskulptur als diagnostische, therapeutische und Ausbildungstechnik. Familiendynamik, 8, 113-128.

SHALINSKY, W. 1983. One-Session Meetings: Aggregate or Group? Small Group Behavior, 14, 495-514.

SOMMER, R. 1969. Personal Space. Englewood Cliffs: Prentice-Hall.

SPRÜNGLI, K. 1981. Evolution und Management. Ansätze zu einer evolutionistischen Betrachtung sozialer Systeme. Bern: Haupt.

TOFFOLI, T. & MARGOLUS, N. 1987. Cellular Automata Machines. A New Environment for Modeling. Cambridge: MIT Press.

TSCHACHER, W. 1990. Interaktion in selbstorganisierten Systemen. Grundlegung eines dynamisch-synergetischen Forschungsprogramms in der Psychologie. Heidelberg: Asanger.

TSCHACHER, W. & BRUNNER, E.J. 1992 (in press). Organization and Self-Organization. In: Wunderlin, A. & Friedrich, R. (Eds.): Evolution of Dynamical Structures in Complex Systems. Berlin: Springer.

TUCKMAN, B. 1965. Developmental Sequence in Small Groups. Psychological Bulletin, 63, 384-399.

ULRICH, H. 1984. Management. Bern: Haupt.

ULRICH, H. & PROBST, G.J.B. 1990. Anleitung zum ganzheitlichen Denken und Handeln. Ein Brevier für Führungskräfte. Bern: Haupt (2nd ed.).

WATZLAWICK, P., BEAVIN, J.H. & JACKSON, D.D. 1969. Menschliche Kommunikation (Formen, Störungen, Paradoxien). Bern: Huber.

WEIDLICH, W. & HAAG, G. (Eds.) 1983. Concepts and Models of a Quantitative Sociology. The Dynamics of Interacting Populations. Berlin: Springer.

WEINERT, A.B. 1987. Lehrbuch der Organisationspsychologie. Menschliches Verhalten in Organisationen. München: Psychologie Verlags Union (2nd ed.).

WIEDL, K.H. & GREIF, S. 1991. Störungen betrieblicher Organisationen: Intervention. In: Perrez, M. & Baumann, U. (Eds.): Lehrbuch Klinische Psychologie (Bd. 2: Intervention). Bern: Huber, 395-407.

WOLFRAM, S. 1984. Universality and Complexity in Cellular Automata. Physica, 10 D, 1-35.

Applicability of Dimension Analysis to Data in Psychology

Arno Steitz, Wolfgang Tschacher, Klaus Ackermann, and Dirk Revenstorf

With 16 Figures

Abstract. This paper is motivated by the question of whether dimension analysis is a valid and practical method for the reduction of data in psychology. The paper presents a short introduction to the analysis of chaotic systems by the Grassberger–Procaccia algorithm. General aspects of this method are demonstrated; we tested the limits of dimension analysis depending on signal–to–noise ratio, length of time series, and resolution of measurement. For this purpose, the Hénon map was used as a basic model. The Grassberger–Procaccia algorithm was also applied to a simulated time series of group processes and an empirical time series of smoking behavior. To compensate for artefacts induced by local correlations a revised dimension analysis was performed with the group simulation data. Results suggest that neither group simulation nor cigarette consumption data can be reduced to a low–dimensional deterministic system.

1. Introduction

The study of various physical and mathematical models has shown that even simple nonlinear systems display very complex behavior within certain ranges of their parameters. Time series of these systems may look like irregular random series but actually are totally determined by only few variables, a phenomenon referred to as deterministic chaos. Therefore, one may search for the fingerprints of simple systems in observational data of noise–like complexity. In the biological and social sciences analyses for chaotic behavior have been applied, for example, to epidemiology (Schaffer & Kot, 1986), chronobiology (an der Heiden, this volume), cardiac electrophysiology (Glass et al., 1986) and, with special relevance to psychology, to EEG activity (Babloyantz, 1985; Graf & Elbert, 1989).

We are basically interested in analyzing dynamical systems within clinical psychology. The most convenient and thus most popular tool to find such fingerprints of nonlinear systems is the evaluation of the correlation dimension by means of the Grassberger–Proccacia algorithm (Grassberger & Procaccia, 1983). We studied some aspects of its application analyzing time series from three systems. The first one is the well–known analytical Hénon map. The second is a time series from a computer model simulating group interaction. The third time series consists of about 1000 observations concerning a person's daily cigarette consumption.

Springer Series in Synergetics, Vol. 58   Self-Organization and Clinical Psychology
Editors: W. Tschacher, G. Schiepek, and E.J. Brunner   © Springer-Verlag Berlin Heidelberg 1992

2. An Introduction to Dimension Analysis of Chaotic Systems

Since a number of papers are available that give a good overview of the dimension analysis of chaotic systems (Simm et al., 1987; Mayer–Kress, 1987), we start with a short introduction to dimension analysis with the Grassberger–Proccacia algorithm.

The set of mutually independent variables necessary to describe the behavior of a system span the system's **state space** (the variable set of a particular state of the system corresponds to one coordinate point in the state space). In general, we do not know the number of variables governing the system nor do we know the effective dimension of its state space. Instead, we measured the evolution in time of a system by one or few projections of the system's state space onto our measure quantities. Time series that show no spectral structure, i.e. no clear–cut periodic pattern, may result either from projections in noisy directions, or from projections of high dimensional state surfaces, or from low dimensional but geometrically complex attractors. Dimension analysis will discriminate the latter from the two former ones (i.e. discriminate systems which are governed by few, from those which are governed by a large number of variables).

From the study of dissipative chaotic model systems we know that their evolution in time — at least on chaotic time scales — is unpredictable due to the exponential divergence of initially close trajectories. Nevertheless, the possible states of the system do not fill up the whole volume of the state space but are confined to the so–called chaotic attractor — a generalization of the concept of state surfaces to which the evolution of a system is bound. By describing the geometry of the attractor, dimension analysis can be one way to recognize and to characterize a chaotic system.

2.1 The Generalized Dimensions

How can we describe an attractor (i.e. the ensemble of state points in the state space of a chaotic system) by its geometrical properties? To begin with, we assume sufficiently long time series of a chaotic system in all its relevant variables. The information of all time series can be represented as an ensemble of state points in a state space of dimension equal to the number of relevant variables. Each point represents a state of the system at a given point in time.

The probability distribution of the attractor points, i.e. the probability of finding one out of N attractor points in cell k under a covering Ω of the state space by $M(r)$ cells of radius r is

$$p_k(r) = \frac{N_k(r)}{N(r)} \tag{1}$$

where N_k is the number of attractor points in cell k. (Every point of the state space lies in only one cell of the covering Ω.)

This enters in the definition of the so-called q-dimensions which describe the 'strange' geometrical structure of a chaotic attractor. The q-dimensions can be seen as the moments of an expansion of the probability distribution of the attractor points:

$$D_q = \frac{1}{q-1} \lim_{r \to 0} \frac{\log(\sum_{k=1}^{M(r)} p_k^q)}{\log(r)} . \tag{2}$$

The Hausdorff dimension D_0 can easily be seen to correspond to the standard dimension for n-dimensional surfaces. It can be shown that in general $D_{q+1} \leq D_q$ (Hentschel & Procaccia, 1983).

2.2 The Correlation Dimension and Its Approximation by the Grassberger–Proccacia Algorithm

For practical purposes we are not able to calculate all q-dimensions (especially for large q) because of the computational burden or lack of sufficiently long time series. For D_2 a practical algorithm to approximate the above definition in (2) has been suggested by Grassberger and Proccacia. From (2) the D_2 or correlation dimension for a dense covering Ω with $r \to 0$ is written

$$D_2 = \lim_{r \to 0} \frac{\log(\sum_{k=1}^{M(r)} p_k^2)}{\log(r)} \tag{3}$$

where p_k^2 is the probability that two arbitrary points P_i, P_j are in cell k.
$\sum_{k=1}^{M(r)} p_k^2$ is the probability to find these points in any cell of the covering. This is approximately the probability $C(r)$ to find two points P_i, P_j with distance less or equal to the diameters of the cells r

$$\sum_{k=1}^{M(r)} p_k^2 \cong C(r) = \lim_{N \to \infty} \frac{1}{N^2} \sum_{i=1}^{N} \sum_{j \neq i=1}^{N} \Theta(r - |\vec{P_i} - \vec{P_j}|) . \tag{4}$$

where Θ is the Heaviside function $\qquad \Theta(x) = \begin{cases} 1 & \text{for } 0 \leq x < 1 \\ 0 & \text{else} \end{cases}$.

Finally, the correlation dimension ν is defined as

$$D_2 \cong \nu = \lim_{r \to 0} \frac{\log(C(r))}{\log(r)} . \tag{5}$$

2.3 The Reconstruction of the Attractor from One–Dimensional Time Series

A prerequisite for the application of the Grassberger–Proccacia algorithm is that we know the attractor points in state space or at least are able to reconstruct the attractor in a way that preserves its interesting geometrical structure (Packard, 1980; Takens, 1981).

For a given dimension of the state space, state vectors of that dimension have to be constructed from the time series. Since these vectors should reveal the effective dimension of the attractor, which is smaller than the dimension of the embedding space, the components of the vector should reflect this in a correlation among them. Several procedures for the construction of the vectors have been proposed. We refer to the one called time delayed reconstruction. The time series of length n yields $(n - m\tau)$ m–dimensional vectors $(x_i, x_{i+\tau}, x_{i+2\tau}, ..., x_{i+(m-1)\tau})$.

The choice of the delay parameter τ is important. A too small delay with respect to the characteristic correlation length of the time series compresses the reconstructed attractor by inducing a high correlation among the components of the vector of the state point, thus yielding a dimension smaller than the real one. A too large delay decouples the components of the vector. The reconstruction then tends to fill up the embedding space resulting in a dimension of the reconstructed attractor corresponding to the number of components of the vector.

2.4 The Evaluation of the Correlation Dimension from $C(r)$

For a genuine embedding of the attractor (dimensions of the embedding space greater than the dimension of the attractor) the correlation dimension $\nu = \lim_{r \to 0} \frac{\log C(r)}{\log r}$ should be independent of the dimension of the embedding state space. In practice, $C(r)$ can only be calculated down to a small r_f because of the limited resolution of experimental data. Also, the finite length of the time series accounts for an incomplete and inhomogeneous reconstruction of the attractor and causes a noisy and unreliable evaluation of $C(r)$ for $r < r_f$. The correlation dimension will show up as a constant ratio of $\log(C(r))$ to $\log(r)$ across a range of embedding dimensions and a range of r. To evaluate $C(r)$ for a given r and a given m–dimensional reconstruction we compute the mean over all attractor points of the number of their neighbor points with distances less than r. Variations of $C(r)$ for different locations on the attractor are neglected.

In the case of reconstructions with large time delays, strongly correlated neighboring points of the time series should be given attention, since they will be neighbors in state space as well. Thus they induce spurious correlations and have to be carefully discarded in the evaluation of $C(r)$.

3. A Simple Analytical Model: the Hénon Map

We examined a 1500 point time series from the two–dimensional Hénon map given by the equations:

$$
\begin{aligned}
x_{n+1} &= y_n + 1 - 1.4x_n^2 \\
y_{n+1} &= 0.3x_n .
\end{aligned}
\tag{6}
$$

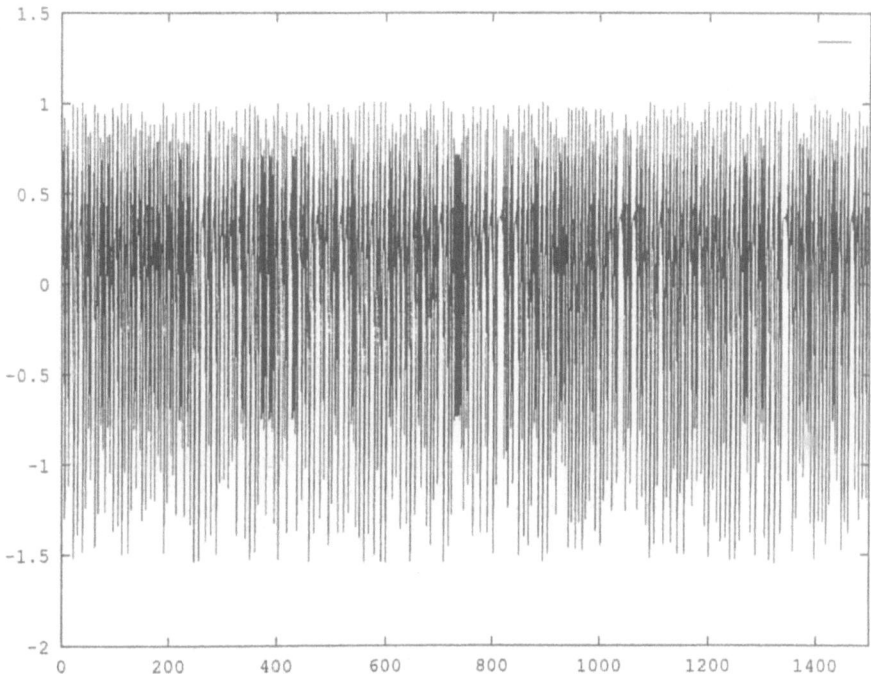

Fig. 1. Time series of Hénon map

The plot in Fig.1 shows clear oscillations around a mean value, which appears to remain constant throughout observation (stationarity). The oscillations have various amplitudes and give no evidence of regularity.

Fig.2 shows the results of a FFT (Fast Fourier Transform) of this time series. By this method, the signal is decomposed to harmonic oscillations at different frequencies. The contribution of each individual wave to the signal is measured by its power. The resulting power spectrum confirms that the frequencies of sine waves are not equally probable — waves of frequency around 300 and between 900 and 1000 (in units of the inverse sampling–time) are represented more prominently; still the power in between these frequency bands does not decrease significantly as it would in a quasiperiodic system.

The autocorrelation of the time series decreases considerably after a few iterations of the Hénon map. In order to reconstruct phase space with the method of time delays, we chose as a rule of thumb the first minimum of the autocorrelation function as suitable. In our case a phase space was reconstructed using a lag of 2. In two–dimensional phase space (Fig.3) the well–known structure of the Hénon attractor appears as expected (i.e. the reconstruction procedure yields the same structure as in a phase space spanned by the analytical dimensions of the map given in (6)).

The Grassberger-Proccacia method was applied with embedding dimensions $1 \leq m \leq 5$ (Fig.4). Quite clearly slopes converge to a value of around 1.25, which is consistent with the dimension 1.26 given in the literature.

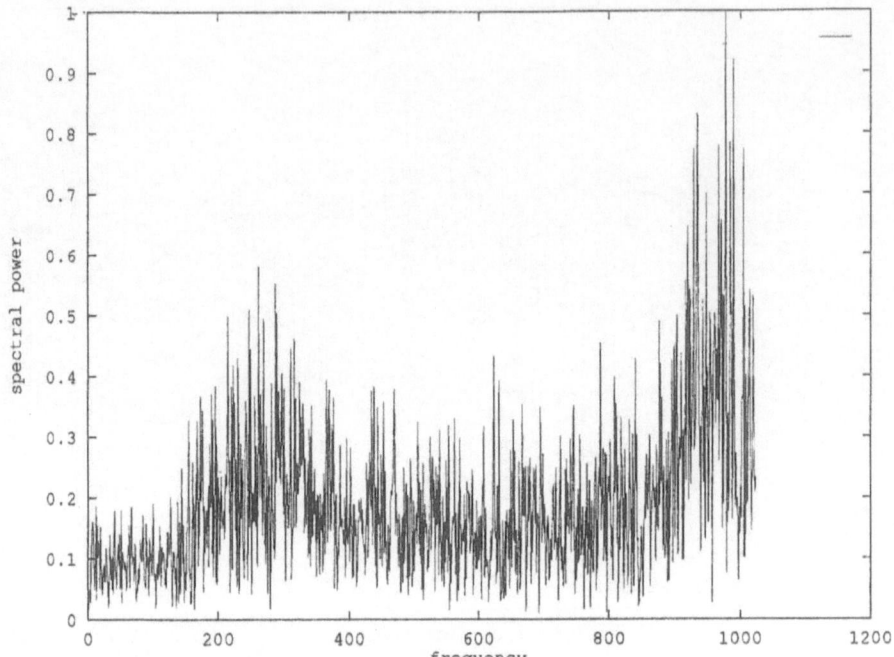

Fig. 2. Power spectrum of Hénon time series

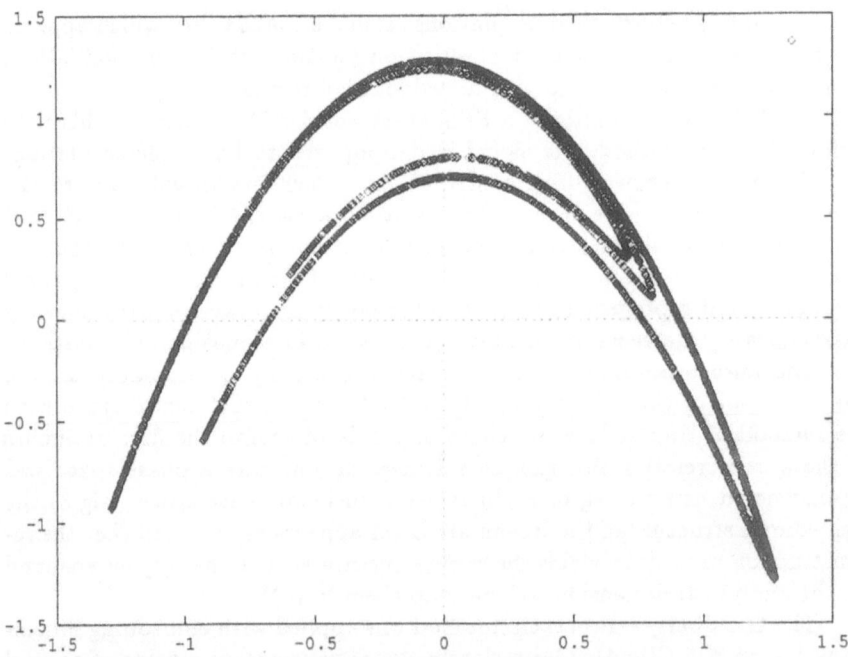

Fig. 3. Reconstructed 2D map from Hénon time series

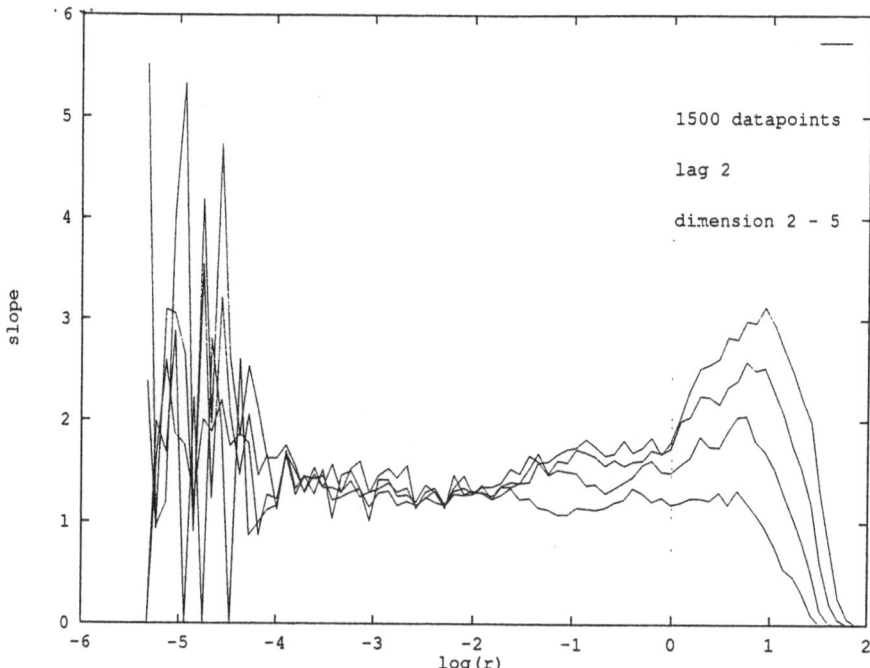

Fig. 4. Slope vs log(r) for Hénon time series with embedding dimensions 2–5, $\tau = 2$

This value appears as a plateau in all curves (except for embedding dimension $m = 1$) within the significant range of r; it is evidently not dependent on m as it would be in the case of white noise.

4. Evaluation of the Correlation Dimension Under Different Constraints for the Time Series

4.1 Experimental noise

Pure noise has no attractor; it fills up the state space. Thus, the correlation dimension of noise just measures the dimension of embedding space. In contrast, the correlation dimension of any reconstructed attractor is independent of embedding space (at least for embedding dimensions near that of the attractor). If noise is added it will tend to smear the attractor in embedding space. As a result the correlation dimension will couple to the embedding dimension. The results of application of noise to the Hénon map are shown in Fig.5. A noise level of 10% destroys the constancy of the correlation dimension for different dimensions of the embedding space, making it impossible to estimate safely the dimension of the attractor.

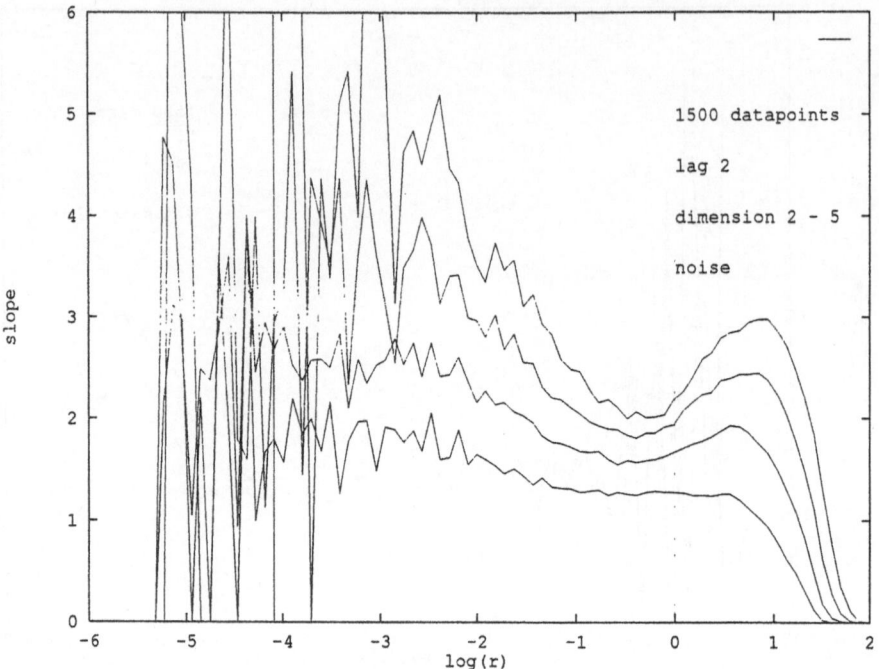

Fig. 5. Slope vs log(r) for Hénon time series (10% noise superposed) with embedding dimensions 2–5, $\tau = 2$

4.2 Length of the Time Series

The length of the time series and the dimension of the embedding state space define the resolution of the attractor reconstruction. As a general rule (see Mayer–Kress, 1987, for a derivation from error estimates on $C(r)$ and the procedure to find the range and the value of constant slope) the length of the time series should be at least $n = b^m$ where m is the dimension of the embedding space and b the minimal tolerable resolution in an coordinate direction. For the Hénon map with dimension $\cong 1.25$ it should be at least $m = 2$ and $b = 10$. In Fig.6 a dimension estimate for $n = 100$ corresponding to the above minimal length is shown. For $m = 2$ a plateau is still visible, but especially for higher embedding dimensions the estimation of the correlation dimension becomes difficult.

It should be noted, though, that in order to observe convergence larger embedding dimensions must be considered as well. The lag τ that was chosen for reconstruction further increases demands for long time series. As was stated in Sect. 2.3, a time series of length n yields $(n - m\tau)$ attractor points. As an example, when the attractor is reconstructed with a lag of 5 and $C(r)$ is tested up to embedding dimension 10, it takes 150 time series points to reconstruct 100 attractor points!

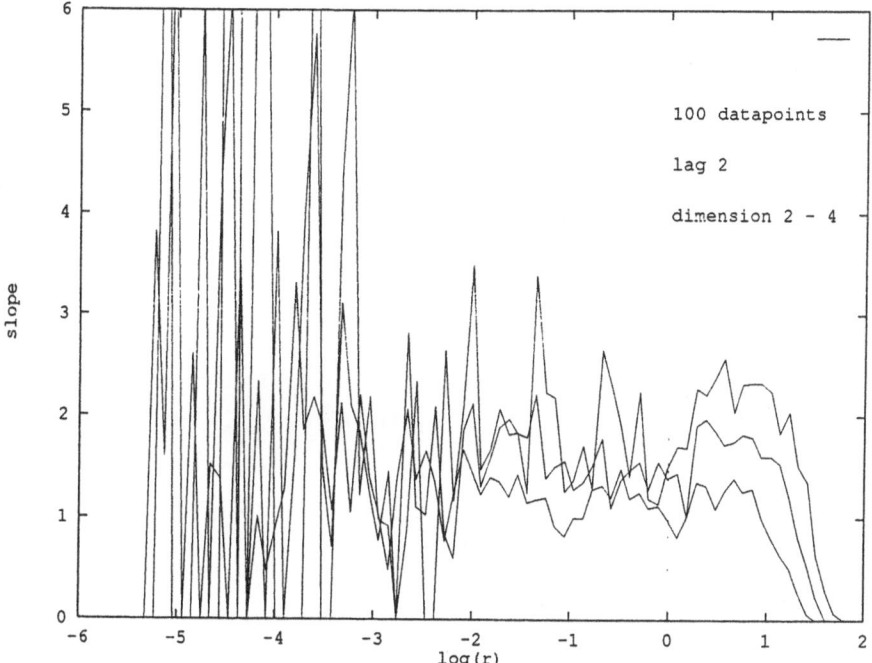

Fig. 6. Slope vs log(r) for Hénon time series (length of series reduced to 100 data points) with embedding dimensions 2–4, $\tau = 2$

4.3 Insufficient Resolution of the Time Series

The resolution of the data from the time series determines the resolution of the reconstructed attractor in the embedding state space. Too low a resolution constrains the evaluation of the correlation to large radii thus making it impossible to measure the filigrane local structure of the attractor. In Fig.7 the resolution of the Hénon time series has been diminished to only 6 bins. This simulates the effects of measuring the Hénon time series with a 6 point rating scale. As can be seen, the plateau in the slope versus log(r) plot has disappeared (with $n=$ 1500 as above).

5. Two Time Series

5.1 Simulation of Group Processes

Figure 8 presents a time series from a simulation system designed in order to model group processes. The simulation computes 'social distances' between members of a group; these may be interpreted like a sociogram (Moreno, 1953) or a 'sculpture' (Schweitzer & Weber, 1983). Further descriptions are given by Tschacher et al. (this volume). The time series maps the spacial expansion of a configuration of eight 'persons' by the sum value of all persons'

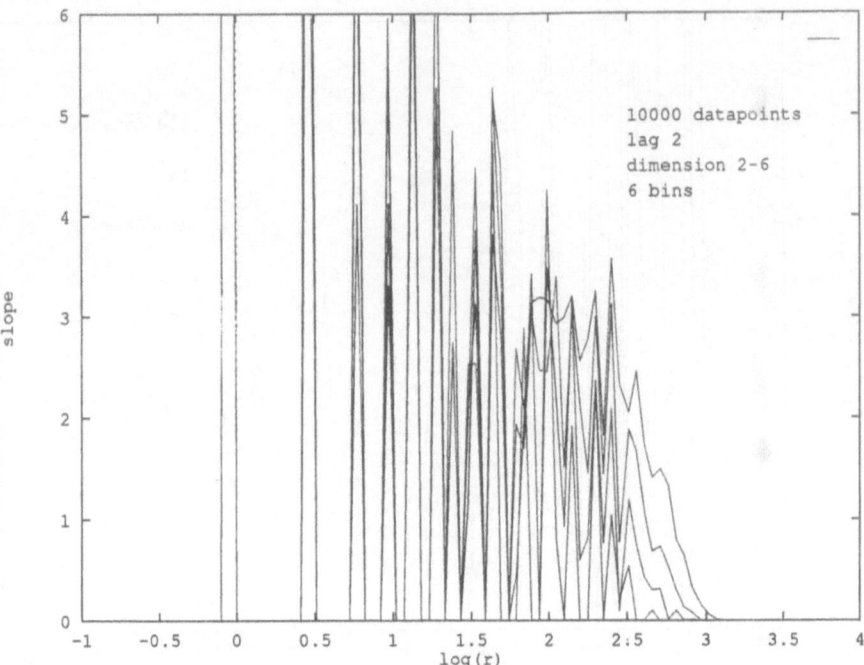

Fig. 7. Slope vs log(r) for Hénon time series (resolution reduced to 6 bins) with embedding dimensions 2–4, $\tau = 2$

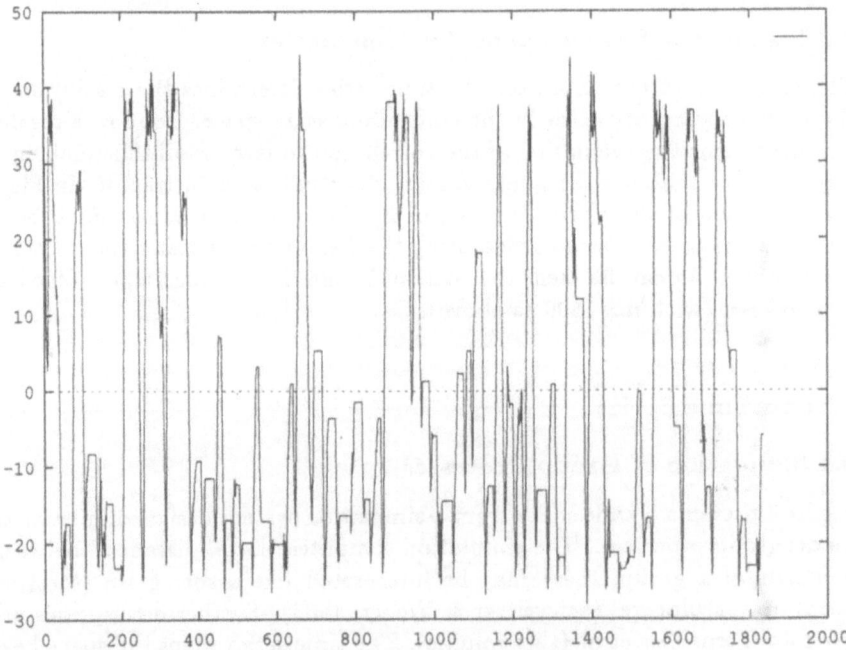

Fig. 8. Time series of simulation of group interaction

distances from a reference point. This value varies from iteration to iteration. In the time series analyzed here model parameters were set in a way corresponding to 'conflict' between group members — the simulation takes a strange course and does not approach any of several stationary states as usual. 'Persons' form a close turbulent unit instead, occasionally building up formations that resemble 'gliders' in cellular automata.

A total of $n = 1800$ values from this process (corresponding to 1800 iterations of the simulation) were sampled for analysis. The series plotted in Fig.8 does not appear to have any recognizable pattern.

The power spectrum is given in Fig.9. Several peaks can be observed, but the spectrum in between is continuous. Thus, the system is not a mere superposition of several different oscillations, even if some oscillations are represented more. As the signal is neither periodic nor white noise, it might be chaotic with some fractal dimension. In order to test for chaos, phase space was reconstructed. To establish coordinates we determined the first minimum of the autocorrelation function which yields an appropriate time delay of $\tau = 40$ (i.e. phase space is spanned by coordinates $x(t), x(t + 40), x(t + 80)$ etc.). The system in 2D phase space is presented in Fig.10.

Fig.11 gives the slopes of the Grassberger–Proccacia method. Any region of constant ratio of the log–log plot of the correlation dimension $C(r)$ to radius r should show up here as a horizontal plateau of slope curves irrespective of embedding dimension. Although the plot is not as clear as in the case

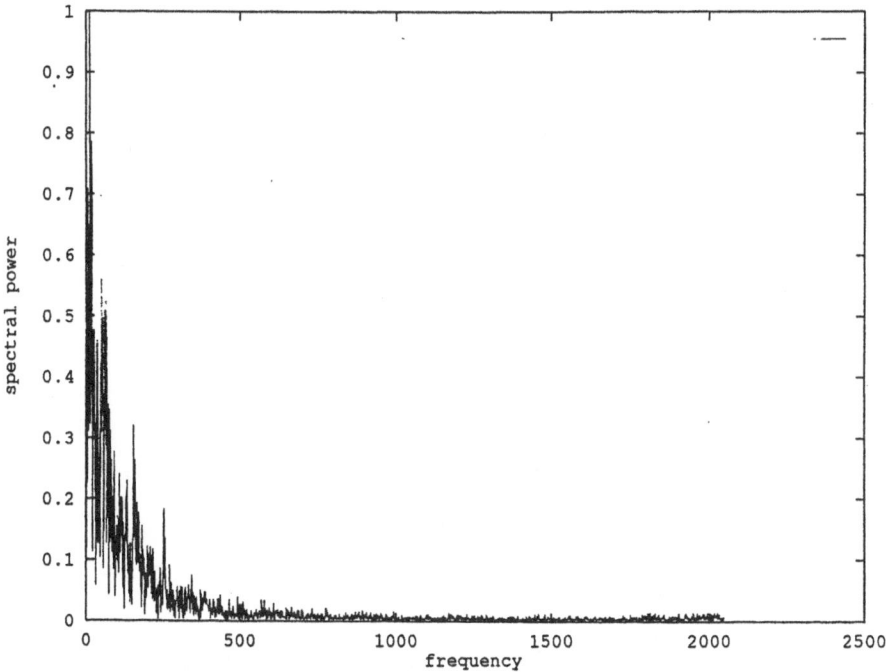

Fig. 9. Power spectrum of simulation data

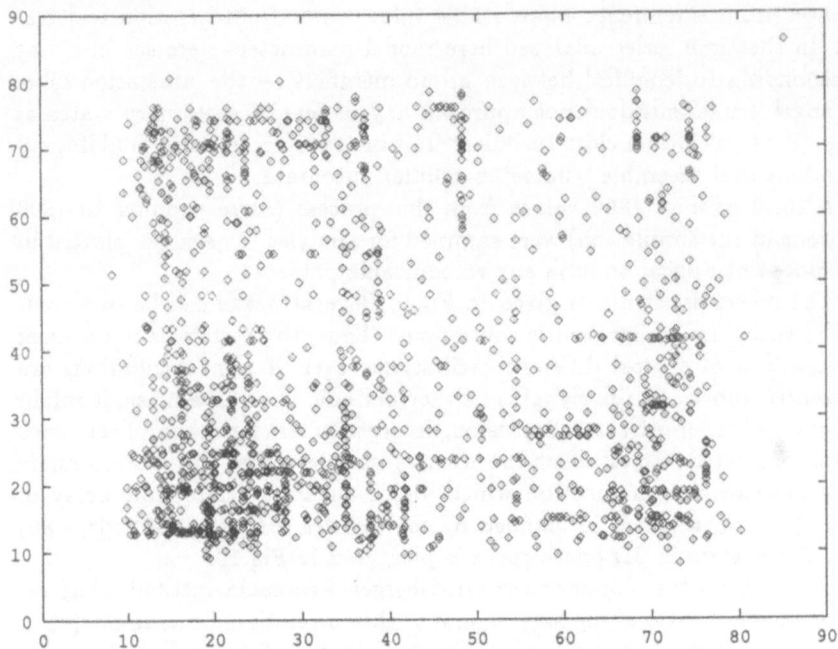

Fig. 10. Reconstructed 2D map from simulation data

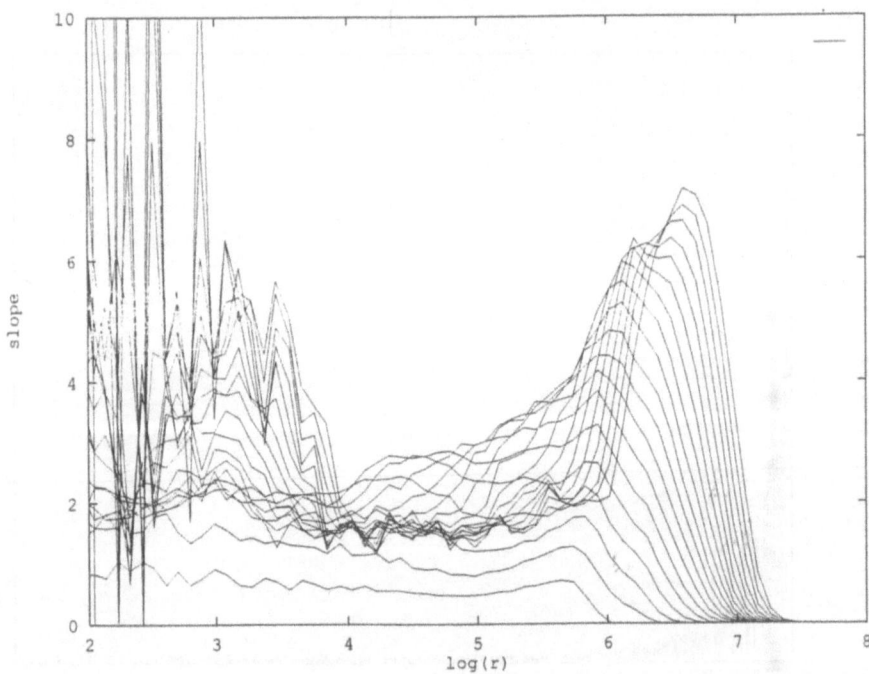

Fig. 11. Slope vs $\log(r)$ for simulation data with embedding dimensions 1–20, $\tau = 40$

of the Hénon map, a plateau of this sort can be derived from the diagram. This points to a fractal attractor of dimension $\nu \cong 1.6$. But peculiarly, the plateau is established only from embedding dimensions $m > 12$. The fact that convergence is not recognizable in lower dimensional embeddings casts doubt on this result (see Sect. 2.4).

Therefore a revised dimension analysis was employed in order to check for artefacts induced by local correlations. For the computation of the correlation $C(r)$ for each attractor point all points with time distances less than τ in the time series were discarded. This procedure yields Fig.12. There is no common plateau left in the slope versus $\log(r)$ diagram. Evidence for a low dimension has disappeared. This indicates that the correlation dimension derived from the Grassberger–Proccacia algorithm was largely due to local effects caused by points on the same segment of the trajectory; the global structure of the hypothesized attractor given by many foldings of the trajectory was not grasped by the standard method.

5.2 An Empirical Time Series of Smoking Behavior

The time series depicted in Fig.13 comprises 1555 observations of an individual's cigarette consumption. Being trained in clinical psychology and behavior therapy the 28–year–old male student took these daily counts to monitor his smoking behavior. Fig.13 indicates that major periods of the present se-

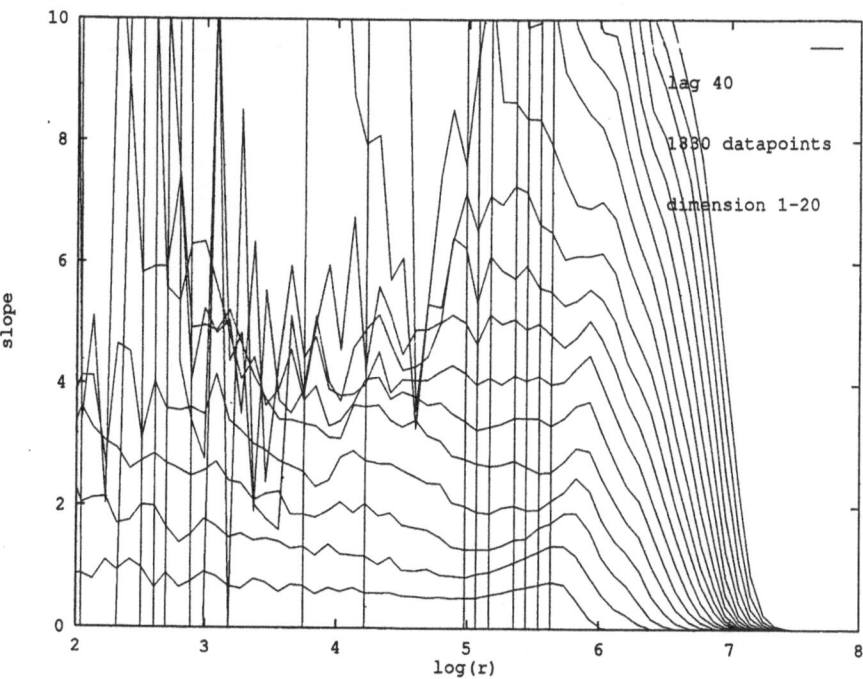

Fig. 12. Corrected slope vs $\log(r)$ for simulation data with embedding dimensions 1–20, $\tau = 40$

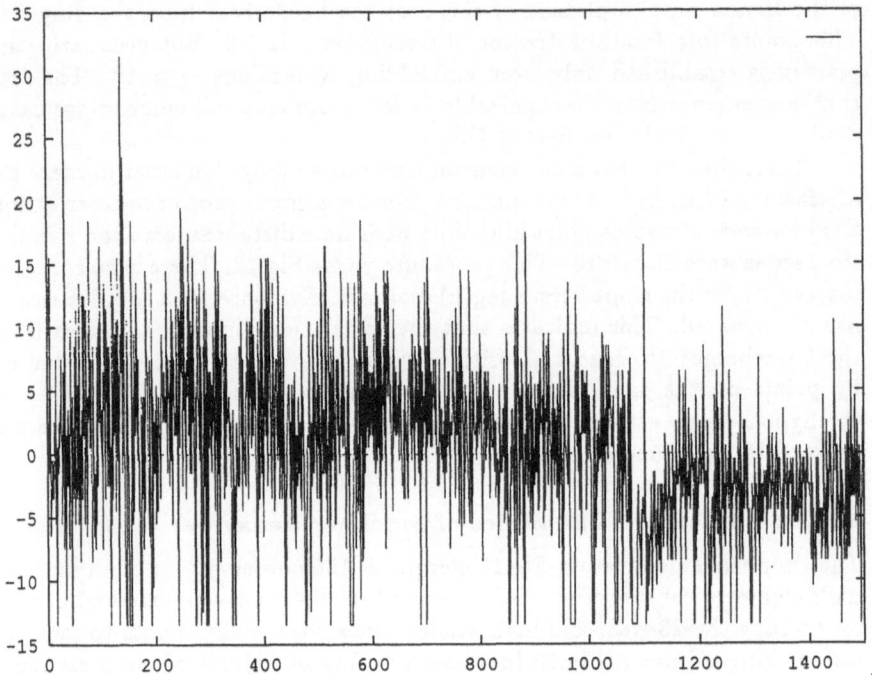

Fig. 13. Time series of smoking data

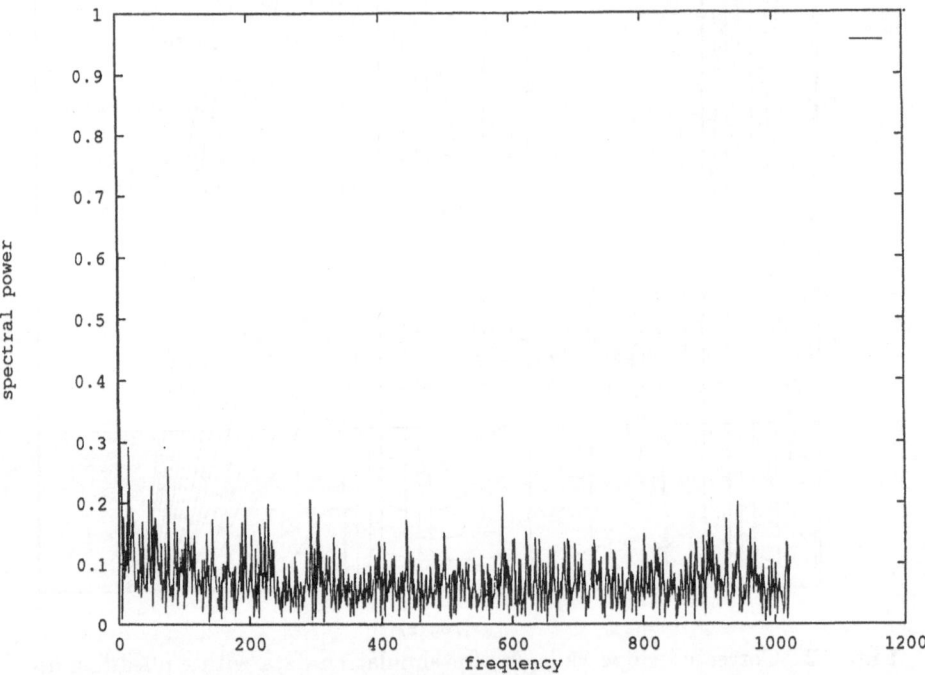

Fig. 14. Power spectrum of smoking data

ries appear to have means different from the means of other periods. There is a distinct level shift after approximately 1100 observations. Though stationarity could have been accomplished by appropriate transformations such as differencing, i.e. the calculation of successive changes in the data values, we preferred to restrict our statistical analysis to the initial 1000 observations and ignored the subsequent data. The power spectrum of the resulting time series given in Fig.14 is smooth and therefore all cycles are assumed to have occurred at an approximately equal intensity. Thus the FFT gives no indication of any hidden periodicity or regularity inherent in the time series.

With respect to the identification of an optimal time delay τ for the reconstruction of an appropriate attractor we again determined the first local minimum of the autocorrelation function, in the present example of $\tau = 4$. Provided that the corresponding embedding dimension is greater than the correlation dimension any range of r with a constant ratio of $\log(C(r))$ versus $\log(r)$ should show up as a common horizontal plateau of various slope curves in the slope versus $\log(r)$ diagram (Fig.16). However, this diagram clearly indicates that there is no such common plateau. Moreover, slopes apparently increase with higher embedding dimensions, consequently the points within successive reconstructions of the attractor exhibit a noise–like pattern.

In general, cigarette smoking is viewed as resulting from the complex interactions of environmental, physiological and psychological processes (Lichtenstein & Brown, 1982). If behavioral time series like the present one could

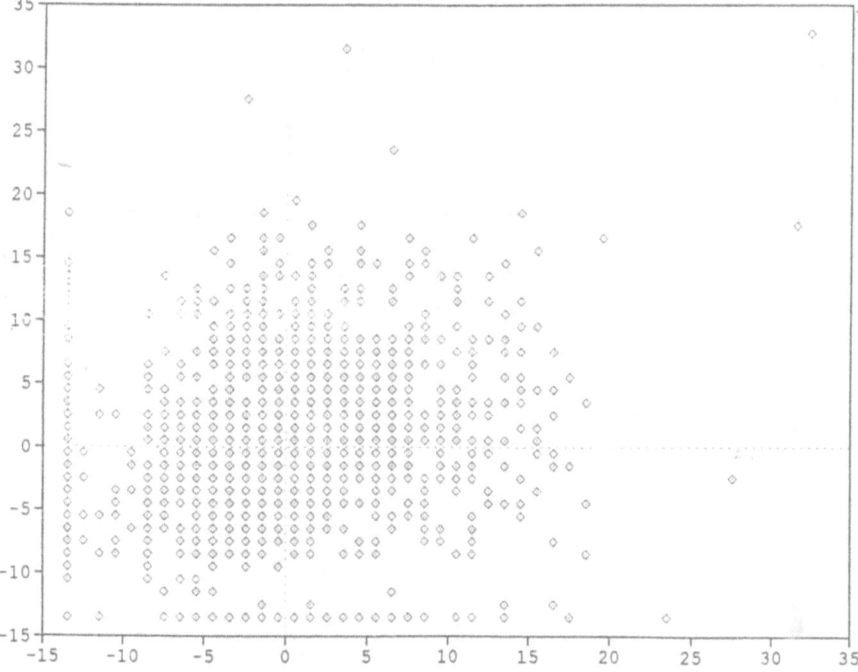

Fig. 15. Reconstructed 2D map from smoking data

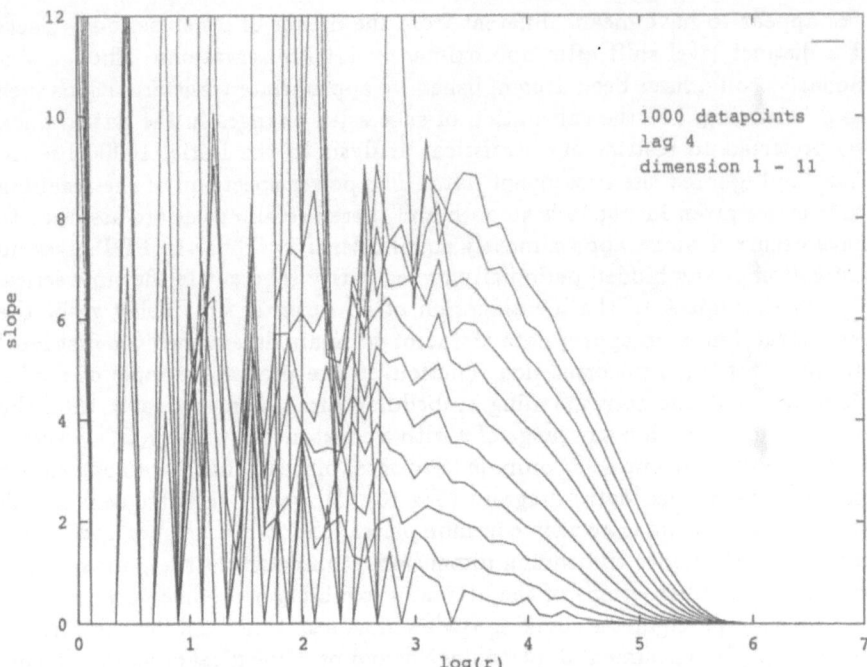

Fig. 16. Slope vs log(r) for smoking data with embedding dimensions 1–11, $\tau = 4$

be reduced to a finite, possibly small number of variables this could be of considerable relevance for clinical theory and therapeutic practice. But this is not yet supported by our data.

6. Options and Restrictions of Dimension Analysis in Psychology

The options of this method have already been stated in Sect. 1: it is possible to evaluate seemingly erratic behavior by determining the dimension of the underlying dynamical regime (if there is any such regime of just a few variables). In psychology, the multicausality of mental and social events has often been emphasized. This is accepted as the reason for either interpreting field data   massively or controlling all circumstances by laboratory methods. In either case, complexity is reduced rather artificially. On the other hand, self–organization theory has shown in many instances that under certain conditions very complex systems are governed by but a few order parameters. These conditions seem to be met by most psycho–social systems, so that self–organizing processes should be expected in the field of psychology, too (Schiepek & Tschacher, this volume). Dimension analysis may be a suitable method to detect self–organized systems in apparently nonperiodic time series generated by psycho–social systems.

Even some basic assumptions about the etiology and maintainance of behavior mapped by the time series may be put to test: if our observations reflect an internally controlled process unfolding no matter what environmental influences exist at the time, a simple attractor or an attractor of a finite correlation dimension should be effective. This result would be compatible with psychological theories stressing cognitive 'inner' dynamics as primary causes of behavior. On the other hand, if behavior is to a high degree steered by 'external' control parameters, the fluctuation of these parameters renders any determination of the system's characteristics impossible. The time series then maps many instantiations of quite different systems, so that no overall values of attractor dimensions can be found. In order to map the system, a rigorous control of environmental parameters should be accomplished. The smoking behavior data rather point to this conclusion.

But methodological restrictions also have to be considered following the above discussion of the influence of noise, length of the time series, and resolution of data points. From the discussion some minimum quality standard for empirical time series must be achieved. The resolution of the variable measured and the length of measurement are independent indicators of data quality. They are also limiting factors, i.e. insufficient resolution cannot be remedied by a longer time series and vice versa. For research in psychology, high resolution of measurement is a demand which is quite difficult to satisfy. It will most certainly not be accomplished by the application of usual rating scales since only very simple structures can be mapped in a low–grained, discrete phase space — fractal attractors for one are not simple structures. The implementation of dynamical methods in psychology is largely a question of finding and measuring appropriate observables (Tschacher, 1990).

All of the factors listed in Sect. 4 severely limit the applicability of dimension analysis, maybe even to the point that while there is evidence of complex low–dimensional homeostatic mechanisms theoretically, the accompanying attractors may never be revealed empirically. Still, in our opinion it may prove worthwhile to test further time series from psycho–social systems, bearing in mind these restrictions. To this end observations of (socio–)physiological variables and variables of spacial behavior in restricted settings (such as therapy settings) may be useful.

Acknowledgment

The research reported in this article was in part supported by a grant from the Deutsche Forschungsgemeinschaft (DFG), Grant No. RE402/9-3.
We would like to thank Dipl.–Psych. Uwe Prudlo for permitting us to make use of the cigarette consumption data.

References

BABLOYANTZ, A. 1985. Strange Attractors in the Dynamics of Brain Actitivity. In H. Haken (Ed.): Complex Systems – Operational Approaches to Neurobiology, Physics, and Computers. Berlin: Springer, 116-122.

GLASS, L., SHRIER, A., & BÉLAIR, J. 1986. Chaotic Cardiac Rhythms. In A.V. Holden (Ed.). Chaos. Princeton: University Press, 237-256.

GRAF, K.-E. & ELBERT, T. 1989. Dimensional Analysis of the Waking-EEG. In E. Başar & T.H. Bullock (Eds.). Brain Dynamics. Progress and Perspectives. Berlin: Springer, 174-191.

GRASSBERGER, P., & PROCACCIA, I. 1983. Characterization of Strange Attractors. Physical Review Letters, 50, 346-349.

HENTSCHEL, H.G.E. & PROCACCIA, I. 1983. The Infinitiv Number of Generalized Dimensions of Fractals and Strange Attractors. Physica 8D, 435-444.

LICHTENSTEIN, E. & BROWN, R.A. 1982. Current Trends in the Modification of Cigarette Dependence. New York: Plenum.

MAYER–KRESS, G. 1987. Application of Dimension Algorithms to Experimental Data. In H. Bai-lin (Ed.). Directions in Chaos. Singapore: World Scientific.

MORENO, J.L. 1953. Who Shall Survive? Foundations of Sociometry, Group Psychotherapy, and Sociodrama. New York: Beacon House.

PACKARD, N.H, CRUTCHFIELD, J.P., FARMER, J.D., & SHAW, R.S. 1980. Geometry From a Time Series. Phys.Rev.Lett., 45, 712-716.

SCHAFFER, W.M. & KOT, M. 1986. Differential Systems in Ecology and Epidemiology. In A.V. Holden (Ed.). Chaos. Princeton: University Press, 158-178.

SCHWEITZER, J. & WEBER, G. 1983. Beziehung als Metapher. Die Familienskulptur als diagnostische, therapeutische und Ausbildungstechnik. Familiendynamik, 8, 113-128.

SIMM, C.W., SAWLEY, M.L., SKIFF, F. & POCHELON, A. 1987. On the Analysis of Experimental Signals for Evidence of Deterministic Chaos. Helvetica Physica Acta, 60, 510-551.

TAKENS, F. 1981. Detecting Strange Attractors in Turbulence. In D.A. Rand & L.S. Young (Eds.). Lecture Notes in Mathematics. New York: Springer.

TSCHACHER, 1990. Interaktion in selbstorganisierten Systemen. Grundlegung eines dynamisch–synergetischen Forschungsprogramms in der Psychologie. Heidelberg: Asanger.

The System Game as a Research Paradigm for Self-Organization Processes in Complex Social Systems

Günter Schiepek and Michael Reicherts

With 5 Figures

In the present contribution the research paradigm of system gaming is outlined and integrated into the theoretical framework of psychological synergetics. Questions concerning the theoretical and empirical conceptualization of the order and control parameters governing the self-organizing processes which occur during the course of a system game will be discussed. A multi-method approach at evaluating intrapersonal processes of coping with stress and stressful emotions as well as the dynamics of interpersonal communication will be suggested. In the second part we will describe the empirical results obtained from the implementation of a system game scenario ("youth home scenario"), revealing not only the differing dynamics of coping with stress within the professional subsystem as opposed to the subsystem of the problem persons, but also the dynamics of the interactions between the participants involved.

1. The System Game as a Research Paradigm

1.1 From Planned Game to System Game

The planned-game method has been familiar within the practical sphere of human science for some considerable time. It has been used in such fields as education and didactics — for example, within the realm of university didactics courses (Reinisch 1975) or educational practice for school teachers (Giesecke 1973) - the world of business (Bleicher 1969), and the realm of military strategic planning, but also in psychosocial practice (Cumpelik 1985; Daigl 1988). Planned games (in German, "Planspiele"; one may compare them to English "simulation games") aimed at emancipating the individual within society were employed particularly in the 1960s and 1970s in various fields of work relating to education (e.g. Rehm 1964; Reimann 1972; Lehmann & Portele 1976). Since then, however, it would seem to us that silence has descended around this method. It is true that it continues to be used as before in industry — within the framework of complex planning strategies or assessment centers, for example; however, it is not a component of standard methods adopted by psychology. Instead, the realm of psychology is dominated by the delimited method of role-playing — the cause of some sensation with the advent of behavioral therapy — or exercises in group dynamics. Those areas where planned games have been taken up for psychosocial practice

Springer Series in Synergetics, Vol. 58   Self-Organization and Clinical Psychology
Editors: W. Tschacher, G. Schiepek, and E.J. Brunner   © Springer-Verlag Berlin Heidelberg 1992

represent more the exception than the rule (with regard to such exceptions, see, for example, the small practical introduction by Daigl (1988), and also the dissertation by Cumpelik (1985) from Bergold's group in Berlin). It is remarkable that even those practitioners evidencing a systemic orientation have hardly dealt with the planned-game method, and this in spite of the fact that those institutional contextual conditions deemed relevant for psychosocial practice can thereby be simulated (Mackinger 1984; Schweitzer 1989; Imber-Black 1988). Research in the field of psychology — and especially in that of social psychology — regards planned games as something meriting no in-depth interest at all, despite the existence of prominent models (see the famous prison experiment by Zimbardo 1979, for example).

In principle, a planned game is concerned with expanded role-playing, with respect both to the number of persons involved and to the temporal perspective. With regard to the number of participants, the literature considers three to fifteen groups of two to five members each to be practicable (see Cumpelik 1985, p.18). Scenarios drawn up on too small a scale often fail to develop the required complexity, especially from the perspective of social self-organization. The participating persons are divided up into groupings or institutions (e.g. members of an educational psychology service or a family), and receive generally brief descriptions of their roles, possibly also an initial problem or conflict forming the starting point for the further development of the game. In other words, these are life-simulations of the interactional dynamic(s) involving several persons and institutions, usually conducted for purposes of innovation (e.g. testing new forms of co-operation, changes within an institution), self-experiences or competence acquisition (acting and deciding in complex social situations), and occasionally as an aid to cadre selection in businesses. A detailed portrayal of the method may be found in Schiepek (1991, pp. 199-220).

It is evident that planned games involve dynamic systems in which psychological processes engage with those requiring interactive communication in a manner that is quite as complex as the respective operational logic of the social subsystems (or institutions or groups) (see Willke 1989). While starting situations and peripheral conditions exist, these can be said to influence the concrete processing dynamics, but not to determine it in its progressive dynamics, since the dynamics is self-organizing. In the light of this background, it would appear advisable to abandon any suggestion of it being possible to plan this type of game. In consequence, we should speak no longer of "planned games", but rather of "system games". The term "system game" describes what is meant in a more precise manner; however, a change in terminology always requires thorough justification:

1. Complex social systems usually lie beyond the possibility of comprehensive planning, especially since the planners themselves are members of the system, in consequence planning and acting in accordance with a certain, particular perspective from within this system (cf. Krohn & Küppers 1990; Schiepek 1991, Chap. III; Willke 1989). It is of course true that one not

only should but also must plan; however, such planning is carried out by many, and the resultant multitude of plans and efforts at implementation clash, interact with each other, and develop an emergent momentum of their own the extent of which cannot normally be calculated. Inasmuch as system games represent heterarchically interrelated systems, they exhibit neither an identified center for planning or decision-making, nor a hierarchical apex.

2. The interest of our research group is focussed not so much on planning processes as on the variety of emotion, cognition and behavior under conditions of complexity. Within this, special interest is devoted to the realm of stress experience and coping with stress, together with the realm of interpersonal dramatization and self-presentation by the participating actors.

3. It is for this reason that we refer less to planning theories than to coping theories (Reicherts 1988), theories of problem-solving in dynamic scenarios (e.g. Dörner et al. 1988), theories of self-presentation and impression management (Laux, 1991) and, in particular, models from the realm of synergetics and other theories of self-organization (e.g. Haken & Stadler 1990; Willke 1989; Schiepek 1991). The center of our attention is occupied by inherent dynamism, dynamic processes, states of order, and phase transitions in highly complex psychosocial systems. The question also arises as to whether it is possible, using empirical means, to identify chaotic dynamic processes within the system game, and, if so, what possibilities it reveals for managing chaos within the social system.

4. It is important to avoid confusion with the concept of "plan analysis" (Caspar 1989), in particular since we are working in another study group at the University of Bamberg on the development of a procedural form of plan analysis. This so-called *sequential plan analysis* (SPA) is to be used in system-game research with the aim of conducting a fine analysis of communicative processes.

1.2 Self-Organization Processes in System Games

If system games are considered from the internal perspective of a participating actor or through the bird's-eye view of the external observer, it is possible to recognize intuitively that there are identifiable phases in the course of system games which are determined by particular interactional constellations (e.g. who has a central role to play, who performs a peripheral role), by particular temporal patterns (possibly rhythms) within the interactions (communicative acts that are hectic or demonstrate a slow rhythm), by predominant topics or problems, and finally by a characteristic emotional climate. This observation is sufficient to suggest the presence within system games of states of order, or "modes", "enslaving" the highly complex cognitive, affective, social and — up to a certain point — biological (e.g. the physiological correlates of stress) course of events at the microscopic level of the game for at least a certain period of time. Such states of order may also be observed in real-life organizations (e.g. businesses) and in networks of interaction between organizations (e.g. in complete branches of industry): here one is impressed by the "working atmosphere" in business or branch, or by the "philosophy" or "image", together with the relevant forms of co-operation between producers and customers. In system games, however — so, at least, runs our working hypothesis — such states of order arise at shorter intervals and also more closely spaced, making them both easier to observe and examinable from an empirical point of view. For the following reasons, this working hypothesis would appear plausible:

1. The cited phenomenology speaks for the manifestation of coherent psychosocial patterns in system games.

2. System games take place over stretches of time (of the magnitude of one to four days) and spatial distribution (in several rooms of a building) that are such as to enable one to maintain a general overview; in addition, they represent not private but rather public situations — despite a high degree of emotional involvement on the part of the participants — which are accessible by means of observation and questioning methodology.

3. System games fulfill every formal prerequisite for the appearance of self-organization as it is understood by synergetics. These involve

a) the existence of a thermodynamically *open system*. That this is the case is obvious from the fact that we are concerned here with interactions between living systems. Each of the subsystems (individuals or groups) is of course open to every possible kind of stimulation from any of the others.

b) the existence of a state far from thermodynamic equilibrium, but in particular *far from motivational equilibrium*: the initial constellation, but also —. to a greater extent — the questions, problems, conflicts, requirements and relational constellations that are constantly being produced, keep the mostly highly involved actors on the go and energize both the psychological and the social course of the game in an intensive manner.

c) the existence of a *biopsychosocial microscopic level*. As has already been asserted repeatedly for psychological synergetics, so too in the case of system games: the persons should not be seen as elements of the microscopic level. Rather, the microscopic level must be introduced as the virtual horizon of high resolution for every psychological, social and biological process of the participating individuals (biopsychosocial system, cf. Schiepek & Tschacher, this volume).

d) the existence of *nonlinear interactions* between the processes of the microscopic level, between system and environment, but also between the macroscopic variables describing the system at the order-parameter level (macroscopic system of variables, cf. Schiepek & Tschacher, this volume).

Inasmuch as conditions (1) to (3) are fulfilled — something which may be tested empirically — and condition 3a-d includes the prerequisites for the appearance of self-organization in complex systems, one is prompted to conclude that use should be made of *the system game as a research paradigm* for the study of self-organization processes in biopsychosocial systems, and thereby for *psychological synergetics*. Experience collected to date indicates

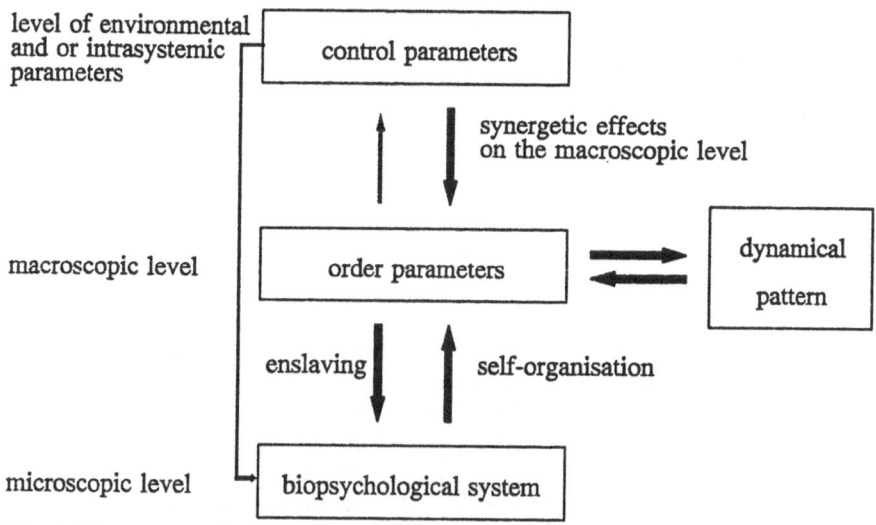

Fig. 1: The synergetic model of self-organization

that this paradigm is highly practicable and efficient, the amount of time and energy required for preparation and execution standing in favorable relation to its use; this is not least on account of the multifunctional nature of system games, since they fulfill research-oriented, educational, self-experiences and innovational purposes simultaneously. This continues to be true even when they are carried out with only one objective in mind: it is impossible, for example, to exclude the element of self-experiences even when conducting a system game for research purposes only. Nevertheless, further experience and, in particular, methodological development is necessary if the system-game paradigm of self-organization research is to be established on a broader basis.

If one visualizes the basic scheme of synergetics (Fig. 1, cf. Schiepek & Tschacher, this volume), then this, together with conditions 1 and 3a-d given above, serves to make it clear that system games represent in principle an "intended application" of the non-statement view of theory (Stegmüller 1973),in the sense of synergetics. In order to conclusively establish the veracity of such a statement, however, the phenomenological evidence of the existence of disorder-to-order and order-to-order transitions, and the principal assumptions of conditions 3a-d, are not sufficient. Empirical proof must be provided that identifiable states of order are actually present and that phase transitions between such states of order are actually taking place. We are thus faced with the task of identifying suitable *order parameters*, or macroscopic variables, for the dynamics of complex psychosocial systems.

1.2.1 The Identification of Order Parameters

One candidate for a suitable macroscopic indicator, meaning one representing the entire interactional event as comprehensively as possible, may be seen in the matrix of interaction between the participating persons or groups. Such matrices of interaction refer to a period of time, one laid down for each instance (e.g. half an hour or an hour of playing time in real terms), and can be drawn up for the entire chronological course of the game. A succession of matrices results, covering the course of the entire game. The matrices can be seen to a certain extent as windows by means of which one may observe the course of interaction taking place in each respective section of the game. These matrices enable one to make statements regarding the density of interaction, the concentration of interaction (are only a few involved and others excluded, or are the events distributed among the whole system game?), and the structure of interaction (who is transmitting to whom, who is receiving from whom?), it being possible to directly express these statements in the form of quantitative indicators (cf. below, Sect. 2.2.4). Continuity in these indicators would indicate the presence of a stable state of order in the system dynamics, while great variations or clear structural differences between successive matrices would suggest disorder (instability) or fluctuations in discontinuous phase transitions between states of order. In addition, the contents of the activities and interactions, which are documented by every player in the form of activity records, could be enlisted as a criterion for states of order: a constant distribution of activity among the groupings, or recurring sequences at least, would indicate relative stability and order within the course of the game.

Such global indicators of interaction may make sense as initial approaches in the search for suitable order parameters. However, closer observation reveals the need for differentiation:

a) The courses of such parameters of interaction often reveal ambiguous changes between stable and unstable phases (fluctuations). This leads to the question as to whether coherent states of order (stable modes) either do not occur, which would contradict the phenomenological experience collected up to now; whether the dynamics of interaction found in system games involve low-dimensional chaos: this is to be tested, but requires longer time-series; whether methodological artefacts are present, although this would call into question the suitability of indicators of interaction as order parameters for system games; or whether the states of order of highly complex psychosocial systems also involve correspondingly more complex dynamic patterns than those expressed within the matrices of interaction.

b) Should the latter be the case, the identification of complex, dynamic states of order must be conducted via the configuration of several parameters of order and macroscopic variables, each of which would then record different aspects of the biopsychosocial system. The consequence of this is the necessity for system-game research of pursuing a multimethodological approach.

c) Self-organization within complex social systems does not only mean the creation of *coherence*, and thereby a reduction in the degrees of freedom regarding experience and behavior enjoyed by the members of the system.

Slaving resulting from simple modes, such as a collective ritual, a decisive topic of communication, an emotional atmosphere that sweeps everyone along with it, a compelling code of speech: all these are forms of uniformity but simultaneously of reduction in complexity. However, self-organization in social systems can additionally mean *differentiation*: new structures and contexts of action develop, role differentiation takes place, different forms of operational logic and emotional relational qualities in individual subsystems coexist. This is still bound up with an enormous reduction in degrees of freedom when compared with microscopic chaos — that is, disorder at the system's microscopic level. However, the states of order themselves cannot be represented by one individual order parameter, but only by a dynamic configuration of several macroscopic variables.

The consequence of this is that the system-game paradigm takes as its theme several aspects of the biopsychosocial system simultaneously, namely

1) the individual experience(s) of the participants, our research project concentrating here on emotions from the field of experiences when under stress and pressure. Accordingly, the coding is conducted at the level of events involving experiencing and coping with stress;

2) the individual regulation of action, the emphasis here, too, lying in the realm of coping with stress ((1) and (2) are recorded by means of a shortened version of the COMES ("Questionnaire assessing the coping with current stress situations", Reicherts 1988; Perrez & Reicherts 1992, even available in the form of a computer assisted recording system);

3) the nature of the activities realized by individual persons and subgroups;

4) the structure of interaction between persons and subgroups (who consults whom and when? - who is alluded to? - who is not involved?), this enabling the derivation of parameters covering the density of interaction, the chronological distribution of interaction, the concentration of interaction or the chronological change in the structure of interaction;

5) the concrete behavior of the participants, something which can be recorded by making a systematic observation of behavior over the course of the game as documented on video film;

6) the emotional relational dynamics between the participants in crises or (so-called) phases of high communicative intensity compression, the latter serving as indications of instabilities within the system dynamics (recorded via the method of "sequential plan analysis", a procedural variant of the hierarchical plan analysis developed by Grawe and Caspar (Caspar 1989)), together with

7) the retrospective description by the participants of the course of the game, together with their individual actions and experience; such descriptions take place in the evaluation session and, where necessary, in separate, semi-structured interviews. Within this process, all of the video films and extracts from video recordings can serve to stimulate the memory (stimulated recall).

Two further elements will come into consideration in the near future, namely

8) methods serving the high-level analysis of nonverbal communication and

9) telemetric assessment of the participants' psychophysiological parameters (see e.g. Fahrenberg et al. 1991).

Accordingly, the states of order of a system game should be considered as a complex, dynamic pattern (configuration) of these various data levels and data sources. Within thus configuration, the likelihood of the appearance of coherence phenomena (e.g. the synchronous courses of the physiological parameters of several persons) will be equal to that of divergence phenomena (different coherent modes in individual subsystems of the data levels).

1.2.2 The Identification of Control Parameters

Questions naturally arise not only with respect to the order parameters but also regarding the *control parameters* for self-organizing processes. We are normally concerned here with magnitudes within the environment of a system (e.g. the temperature gradient between the bottom and the surface of a liquid in connection with Bénard instability, or the stimulus of energy resulting from radiant exposure in the context of a laser), the behavior of the system reacting to their linear change in a nonlinear manner. It is true that such external magnitudes of influence are conceivable in the case of the system game, e.g. through increasing the pressure of time on an experimental basis, through the person(s) coaching, monitoring, observing the game presenting the players with additional kinds of problems, through the infiltration of additional participants, or through informed behavioral change on the part of one or several players. Such interventions enable experimental manipulation during the course of a system game. However, the external milieu usually remains constant, processes of self-organization nonetheless occurring. For this reason, other possibilities must be taken into consideration:

a) We can assume the existence of the system's internal control parameters, comprising, for example, the internal activation or intensity of emotion of the entire system or a form of net motivation of all the participants. It would be possible to measure such parameters, for instance via the participants' emotional or stress ratings. In analogy to thermodynamic (physical) systems, this would indicate the distance of the system from motivational equilibrium.

b) It is precisely in the case of nonlinear systems that the parameters themselves can be conceived of as a function of the structure and dynamics of the system. Internal changes in the game's conditions (e.g. changes in stress experience, the activation of particular emotions, an increase or decrease in complexity, a change in the density of interaction) thus function as parameter changes. They can be utilized to mediate the further processing, dynamics of the subsystem in which the change in conditions takes place; alternatively, the change in conditions within subsystem A can function as a control parameter for the further dynamics of a different subsystem B. In this way, parameter dynamics and system dynamics are coupled to each other. The prerequisite for this coupling and the nonlinear processes thereby made possible is merely that the system is kept beyond a critical distance from motivational equilibrium, that is the total activation of the system game should not fall below a critical limit (the players should not lose their motivation or abandon their role identification, for example).

c) As mentioned above, the course of events in a subsystem of the system game can become the external control parameter for the internal dynamics of another subsystem. The individual subsystems alternately constitute the structural conditions for one another under which their internal processes develop. This leads to the emergence of a *hypercyclic coupling* between the subsystems, the latter providing and also altering the relevant control parameters for one another (Fig. 2). In this way, it is possible for emergent states of order to develop, including the individual subsystems (cf. Eigen & Schuster 1977; 1978).

1.3 Prospects for Research and Application

The paradigm of the system game shows itself to be useful and multifunctional in a considerable number of respects:

1) It embraces the possibilities offered by both *experimental* and *field research*.

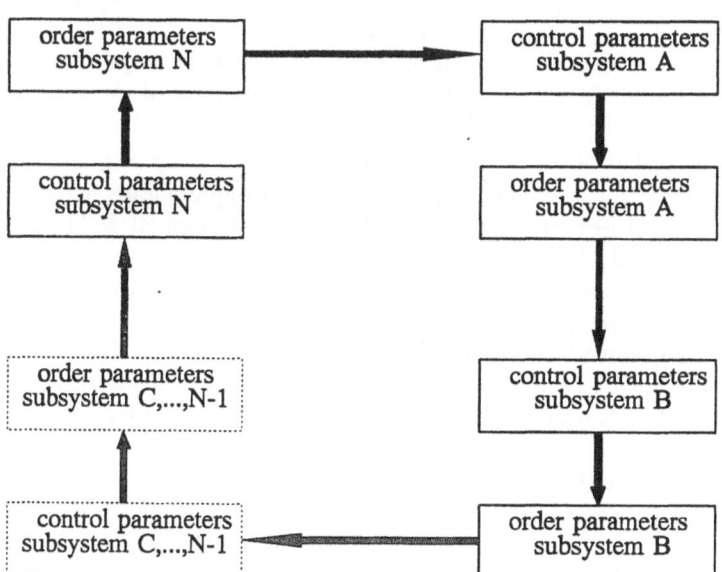

Fig. 2: The hypercycle of order- and control parameters: The dynamics of order parameters within one subsystem provide the control parameter(s) for the dynamics within another.

System games include the advantages of experimental research inasmuch as the course of the game can be well documented (e.g. via records of action and stress, follow-up studies, and interviews) and even directly observed by means of video cameras installed in the rooms. The paradigm fulfils the experimental criterion that it should be repeatable, since the same scenario can be played by different, comparable (e.g. parallelized) groups of test subjects under the same experimental conditions. Once a scenario has been drawn up — in other words, as soon as the role descriptions and starting situation exist — this scenario can be used and the results obtained by several research groups compared. Since one must reckon with chaotic dynamic processes in complex social systems, however, it is to be expected that the course of the various games will not be similar, even where the role instructions are identical (sensitive dependence upon the initial conditions). Finally, the system-game paradigm also fulfils the experimental criterion of variability, since the scenarios and structural conditions can be altered systematically according to the research interest or hypotheses (e.g. different number of participants, different playing-time/real-time relation, conflict given or not given in initial scenario, etc.). The person(s) preparing and observing the game can even intervene on an experimental basis in the course of the game, for example by issuing further instructions or tasks such as would induce or reduce stress or conflict. It is even possible in an extreme case to work with previously instructed participants, enabling the observation of behavior changes of individuals or of systems and systemic behavior in response to specific behavior on the part of a participant, for example.

Inasmuch as the criteria of repeatability, controllability (of certain environmental conditions) and variability within the framework of the system-game paradigm are at least partly fulfilled, it becomes clear that this offers access on a basis that is at least quasi-experimental to the study of nonlinear psychosocial systems (for a detailed comparison of the requirements made of experimental and quasi-experimental methodology, see Campbell & Stanley 1966; Cook & Campbell 1979).

The system-game paradigm includes the advantages of field research in that situations of considerably greater complexity and dynamism can be studied than would be possible with the usual laboratory research employed by social psychology. If the scenarios are conceived in a sufficiently realistic manner, then role identification is normally high and ego-involvement considerable. It is even possible to create scenarios in which institutions and their role-takers play themselves. Despite this considerable approximation to reality, one which is of advantage for the external validity of such studies, the difference between game and reality remains upheld, so that the ethical problems regarding research are less serious than is the case with scientific interventions in the sensitive, and equally intimate, zones of real-life. (It is necessary to conduct further, detailed analysis of the relationship existing between internal and external validity within the context of system games).

2) The system game as *research* and *training method*.

Possible concerns relating to the ethics of research are further reduced if one considers the fact that it is only of secondary importance that the participants fulfil the function of test subjects; what is of primary importance is that they profit from their participation in the system game. This can be seen from the fact that system games are employed within the context of practical education and university seminars, for purposes of self-experiencing, as a means of illustrating the knowledge in the realm of system theory, action theory, and research into experiencing and coping with stress, and also by social, industrial and organizational psychology, as well as for purposes of innovation testing and conflict resolution in institutions. In this way, participation is made attractive for highly qualified practicioners or senior employees, people who would not normally be willing to take part in the usual kind of psychological study. This is a further argument in favor of the high practical relevance and external validity of the system-game paradigm. Additionally, the fact should not be forgotten that the collection of data (e.g. the compiling of reports regarding actions, experiencing and coping with stress) can in fact assist the participants in immediate feed-back in the evaluation session. This means that (a) the collection of data is an organic component of every system game; (b) personal motivation can be expected when compiling reports, instead of the usual reactance; (c) reporting back to the test subjects — something often asked for in scientific studies but rarely fulfilled — not only ensues immediately but in turn itself becomes a data source as a detailed evaluation phase. The system-game paradigm thereby approaches the ideal of actional research and systemic methodology through the coinciding of practice and research, with no concessions as regards the high standards of research methodology being necessary (cf. Schiepek 1988).

3) The connection between the *individual* and *"system" perspectives*.

Participation in system games does not only require a complex cognitive performance: processes of problem solution, decision and planning often take place under considerable pressure of time. Social competence (e.g. negotiation, impression management) is also in demand, interpersonal conflicts can appear, and intensive emotions and stress are experienced, something in turn requiring intrapsychic and social coping processes. For this reason, the interest of the authors' research is directed less towards cognitive performance and error in complex problem-fields, something examined in detail by the study group around D. Dörner, for example (see e.g. Dörner 1989); rather, we are concerned with the emotional experiences of, stress upon, and coping with stress by the individuals involved. Within this context, the system-game paradigm offers a possibility that is of particular interest for research into experiencing and coping with stress, namely that of comparing the subjective experience of stress, the subjective appraisal of the situation, and the subjective coping strategies, on the one hand, with an "objective" observation of the stress-relevant situation, with the observable behavior of the subject and his interactional partner, and finally with this interactional partner's assessment of the situation and observation of the his/her partner, on the other hand. It is thus possible to relate subjective experience mutually to each of the partners and to set this in relation to the interpersonal interactive course of events. In other words, an insight is obtained into the co-processing of psychological and social systems, including their mutual interpenetrations (Luhmann) and irritations. Accordingly, for the identification of collective order states temporal progressive patterns of individual experience-processing may be employed, as may be the characteristics of interpersonal communication (relational dramatization) and those of the pattern of social (inter-)action.

4) The great *variety of possible scenarios.*

System games can be devised for every interpersonal field of action and problems. The customary procedure here is that the person(s) preparing the game give(s) the starting scenario and role descriptions. However, it is also possible for the organisators and the participants to develop the scenario together (e.g. when system games are used within the context of a "future workshop", see Jungk & Müllert 1981). Another possibility is to have the participants (e.g. students) conduct their own investigations into the practical sphere in connection with which they are planning a system game, offering the opportunity of acquiring relevant on-the-spot occupational and role descriptions from practitioners. The participants may also play themselves in their institutional functions or exchange functions in a deliberate manner, as when a system game is played at an institution with the aim of crisis management or innovation planning.

The experience gathered by the authors to date lies in the realm of psycho-social scenarios: a "multigeneration-family/educational-psychology scenario", ⓒSchiepek 1987, Schiepek 1991, Chap. VI, and a "youth home scenario", ⓒGisi & Schiepek 1989, see below Sect. 2, have been developed. It is instructive in this area to observe not only how problems either are solved, or remain open and unresolved, or are put off until later, but also how problems and problem sources come into existence. It is possible to trace the course whereby problem persons and "problem systems" (Ludewig 1987) are constituted, sometimes resulting from predisposed role descriptions, sometimes without any role-immanent predispositions. The "social construction of patients" could become a research topic in its own right within the system-game paradigm, for example as social psychology research into etiology or "experimental sociopsychopathology".

In addition to psychosocial scenarios, there are of course also business-administration scenarios intended for management-training purposes. *Environmental-policy scenarios* would appear to be of particular relevance for the future: within the context of environmental problems that are becoming increasingly apparent in nature, the conflict of interests is becoming more explosive, the possible solutions more complex (more "systemic" — one need only think of the emotive word "traffic"), and at the same time communication between the groups involved — whether citizens' action groups, political groups, scientists or industry — more necessary, but no easier. The authors are currently working on a study of scenarios related to environmental policy.

2 The Study of a System Game

2.1 Method

The Scenario

In the following we would like to describe the realization of an a system game, together with the preliminary results thereby obtained. The scenario is that of a "youth home", developed during a university course on system games (© Gisi & Schiepek 1989) at the University of Bamberg. We refer here to an implementation of this scenario which took place at the University of Bern. The following parts were taken:

Staff of the youth home: Franz Rupprecht, director; Monika Seibold, head of education department; Veronika Maier, educator; Karin Huber, psychologist; Otto Kaltenegger, head of workshop

Adolescents at the home: Peter Köhler, Freddy Müller, Sabine Sauer

The parents of Peter Köhler: Helga Köhler, Manfred Köhler

The psychiatrist of the psychiatric ward for children and adolescents at the hospital: Dr. Hans Weinberger

Staff member of the youth welfare office: Julius Jelloncek

The scenario simulates a psychosocial care system comprising a home for adolescents, a youth welfare office and the psychiatric ward of a hospital. From the perspective of the professionals working within these institutions, all other persons are viewed as "problem persons", or, put differently, as "having problems". This leads to the whole system being divided into two parts, the psychosocial care system and the problem persons (see Fig. 3). In addition, the planning of the scenario setting has designated a special role for Peter Köhler, one of the adolescents living in the home. At the present, he is in the psychiatric ward for children and adolescents for purposes of observation, so that the resulting expert opinion can be used to clarify the question as to whether Peter is to remain at the youth home in the future. As indicated above, Peter's parents are also considered "problem persons", since they are living under difficult social and economic circumstances.

2.2 Implementation of the Study

Procedure

The system game took place during a university course on "Systems theory and clinical psychology" at the Department of Clinical Psychology of the University of Bern (Prof. Grawe). The authors were responsible for the implementation and documentation of the system game.

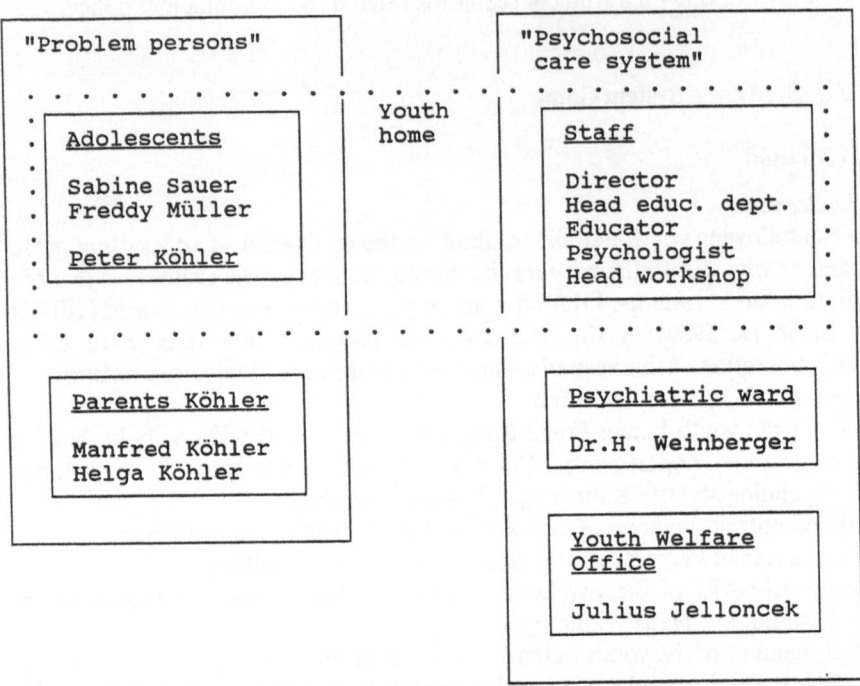

Fig. 3: Subsystems and role players within the system-game "youth home scenario"

The *participants* were advanced students of psychology and psychologists experienced in practical therapeutic work. Five of the twelve participants were women.

Rooms: The premises of the Psychologische Praxis- und Beratungsstelle of the University of Bern were placed at our disposal. Each of the rooms was equipped with a telephone, with the exception of the adolescents' communal room. The doors were marked with nameplates and had been arranged so as to suit the requirements of the role players; the parents' apartment was furnished with several chairs, for example. Separate, suitable rooms were provided for each subsystem or participant actor.

Time schedule: The experiment began at nine a.m. with preparatory instructions. In addition to a few general comments, the subjects were given instructions regarding the recording of planned actions and stressful events (see also below). Every participant received a written role description giving a short characterization of his/her personality (gender, age, biographical information) and his/her formal and informal position within the institution to which he/she was allocated. Otherwise, the description only offered information about certain details regarding the whole system (for example, who were the most important persons to contact etc.). The roles were assigned to the players on a random basis.

The system game itself started at 10 a.m., after the participants had been shown to their rooms. The game simulated three fictitious days of the system, taking up two hours of real time each, making it last six hours altogether. The first "game" day was followed by lunch. The actors were asked not to exchange information about the game (in order to avoid meta-communication), yet they were allowed to contact each other within their roles. Two further "game" days followed after lunch and the game ended at 6 p.m. Following a short concluding discussion, the participants were formally released from their roles. For assessment purposes, the complete game was subdivided into six periods of one hour, each thus corresponding approximately to half a "game" day. A detailed follow-up discussion took place on the following morning (evaluation session), during which the group exchanged and reviewed the experiences acquired by the participants. The evaluation session was concluded by lunch.

Measures: Two different self-report instruments especially designed for the purposes of the present study were applied for data collection.

1) The so-called "advance notice of game activity" consists of a semi-standardized record sheet which the participants are asked to submit for documentation prior to any step taken in the game. In addition to the time and the persons involved, it includes an indication in few key words of the intended action.

2) Stressful events occurring in the course of the game are recorded by means of the COMES self-observation method (Perrez & Reicherts 1987). This includes a detailed description of the stressful situation as experienced by the subject, together with the time of occurrence, by means of different variables involved in the coping process (appraisal of the situation, emotions, coping efforts and results). This instrument, which is based upon the "situation-behavior approach" proposed by Reicherts (1988) and Perrez & Reicherts (1992), has already been successfully applied in different previous studies. The data on coping with stress obtained during this system game can thus be compared with data collected within other studies and subject populations.

While the data obtained by means of the "advance notices of game activities" reflect intended (i.e. anticipated and attempted) action, the data collected by the COMES self-observation method yield information concerning stressful situations after they have in fact occurred. An event taking place during the game, e.g. a stressful team session, may be perceived as subjectively stressful by several participating persons and recorded accordingly. Of course, the situation may be perceived differently by the various persons involved.

Rating and Scoring of Measures

Only those data generated by the subjects during the course of the game, i.e. in close association with behavior and experience during the game ("hot" cognitions, emotions and actions), are included for analysis.

The subjects' responses concerning their stress and coping as gathered by means of rating scales were evaluated directly. Following completion of the game, it was additionally determined by means of the stressful-situation descriptions which of the other participants had been involved, regardless of whether this had only been anticipated (such as an invitation by the youth welfare office to the index-child's parents), or whether the event had in fact occurred (such as an argument between husband and wife). One or more other role players were involved either directly or indirectly in every stressful event recorded by the participants during this game.

The game actions ("advance notices of game activities") were assigned ex post facto by means of content analysis into eight categories of a category scheme for the evaluation of system games developed on the basis of the IPA suggested by Bales (1950; compare Nowack 1989) as well as the work of Dörner and his research group (Badke-Schaub 1988). It consists of the following categories:

1) "Self-centered activity" (e.g. working, relaxing);
2) "Informal search for contact" (e.g. inviting someone for supper);
3) "Search for information" (e.g. asking questions, making inquiries);
4) "Providing information" (e.g. giving notice of something, informing someone of something);
5) "Personal assistance" (e.g. counseling, therapeutic assistance);
6) "Planning, decision-making, designing plans of action";
7) "Initiating, persuading, achieving objectives" (e.g. initiating or arranging steps to be taken or achieving objectives);
8) "Addressing and solving conflicts".

These categories refer to the designated intentions, which means that an intended behavior is encoded as belonging to a certain category even if the aim was, in fact, not accomplished (for example, a telephone call with which the person intended to gather information will be encoded as belonging to category (3) even if the call was not successful). The encoding specifications allowed for an assignment of one activity to two different categories. The inter-rater reliability of two independent raters amounted to 85%.

In addition to the registration of the total number of activities and stressful situations (per subsystem and time period), simple measures of interdependence were computed referring to (1) the activities within each subsystem and to (2) the activities exchanged between the subsystems. These measures serve as formal indices for the system dynamics.

The indices designating the changes in the structure of the interactions were obtained by calculating the differences between the matrices of interaction. These values reflect the number of completely changed cells within the interaction matrices (the number of additional or cancelled cells), as well as their numerical/intracell/intragroup difference in relation to each previous half-hour interval.

The statistical analysis was carried out (a) with respect to the maximum aggregation/accumulation of activities and stressful events over the complete course of the game (i.e. static considerations) and (b) concerning the two subsystems "psychosocial care system" and "problem persons". The statistical tests and analyses were carried out on the basis of the events within the subsystems, instead of the events reported by the individual participating actors.

For further analysis, the data reflecting the course of the game were subdivided into the three fictitious game days and also into six one-hour intervals. In view of the fact that a total of 191 events (activities and stressful events) had been recorded, an even finer differentiation did not seem appropriate. Also, the preconditions which must be fulfilled if a statistical analysis of temporal evolution is to be carried out were hardly met. Although the analyses carried out yielded different formal/statistical indices, they are, essentially, of an heuristic and descriptive nature.

2.2 Results

In presenting the results, we wish first of all to demonstrate to what extent this systems-game scenario enabled processes of system dynamics and self-organization, such as can be inferred by examining the experience and behavior described by the participants, to take place.

Second, the dynamics of this complex scenario evolving out of the game events as indicated by the activities and stressful situations within and between the subsystems are to be illustrated — at least in rough outline — by means of selected variables and their interactions.

Third, we would like to point out possible ways of presenting data indicating synergetic or self-organizational processes.

2.2.1 The Validity of Role Identification

Each participant was asked to indicate — separately for each of the three fictitious game days — to what extent he/she was able to identify himself/herself with the role description during the game. The rating scale applied ranged from 0 ("not at all") to 10 ("extremely"). With marked differences between the participants, the mean was 6.7, which is regarded as a medium to good identification. No distinct differences were observable between the fictitious game days or between the subsystems. The lowest value (4.3) was recorded by the participant acting as the "head of the workshop", who had had difficulties entering into the game despite trying hard. The highest value was reported by the participant acting as the "mother of the index client", Mrs. Köhler. She had experienced the game as very realistic, something which also became evident during the evaluation session. (Incidentally, she was at an advanced stage of her therapist training.) In general, the data obtained on role identification indicate that the participants experienced the game as sufficiently authentic and realistic, a fact which was also emphasized during the evaluation session.

2.2.2 Activities

A total of 147 activities or interactions were documented by means of the "advance notices of game activity" (see Table 1). These include such events as the director calling in the staff for the morning report (category 7 of game actions); the parents of the index client making inquiries at the youth welfare office (category 3); the psychiatrist having a talk with the index client (categories 3 and 4); the adolescents visiting the index client at the hospital (category 2); or the parents having an argument (category 8).

There is a striking difference between the number of activities initiated by the psychosocial care system (consisting of the actors representing the youth home, the youth welfare office and the psychiatric ward of the hospital) and that from the subsystem of the problem persons (represented by the adolescents together with the client Peter Köhler and his parents), the former being much more active than the latter with 102 as opposed to 45 activities respectively.

Even when the different number of actors within the psychosocial care system and the subsystem of the problem persons, $n = 7$ and $n = 5$, respectively, is taken into account, this marked difference still persists.

The following results are obtained when the activities are rated as to content in accordance with the above-mentioned categories (see Table 2): As a

Tab. 1: Activities and stressful events within and between the subsystems during the course of the game

Subsystem			1st day 1	2	2nd day 3	4	3rd day 5	6	Total
Activities:									
(1) Problem persons									45
with reference	to	(1)	1	1	5	0	1	4	12
	to	(2)	4	2	4	6	8	9	33
(2) Care system									102
with reference	to	(1)	3	4	2	1	5	9	24
	to	(2)	8	23	15	6	20	6	78
Total			17	30	26	13	34	28	147
Stressful events:									
(1) Problem persons									32
with reference	to	(1)	2	1	2	0	0	1	6
	to	(2)	5	9	1	5	4	2	26
(2) Care system									12
with reference	to	(1)	0	4	1	1	1	1	8
	to	(2)	0	0	1	0	3	0	4
Total			7	14	5	6	8	4	44

Fictitious game day/period

Tab. 2: Categories of activity (advance notices of game activity)

Cat	Total	Subsystem		Fictitious game day/period					
				1st day		2nd day		3rd day	
		(1)	(2)	1	2	3	4	5	6
1	3	3	0	1	0	0	1	0	1
2	22	14	8	1	4	4	0	0	13
3	30	10	20	5	11	3	2	5	4
4	12	1	11	2	6	0	1	3	0
5	8	3	5	1	1	1	0	1	4
6	46	6	40	4	7	16	5	8	6
7	22	6	16	3	1	2	0	16	0
8	4	2	2	0	0	0	4	0	0
Total	147	45	102	17	30	26	13	33	28

consequence of the activities initiated by the psychosocial care system, the category "planning, decision-making, designing plans of action" (6) was the most frequent one, followed by "search for information" (3), "initiating, persuading, achieving objectives" (7) and "informal search for contact" (2). The remaining categories "self-centered activity" (relaxing, playing; 1), "addressing and solving conflicts" (8) and "personal assistance" (counseling, therapeutic assistance; 5) were very rare.

However, the frequency with which the categories occur varies distinctly over the course of the game. While the category "search for information" is initially predominant, especially during the second period, the third period is characterized by the predominance of the category "planning, decision-making". The categories "initiating, persuading, achieving objectives" and finally "informal search for contact" occur more frequently towards the end of the game, i.e. during the fifth and sixth period, respectively.

Also, the two subsystems exhibit different activity profiles. While the categories "informal search for contact" and "search for information" occur most freqently within the subsystem of the problem persons, the categories "planning, decision-making" followed by "search for information" and "initiating, persuading, achieving objectives" are found most frequently within the psychosocial care system, as these are task-oriented activities which are required to be carried out by these institutions. We will return to a more detailed analysis of the changes in activity profile in the presentation of the configurational patterns of the system dynamics, which will be interpreted in terms of phase transitions.

Interactions between the two Subsystems
A quite remarkable phenomenon is the fact that the number of activities by the psychosocial care system taking place *within* this subsystem by far exceeded the number of activities directed at the problem persons actually to be taken care of (see Table 3). This relationship is reversed with respect to the subsystem of

Tab. 3: Interdependence of the subsystems concerning activities

			Activity initiated by	
Subsystem			(1) Problem persons	(2) Psychosocial care system
Activity with reference to	(1)	Problem persons	12	33
	(2)	Psychosocial care system	23	79

Chi^2=53.06; p=.000

the problem persons: that is, activities directed at the psychosocial care system by far exceeded those activities taking place within the subsystem. The corresponding statistical test yields a high significance (χ^2 = 53.06; p < .000).

2.2.3 Stress

A quite different picture results when we turn to the extent of perceived stress (see Table 1). In contrast to a total of 32 stressful events reported by the group of the problem persons, only 12 events were reported by those participants acting as members of the psychosocial care subsystem. This corresponds to a mean of 6.3 stressful events per problem person, as opposed to only 1.7 stressful events per member of the institutional staff.

This disproportion may well be understood as reflecting in a realistic way the psychological strain suffered by clients and other persons involved in problematic situations. To a certain degree, this strain is a result of their interactions with members of the institutions concerned. This view is also substantiated by the specific differences in the content of the stressful situations reported.

Characteristics of Coping with Stress
Further analysis reveals remarkable differences between the problem persons and the psychosocial care system with regard to important characteristics of coping with stress (all differences are significant; p < .05).

Not only were stressful situations reported more frequently by the problem persons; they were also perceived as less controllable (t = −4.18; p < .001). Moreover, the problem persons reported more intense negative emotional reactions to stress (t = −2.72; p < .01): the degree of anxious and depressed stress-related emotions was higher.

The problem persons' attempts at coping are characterized by an increased suppression of information (t = 3.41; p < .001) and a decreased search for information about the situation (t = −3.38; p < .01). Their coping operations are notably evasive (avoiding, leaving the field; t = 3.21; p < .01). Their

Tab. 4: Interdependence of the subsystems concerning stress

| | | Stress experienced by | |
Subsystem		(1) Problem persons	(2) Psychosocial care system
Stress in relation to	(1) Problem persons	6	8
	(2) Psychosocial care system	26	4

Chi²=9.24; p=.002

attempts at actively influencing the stressful situations are drastically reduced ($t = -6.39$; $p < .001$).

Accordingly, the problem persons perceive the results of their coping efforts as unfavorable. Both problem solving and emotional adaptation during the stressful episodes are described as less effective by the problem persons than by the actors of the psychosocial care system ($t = -2.30$; $p < .05$).

The pattern of the characteristics of coping with stress described here is a typical example of the patterns found in "real-life" clients, for instance in studies on depressed persons or phobic patients (Perrez & Reicherts 1986; Reicherts et al. 1987; 1991).

Interdependence of the Subsystems Concerning Stress
A highly significant ($\chi^2 = 9.24$; $p = .002$) and very remarkable result becomes evident when the stressful events are categorized into those occurring within the relevant subsystem or in connection with members of the other subsystem (e.g. who else was involved in the stressful situation, who caused it?) (see Table 4). Members of the psychosocial care system are involved in more than 80% of the stressful events reported by the problem persons, while only 30% of the stressful situations encountered by the members of the psychosocial care system occur in connection with the problem persons. In other words, not only is a lower extent of stress reported by the members of the psychosocial care system; a considerable portion of that stress is also caused by the system itself. At the same time, the psychosocial care system is the predominant source for the stressful situations encountered by the problem persons who are actually to be taken care of. However, the differences between the two subsystems regarding their interactions and coping with stress are due mainly to the activities during the first part of the system game.

2.2.4 Dynamics of the Evolutionary Pattern

In the following we would like to address the questions as to which sections may indicate the occurrence of dynamic patterns of self-organization during the course of the system game, and the extent to which any characteristic features and their specific configurations may suggest the occurrence of phase

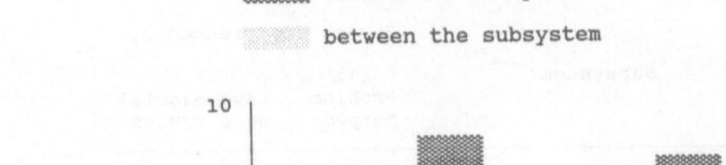

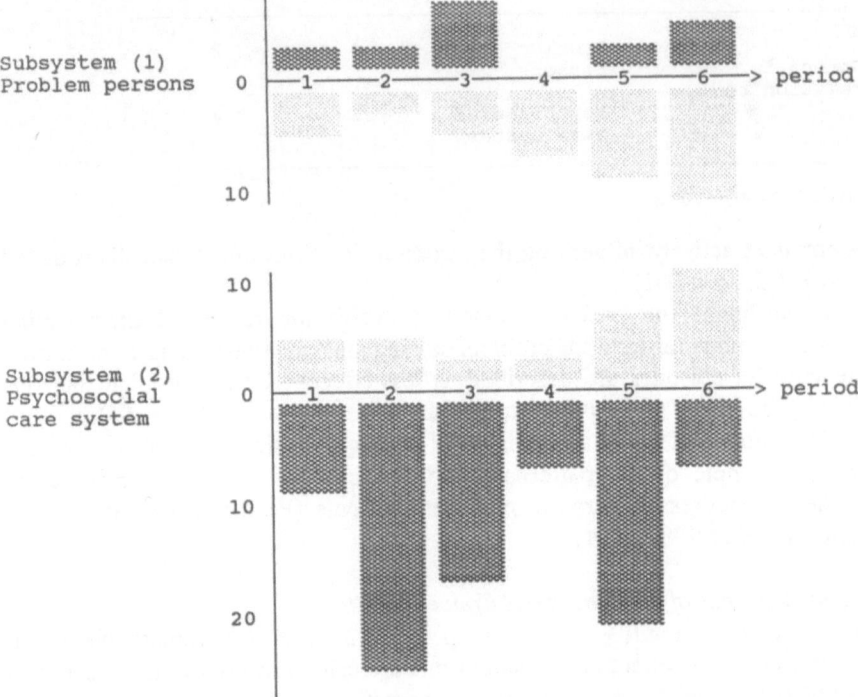

Fig. 4: Activities within and between the subsystems during the course of the game

transitions and periods of stability and instability. To this end, we will consider from a heuristic point of view the sequence of the six one-hour periods. (See also Figs. 4 and 5, as well as the configurational pattern presented in Table 5).

After the initial "warming-up" period, with relatively few activities and a low degree of stress, the second one-hour period is characterized by a high level of activities on the part of the psychosocial care system, being directed mainly at role players within the subsystem. These self-referential activities predominantly concern the exchange of information, especially the search for information. The problem persons, on the other hand, encounter the highest level of stress during this period while at the same time performing only a very few activities. Also, we can observe a climax in the degree of interrelatedness, as indicated by the index designating the changes in the structure of interactions obtained by means of the interaction matrices. High values are indicative of changes in the structure of interactions due either to activities being directed at a lower number of partners than during the previous period

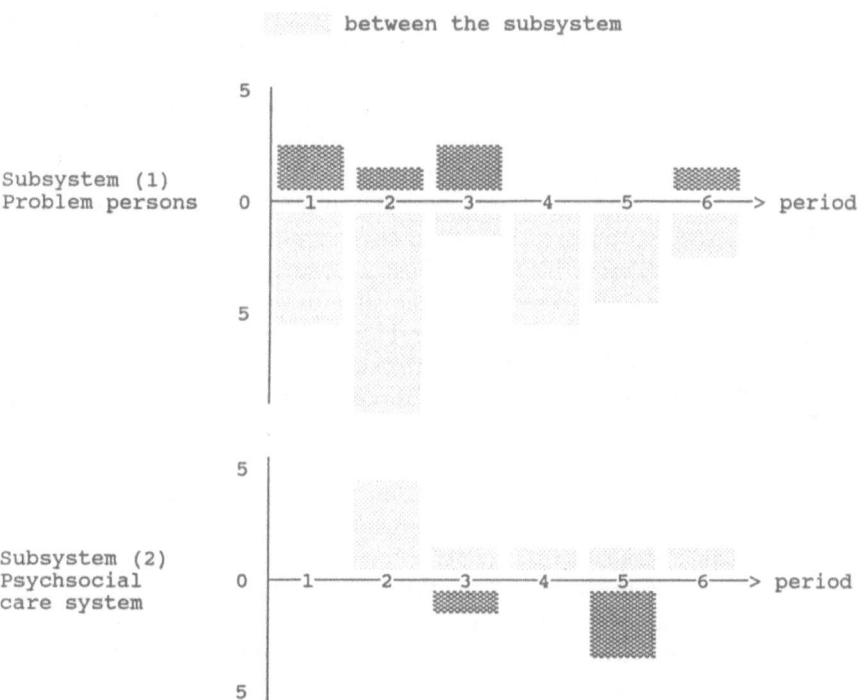

within the subsystem

between the subsystem

Fig. 5: Stressful events within and between the subsystems during the course of the game

Tab. 5: Configurations of different indicators of system dynamics (positive or negative z-values)

Indicator of systems dynamics	Fictitious game day/period					
	1st day		2nd day		3rd day	
	1	2	3	4	5	6
Activities total	−	+	+	−	+	+
Activities between the subsystems	−	−	−	−	+	+
Changes of interaction matrices	−	+	+	−	−	−
Stressful events total	−	+	−	+	+	−
Stressful events between subsystems	−	+	−	+	−	−

and/or to an increased number of different role players being addressed than before. It is mainly the activities within the psychosocial care system which contribute to this increase in value.

With regard to the psychosocial care system, this configuration of characteristic features may be interpreted as indicating a period of instability together with an intensification of communication processes. The unpredictable complexity and the absence of familiarity with the actors and the structure of the different institutions, especially within the youth home (distribution of competence, type and structure of the problem), are characteristic of a system game's initial phase and require a high degree of information to be exchanged between the different players. Apart from clarifying the respective tasks, this should also serve to build up an identity for the different institutions within the psychosocial care system.

The analysis of the third section of the game reveals changes which might indicate the occurrence of a phase transition (compare Figs. 4 and 5). The number of activities is increased within the subsystem of the problem persons, exceeding the number of stressful events reported, meaning a "switching" in the ratio of activities and stress. At the same time, the number of activities performed by the psychosocial care system decreases, yet these activities are still mainly addressed at members within the subsystem. The previously predominant category of "search for information" is now replaced by "planning, decision-making, designing plans of action" as the principal category.

The fifth section (beginning of the third game day) in a way designates the onset of a new state of order, as it represents that section characterized by the highest amount of activities overall (compare Fig. 4 and 5). There is a distinct increase in interaction between the two subsystems. Nonetheless, the psychosocial care system exhibits a high level of activity within the subsystem, but there is increasing exchange with the problem persons. Now the main category of activity is "initiating" agreements to be put into practice and "achieving" anticipated goals (category 7). Also, the degree of stress perceived by the participants is increased.

This state of affairs is maintained during the final sixth section of the game (compare Fig. 4 and 5). We continue to find a relatively high level of activities, accompanied by a high rate of exchange. In contrast to the increase during the fifth section, the rate of stress now returns to a relatively low level, both within and between the two subsystems. Regarding the content of activities performed, the category of "informal search for contact" now dominates the scene. Among other activities, this is due to a party being organized on behalf of the index client's return from the psychiatric unit to the youth home.

The indicators mentioned above suggest that the two subsystems converge during the fifth and sixth section. In addition to various other examples, this convergence becomes evident through an interactional exchange between the subsystems appropriate to their respective functions, together with constructive efforts at solving the problem initially introduced into the scenario. The dynamics may be understood as representing a relaxed state of order facilitating a potentially productive interplay between the different participants in the game.

In principle, this heuristic analysis can be made more formally precise by means of configuration—frequency analysis, which would allow for statistical testing especially in the case of longer time series with a higher level of resolution (and, accordingly, a higher number of events to be included).

The results obtained in the detailed evaluation session have not been considered here. As mentioned before, only those data reported by the participants in close connection with their behavior and experience during the course of the game were analyzed. However, we should like to mention here that during the evaluation session latent conflicts within the psychosocial care system were indicated which were not yet conceived as overt stressors, especially by the staff of the youth home. They were based upon the staff member of the youth welfare office being dissatisfied with the proceeding of the staff of the youth home concerning the client, Peter Köhler. However, these problems had precipitated merely one stressful situation reported by the staff member of the youth welfare office.

Specific differences are discernible between the various institutions and actors of the psychosocial care system concerning their individual coping with stress. These have to be considered when interpreting the individual coping characteristics as indicators for the system dynamics. Strictly speaking, only the staff members of the youth home reported extremely few and rather mild stressful situations, whereas the psychiatrist and the staff member of the youth welfare office reported far more serious stressful events. The reasons for this need to be clarified by means of further analyses which take into consideration the specific situations and environmental contexts.

3. Discussion and Prospects

The present study shows that by employing a system-game scenario of psychosocial services, authentic and very intense processes were stimulated which may be considered to depict real system dynamics in a plausible way. The "problem persons" performed activities and encountered stressful episodes typical for real clients. The predominant portion of the activities performed by the psychological care system showed it involved with itself, thereby reflecting the complexity of intra- and interinstitutional structures. This is especially true for the initial period of the present life simulation. Remarkable changes took place during the system game, typified both by individual characteristics of the system dynamics and by the configuration of these characteristics. Thus we find characteristic variations in the temporal density of activities and in the ratio of activities exchanged between the subsystems. Also, the pattern of the interactions by which the individual actors and institutions relate to each other varies in the course of time. The changes indicated by these formal criteria are accompanied by changes regarding the content of the activities. Significant features of the system dynamics are marked not only by changes at the level of social processes, but also by processes at the individual personal level, such as the subjective perception of stress and subsequent coping efforts. Unlike the members of the psychosocial care system, the problem persons' stressful experiences are characterized by situation appraisals, emotions and coping behavior typical for mentally disordered clients. Over the course of the game, a decrease in stress and an enhanced degree of interdependence are to be noticed. Together with the activity profile, these characteristics are seen as indicating an increasing convergence of the overall system towards a new state of order. By way of this newly reached state of order, conditions are at the system's disposal which allow for an effective solution of the problem initially introduced into the scenario setting.

We would like to emphasize once again that the results described here are based on data collected via self-reports and self-observation; thus, they only reflect processes as they are subjectively perceived and reconstructed by the participants in the game. It is both conceivable and desirable, however, that these self-reports be compared with physiological recordings (see Fahrenberg et al. 1991) or with data gathered by direct behavior observation, as the paradigm of system gaming offers controlled conditions which make it possible in principle for such data to be collected. This would allow for a comparison of the subjective perspective of participants acting within a highly emotional atmosphere with the level of objectively and systematically observable processes.

The method of analysis described here is of an heuristic nature. A more differentiated analysis and interpretation of the data is possible, for instance with regard to more refined and various other indices of the inherent system dynamics (such as interactional concentration measures) or in connection with the qualitative results of the evaluation session.

Clearly, the realization of this particular system game and the system dynamics resulting from it should not be regarded as representing the sole general principle of self-organizational dynamics that can occur within the paradigm of system gaming. The present study depicts a highly complex, but unique, chain of events, on the basis of which hypotheses concerning the underlying synergetic dynamics are to be formulated and tested in further studies. As already mentioned, the system-game paradigm offers the advantage of enabling controlled experimental research to be carried out. In addition to replicating the research, it is possible to systematically vary research conditions such as game duration, composition and participants' previous experience, varying the scenario setting as to structure and content. There is still a long way to go.

Nevertheless, we believe that steps have been made in the direction of an improved understanding for synergetic processes of self-organization within complex psychosocial systems, both in terms of the methods of data collection and analysis and also with respect to the initial results presented here.

The research approach adopted here can be characterized by the following prospects:

1. *Replications*. The intention here is to create a systematic comparison of the course taken by several system games either already conducted or planned for the future within the scenario employed above. A necessary stage in this respect is the development of a readily usable data-management system, one enabling the direct comparison between data of different types, from different data sources, and stemming from different temporal degrees of resolution. Finally, systematic replication makes possible the specific variation of relevant control parameters within the course of the game (e.g. variations in the playing-time/real-time relationship, in the induction of conflict from without, in role description etc.); it would also be possible within this framework to conduct hypothesis-testing research via theory-based variations in the parameters. However, hypothesis-testing in the realm of complex, dynamic systems assumes the existence of computer simulations for appropriate theoretical models (for an overview of various simulation methods which may also be used in a hybrid manner, see Schaub & Schiepek, this volume).

2. *Chaos and Chaos Management*. Once time series are available that are of adequate length (comprising at least 500 to 1000 measuring points) and use scales of measurement manifesting adequately fine resolution, it should be possible to carry out analyses of dimensionality, resulting in indications of the degree of chaoticity within the course of a game (for a practicable algorithm with regard to ascertaining the dimension of correlation D2, see Grassberger & Procaccia 1983; Theiler 1987; Steitz et al., this volume). The identification of chaotic attractors in psychosocial system dynamics would not only be of eminent theoretical interest; it would also open up to empirical treatment a question which has already been raised theoretically a number of times, that of the possibility of planning and the rationality of action within the context of macroscopic chaos (e.g. Krohn & Küppers 1990). Rather than forcing one to defeatism, the existence of chaos opens up chances for new basic positions, management philosophies and co-operative methods of planning. System games can offer territory for research and experimentation in this respect, with attempts at chaos management being undertaken by the scientists organizing and observing the game, on the one hand (necessitating — among other things — efficient and rapid feed-back on the current state of events within the game), but also by individual players or groups within the system game, on the other. Possible courses of action to be taken in chaos management include the alteration solely of structural conditions, giving a system the possibility of self-management and self-organization ("context management", see Willke 1989; Schiepek 1991); minimal intervention at short intervals (microtuning), rather than one or more massive "ballistic" interventions (Dörner 1989), enabling one to take into account the sensitive dependence of the dynamics on the system's current state (in formal terms: the existence of at least one positive Lyapunov exponent); the creative usage of unforeseen events, signifying greater flexibility and an increase in actional leeway; finally, should self-similar patterns be present, the variation of the most easily accessible one (principle of minimal

intervention on the basis of "scale invariance" and "sensitive dependency on the current state").

3. *Ecological Scenarios.* A scenario concerning environmental policy is currently being developed in order that progress be made regarding research and actional competence in one of the most relevant fields of conflict within social policy. In this connection, it is also intended to look into the question of whether there is a connection between the "quality" of the entire system performance and the individual condition of the participants (in particular, experiencing and coping with stress). This question is of theoretical interest, first of all since it is not posed with respect to one single functional system (e.g. a business concern, involving criteria such as productivity, profit, financial reserves, or employee work-satisfaction), but presupposes a quality evaluation for co-operation or conflict between several functional systems following quite different kinds of operational logic and criteria of evaluation (in accordance with the model of functionally differentiated societies). Secondly, the connection between "system performance" and individual condition is of practical relevance, since the latter is itself a quality characteristic, or at least an important prerequisite for the "quality" of system processes (one negative example of this may be seen in the consequences for the quality of care and treatment resulting from personnel "burnout" in psychosocial institutions).

4. *Socioaffective System Competence.* Where the acquisition of actional competence in dealing with complex dynamic systems is a matter of discussion, it is usually a question of cognitive performance (planning, making decisions, evaluating, solving problems). As has been set out above (Sect. 1.3, Point 3), however, social competence and social intelligence, experiencing and dealing with (intense) emotions, experiencing and coping with stress, all play a role that is at least as important. In social reality, it is not via the push of a button that decisions are effected: their realization requires personal commitment, feeling, conviction and the toleration of frustration. Emotions run through and integrate everything involving human perception, thought and action, as has been verified by numerous psychological and biological findings (Ciompi 1991). It is for this reason that the acquisition of *socioaffective system competence* is decisive, and suitable training units should be developed for this (e.g. a system game is played, followed by a discussion of aspects relating to self-experience, along with the communication of suitable coping strategies and problem conceptions such as are sound with respect to system theory; specific role-playing is performed, variations are introduced in particular sections of the game; finally, the system game is played again and intensively analyzed and evaluated afterward). It would appear, therefore, that the perspective for the future of the system-game paradigm is indeed extensive and complex.

Acknowledgement. We would like to thank Mr. Michael Claridge (Sect. 1 and 3) and Dipl.-Psych. Martina Weisensel (Staatl. gepr. Übersetzerin) for the translation of the present contribution.

References

Badke-Schaub P (1988) Protokollierung in der AIDS-Simulation. Unveröff. Arbeitspapier, Universität Bamberg

Bales RF (1950) Interaction Process Analysis. University of Chicago Press, Chicago

Bleicher K (1969) Entscheidungsprozesse in Unternehmensspielen. Bd 1: Die Darstellung von Unternehmenspolitik und -planung an Idealmodellen. Baden Baden

Campbell DT, Stanley JC (1966) Experimental and Quasi-Experimental Designs for Research. In: Gage NL (Ed) Handbook of Research on Teaching. Rand McNally, Chicago

Caspar F (1989) Beziehungen und Probleme verstehen. Eine Einführung in die psychotherapeutische Plananalyse. Huber, Bern

Ciompi L (1991) Affects as Central Organising and Integrating Factors. A New Psychosocial/Biological Model of the Psyche. British Journal of Psychiatry 159: 97-105

Cook TD, Campbell DT (1979) Quasi-Experimentation. Design and Analysis Issues for Field Settings. Houghton Mifflin Company, Boston

Cumpelik K (1985) Das Planspiel. Versuch einer handlungstheoretischen Analyse. Unveröffentlichte Diplomarbeit, FU Berlin

Daigl KA (1988) Kleine Planspiele für Helfer. Lambertus, Freiburg

Dörner D (1989) Die Logik des Mißlingens. Strategisches Denken in komplexen Situationen. Rowohlt, Reinbek

Dörner D, Schaub H, Stäudel T, Strohschneider S (1988) Ein System zur Handlungsregulation. Sprache & Kognition 7(4): 217-232

Eigen M, Schuster P (1977) The Hypercycle. A Principle of Natural Self-Organization. Part A: Emergence of the Hypercycle. Die Naturwissen-schaften 64: 541-565

Eigen M, Schuster P (1978) The Hypercycle. A Principle of Natural Self-Organization. Part B: The Abstract Hypercycle. Die Naturwissen-schaften 65: 7-41

Fahrenberg J, Heger R, Foerster F, Müller W (1991) Differentielle Psycho-physiologie von Befinden, Blutdruck und Herzfrequenz im Labor-Feld-Vergleich. Zeitschrift für Differentielle und Diagnostische Psychologie 12: 1-25

Giesecke H (1973) Methodik des politischen Unterrichts. Juventa, München

Gisi M, Schiepek G (1989) A Children's Home Scenario. © University of Bamberg. For requests please adress to the author.

Grassberger P, Procaccia I (1983) Measuring the Strangeness of Strange Attractors. Physica 9 D: 189-208

Haken H, Stadler M (Eds)(1990) Synergetics of Cognition. Springer Series in Synergetics, Vol. 45. Springer, Berlin

Imber-Black E (1988) Families and Larger Systems: A Family Therapist's Guide through the Labyrinth. Guilford Press, New York

Jungk R, Müllert NR (1981) Zukunftswerkstätten. Hoffmann & Campe, Hamburg

Krohn W, Küppers G (1990) Selbstreferenz und Planung. In: Niedersen U, Pohlmann L (Hrsg) Selbstorganisation. Jahrbuch für Komplexität in den Natur-, Sozial- und Geisteswissenschaften. Duncker & Humblot, Berlin, S 109-127

Laux L (1991) Selbstdarstellung und Selbstinterpretation: Herausforderungen an die Persönlichkeitspsychologie. Manuskript, Universität Bamberg

Lehmann J, Portele G (Hrsg) (1976) Simulationsspiele in der Erziehung. Beltz, Weinheim

Ludewig K (1987) Vom Stellenwert diagnostischer Maßnahmen im systemischen Verständnis von Therapie. In: Schiepek G (Hrsg) Systeme erkennen Systeme. Psychologie Verlags Union, München Weinheim, S 155-173

Mackinger H (1984) Sind Rahmenbedingungen Randbedingungen? Überlegungen zum Bereich stationärer Psychotherapie. Verhaltenstherapie und psychosoziale Praxis 16(4): 543-552

Nowack W (1989) Interaktionsdiagnostik. SYMLOG als Rückmelde- und Forschungsinstrument. Dadder, Saarbrücken

Perrez M, Reicherts M (1986) Appraisal, Coping, and Attribution Processes by Depressed Persons: An S-R-S-R Approach. The German Journal of Psychology 10: 315-326

Perrez M, Reicherts M (1987) Coping Behavior in the Natural Setting: A Method of Computer-aided Self-observation. In: Dauwalder JP, Perrez M, Hobi V (Eds) Controversial Issues in Behavioral Modification. Swets & Zeitlinger, Lisse, pp 127-137

Perrez M, Reicherts M (1992) Stress, Coping and Health. A Situation-behavior Approach. Hogrefe & Huber Publ., Toronto

Rehm M (1964) Das Planspiel als Bildungsmittel. Quelle und Meyer, Heidelberg

Reicherts M (1988) Diagnostik der Belastungsverarbeitung. Universitätsverlag, Fribourg und Huber, Bern

Reicherts M, Christensen S, Gronlykke S (1991) Analysis and Treatment of Phobic-panic Disorder as Deficient Stress and Coping Behavior. Forschungsbericht, Universität Fribourg

Reicherts M, Käslin S, Scheurer F, Fleischhauer J, Perrez M (1987) Belastungsverarbeitung bei endogen Depressiven. Zeitschrift für Klinische Psychologie, Psychopathologie und Psychotherapie 35: 197-210

Reimann HL (1972) Das Planspiel im pädagogischen Arbeitsbereich. Bonn

Reinisch H (1975) Planspiele. In: Reinisch H (Hrsg) Hochschuldidaktische Arbeitspapiere 7: Planspiele. Interdisziplinäres Zentrum für Hochschuldidaktik, Hamburg, S 1-11

Schiepek G (1987) A Multigeneration Family Educational Psychology Scenario. ©️ University of Bamberg. For requests please adress to the author.

Schiepek G (1988) Beitrag zur Diskussion im Vorfeld systemischer Methodologie I. Zeitschrift für systemische Therapie 6(2): 74-80

Schiepek G (1991) Systemtheorie der Klinischen Psychologie. Vieweg, Braunschweig

Schweitzer J (1989) Professionelle (Nicht-)Kooperation: Ihr Beitrag zur Eskalation dissozialer Karrieren. Zeitschrift für systemische Therapie 7(4): 247-254

Stegmüller W (1973) Theorie und Erfahrung, Zweiter Halbband: Theorienstrukturen und Theoriendynamik. Springer, Berlin

Theiler J (1987) Efficient Algorithm for Estimating the Correlation Dimension from a Set of Discrete Points. Physical Review A 36(9): 4456-4457

Willke H (1989) Systemtheorie entwickelter Gesellschaften. Juventa, München

Zimbardo PG (1979) Essentials of Psychology and Life. Scott, Foresman & Co., Glenview/Ill

The Systemic Character of the Psychosocial and Psychiatric Health Services[1]

Jarg B. Bergold

With 4 Figures

Abstract. In this paper I try to examine the basic mechanisms underlying interdependency between psychosocial and psychiatric institutions in one region, which is therefore also responsible for the systemic effects. Using the findings from a survey on the work of an outpatient crisis centre we are able to show that both interdependency as well as different network patterns result from the decisions made in the selection procedure as to the assignment or further referral of clients.

1. Introduction to the Problem

The term system is frequently used in connection with the psychosocial and psychiatric health services. When we speak of the "health care system" we mean that the various institutions are linked in different ways, this having effects that cannot be explained by any one single link, but rather as a result of the combination of the different institutions and their employees and clients or patients. Mitzlaff (1987) and Schiepek (1986), who have most recently referred to the systemic character of the psychosocial and psychiatric health services, based their conclusions on their own experience and theoretical reflection. It would therefore make sense to examine the elements that explain the systemic effects more closely.

In this paper I will try to examine to what extent it is justified to speak of a "psychosocial/psychiatric health care system" in a limited regional area, a district of a large city. This will be theoretical on the one hand, and based on the results of an empirical study on the other. First, however, I would like to explain the terminology. I will distinguish between the terms system and network. The term network will be used to mean the links between independent units as seen from the observer's point of view. The term system, however, will be used to describe varying units, which are restricted temporally and spatially to different levels, when their day-to-day processing and its consequences are the object of the analysis. One can in this way, for example, see a psychosocial district work team as a transitory system, when focussing on the processes during a session or, on the other hand,

[1] The empirical survey was supported in part by a research commission on behalf of the Bundesministerium für Jugend, Familie, Frauen und Gesundheit (Federal Ministry for Youth, Family, Women and Health) and by the Freie Universität (Free University) of Berlin within the scope of a research project focussing on "Researching Community Psychology in a City District".

416

as part of the network when it is analysed in its role as negotiator between the various institutions. Schiepek (1990) makes a similar point when he distinguishes between two concepts of the term system, firstly the term self-reference, which includes temporal components only existing in relation to the system and the productive interaction between them, and secondly, the term network, covering the structure of elements which are in principle independent of each other, but linked as a result of specific relations.

2. Selection as a Basis for Interdependency

Castel et al. (1982) were among the first to refer to the systemic characteristics of psychiatric health care. Within the framework of an analysis of the development of psychiatric health care in the USA they were able to show how changes in the clientele and the resources available to them led to changes of a systemic character in the landscape of the institutions. Castel et al. do not, however, go into great detail on the microgenetic origins of this system effect, i.e. how it is reproduced on a day-to-day basis. Let us now reflect upon this point.

In this paper I assume that the interdependency effects at an institutional level emerge as a result of each institution having to recruit clients in a field which is characterized by competition between the institutions. This means to say that each institution does not accept any client available but that, according to a specific selection process, only certain clients are accepted. It is in the interests of each institution to have a sufficient number of clients to choose from and as a result of this, the selection criteria are continually changing according to the conditions in the welfare scene and in each institution itself. The process of selection takes place in all of the institutions involved, which means that it is not possible for a particular institution to include any client in its own selection process, only those who are not already tied to another institution, or those who cannot be sufficiently provided for by another institution.

Each institution in the "ensemble"[2] of welfare services has its own specific clientele, which is a result of a system effect caused by the selection process within the context of political and administrative decisions. It is not possible for one institution alone to determine this effect, nor predict it. An unexpected result of changes within the other institutions or on the part of the clients could be that "suitable" clients are either not to be found or abundant. Here we see typical interdependency effects, which are characterized in the experience of the individual by his actions having unforeseen consequences, or that there are surprising changes in his environment that contradict existing expectations.

---

[2] I follow on here from Norbert Elias (1983), who based his concept of figuration on interdependency, which he also describes as having systemic characteristics. We can view institutions as figurations of interdependent people, and at the same time, on a second level, study the interdependencies between the various institutions. It is primarily the latter that we are concerned with in this paper.

Cramer (1981) described in detail the development of a welfare sequence based on interviews with professional helpers; this serves here as evidence of such interdependency effects. Welfare sequences, or a continuing cooperation of this kind, can be considered as patterns which emerge within the field of psychosocial psychiatric welfare. On the basis of the chosen literature and my own survey I would like to describe several of these patterns.

Based on the findings of his interviews Cramer suggests that one can see psychosocial work as being dependent partly on the client and partly on the welfare services, and that it is the latter which is more essential to the maintenance of the institution. The survival of an institution is, in his opinion, dependent on two factors. Firstly, the institution is an asset to other social organisations, it has a utility value. Concentrating on the utility value causes the institutions to become interdependent and they are forced to carry out their work in dependency of the welfare service, with a view to safeguarding supplies.

It is on this basis that he describes the development of different welfare sequences, which he attempted to classify according to the degree of specialization:

– **ascending**, becoming more specialized: everyday problems are defined as being specific and requiring expert help.

– **descending**, becoming more concrete: specific problems (as defined by experts) take on a more general/common appearance.

– **linear**: sequence seen as being accumulative, view of problem remains the same, approach successive

– **non-linear**: sequence diversifies as a result of varying definitions of problem, approach not successive

Secondly, client recruitment takes place according to "case production". The starting point of the sequence and therefore also of the beginning of cooperation between the institutions are so-called motives, which are determined by the structure of the institution. Referral occurs as a function of the institution's assessment of the problem, related to its own disposing capacity and with a view to safeguarding supplies. The client's problem is "only" the motive for this assessment of its disposing capacity. This can be viewed as a sign that the institutions react to the demands from without, in a way that is inherent to the system.

Without a doubt, Cramer's analysis is interesting and expansible. It was empirically based on interviews that the author conducted with welfare professionals in three districts. The findings reflect the theoretical ideas and the viewpoint of those interviewed and also clearly refer to the systemic character of the psychosocial and psychiatric welfare service. This raises the question of whether it is possible to describe more exactly the interdependencies and patterns and, above all, their process of development using other data.

3. A Survey of the Work of an Outpatient Crisis Centre and Its Cooperation with Other Local Institutions

We evaluated the work of an outpatient crisis centre, in existence since 1987, over a period of two years, working under a current research project concerned with the development of a district welfare system. This institution came into being as a result of a detailed debate within the district on the prevailing high suicide rate and the shortcomings in the welfare service. Conflicting demands arising from this debate were then made on the new welfare institution[3]. There was a call for a traditional medical/psychiatric inpatient emergency service on the one hand, and for a psychosocial outpatient crisis intervention centre on the other. While the concept was being developed and the application filed, this conflict was "solved" after a lengthy process of negotiation by entrusting the new centre with the care of both groups of clients.

The outpatient crisis centre's concept entails short-term counselling (1-5 sessions) but also long-term counselling in individual cases. The staff carry out home visits (which also includes locating clients, e.g. in hospitals) and counselling by telephone, but mostly counselling takes place in the centre itself. The opening hours are from 3pm till 12 midnight. Because the centre's work also includes short-term intervention it cooperates with many other institutions in the district, as it transfers clients to them and also receives referrals itself.

We have collected a series of data with this study which gives us direct or indirect information on interdependencies. We would like to distinguish between two basically different types of data at this point that lead to different designs of the networks. The two types are, firstly, the facts noted in actual situations which give us information on the course of events and the frequencies (main data registered in or after the first interview) and secondly, statements made by the participants about these procedures. The first type of information allows us to

---

[3] see Handrack (1988)

Table 1. Data on interdependencies within the network

survey method	aim/ data	number
report register	contact to other institutions	1375 contacts[4]
main documentation	description of clients, contacts with other institutions, type of counselling	964 clients[5]
qualitative interviews with employees of various institutions	crisis concepts, treatment methods, development of crisis work, acceptance and approval of outpatient crisis centre	10 members of staff[6]
questionnaire	acceptance, contacts between the institutions, type of work and crisis concepts of the institution	69 members of staff (rate of response=51%)[7]

identify the networks from the viewpoint of the observer, the latter through the eyes of the participants. The data sources are shown briefly in Table 1.

3.1. The Selectivity of Contacts

We can reach conclusions about the selectivity of the interaction and the most relevant details seen from the viewpoint of the observer by looking more closely at the frequency scores.

The staff of the outpatient crisis centre make informal and handwritten notes of all events during any one shift in the report register, and so all contacts with other professionals, mostly colleagues in other institutions, are also recorded there[8], As it is used daily, the report register is good evidence of actual contacts made between the outpatient crisis centre and other institutions. By counting the categorized records we can conclude that the outpatient crisis centre contacted 150 different institutions in some form or another since it came into existence. Of these 61 were within the same locality and 86 in other districts. All in all 1375 contacts were recorded. We can group the institutions according to the frequency of contacts as shown in Table 2.

This shows that roughly three quarters of the total contacts are made with ca. 10 % of the institutions in question. There were less than 10 contacts made to the other institutions. If we assume that there are probably more than 150 institutions that could possibly be contacted, we can clearly see a strong selective approach here. Only 16 institutions remain which are contacted with any significant frequency. Of these contacts 73% directly concern clients, and roughly 25% are

[4] see Holz (1991)

[5] see Bergold et al. (1990)

[6] see Möller & Schürmann (in press)

[7] see Leferink (in press)

[8] see Holz (1991) for a detailed analysis of the results

Table 2. Distribution of contacts among institutions

number of institutions	frequency of contacts	percentage of total contacts
4 9	100 and >200	50 %
4 10	ca. 50 :	14 %
8	10 - 20	10 %

general exchange of information. Therefore, it is the clients that present the basis for the interaction.

We can see similar findings in the main documents, where the cases assigned to the crisis centre and the referrals made to other institutions are recorded. We can again reach conclusions from these documents as to the cooperation between the institutions, resulting from the referral of clients. Here we see cooperation as a direct result of a client being referred from one institute to another. Here, too, is the finding that the outpatient crisis centre cooperates with 6 or 7 institutions from a total of a possible 40 within the same district particularly worth noting.

Let us then first conclude that, as a rule, the large number of possible connections is greatly limited. The limitations are not only to be seen in the institutions assigning cases, i.e. only a small number cooperate with the outpatient crisis centre in such a way that they refer clients there, but also on the part of the crisis centre and its cooperation with and referrals to other institutions.

3.2. Identification of Three Patterns of Cooperation

This finding that the institution surveyed only refers clients to a limited number of other institutions and also for its part obtains referred cases from very few, and that the data on the contacts corresponds to that on the assigned and referred cases, suggests that it could be justified to assume that the interdependency between various institutions is based on the allocation and referral of clients. Secondly, we would like to examine the main documents for possible patterns in the cooperation or the referrals of clients, which could be connected to the process of selection.

There are two categories in the main documents which could help us to identify such patterns. There is a record made for each client stating the institution that referred him to the centre and to whom he was then referred. We can clearly define three groups of clients with varying referral patterns based on a simple graph of the frequencies and describe them in more detail using other categories from the main documents. The first group contains the 62 clients who were referred to the outpatient crisis centre by three psychiatric institutions (the local mental health clinic, the district social and mental health service, the district day

[9] General Hospital with an emergency department (Jüdisches Krankenhaus), social and mental health service, crisis clinic, police

[10] Day centre, General Hospital with an emergency department (Rudolf Virchow Krankenhaus), research group

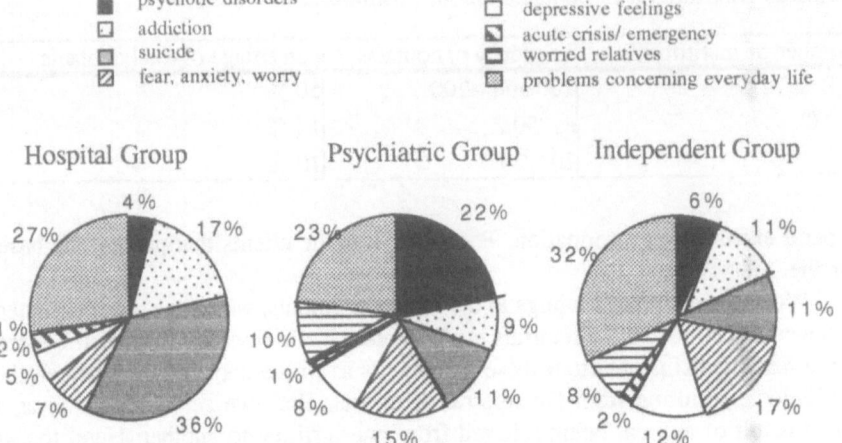

Fig. 1. Patterns of distribution showing the events leading to contact with the outpatient crisis centre

centre. The second group comprises the 114 clients from a general hospital with an emergency service and lastly, 437 clients who had come to the outpatient crisis centre on their own initiative, independent of any institution (independent group).

I would like to describe these three differing groups in order to achieve a more exact characterisation of the various patterns. They will be described according to the events leading to the first contacts with the outpatient crisis centre, the diagnoses and some personal charateristics of the clients. The pattern of distribution of these events (see Fig.1) shows how the institutions react selectively to external events.

I will draw upon quotations from the interviews with the employees of these institutions in order to explain the selection procedures of the referring institutions and some of the outpatient crisis centre's principles of referral more clearly. The outpatient crisis centre's selection procedure will then be presented in the next two figures (Fig.2 and Fig.3) showing the agreements made at the end of the first session and the referrals made to other institutions.

This procedure compares different technical data. In doing so, the characteristics of the various networks can be described. The networks will be described with the main focus of analysis on the outpatient crisis centre. We are concerned, therefore, with a small section of the networks, on the basis of which we are able to reach our conclusions.

3.2.1. The Hospital Group

The group of clients from a General Hospital with an emergency department is most clearly characterized by the fact that the incident leading to the first contact in the case of 36% of clients[11] is registered as being attempted suicide, addiction covers 17%. The incidents of "psychotic disorder" 4%, "fear, anxiety, worry" 7%, "worried relatives" 1% are distinctly lower. On examining the diagnoses it is

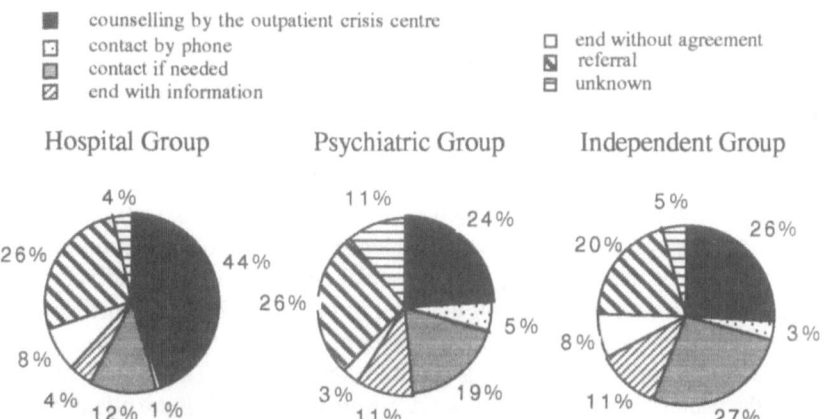

Fig. 2. Patterns of distribution showing the agreements made at the end of the first session

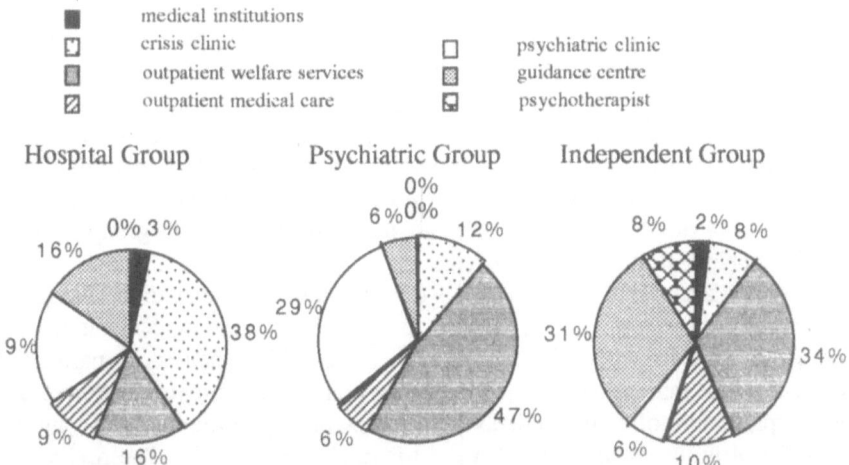

Fig. 3. Patterns of distribution showing referrals by outpatient crisis centre

noticeable that those clients without a specific diagnosis are the most frequent (50%). Within the diagnostic categories only alcohol and drug addiction are more frequent (24%). Almost all the other diagnostic categories are smaller than in the other two groups, including the diagnosis "neurosis".

We can distinguish these clients sociologically in comparison to the other groups as more likely to be married, earning a wage or a salary and as more likely

[11] Interview statement: "Well, I think it's kind of, it was very clearly, that is, vice versa, basically the therapists don't need those of us in crisis intervention, so it seems, well, I don't know, maybe they don't have any crises to deal with, I guess, 'cause they don't need us but we need them so to speak, see, 'cause, well, quite a few want to have therapy."

to have completed some form of training. This suggests that this group of persons are socially slightly better off than those in the other two groups. Furthermore these patients differ in that their previous history shows few contacts with institutions in the psychosocial or psychiatric field.

During an interview which, among other things, covered the procedure immediately after admittance to hospital, one doctor from the emergency department underlined that the first step was to carry out a detoxication and then further practical steps were taken, which included involving other institutions, the hospital's own consultant neurologists, the social and mental health service or the outpatient crisis centre. A consultant neurologist or the public health officer were called in if the client was seen to be in danger of making a new suicide attempt, which had obligatory admittance as a possible consequence. The outpatient crisis centre or the social and mental health service were called upon if the client wanted to come to terms with and was capable of coping with the suicide attempt. The outpatient crisis centre was usually preferred because there were no files kept and the anonymity of the clients was ensured.

These findings can be summarized thus: most of the patients transferred from the hospital to the outpatient crisis centre have experienced a crisis situation (attempted suicide or drug/alcohol addiction) but are still within the spectrum of being "normal". They have little or no experience of psychosocial or psychiatric care and it is important to ensure that they remain within the boundaries of normality as far as possible. It is not possible for the hospital to help them with their social and psychological problems; for this the outpatient crisis centre seems to be the most suitable institution. We can, therefore, describe this group of clients as "newcomers" to the welfare system.

We can also see the characteristics of each network by examining the further treatment of the clients. Let us then do so by looking at the agreements made at the end of the first session and the type of referrals made. It is particularly noticeable here that the outpatient crisis centre counsels the clients transferred from the hospital more often than clients from all the other institutions (44%) and that the offer of further counselling if needed is rarely made (12%). The first session ends less often with information on further counselling in comparison to the psychiatric group (4%). The number of transfers here is similar to the clients from other institutions. Clients are referred to 25 different institutions, and the number of referrals to a crisis clinic is clearly higher. Referrals to a psychiatric clinic are also higher in comparison to the independent group. This shows that transfer to a receiving clinic for resident treatment is necessary for some of the patients.

These findings suggest that those clients referred from the hospital are seen as a special group by the employees of the outpatient crisis centre. As mentioned above, this group of clients are persons who have had little or no previous contact with the psychosocial or psychiatric welfare service and now, as a result of their suicide attempt, are confronted with psychological counselling for the first time. The guidance counsellors expect them to refuse offers of treatment if they are only recommended. This corresponds to the current discussion to be found in the literature on crisis intervention, where it is emphasized that it is easier to establish contact to persons who have attempted suicide during the first few days after the attempt, and that after this time the willingness to talk about their problems is greatly reduced.

The outpatient crisis centre sets the course of further action at this point. Some of the patients will be able to solve their problems through the counselling at least to such an extent in that they will be able to carry on with their lives without any form of professional help. Here the outpatient crisis centre is able to normalize the crisis situation, which was the cause of the patient being admitted to hospital in the first place. It simultaneously prevents the patient being caught up in the psychosocial or psychiatric welfare system.

Some of the patients are referred elsewhere, which can occur immediately after the first session. Here we can distinguish between clients who are in need of inpatient care – they are referred either to a crisis clinic or to a psychiatric clinic – and clients who are referred to various psychosocial institutions. Sometimes these referrals occur after a period of counselling by the outpatient crisis centre. In such cases, the outpatient crisis centre takes on the task of "motivating" the clients to be interested in further professional help, or, seen from a different perspective, to socialize them for the professional welfare service.

3.2.2. The Psychiatric Group

As was expected, the incident leading to the contact with the outpatient crisis centre in the group of clients from the three psychiatric institutions was in 22% of the cases a psychotic disorder. Attempted suicide, in contrast, is less common (11%). Other important motives are the client's anxiety, fear or worry, or worried relatives. There are a higher number of diagnosed cases of schizophrenia, endogenous neurosis, other psychotic disorders and depressive neurosis. It is noticeable that diagnoses are missing less frequently with this group of patients (19% vs 40% in the overall sample).

With regard to sociological characteristics these patients are more likely to be single, have fewer social contacts, are less likely to have steady, reliable employment and are, as a rule, mostly male. Their previous history shows substantially more experience with psychiatric institutions, whereas their experience of psychosocial institutions is remarkably small.

The employees from the psychiatric institutions emphasized in their interviews the outpatient crisis centre's task as primarily offering a supplementary service. This was especially so in the case of the social and mental health service, whose employees see the outpatient crisis centre as an additional service outside of their own working hours. If they expect a patient to become difficult or in need of help towards the end of a duty period the outpatient crisis centre is contacted. By taking this action it is sometimes possible to avoid inpatient treatment. The clinic, where patients are only usually informed of the outpatient crisis centre, also emphasizes its role as being additional to the social and mental health service. The psychiatric clinic occasionally refers patients to the outpatient crisis centre to bridge the gap until the social and mental health service reopens the next day.

It becomes clear from the information available that the selection criteria and, accordingly, the type of patients referred are fundamentally different from the group of patients from the General Hospital. In most cases, these patients have been in the care of various institutions within the psychiatric welfare service for some time. The outpatient crisis centre was included in this network of institutions because it is the only institution able to offer outpatient care at a time of day when every other outpatient institution within the network is closed. Those patients who circulate within this network are well known in the various institutions and have accepted the fact that they themselves are part of it. This is suggested in the finding that even the outpatient crisis centre ascribes a diagnosis to the majority of these patients. We can interpret this to mean that they have previously been given this diagnosis and bring it with them to the new institution.

On examination of the outpatient crisis centre's process of selection it becomes clear that this is an institutional network characterized by a relatively steady clientele and firm referral practices. The outpatient crisis centre is less likely to counsel these clients itself, in contrast to the group from the hospital. The first contact, however, rarely ends without some form of agreement. Patients are mainly transferred to outpatient services (47%), i.e. institutions in the psychiatric field, and to psychiatric clinics and, in comparison to the other two groups, much less often to guidance centres or psychotherapists. The whole range of institutions to whom clients are assigned numbers only 11, with 50% of all transfers being made by the social and mental health service, the clinic and the day centre.

The data on the outpatient crisis centre's selection procedure and referrals suggest similar conclusions to those made from the findings on the referral criteria of those institutions referring patients to the outpatient crisis centre. Here we see a network of a limited number of institutions closely linked as a result of referrals and transfers of a relatively steady number of clients. The outpatient crisis centre is not a typical institution but is included in the network because it offers additional help to those clients who represent the base of this network. The most important factor behind these short-term interventions with offers of further help is the transfer of clients to the particular network, as the outpatient crisis centre has no interest in long-term counselling and a basic welfare service is guaranteed.

In contrast to the outpatient crisis centre's role in setting the course for further action as described above in the network for clients from the General Hospital, its main task here is one of collecting up and returning. The outpatient crisis centre collects up those people who need help at a time when their usual outpatient institution is closed, and brings them back into the network.

3.2.3. The "Independent" Group

Finally, the third group represents those people who contacted the outpatient crisis centre on their own initiative, independent of and without having being referred by any institution. These clients differ very little from the average client. It is, however, noteable that attempted suicide is not a prevalent reason for contacting the outpatient crisis centre, in contrast to the group from the hospital (11%). The main reasons here are the problems they have in their everyday lives. The number of non-diagnosed cases is not as high as in the group of clients from the hospital but, however, very much higher than the clients from the psychiatric institutions. The most common diagnosis is "neurosis" (ca. 30%).

Again, this group differs very little from average ratings in the sociological categories. As with the clients from the hospital the majority are female. These clients have had experience of a number of very different psychosocial institutions. They gave 65 different sources in answer to the question where had they heard about the outpatient crisis centre. These cover the conventional psychiatric institutions and various psychosocial services as well as the telephone hot line service and articles in the press. The main focus, however, was on a large number of psychosocial institutions. Similarly, the clients specified various and very different institutions when asked about their previous experiences with the psychosocial and psychiatric welfare services.

Firstly, we must conclude that this group of people is very heterogeneous, having gathered their information about the outpatient crisis centre from very different sources. Some of them (14%) also learned of its existence from the three psychiatric institutions who have information leaflets readily available. It is certainly interesting, with a view to interpreting the results, to note that this group first heard of the outpatient crisis centre in a large number of very different psychosocial and state-run institutions. We may therefore conclude that they had previous contact to these institutions. This is, then, a group of people, the majority of whom are familiar with what is on offer in the psychosocial field. Here is a fundamental contrast to the other two groups of clients. This group also differs in that it is the client who decides to go to the outpatient crisis centre and so determines the selection process, rather than another institution by making a referral. In this sense it will be necessary to examine whether we may speak here of an institution network at all.

With regard to the outpatient crisis centre's selection process it is noticeable that, on the whole, fewer of these clients see the necessity for further treatment after the first session (20%). This is particularly clear in comparison to the group of clients from the hospital who, at first glance, seem to be quite similar. 70% of the hospital group are counselled by the outpatient crisis centre or referred elsewhere for further counselling, whereas this is only 46% in the independent group. There is no further counselling for 50% of these clients after the first session and it is left up to them to take any further steps.

If they are referred elsewhere it is mostly to outpatient services, outpatient medical care and to guidance centres and psychotherapists. Again we can see a clear contrast to the hospital group. This finding that clients are referred to a large number of institutions (58 in all) suggests that here, decisions are being taken in tune with the varied needs and wishes of people who are credited with the ability to take up these offers and check their suitability for themselves.

In this third group we can identify a collection of loosely connected institutions, each having services on offer within very different frameworks. It is the clients in the various institutions who make up the connection between them, and who are thus far more actively involved in forming the network. The services on offer meet different needs on the part of the client, who in turn is under various lasting influences and also subject to short-term fashions propagated by the media, cultural and subcultural standards, particular predicaments caused by society (e.g. unemployment) and so on. The outpatient crisis centre's task here is to be one of the many services on offer. Furthermore, it works as a distributor, showing the client the wide range of services available and helping him choose a service to suit his needs.

In contrast to the psychiatric network this is a loosely linked network. The only reason the links come into being are as a result of the clients utilizing several institutions, either consecutively or at the same time. Cooperation between the institutions on another level, e.g. interinstitutional discussions or an exchange of information on clients, rarely occurs. This also follows from the survey of the frequency of contacts in the report register as shown above. It is also confirmed by one of the outpatient crisis centre's employees in an interview, where she describes how clients are indeed referred to them from the other institutions, but that there was otherwise little communication. Direct contacts to regular G.P.'s and those involved in therapy were especially rare[12].

3.3. Rules of Pattern Formation

On examination of the presented data on referrals to and from the outpatient crisis centre, the institutions involved in this process and the particular characteristics of the clients, we can therefore identify an interdependency between the institutions and, in addition, define and describe three different patterns of institutions as shown in Fig.4.

– Pattern 1 shows the hospital's referral of people from the everyday world, or clients who have already been socialized, to the outpatient crisis centre who are then discharged from there back to the outside world or assigned to professional institutions.

– Pattern 2 shows the realitively closed circuit of psychiatric patients and

– Pattern 3 is created by clients who are self-reliant to a certain degree in choosing between the various institutions.

In order to characterize the pattern formation more clearly I would like to define which rules influence how the field is organized and could possibly explain the basis of the different networks. The term rule should be understood here as characterizing the basic principles according to which a social system functions, as seen from the viewpoint of the observer. The term rule is used to underline the fact that each participant behaves, as a rule, in a certain way which has an effect on the system, but that it would also be possible for them, with a certain amount of effort, to behave differently. We can define these interdependent rules in the area surveyed for the participating groups, the institutions and their users.

[12] Due to the different numbers of patients and the missing values only percentages are given here.

428

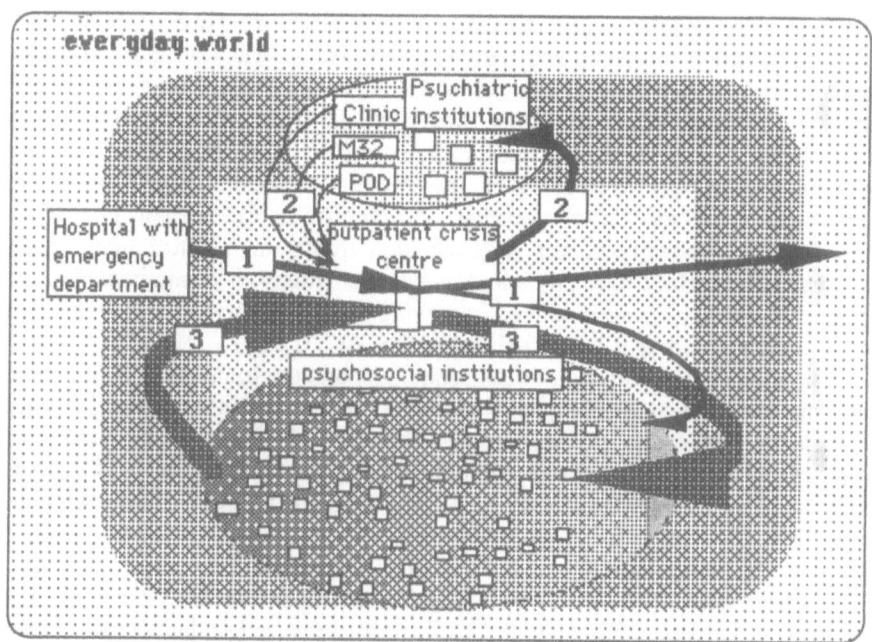

Fig. 4. Diagram of the movement of clients into, out of, and between institutions

Table 3. Rules and orientations for different client groups

client group	rule	main orientation
Independent Clients		
client	choose according to your needs	need- oriented
crisis unit	supply an attractive service	supply- oriented
Hospital Patients		
client	allow yourself to be helped	support- oriented
crisis unit	cope with temporary crisis	normalisation- oriented
	interest client in further treatment	socialisation- oriented
Psychiatric Patients		
client	handover responsibility	provision- oriented
crisis unit	act on behalf of patient	oriented towards incapacitation of patient

Moreover I will try to determine the main orientation resulting for each participant from the rule, which represents the context and so the general orientation of each individual's action (Table 3).

We would now like to characterize the networks in more general terms, on the background of this analysis focussing on the outpatient crisis centre. We can best describe the institutional network connected by the clients who seek out the outpatient crisis centre on their own initiative, with the term free market. The different institutions have little contact, but are connected, above all, by the sum of their selection processes and the clients' decisions. They are able to secure enough clients to ensure their own survival if they can supply an attractive enough service for them. As it is almost impossible to create such an appeal they concentrate on one particular small group of clients, for whom they can then develop an attractive service. We can assume that the institutional distinction in this field is based on competition and limited governmental controlling mechanisms. Furthermore, this differentiation results in the institutions rarely communicating. It is not possible to maintain extensive communication and cooperation whilst concentrating on improving one's own services within this type of competition among the multitude of institutions.

In contrast, the outpatient crisis centre supports the basic medical principles and the structure of the hospital in the case of the patients coming from the emergency department of the hospital. The General Hospital concentrates on physical illnesses, and, due to staff shortages and financial cutbacks, is scarcely able to cope with psychological problems. Through the cooperation with the outpatient crisis centre it is possible to justify a swift discharge without having to consider the psychological side of the matter. Through its own extended welfare the outpatient crisis centre sets the course for further action. On the one hand, it channels the clients swiftly on to the free market of the counselling services and/or socializes them in such a way that they are able to take up these offers. On the other hand it offers them the chance to leave the network of professional help. This has a normalizing effect on the reactions to a psychological crisis, e.g. a suicide attempt.

In the case of the latter we are not so much describing an institutional network, rather an institutionalized door to everyday life. This task bonds both institutions – but only as long as the hospital, which is not responsible for the psychological health of its patients unless it is directly connected to the illness, is committed to and establishes such a bond. This may come about as a result of one of the staff feeling morally responsible. This again calls for communication and so shows us another level of the interrelations between the institutions.

The network formed by the psychiatric patients is relatively firmly in place, in contrast to the other two patterns. It leaves both the institutions and the patients little room for choice. The data clearly shows us how the patients pass on from one institution to another, remaining within a small circle. This small, clearly defined group of patients closely links these institutions. As they communicate about these patients a further bond is established, e.g. in the form of doctors' letters or telephone calls during the referral procedure. Thirdly, they are linked as a result of a mutual norm or ideology, that is the so-called medical model on mental illness. This means that they view people, first and foremost, from the point of view of long term or possible chronic illness and the necessary medical, social and mental care resulting from that. The institutions themselves prefer a close network here, which should serve the well-being of the patient group. We can assume that, resulting from its close connections and in addition to the desired welfare effect, this network produces a system effect, which compels involvement in the network itself. The communication and, above all, the mutual ideology result in the attribution of joint significance between patient and institution, which define the patient as a part of the network.

Based on these considerations, it seems justified to assume that the more often a person is admitted to an institution within the psychiatric network, the more probable it is that he/she will experience further contact to a similar institution and receive a psychiatric diagnosis. This is certainly the case in the conventional psychiatric hospital where this effect is already known under the name of "revolving door psychiatry" but also, as this analysis suggests, likely in the case of an outpatient psychiatric welfare service as is currently being called for by the more progressive elements in psychiatric welfare, the social psychiatric welfare service. In contrast to the hospital patients, these people have few exit doors into the everyday world.

4. Conclusion

The present survey indicates that the psychosocial welfare institutions in one area are linked together in such a way that system effects can occur. At this point I must qualify this by stating that the data on which this finding was based is very limited. We reached our conclusions on the whole "ensemble" of institutions from our findings in the patterns of referral to and from just one institution. The data presented nonetheless seem to overthrow the theory that the decisions made in the selection procedure as to the assignment or further transfer of clients create a basic interdependency between the institutions, and also raise a number of implications for the whole institutional "ensemble".

One of these implications is the development of different patterns of institutions which are linked in a special way. We can roughly distinguish three patterns. First of all we are concerned with a kind of market design, whereby the patients approach the institutions which then primarily decide on their admission or rejection. The following two patterns are similar to begin with. Referrals come about in most cases through institutions which are authorized by society to use constraint, that is the police force, the fire department, the medical emergency services etc. In one of these cases referral occurs as a result of psychiatric deviations, in another because of risks, mostly in a medical sense. In the case of the former the institution makes the decision on the person's admittance and also further referral in a relatively tight network of institutions, characterized by a strong pattern of perception and action, that is the medical, psychiatric pathological model. In the latter case, referral occurs as a result of the medical system. We can, however, define two principally different eventualities in the case of further referrals, which are again determined by the institutions: on the one hand we can see help in normalizing the problem and so helping to reintegrate the client back into the everyday world, on the other hand the client is settled into one of the previously mentioned networks.

There are not only different institutions linked together in these three networks, but also the type and intensity of their interdependency differs. Based on Aldrich & Whetten (1981) we can determine three types of interactional content: these are first a "material" exchange, i.e. referral of clients, second "communication", i.e. an exchange of information and third, "normative content" in the sense of a system of mutually expected behaviour coherent to theoretical/conceptual ideas. We can thus classify these three networks according to their content. There is primarily a material exchange in the network of the independent group, the clients visit the various institutions consecutively. In the hospital network there is also communication about the clients and, in addition to this, the psychiatric network is linked by a mutual normative content, the psychiatric pathological model. We can therefore assume that the degree of coherence of the networks also differs.

With this description we have been able to show network structures by analysing the decisions made over referrals. It is a matter of what the observer can see from a distance, as it were. It is possible to identify patterns, but not possible to reach conclusions as to how these patterns come into being, nor which route any individual might take through this labyrinth. More information is necessary in order to be able to comment on the regular creation of, and in this sense, the dynamics of this process. What is needed here are additional analyses of the processes of coordination and cooperation between the institutions and the public and administrative authorities, and, above all, registration and study of the selection processes in the various institutions. This must be reserved for further surveys.

This is also the reason why there has not yet been an explicit attempt to introduce synergetic concepts here, although this would be a reasonable approach. In my opinion, we are in need of a great deal more empirically sound data in order to ensure that these concepts, and possibly even simulations, really do yield adequate findings. As we can see from a series of surveys in the field of the social sciences, it is certainly possible to carry out simulations on the basis of fewer data. The findings, however, then tend to take on the form of metaphors, whereby very little can be said about their relation to the phenomena in question. In my view it is necessary, especially within the field of contemporary social sciences, to not only adopt the concept of "system" as a metaphor, but to demonstrate the basic processes generating the systemic manner of social occurrences.

References

ALDRICH, H. & WHETTEN, D. 1981. Organization-sets, action-sets, and networks making the most of simplicity. Handbook of Organizational Design, 385-408.

BERGOLD, J. & FILSINGER, D. (in press). Einleitung. in: Bergold, J. & Filsinger, D. (Eds.) Vernetzung psychosozialer Dienste. Weinheim: Juventa.

BERGOLD, J., ZAUMSEIL, M. & FILSINGER, D. 1990. Forschungsprojekt-schwerpunkt "Gemeindepsychologische Forschung im Großstadtbezirk". Zwischenbericht und Antrag auf Verlängerung. FU Berlin: Forschungsantrag.

CASTEL F., CASTEL R. & LOVELL A. 1982. Psychiatrisierung des Alltags. Frankfurt: Suhrkamp.

CRAMER M. 1981. Wie psychosoziale Arbeit gemacht wird – eine qualitative Studie. München: dissertation.

ELIAS, N. 1983. Die höfische Gesellschaft. Eine Untersuchung zur Soziologie des Königtums und der höfischen Aristokratie. Frankfurt: Suhrkamp.

HANDRACK, A. 1988. Die Geschichte der Krisenambulanz Wedding. FU Berlin: unpublished diploma thesis.

HOLZ, M. 1990. Interinstitutionelle Kontakte bei der professionellen Bearbeitung individueller Krisen in einem Großstadtbezirk. Eine quantiative und qualitative Untersuchung am Beispiel der Krisenambulanz Wedding. Universität Heidelberg: unpublished diploma thesis.

LEFERINK, K. (in press). Bewertung psychosozialer Einrichtungen in ihrem Entwicklungskontext. in: Bergold, J, & Filsinger, D. (Eds.) Vernetzung psychosozialer Dienste. Weinheim: Juventa.

MITZLAFF, S. 1988. Kooperation und Konflikte in der Reform psychiatrisch-psychosozialer Arbeit. Sozialpsychiatrische Informationen **3**, 2-9.

MITZLAFF, S. 1987. Zur Machbarkeit psychosozialer Arbeit. Systemische Ansätze und Erfahrungen wider die Gradlinigkeit. Bonn: Psychiatrie-Verlag.

MOLLER, H. & SCHÜRMANN, I. (in press). Interinstitutionelle Kooperation bei der Krisenbearbeitung in einem Großstadtbezirk – aus der Sicht von Mitarbeitern unterschiedlicher Einrichtungen. in: Bergold, J., & Filsinger, D. (Eds.) Vernetzung psychosozialer Dienste. Weinheim: Juventa.

SCHIEPEK, G. 1990. Selbstreferenz in psychischen und sozialen Systemen. in: Kratky K.W. & Wallner, F. (Eds.) 1990. Grundprinzipien der Selbstorganisation. Darmstadt: Wissenschaftliche Buchgesellschaft, 182-200.

SCHIEPEK, G. 1986. Systemische Diagnostik in der Klinischen Psychologie. Weinheim/München: Psychologie Verlags Union.

Recursive Interaction and the Dynamics of Knowledge Production in Research Groups: An Empirical Simulation of Knowledge Production[1]

Wolfgang Krohn, Günter Küppers, and Wolf Nowack

With 1 Figure

1. Introduction

The last 10 years have seen a change in the study of the processes of scientific knowledge production in research on science:[1] They are no longer only being studied indirectly through interviews and qualitative and quantitative analysis of written documents (publications) but directly through observation of research activity in laboratories. These so-called "laboratory studies"[2] have lastingly shaken some of the received views on scientific development, particularly those based on the existence of a special scientific rationality. Empirical observation of communication and cooperation in day-to-day research has reported complex negotiation procedures (Knorr-Cetina, 1984, p. 197), flexible opportunism (Knorr, 1984; Krohn, 1988), and rhetorical skills (Myers, 1985, 1986), while stringent procedures that could form the basis for confirming and rejecting rational discourse have rarely been reported.

This new approach of empirical research on science has condensed into the formula of the "social construction of scientific knowledge." This has proved to be a strong formula in the campaign against philosophical positions that are not supported by empirical research but by reconstructions (such as critical rationalism and logical empiricism). However, when its opponents have vanished because, at least in a sociological context, hardly anyone assumes a rationality and knowledge that is independent of culture and history, it becomes an empty statement: universally applicable while simultaneously fragile when applied to itself.[3]

---

[1] The text was translated from German by Jonathan Harrow at Bielefeld University.

[2] See B. Latour and S. Woolgar (1979), K. Knorr-Cetina (1984), and Traweek (1987).

[3] Social constructivism has to tolerate being asked what is the meaning of saying that "knowledge is a social construction" is a social construction. For a treatment of this problem, see the articles in Steven Woolgar (1988). For an attempt at a constructive solution of the paradox problem, see Niklas Luhmann (1990).

434

Springer Series in Synergetics, Vol. 58
Editors: W. Tschacher, G. Schiepek, and E.J. Brunner

Self-Organization and Clinical Psychology
© Springer-Verlag Berlin Heidelberg 1992

The metaphor[4] of "construction" particularly suggests two questions: 1) What constitutes the quality of a construction? 2) How is a construction established? Both questions – on the structure of social construction and the process of the social constructing of knowledge – have only recently become issues in sociological constructivism.[5]

Our interest starts at this point. In contrast to the research practice of "laboratory studies", we are working with a theoretical model and are also prepared to use[6] the laboratory for laboratory studies, that is, not just observe but also generate processes under our own conditions. Compared to actual research processes, these processes are idealized in a way that is typical for experimental research (e.g., standardized task presentation, shortened time span, disengagement from a vocational context). The purpose of these idealizations is to generate transparent conditions for the observation of the process and structures of knowledge construction. How far results obtained with such simulation can be used to interpret actual research processes must initially remain an open question; it cannot be answered within the present approach at the same time.

2. What Are Research Groups?

In the science system, research groups are those social systems that are the smallest units responsible for the production of new knowledge.[7] By research, we

[4] The metaphorical content relates to three associations:
  i) Within the history of language, to the link to the construction (*constructio*) of buildings and to the constructing activity (*construere*). Naturally, every building is a social construction.
  ii) Within the history of science: to the construction and design sciences whose domains are the formal aspects of construction independent of content.
  iii) Within the history of logic: to logical constructivism that is characterized by a particularly rigid standpoint regarding the rejection of nonfinite procedures in proof theory.

[5] In Amann's (1990) publication "Natürliche Expertise und Künstliche Intelligenz" (natural expertise and artifical intelligence) the central idea of a "constructivist model" for analyzing laboratory activity is the formation of analogies between the concept of the expert system in artificial intelligence and the "natural" expert system in the laboratory. The metaphorical transformation is certainly astonishing here: That which is traditionally treated as "cultural" in social science is now labeled "natural" and contrasted with "artificial."

[6] Because of their ethnomethodological background, proponents of laboratory studies are not prepared to carry out laboratory studies themselves in their approach to the study of the laboratory.

[7] For a more detailed discussion, see Krohn and Küppers (1989).

understand activities that are systematically directed toward the production of new knowledge.

As a rule, research groups are embedded in more comprehensive institutional frameworks such as laboratories, institutes, or, in general, formal networks of cooperation. These networks do not necessarily have to be scientific; industrial or government organizations can also be included as long as they possess research departments.

Organizations are normally the source of boundary conditions[8] for research groups. This is because work contracts generally set work targets that restrict the forms of interaction in research groups. To the extent that institutional conditions are insufficient as boundary conditions, the system's function – in the case of science, the production of knowledge – is set by the boundary conditions for interaction within a research group. These external conditions are internal selection criteria with which the group selects the forms of cooperation that it views as being functional from all possible forms of cooperation. External boundary conditions make the interaction recursive because they force interactions within the frameworks set by these conditions to link together or network. In the formal language of dynamical systems theory, this networking is equivalent to the existence of an autonomous system of equations (an der Heiden, 1991).

A second step generates the reproduction (self-organization) of the recursive network through its internal interaction. This is because the external boundary conditions initially permit a margin of freedom in the final determination of the ways and means by which work targets are to be achieved by the research group. By using up this freedom – in the complex network of task-specific interactions, certain forms of interaction are considered to be more important than others – specific interaction rules develop that are considered necessary to achieve the set target. Decisive for this (internal) boundary formation is that this system of rules is reproduced by the continuation of the interaction and thereby establishes the research group as a permanent social system.[9]

To further understand the activity of research groups, it is useful to characterize their tasks with systems theory. Research groups are – as we mentioned above – the smallest social units of the science system that generate the function-specific production of this system: the production of new knowledge. Which areas of knowledge are considered relevant – nature, society, technology, or God – which qualities are to be demanded of them – replication,

---

[8] In this context, boundary conditions are understood as conditions of framework formation. The term, constraints is also frequently used in literature on the natural sciences. It signalizes that the environment sets conditions that have to be met by the system.

[9] Not only the expectations of the environment but also the expectations of the individual members of the research groups are encoded in this system of rules. Elsewhere, we have called this the cognitive-emotional group matrix (Krohn & Küppers, 1989, p. 37).

generalizability, comprehensibility, objectivity (or value orientation) – is regulated by the science system through a comprehensive set of institutions in which research is embedded.[10] This set of institutions includes publishers, scientific associations, training institutes, developmental departments, and review bodies. How far such institutions should not only be assigned to the science system but also represent components of business, policy making, and administration can be ignored here;[11] only the fact that they are dependent on the results of research is decisive. Research groups that produce research findings are thus subject to two general functional demands:

1) A necessary condition for the functional success of a research group is for the research finding to be perceived as new or innovative outside the research group.

2) A necessary condition for functional success is for the research finding to be perceived outside the research group as being consistent with the knowledge constituting the field of research. Research groups must ensure that a finding conforms to the theoretical, methodological, and empirical standards of their field of research so that it is broadly compatible with previously accepted findings.

These two conditions – one innovative and the other conservative – must be met simultaneously, and this contains the core of the entire further problem. Not everything that is new can be attached to accepted knowledge and certainly every attachable statement cannot be regarded as innovation. A product that meets both conditions is therefore not easy to produce.[12]

3. Similar Organizations in Other Subsystems

Before discussing the process of producing new knowledge in the science system, we will take a look at other systems. It can be seen that groups also exist as the smallest unit in other functional systems. In the sense of self-organization, these groups are characterized by external boundary conditions (functional duties) on the one hand, while, on the other hand, they set the rules by which specific results are produced through the development of an internal boundary. The results vary across different systems. However, their common point of reference can be summed up

[10] For a presentation of these reactive mechanisms, see Krohn & Küppers (1989, p. 66).

[11] Of course, this relation between scientific institutions and nonscientific contexts cannot be ignored when analyzing criteria of relevance that determine the directions of growth in research fields. However, they can be disregarded as far as the production and the structure of an individual research finding is concerned.

[12] New research findings frequently contradict accepted knowledge and are therefore not consistent. Precisely for this reason, it is a part of normal rhetorics to withdraw acceptance from that knowledge that is in contradiction to the new.

as construction of new information. Legal rulings, contracts, policy decisions, and business decisions are tenable insofar as they take account of a series of general conditions (consistency) and respond to the perceived situational expectations and constraints (innovation). Depending on the system, productions differ in context: A majority opinion can be generated, a decision justified, a fact confirmed, or an action motivated. In addition, consistency and innovation are not always balanced. The anticipated achievements can be directed more strongly toward routines or toward new solutions. Nonetheless, whenever a regulated procedure (algorithmic information processing) does not promise a solution, groups arise that produce relevant information through their recursively closed mode of operation.

The performance generated by groups in the political system is primarily to collate background information into a report for a decision-making body. By appointing commissions (e.g., committees and subcommittees), operational units in political bodies are reduced to an appropriate size for group dynamics. The documentation of information must be innovative; that is, it must offer a solution to the situation created by the political need for action; it must be consistent with both policy programs and tradition.

In the economic system, new products and services are generally viewed as system-specific performances. However, before these can in any way be produced or offered, inservice agreement has to be achieved on their design. This design is the initial product in the economic system. It is the responsibility of committees that are organized in different ways depending on the size and structure of the company: management, departments, specialized development units, and so forth. The development of a design for the production of goods and services must also not only satisfy the requirement of novelty of solution but also be adjusted to the technological specialization, the sales potential, and the image of the company (consistency).

In the legal system, the appearance of groups is formally regulated at a decisive point: in the adjudication process. There are (legal) boundary conditions for the recursive procedures of those involved (e.g., judge, jury, defense council, public prosecutor, expert witness, defendent, witness). It is also obvious why a "good solution" is hard to achieve. The entire interaction process is based on dispute between opposing interests who are awarded equal influence through the formal boundary conditions. Although in adjudication, the dimension of "innovation" is not normally effective, the typical tension between the treatment of a new (and strictly speaking always singular) case and its allocation within recognized law can be observed. Theoretically, it has also been described as the source of a permanent "creation of statutory law by judges" (Esser, 1964). As in science, the construction of information is formally documented in law: Each adjudication and, in case of doubt, each legal decision is linked to a written justification.[13] As well as courts, there are also further legal institutions for solving complex legal problems: drawing up a contract, settling out of court, and so forth.

4. Complex Problems

In all these cases, which only serve as illustration and possibly need to be considered much more factually, the potential difficulty in the construction lies in the fact that two fundamentally opposed requirements have to be met:

1) The linking of new information to the global set of rules applying to the institution (group) as a component of an organization and furthermore as a component of a subsystem.

2) The treatment of a concrete demand (local disorder) that is given by a completely specific task specification and is the reason for action.

The simpler the system of rules and/or more habitual the concrete case, the fewer the difficulties; but the self-organization of functionally differentiated areas of society is built to deal with complex problems.

This complexity is not (solely) due to the breadth of variables but to the contrasting nature of the two major features of construction: "adjustment to a 'global set of rules' " and "dealing with a 'local disorder'." However, these only become opposites from a dynamic perspective, that is, when the modification of rules depends on cases, and the interpretation of cases (disorders) depends on rules. From a static perspective, a new case would either be rejected because it was irreconcilable with the rules, or a rule would be exchanged or introduced in order to deal with the case. In contrast, the institutions concerned are also particularly expected to shoulder responsibility in questionable cases, that is, in cases in which the respective set of rules needs to be administered flexibly to encompass an unusual case, or a case has to be reinterpreted to make it fit the rules.

The opposing effects of the two postulates – maintaining the set of rules for action and safeguarding the responsibility for new cases – is the reason why all products of systemic information processing have a complex internal structure. This is always spanned between the retention of the systemic consistence on which the global competence of the system is founded and the integration of a problem state that is based on a singular situation in which special competence is maintained through innovation.

The dual character of this complex internal structure of integration (new cases under given rules) and extension (rules on new cases) corresponds to two completely independent but dynamically linked mechanisms in the process of generating these performances. We label these in general terms as the process of the operationalization of rules (laws, theories, etc.) and as the process of the interpretation of actualities (facts, data). In constructing an achievemnt, the rules to which the activity is subject (e.g., legal laws, market laws, scientific laws) must be more or less explicitly operationalized to fit a problem. In contrast, the facts of

[13] As in science, the justification is frequently constructed de facto after the decision (the outcome). For the regulation of proceedings (appeal, revision), however, the constructed appearance that the decision follows from the information premises is sufficient.

the problem must be more or less explicitly interpreted far enough to correspond to the rules. Operationalization and interpretation are thus means of modification. These modifications must be performed until an informational context is given that is regarded as a satisfactory solution to the problem – initially in the eyes of those involved, but in anticipation of acceptance by the environment. We have termed the construction of such a solution as the construction of an eigensolution.

Using this model, we attempt to present a testable, empirical version of epistemological constructivism as it has been developed by, among others, Foerster, Maturana, Glaserfeld, and Luhmann (Schmidt, 1987). The demand that cognitive constructions must meet at least two independent, though dynamically linked, validity criteria – the operative part must work, the interpretive part must apply – makes it apparent why "good" solutions require the effort of a recursive operation and why institutional effort and expert skills are required for their generation.[14]

5. The Generation of Eigensolutions

A description of system-specific performances as the eigensolution of a process of self-organization requires not only the introduction of suitable variables as elements of this description but also the definition of system-specific operations as mechanisms of their change. In this context, we understand variables as changeable elements of a complex description of the performance[15] expected by the environment that are typical for the institution under consideration. The starting point for determining the variables is the bipolar structure of systemic performances, that is, the combination of interpretive and operative components and their dependences. The term eigensolution indicates that the product to be

[14] In a rudimentary form, legal, economic, political, and epistemological problems also occur in real life at a level below functional differentiation and institutional self- organization: adjudications outside the legal system (e.g., when the head of the family lays down the "law"), the exercise of power outside of politics (e.g., the use of force to break somebody else's will), knowledge outside of science (e.g., searching for a good traffic connection), or investments outside the economic system (e.g., looking after one's own front garden). However, activities such as these that are not embedded in functional systems are free from the construction problem addressed here. Ad hoc solutions, which only serve to cope with a situation, are just as permissable as the conservation of traditions that are blind to new tasks. Improbable and therefore brief are performances that simultaneously provide consistency and innovation.

[15] A functional system does not export a performance in the true sense but only a description that suffices to represent the performance in the environment. Law produces judgments for society or its subsystems as legally binding descriptions of legal content that are realized as just decisions in society. Science produces knowledge whose description can be used for various purposes in the environment.

Table 1. The variables of the research process

VARIABLE	VALUE DOMAIN
PROCEDURES	{EXPERIMENTAL DESIGNS, SIMULATION MODELS, NUMERICAL CODES, FIELD RESEARCH ....}
DOCUMENTS	{MEASUREMENT PLOTS, PROTOCOLS, FILM RECORDINGS, INTERVIEWS ....}
DATA	{NUMERICAL SCORES, GRAPHS, EMPIRICAL EXAMPLES ......}
PATTERNS	{CORRELATIONS, REGULARITIES, CAUSAL RELATIONS, INTERPO-LATIONS ....}
THEOREMS	{DERIVATIONS, CONFIRMATIONS, REFUTATIONS ....}
HYPOTHESES	{STATEMENTS, QUESTIONS ....}

produced – the anticipated performance – is a structure that is generated in a process of recursive operation that reproduces itself in this process.[16]

The concretization we are attempting relates to the science system. On the basis of empirical and epistemological studies, it can be postulated that research findings in the science system are determined by the coupling of six variables that can each adopt different "values." We do not mean quantitative values here but various representations of the respective variables.[17]

Each scientific finding contains a selection of appropriate "values" of the individual variables, whereby appropriateness is a result of their reciprocal interdependence: The values must form an eigensolution of the dynamic process of "knowledge production." As a research group cannot fully estimate this appropriateness in advance, it commences by selecting initial values that are considered to be plausible though finally arbitrary. Through recursive operating, it is then revealed whether these either endure as values of a stationary solution, require slight modification, or have to be completely replaced.

The reciprocal interdependences that link the variables to one another are defined by the "functions" listed in Table 2. In summary, we can state:
Hypotheses = Derivation of (networking of (interpretation of (selection/evaluation of (application of (operationalization of (hypotheses))))));
or:

$$H = OP(H)$$

[16] Viewed formally, the desired eigensolution is an attractor of a dynamical system. As we wish to stress its social construction, we have chosen the label eigensolution in preference to the usual terms.

[17] This somewhat unusual use of the term "variable" follows the idea that one can introduce color, for example, as a variable with red, green, and so forth as values.

Table 2. The functions of the research process

HYPOTHESES	=	DERIVATION OF (THEOREMS)
THEOREMS	=	NETWORKING OF (PATTERNS)
PATTERN	=	INTERPRETATION OF (DATA)
DATA	=	SELECTION/EVALUATION OF (DOCUMENTS)
DOCUMENTS	=	APPLICATION OF (PROCEDURES)
PROCEDURES	=	OPERATIONALIZATION OF (HYPOTHESES)
HYPOTHESES	=	DERIVATION OF (THEOREMS)

in which H stands for hypotheses and OP labels the chain of operations in the process of knowledge production. H is the eigensolution of the complex knowledge operation sought.[18]

Thus, the construction of an eigensolution consists in finding a stationary state for this system of networked operations. This requires not only a decision on the relative weights of the individual links but also "values" for the different variables selected in this way, so that these no longer change if the process is repeated. In a recursive procedure, the initially arbitrarily (though with plausibility considerations) chosen values are modified: Measurement procedures are altered so that undesired data are no longer obtained; hypotheses are changed so that deviant data can be integrated; the relations to theories are qualified; and so forth. If a combination of values finally reveals that the complex knowledge production has an eigensolution, then the recursion can be broken off; a research finding is gained and can be transported to the exterior as a product. Recursion as procedure has the effect that each new value of a variable serves as the input for the next operation, and coherences are generated in this way. In addition, the stationariness of a recursive process signifies the finding of an eigensolution.

The intention behind this approach is to study the construction of eigensolutions empirically, not only as an empirical contribution to constructivistic epistemology but also as a contribution to empirical research on science. In developing our empirical design, we have broadly excluded two – very important – aspects: 1) the relations between individual strategies of researchers and group dynamics; and 2) the internal-external relations of groups, particularly with a view to the leadership and management problem. We have concentrated on the isolated world of self-organization within the group in a more or less diffusely set task or theme and have followed the steps in the construction of a research finding.

---

[18] At this point, the text follows the custom of assigning a factual priority to the formulation of hypotheses. This priority is not supported by empirical observation of research processes; it presumably results from post facto reconstruction or from the requirements of a funding application. In the sense of recursive operating, any other relation can be chosen, such as Procedures = OP(Procedures).

6. A Simulation Experiment

As there is no realistic possibility of observing, documenting, and evaluating research processes that normally extend over several months in sufficient detail to completely reconstruct the construction process, we have chosen the method of simulation in order to achieve laboratory conditions. We have compiled groups whose task is to invent a new board game from a set of partially given and partially modifiable materials. We consider the invention of a new game to be sufficiently related to the generation of a research finding. In detail, we see the following analogies:

1) As in research, the development of a new game requires a linking together of the variables that are considered to be relevant and a determination of their "values" in such a way that a stationary solution appears. As in the research process, the variables are not given externally but are invented or, more precisely, constructed by the members of the group as part of the task.

2) The variables that determine a game can, as in research, be assigned to a more conceptual and to a more operative part of the total construction: the idea of the game and the course of the game. The structure of a game can be placed in the eigensolution scheme of Fig. 1.

3) As in research, the materials also set conditions that either cannot be changed or can only be changed slightly: the "universe of discourse" and the "instruments."[19] Abandoning well-defined terms and theories is just as difficult as changing the given materials, even if changes (according to the instructions) are permissable. A pre-established separation between "theme" (What do we want to do?) and "resource" (What is available?) is neither a prerequisite of research nor of our simulation, although it is almost automatically set up.[20]

4) The construction of a game is a favorable problem for group processes because the introduction of ideas and partial solutions as well as the detection of weaknesses and gaps by the individuals balances out the disadvantage of having to seek agreements. It goes without saying that the psychological boundary conditions of the participants in a laboratory procedure differ in their effectiveness from those in occupational research groups. Above all, the conflict between individual strategies and collective goals emphasized so strongly in laboratory studies is not pronounced. With regard to the central issue, however, that individual ideas and intentions have to be transformed into ones that are shared collectively, these differences do not play a decisive role.

---

[19] The former means a differentiated understanding of a board game; the latter the hardware, that is, board, counters, dice, and so forth.

[20] The usual practice of a research group is mostly to adopt instruments, methods, theories, and apparatus and only perform slight modifications in order to generate its own finding. This follows from the above-mentioned discussion on the tension between innovation and consistency.

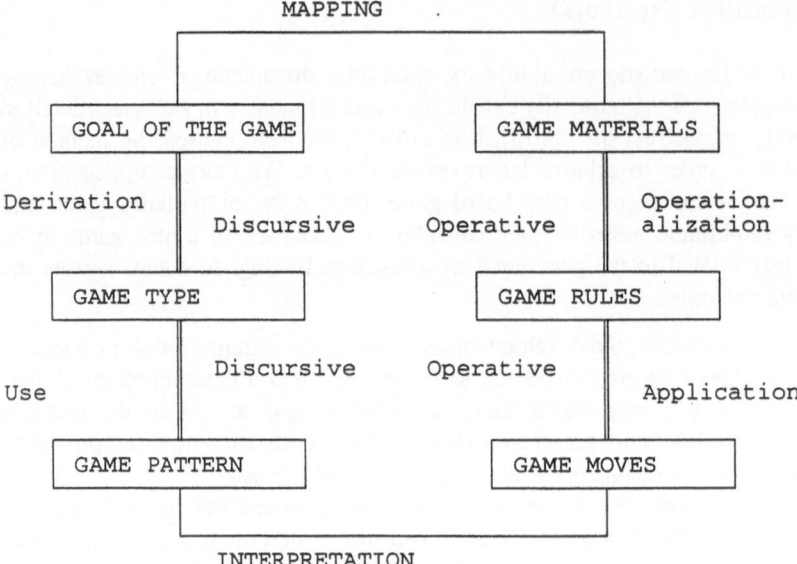

MAPPING

GOAL OF THE GAME   GAME MATERIALS

Derivation   Operation-
              Discursive   Operative   alization

GAME TYPE   GAME RULES

              Discursive   Operative
Use   Application

GAME PATTERN   GAME MOVES

INTERPRETATION

Fig. 1. The functional circle of the game

5) The external boundary conditions in our simulation are generated by presenting them as instructions about the construction: The game must have at least a minimum of novelty and marketability. These instructions are possibly too weak. However, we want to place more emphasis on the study of processes of self-organization than on aspects of the solution of a set task. For the same reason, we have excluded the problem of formal leadership and external controls.

6) Finally, in order to capture some typical problem situations from research processes, we included a difficulty with the board that would simultaneously stimulate interest and also lead to thinking blocks: We have used a digital grid imposed on an unrelated pattern of colors with smooth transitions. Further materials are: tokens, dice, colored pens, and auxiliary aids.

7. Observation Methods: Pilot Study

7.1 Can the Study Situation Be Operationalized Credibly?

We first had to find out whether the task in the simulated setting was taken seriously and was effective. For this purpose, we assembled three groups of between five and six male or female college students. In one group, all members were advanced sociology students who were highly familiar with one another. The second group also included students from other departments. The members of the third group were recruited in the university library. They were students from the Departments of Education, History, Law, Chemistry, and Economics, so that this group could draw on various "intellectual resources." The different forms of group

444

formation should test the independence of the production process from group membership.

Each group was given 90 minutes to develop the game as sketched above. They had to complete their task by writing a short report addressed to the production management and marketing department of a fictitious company. This final report should present the invented game in its current state of development and also point out unsolved subsections ("product").

All groups were videotaped during the 90 minutes of their existence. The videotapes formed the basis for 1) estimating whether the intended empirical operationalization was in any way realizable; and 2) developing a systematic observation procedure for analyzing the self-organization process and the construction of eigensolutions in groups.

Qualitative evaluation of the material showed that the setting and the group task were accepted and taken seriously. All groups worked on their assigned tasks in a concentrated way. Unstructured interviews with the participants after the end of the sessions supported this impression. However, they also indicated that all participants had reached their stress threshold by the end of the 90-min session.

None of the three groups were able to present a finished product in the given time. It is notable that a group comparison revealed that noncompletion was always due to components that either remained unsolved or could not be integrated into the provisional product.

7.2 The Selection of the Observation System and Its Properties

To the best of our knowledge, no previous instruments are available that could be applied directly to the analysis of self-organization processes or the empirical study of the construction of eigensolutions in the sense presented here. The procedures for analyzing social interactions reported in the literature on observation methodology can be divided into three groups (Futoran et al., 1989): activity-centered, process-oriented, and those that are designed for completely specific task situations, such as the analysis of a cockpit crew or a football team.

In activity-centered procedures, very precise data are usually obtained on speech frequencies and speech duration that can be interpreted as indicators for the contribution of a participant to the group. Both content-related aspects – for example, with regard to the group task – and process interpretation of a behavior (behavioral act) remain unconsidered.

The classical example of a process-oriented procedure is the Interaction Process Analysis (IPA; Bales, 1950). As the name indicates, this approach concentrates on the significance of an individual behavioral act for the group process, and this with regard to the group's task processing and in its function for the socioemotional climate. The introduction of time as a category dimension should be noted here.

The third group of approaches to interaction analysis contains all procedures to assess the processing of the most varied kind of task. A typical example is sociometric studies (Moreno, 1954). A frequently reported central problem of this group of procedures is their lack of comparability. This criticism is not so important for the present program, as several groups are each being studied in one and the same situation using the same task.

The SYMLOG system (SYMLOG: A system for the multiple level observation of groups) was developed from the IPA. Bales (1970, 1979) expanded the two IPA dimensions of "socioemotional" and "task-oriented" by adding a third dimension "striving for dominance." All three dimensions were developed on the basis of direct observation of small groups. As in the IPA, the time dimension is included in the classification of behavior although it has rarely been used in a differentiated manner in published research. The two most widespread instruments that have been developed within the framework of SYMLOG are the retrospective rating procedure (Nowack, 1989) and the signifying procedure containing a very complex category system.

Another observation instrument based on much experience with the procedures presented here is the TEMPO approach (McGrath, 1984; Futoran et al., 1989). It also appears to be most suitable for an empirical representation of the major theoretical concepts for the analysis of processes of self-ordering and the construction of eigensolutions. Therefore, foundations and postulates of this approach will be presented in more detail in the following and related to the research question.

TEMPO is an acronym for the time by event by member pattern observation system. It was developed with the goal of analyzing the interaction process in groups that have to process tasks while additionally registering socioclimatic aspects. The following demands are placed upon the performance of the system:

1) The temporal aspect of the interaction process must be made clearly visible. 2) It must be possible to take into account simultaneously both the contribution to the process and the internal significance of a specific interaction (a behavioral act). 3) It should be possible to link the temporal pattern and the content of the interaction process to the extent and quality of the group's task processing or its outcome. 4) The observation system should be universally applicable, that is, to groups of varying size and membership as well as to different tasks. 5) Finally, the system should differentiate the contributions of an individual member of the group to the group process from his or her contribution to the group's task processing.

In addition to these points derived from experience with the published observation instruments, three further postulates that underlie the theory of functional groups (McGrath, 1984)[21] are made for the development of the TEMPO categories.

---

[21] The emphasis on the task orientation is first of all justified by the need to restrict the subject matter. In addition, the situation initiated in this research is a group discussion and decision-making situation. In these types of group session, it is well known for emphasis to be placed on aspects of the task (e.g., Morguet & Boos, 1989).

Fourteen production function categories		
	Content(C)	Process(P)
Propose (P)	<u>PC</u> 1.New task content 2.Prior–previously proposed task content 3.Dictate–repeat content as in dictate to writer or if others didn't hear	<u>PP</u> 1.Goals–quality/quantity production standards 2.Strategies–how to do the task 3.Acts–keep task moving: process acts not 1. or 2.
Evaluation (E)	<u>EC</u> 1.Agree with/accept– proposed content 2.Clarify/modify–the proposed content 3.Disagree–with content that is otherwise accepted 4.Reject/veto–content that is poposed.	<u>EP</u> 1.Agree with/accept– proposed process 2.Clarify/modify–the proposed process 3.Disagree–with process that is otherwise adopted. 4.Reject/veto–process that is proposed.

Seven nonproduction function categories

T. Task digression – talking about content or process that has already been accepted. The discussion or comment does not move the task forward.

P. Personal comments – comments about oneself that do not move the task forward.

I. Interpersonal comments – responses to others' personal comments that do not move the task forward.

R. React to experiment – look up recording booth, react to the time limit, comment about being recorded, etc.

D: Digressions – comments about the specifics of the situation, that don't fit anywhere else.

U. Uninterpretable – you can tell who spoke but not what they said: or you can't tell who spoke.

S. Silence – periods of no overt activity (no one is speaking or writing).

Table 3. An overview of the TEMPO observation system

1. Groups have three simultaneously present, separable, yet interlinked functions: production, support of group members, and maintenance of the well-being of the group.

2. The processing of group tasks covers four phases: (a) start; (b) technical problem solution; (c) conflict solution; and (d) production. The first and fourth phases are always necessary; cycles can arise between the two middle phases.

3. Activities related to the function of supporting members and maintaining the well-being of the group take a comparable course to the phases of task processing (production function).

Two types of function categories are differentiated in the TEMPO observation system: productive and nonproductive behaviors in relation to the task processing of a group (production- and non-production-function categories). The categories for analyzing task-related behavior result from the differentiation between content and process on the one hand, and proposal and evaluation on the other. These four main categories contain 14 further subcategories that can be modified or extended depending on the problem to which they are being applied.

This is a universal observation approach insofar as it is assumed that the four main categories can be found in any group. The personal and interpersonal needs of the members of a group are assessed with the help of the fifth main category. Its seven subcategories refer to behaviors that are not linked directly to task processing (non-production-function categories).

Table 4 illustrates the intended coding procedure. It contains five columns. The first column presents the time unit in which a specific behavioral act is registered. It still has to be decided whether the time unit of 1 sec proposed by Futoran et al. (1989) is not too narrow. A broadening to 3 or 5 sec needs to be tested empirically. The second column of this coding scheme is used to register the sender of the message, the active person. In a five-member group, for example,

Table 4. The coding scheme

Time interval (i.e. 1 sec)	Target	Observation category	Cycle	Comments
1	8	PC1	1	Colors for the
2	8	PC1	1	game board
.	.	.	.	Grid
.	.	.	.	.
.	.	.	.	.
.	.	.	.	.
.	.	.	.	.
60	2	EP3	3	.
61				That doesn't work Silence

this would use the numbers 1 to 5. In the third column, the behavioral act is described in terms of the 14 task-related, "product-oriented" categories of the TEMPO system. In the next to last column, the cycles are counted (this concept will be discussed in more detail in the following section). Their temporal duration can be determined with the "time register" in column 1. The last column finally either registers a simple label for the behavioral act or the occurrence of one of the seven non-production-oriented behavioral categories is entered.

7.3 Adjustment of the TEMPO Categories to the Self-Organization Process

The observation instrument should empirically reflect major theoretical concepts involved in self-organization processes and in the construction of eigensolutions. These perspectives are sketched in the following.

The boundary conditions of the simulated research situation contain all types of preconditions: (a) the physical environment of the "constructed" laboratory in which the group should develop their product (the board game); (b) the material resources (counters, board, etc.) made available; and (c) the intellectual resources of the participants. This situation represents the "isolated world of self-organization within the group" discussed above.

A procedure now has to be developed that is able to identify the stationariness in group interaction that is linked to "completed" eigensolutions. Eigensolutions can refer to subsolutions and the total solution of the group product. For this purpose, the concept of behavioral cycles elaborated within the framework of the TEMPO system is introduced. Cycles are defined as a series of behaviors that refer to one and the same task content or a specific contribution to the process. Experience has shown that the first behavioral act in a cycle, such as, for example, the proposal to rate a game counter in a certain way, often appears as a contribution to its content (Futoran et al., 1989). The response to this proposal can be an evaluation or a modification of the idea (new proposal; evaluation variant). A cycle is completed if the idea is accepted and adopted into the product of the group, or if it is finally rejected. However, a cycle can also be completed by any other task content introduced into the group before the previous content has been finally processed. This is an example of a content-related interruption as compared to distracting or task-independent interruptions introduced by members of the group that belong in the fifth main category. Thus, previous cycles can be taken up again at a later time.

At present, we are trying to apply the cycle concept specifically to the situation of the game development group. The empirical validity of the concept cannot be judged at present. However, it is conceivable that it will be possible to demonstrate a number of characteristics of cycles that can be interpreted as indicators of the ease (or otherwise) of the flow of the transformation of the group task (group's task performance). If, for example, many cycles contain negative evaluations or are more closely related to the group process than to the group task, this might suggest interpersonal activities to cope with conflict rather than technical problem solutions. On the other hand, if many cycles contain positive evaluations and modifications, processing proposals are often received ("evaluated") positively, and/or if they relate to strategies and actions (process orientation) more than to goals or the search for goals, then this might indicate that the group is on the path toward producing a qualitively improved product.

The nonoccurrence of new cycles and perseverence in one cycle could represent an intermediate step toward an eigensolution. If such a cycle enters into the fourth phase of task processing, which is characterized by a similar pattern of cycles, this could be interpreted as an empirical representation of the eigensolution. Using the terms in the categories of the observation system, a higher frequency of "affirmative process statements" is to be expected in such a phase (see Table 3). The concepts of innovation and proof (expansion of rules and consistence) could be indicated in a similar way.

7.4 The Next Steps

First, a precise and manageable final version of the category system must be constructed for the analysis of groups working on board game development so that it can be tested on the three single-case groups already available. This step will then provide information on two aspects: 1) the technical simplification and/or specification of the observation instrument, and 2) whether this empirical approach is fundamentally applicable to phenomena of knowledge production.

References

AMANN, K. 1990. Natürliche Expertise und Künstliche Intelligenz. Eine empiri sche Untersuchung naturwissenschaftlicher Laborarbeit. University of Bielefeld: Dissertation.

BALES, R.F. 1950. Interaction Pess Analysis. Chicago: University of Chicago Press.

BALES, R.F. & COHEN, S.P. 1979. SYMLOG: A System for the Multiple Level Observation of Groups. New York: The Free Press. (Deutsch: 1982).

BALES, R.F. 1972. Die Interaktionsanalyse: Ein Beobachtungsverfahren zur Unter-suchung kleiner Gruppen. In: König, R. (Ed.): Beobachtung und Experiment in der Sozialforschung. Köln: Kiepenheuer und Witsch.

BALES, R.F. 1970. Personality and Interpersonal Behavior. New York: Holt, Rine-hart & Winston.

ESSER, J. 1964. Grundsatz und Norm in der richterlichen Fortbildung des Privat-rechts. Tübingen: Mohr.

FLANAGAN, J.C. 1954. The Critical Incident Technique. Psychological Bulletin, 51, 327-358.

FUTORAN, G.C., KELLEY, J.R. & MCGRATH, J.E. 1989. TEMPO: A Time based System for Analysis of Group Interaction Process. Basic and Applied Social Psychology, 10 (3), 211-232.

HEIDEN AN DER, U. 1991. Selbstorganisation in dynamischen Systemen. In: Krohn, W. & Küppers, G. (Eds.): Emergenz – die Entstehung von Ordnung, Struktur und Bedeutung. Frankfurt: Suhrkamp.

KNORR-CETINA, K. 1984. Die Fabrikation von Erkenntnis. Frankfurt: Suhrkamp.

KROHN, W. 1988. Rationalität in der Wissenschaft – empirisch betrachtet. In: G. Pasternack (Ed.): Rationalität und Wissenschaft. Bremen: Zentrum Philosophische Grundlagen der Wissenschaften, Schriftenreihe Band 6, 66-80.

KROHN, W. & KÜPPERS, G. 1989. Die Selbstorganisation der Wissenschaft. Frankfurt: Suhrkamp, 66ff.

LATOUR, B. & WOOLGAR, S. 1979. Laboratory Life. The Construction of Scientific Facts. Beverley Hills: Sage.

LUHMANN, N. 1990. The Cognitive Program of Constructivism and a Reality that Remains Unknown. In: Krohn, W. & Küppers, G. (Eds.): Self-Organization – Portrait of a Scientific Revolution. Dordrecht: Kluwer.

MCGRATH, J.E. 1984. Groups, Interaction and Performance. Englewood Cliffs, N.J.: Prentice Hall.

MEYERS, G. 1985. Texts as Knowledge Claims: The Social Studies Constructions of two Biological Articles. In: Social Studies of Science, 15, 593-630.

MEYERS, G. 1986. Writing Research and the Sociology of Knowledge. A Review of Three Books. In: College English, Vol. 48, No. 6, October, 595-610.

MORENO, J.L. 1954. Die Grundlagen der Soziometrie. Köln: Westdeutscher Verlag.

MORGUET, M. & BOOS, M. 1989. Kodieranleitung für die Auswertung von Planspielsitzungen mittels Cognitive Mapping und Konferenzkodierung. SFB 221, Universität Konstanz, Arbeitsbericht Nr. 18.

NOWACK, W. 1989. Interaktionsdiagnostik. Saarbrücken: Dadder.

SCHMIDT, S.J. 1987. Der Diskurs des radikalen Konstruktivismus. Frankfurt: Suhrkamp.

TRAWEEK, S. 1987. Buying Time and Taking Space: The Culture of the Particle Physics Community. Boston.

WOOLGAR, S. 1988 (Ed.). Knowledge and Reflexivity. London: Sage.

Author Index

Subject Index

Relationship schema
–, change in a 268-272, 275, 278, 279
Reliability 107, 112, 114, 197, 248,
  349, 353, 400
Religion 35
Repetition compulsion 242
Representation 5, 40, 76, 104, 122,
  124, 128, 158, 207, 214, 231, 235,
  243, 245, 255, 259, 260, 262, 299-
  302, 304, 311, 355, 446, 450
– of knowledge 122, 124
Resistance 252, 273, 284, 288, 306,
  309, 362
Resonance 291-293
Resource 156, 180, 181, 443
Respiration 75, 166
Rhythmic activity 11, 55, 280
Ritual behavior 102, 114
RNA 209
Role-playing 385, 386, 412
Roles 14, 342, 347, 348, 386, 399

Saddle-point 276
Saturation 6, 46, 51
Scale invariance 412
Schema theory 136, 296-298, 310
Schemata 12, 13, 115, 146, 172, 173,
  175-177, 188, 240, 242, 265, 280,
  288, 297
Schizophrenia 8, 22, 30, 71, 163-167,
  169-174, 177-182, 186, 187, 189-
  196, 199, 201, 202, 207-212, 314,
  317, 425
–, long-term evolution of 165, 169,
  186
–, pathogenesis of 163
School 35, 385
Schrödinger equation 93
Science system 435-437, 441
Scientific knowledge production 434
Sculpture 14, 350, 351, 375
Self-concept 12, 305, 306
Self-organization 3, 7, 12, 15, 18-21,
  24, 26-28, 30, 36, 38, 53, 88-90, 98,
  99, 100-103, 105, 107, 112-117,
  151, 191, 239-241, 245, 246, 250,
  257, 264, 266, 269, 280, 281, 337,
  341, 344, 346-349, 359-362, 364,
  366, 382, 386-392, 401, 405, 410,

411, 413, 436, 437, 439, 440, 442,
  444, 445, 449, 451
– theory 102, 103, 105, 112, 113,
  115, 341, 344, 360, 361, 382
Self-reference 250, 257, 293, 294,
  317, 322, 360, 362, 417
Self-reproduction 235, 289, 354
Self-similarity 14, 243, 245, 246, 411
Separatrix 189, 190
SPA (sequential plan analysis) 245,
  255, 258, 259, 387, 391
Simulation 26, 27, 40, 121, 122,
  132-135, 138, 139, 141, 155, 159,
  163, 166, 169, 174, 179, 181, 182,
  186, 187, 189, 231-233, 258, 266,
  299, 309, 324, 341, 350, 354-358,
  363, 367, 375-377, 379, 385, 409,
  411, 413, 434, 435, 441, 443, 444
Singularity 275-277
Slaving 4, 32-35, 38, 47, 293, 387,
  391
– principle 4, 32, 33
Sleep deprivation 206, 207, 210, 212
Slot 272
Smoking behavior 379, 383
Social construction 396, 434, 435,
  441
Social drift 7
Social field 358
Social psychology 136, 285, 342,
  345, 349, 354, 364, 365, 386, 394,
  396, 450
Social skills 173, 175-177, 188
Socialization 234, 290, 359, 364
Society 13, 28, 115, 147, 150, 151,
  160, 235, 280, 281, 312, 324, 346,
  384, 385, 428, 431, 436, 439, 440
Sociogram 375
Sociology 3, 31, 169, 234, 235, 365,
  366, 444, 451
Solipsism 288
Spatial behavior 171, 351, 353, 358,
  364, 388
Species 66, 153, 154, 161, 180
State
– of mind 11, 12, 29, 168, 192, 243,
  326, 327
– space 55, 61-64, 68, 81, 82, 231,
  368-370, 373-375
–, steady 62, 63, 68, 69, 94-100

Springer Series in Synergetics

Editor: Hermann Haken

Synergetics, an interdisciplinary field of research, is concerned with the cooperation of individual parts of a system that produces macroscopic spatial, temporal or functional structures. It deals with deterministic as well as stochastic processes.